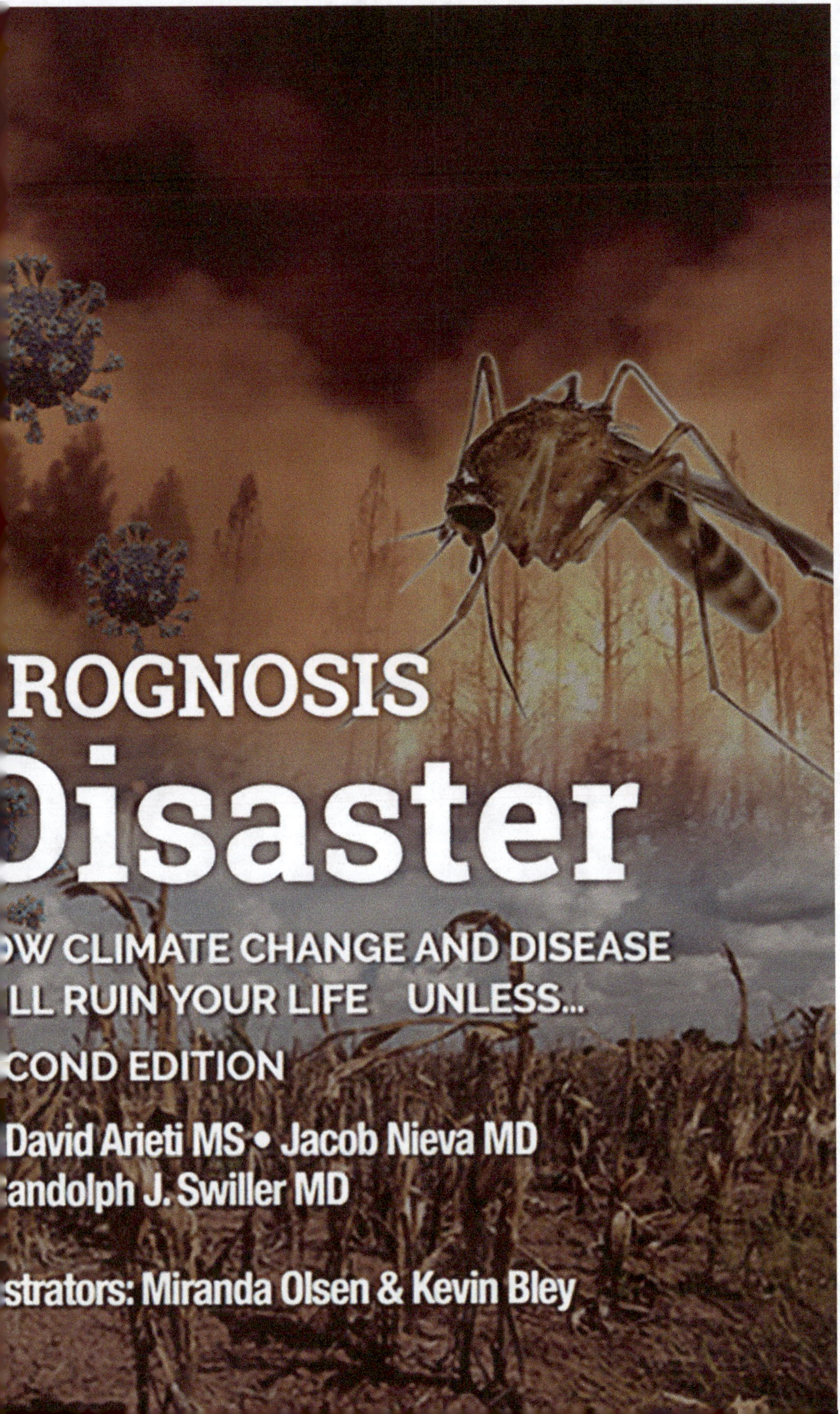
ROGNOSIS
isaster
W CLIMATE CHANGE AND DISEASE
LL RUIN YOUR LIFE UNLESS...
COND EDITION
David Arieti MS • Jacob Nieva MD
andolph J. Swiller MD
strators: Miranda Olsen & Kevin Bley

PROGNOSIS
Disaster

OW CLIMATE CHANGE AND DISEASE
ILL RUIN YOUR LIFE   UNLESS...

ECOND EDITION

David Arieti MS • Jacob Nieva MD
Randolph J. Swiller MD

ustrators: Miranda Olsen & Kevin Bley

ARPress

ARPress
45 Dan Road Suite 5
Canton MA 02021
Hotline: 1(888) 821-0229
Fax: 1(508) 545-7580

Ordering Information:
Quantity sales. Special discounts are available on quantity purchases by corporations, associations, and others. For details, contact the publisher at the address above.

Printed in the United States of America.

ISBN-13:   Softcover   979-8-89389-137-9
           Hardcover   979-8-89356-601-7
           eBook   979-8-89356-602-4

Library of Congress Control Number: 2024912147

THE MAJORITY OF HUMAN SOCIETIES HAVE NOT LIVED SUSTAINABLE LIFESTYLES AND ATTEMPTS TO LIVE ENVIRONMENTALLY FRIENDLY LIFESTYLES NOW DOESN'T CHANGE THE FACT THAT HUMAN EXISTENCE HAS ULTIMATELY BEEN DESTRUCTIVE TO THE EARTH AND ALL ITS NON-HUMAN ORGANISMS.

Les Knight-Founder of the Voluntary Extinction Movement[1]

---

[1] Given to me by Madison Houk

AI STATEMENT

NONE OF THE TEXT WITHIN THIS BOOK WAS CREATED WITH THE HELP OF ANY ARTIFICIAL INTELLIGENCE. All OF IT CAME FROM THE MINDS OF ITS AUTHORS. EVERY ADVANCE WITH NEW TECHNOLOGY OVER THE PAST THOUSANDS YEARS HAS HAD TO FACE A FUNDAMENTAL TEST: DOES IT ADVANCE THE HUMAN MIND (OR VOICE) OR DOES IT REPLACE THE HUMAN VOICE? WE LEAVE THE DECISION UP TO YOU, THE READER.

CONTENTS

BOOKS WRITTEN BY DAVID ARIETI

The Earth is my Patient (2005) Authorhouse

Prognosis Disaster: The Environment, Climate Change, Human Influences, Vectors, Disease and the Possible End of Humanity? (2011) Authorhouse

DEDICATION

FROM DAVID

I would like to dedicate this book to the memory of Randolph J Swiller MD, who was a coauthor of our first edition who died around Thanksgiving in 2015. I have known Randy since meeting in high school in 1959 and we had been best friends ever since.

I would also like to dedicate this book to all the Earth's organisms, especially the copepods, fish, crabs, and birds that were killed because of the oil spill in the Gulf of Mexico which started on 20 April 2010. I would also like to dedicate this book to all those humans who died of COVID 19 during the past couple of years.

I would also like to dedicate it to my children Aviva and Amiel, to my parents Jane and Silvano Arieti, and of course to Maggie my significant other for over 30 years. To my friends and relatives who have passed away: Elde, Lorie Carrol and my Aunt Mary and Uncle Phil Jaffe, Annabelle and Vincent, Marcella and Gulio as well as their pets to include my mother's cat Rusty and my uncle's cat who died prematurely.

FROM JACOB

I dedicate this book to the millions of people who perished from diseases brought about by humans. Their deaths were not in vain. Their deaths served as a wake-up call for us, users of the earth resources, to take good care of the environment. I also dedicate this book to my children, Christian, Marianne, and Jeffrey, who are the precious jewels of my life.

Lastly, I dedicate this book to my special someone, Evelyn.

FROM KEVIN

Kevin would like to dedicate this book to the students of not only Columbia, but the world. May you keep yourselves not simply attuned to environmental changes occurring about you, but ones that occur later as a result. Think, learn, and spur change.

FROM MIRANDA

I dedicate this this book to Mouse, Lola, my friends and family

PREFACE

The day that this sentence was written, (June 7, 2023) New York City suffered the worst air pollution in its history due to a forest fire in Quebec, Canada. The fire is believed to be caused due to climate change. The air quality index (AQI) was over 200 while normal is 50 and under. That puts the description as VERY UNHEALTHY which means that the health risk is increased for everyone with no exceptions. This book emphasizes the issue of climate change and its effects on human and non-human health.

On March 20, 2023, the IPCC (Intergovernmental Panel on Climate Change) came out with a report about the Earth's climate. The report stated that the Earth will surpass 1.5 degrees Celsius (2.7 degrees Fahrenheit) above preindustrial levels by the 2030s. We are looking at heat waves, famine, disease, melting glaciers, human and other organism deaths, economic damage, and a whole lot of other environmental disasters.

It's not a coincidence that this book is about the effects of climate change and the fact that we just got the IPCC report. It seems that every day we are inundated with environmental news which I'm sorry to say is mostly bad, but some articles do mention good news such as another organism has been brought back from the brink of extinction.

It never stops to amaze me that due to greed and stupidity humans KNOW that they are heating up the planet but do it anyway because of the fear of losing money. As mentioned in Chapter 3 the CEOs of the top oil companies and presidents of large countries like Brazil and The United States knew what they were doing to the Earth's environment, but they came up with lies and excuses to continue most likely due to fear of a downgraded economy.

At the time of the first edition of this book, published in 2011, there was an oil blowout off the coast of Louisiana releasing 3.19 million barrels of oil or 134 million gallons. (A barrel of oil contains 42 gallons). I mention this because we are still using oil even though almost every car manufacturer produces electric cars. But it may be too late. The readers of this book should be very angry at the stupidity of humans.

Just as this book is going to press there was a Republican debate with candidates wanting to be the Republican nominee for the 2024 presidential election.

One of the contenders was Vivek Ramaswamy. He is a climate change denier. He mentioned during the debate that climate change agenda is a hoax. Using climate and hoax in the same sentence is quite alarming. In July 2023 Florida ocean temperatures reached 100°F(37.7°C). A most serious consequence of this is lower dissolved oxygen in the water thus leading to suffocation of the fish. Fish need between 5-6ppm (parts per million) of oxygen. Below 5-6ppm they will be stressed.

Climate denial is a hazard for the planet as we see now.

Remember these famous words from Antonio Guterres, the Secretary General of the United Nations who said at the COP27 in Egypt on November 7, 2022: "Cooperate or Perish". (See the link below.)

https://www.washingtonpost.com/climate-environment/2022/11/07/cop27-climate-change-report-us/

ACKNOWLEDGEMENTS

When one writes of a book of this magnitude it is virtually impossible to do it alone. The authors wish to thank the following people who helped make this book possible. We especially want to thank my colleague and friend Valeriy Lyubanov for giving me the impetus for writing the second edition of this book by reminding me that we left out the most important characteristic of life in the first edition: reproduction. We want to thank Jordan Buck for suggesting the eye-catching title of the book. Christine Pfeiffer for adding wonderful eloquence and ideas to the manuscript as well as editing the manuscript. Steve Bithos who helped me write some parts of the book for the first edition. We want to thank Professor Paul O'malley and Lynn Levy-Murray and Caryl Danguilan for wonderful editing jobs. We would like to thank the wonderful librarians at Oakton College (Skokie Campus); they are Tricia Collins, Ted Ramus, Rose Novil, Gretchen Snyder, Marisa Walstrum, Elizabeth Sanderson, Barbara Joy and Kevin Purtell. They helped us get needed articles, books, and journals as references; Amiel Arieti for the cover design for both editions; Duane Gubler, one of the world's experts on Dengue fever who took time from his busy schedule to discuss with me (David Arieti) certain aspects of Dengue fever. Peter Winkler who as a colleague and friend helped me add insight and compelling ideas to this book. We would like to give a special thanks to A.L. Baker from Phycokey who gave us permission to use wonderful algae photos. A giant thanks to Ismir Softic who made sure that the computers used to write this book were all working. A hearty thanks goes to Sam Pudi and Gaya Lasam, heads of the computer center at Oakton College Skokie Campus. A special thanks to John J. Murray and Clara Carr of the Science and Math Department of Columbia College Chicago, and to Byron Bell also from Columbia College for his support.

WANING MOTHER

Progress

Advancement of the human race

Technology

Science

Make use of all resources

The land is ours

The plants are ours

The beasts are ours

The very oceans that spawned life are ours

Or so we'd like to think.

The earth is not our property. She is our mother.

Giver of life

Provider

We are her children

She cares for us

Yet we conspire against her

 In the most heinous of matricides

For we are doing our best to kill her.

We fill her oceans with sludge

And the hazardous byproducts of human progress.

We bury the same deep in her fertile soil.

Her beautiful landscapes are ravaged

To slake our thirst for power.

Her forests and plains and hills and even deserts

We feel are better suited to become factories and refineries and stripmalls and skyscrapers. From these buildings we pour toxic and harmful smoke into the air which supports all life.

We spray CFC's and other chemicals which deteriorate the shield

She has made to protect us from the strong rays of the mighty sun.

We stockpile weapons and wage war destroying our brothers, unbalancing entire ecosystems though she supports us unconditionally. We, her selfish children, only seek to take advantage of her; when will we realize the sheer folly of this incestuous rape?

One day, we will exhaust her infinite generosity.

Then we will suffer for our crimes against nature.

We will then miss our life-giving mother and what use will this technology be then?

This is progress?

In killing the earth, we are in the end killing ourselves, stabbing the breast that feeds us.

Brent Finnel--1994

BRILLIANT STATEMENTS

Homo sapiens, the only deliberate fool that ever evolved, is back tending shop in the good old way![1]

Destroy nature, Destroy yourself.[2]

Microbes pursue every possible avenue to escape from the barriers that are erected to contain them, and we must be forever on our guard. They will seek undercurrents of opportunity and reemerge.[3]

Nature is emitting signals saying that we cannot continue our attempts to ruthlessly dominate her and if we persist, disaster is in the offing….[4]

There is reason to believe that, sooner or later, one or more known or unknown pathogens will cause devastating epidemic disease.[5]

…for although our destiny is not entirely in our hands, we ourselves are among the important forces in its creation.[6]

----

1   Robert Van den Bosch, *The Pesticide Conspiracy.* New York: Doubleday, 1978, p.35.
2   https://mountainsandminds.org/destroy-nature-destroy-yourself/
3   Emerging infections: biomedical research reports Krause, Richard, editor. Academic Press, 1998. Introduction to infectious diseases: stemming the tide, p. 2.
4   Op. cit. Bosch.
5   In Understanding the Environmental, Human Health and Ecological Concerns. Workshop Summary. Stanley M. Lemon et al. The National Academies press. Washington D.C., p.62. Vector-borne diseases. 2008.
6   Silvano Arieti, in D. Arieti *The Earth is my Patient*, Authorhouse Books, 2005, p. iii.

AUTHOR BIOGRAPHIES

David Arieti

David Arieti has been involved in environmental issues for almost forty years. He went to the University of Denver for his BA degree in Science Area Major. He worked on The Sea of Galilee where he did algal research. He returned to the US where he earned his MS degree in Marine Science from Long Island University. He worked studying the effects of chlorine produced oxidants on Chesapeake Bay in Maryland. He then worked at consulting firms in Washington D.C. where he worked on projects that dealt with the fate of pesticides in the soil; health effects on people working with the dyeing and finishing of textiles, and food additives. He then worked at the Baltimore Environmental Center as research director where he worked on hazardous waste issues. He also worked on the Hudson River studying the fish that were prevented from entering the waterways used to cool the condensers from the Indian Point Nuclear Power Plant in New York.

David also began teaching Environmental Science and Biology as an adjunct professor at various colleges in Maryland and Illinois, where he is today. He has been the recipient of three best Teacher-of-the-Year awards at three different colleges. In 1996 he won at Columbia College, in 2002 he won at Oakton College, and in 2005 he won at Daley College; all three colleges are in the Chicago vicinity.

He wrote his first book, entitled The Earth is My Patient in 2005. He has gathered information through newspapers, books, and magazines, as well as the electronic media (radio and TV) and witnessed in real life the havoc that humans have wrought on the planet. In his first book, he lists the real causes of environmental pollution and its effects on the planet, as well as possible solutions. It is due to his witnessing the increasing deterioration of the environment due to humankind's stupidity that led him to continue writing. Now, together with two doctors and an illustrator, he is putting together a book loaded with the effects of humankind's slide toward extinction based on its inability to do something extreme and constructive.

By constructive we mean ending the use of fossil fuels to include a permanent moratorium on off shore oil drilling; limiting human overpopulation (this will be a hard one because certain religions encourage many children), investing tremendous amounts of money in solarizing the planet (getting all our electricity from the sun); and demilitarizing the planet by getting rid of weapons such as planes, tanks missiles, etc. These consume tremendous amounts of money which could be used for building necessary structures like water treatment plants and schools instead of destroying these structures during war. And perhaps doing the most difficult thing possible: GETTING RID OF MONEY AND FINDING A BETTER ECONOMIC SYSTEM so that poor people have a chance at health and a decent life.

Jacob Nieva

Jacob O. Nieva, M.D. is a medical doctor who specialized in Internal Medicine in the Philippines. He graduated from the University of Santo Tomas, Faculty of Medicine and Surgery in 1980. He took five years residency training in Internal Medicine in the Rizal Medical Center, Pasig, Metro Manila, after which he practiced for 18 years before coming to the United States.

He has been an adjunct professor at Oakton College, Des Plaines, Illinois, teaching Human Anatomy and Physiology since 2004.

Being from the Philippines, Jacob knows firsthand about rainforest destruction and its effects on disease transmission. Jacob has written the bulk of the section dealing with diseases in this book as well as a chapter on diseases in David Arieti's first book, The Earth is My Patient.

RANDOLPH J. SWILLER

(In memoriam-1946-2015)

Randolph Swiller, M.D., FACP has practiced Internal Medicine in the Fort Lauderdale area of Florida for the past 28 years. A native New Yorker, he earned his M.D. degree at Chicago Medical School in 1972. His medical internship was at Long Island Jewish Hospital, in New Hyde Park, N.Y., from 1972-'73. He then took a Psychiatric Residency at SUNY in Brooklyn at Downstate Medical Center from 1973-'76. He was an Attending Psychiatrist from 1976-'78 at Maimonides Medical Center in Brooklyn N.Y. Following this he took his Medical Residency at The Jewish Hospital and Medical Center in Brooklyn, N.Y. from 1978-'80. He took one year of Hematology Fellowship at Cornell University-North Shore University hospital in Manhasset, N.Y. (1980-'81). Then he and his family moved to Coral Springs, Florida, where he began a solo Internal Medical practice. He has been in medical practice overall for 34 years. Randolph's further interests include Infectious Diseases, Nephrology, Cardiology, Oncology, and overall, preventive medicine, with public health.

Randolph has a great concern over the emotional apathy of our society about the environment. He believes that this is contributing to the eventual destruction of our environment and that materialism dominates over humanitarianism and common sense. If we fail to recognize the seriousness of the "decay" of our environment, the following generations will end up paying the price for our mistakes. These include more tropical ailments, CO_2 accumulation from the combustion of fossil fuels, global warming and a rise in sea levels with concomitant flooding of our coastlines. Large cities in the northeast will cease to exist, and low-lying land such as Florida, southern Georgia, North Carolina, and Virginia tidewater country will become submerged, along with New York City and New England. Illnesses such as Malaria, Yellow Fever, and Tuberculosis will spread northwards. He believes that only through careful vigilance and positive actions can our generation put an end to this destructive process. There is only one Earth, and if not dealt with wisely and properly, it will be lost. Where will mankind live with such a catastrophe? Randolph contributed to Section 2 on diseases, as well as to Chapter 3 dealing with how humans influence disease spread. He has a great concern over the emotional apathy of our society about the environment. He believes that this is contributing to the eventual destruction of our environment and that that materialism dominates over humanitarianism and common sense.

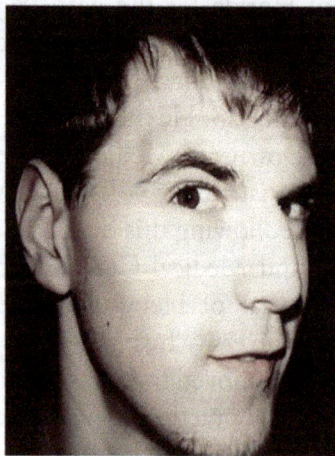

Kevin Bley

Kevin Bley is a graduate of Columbia College in Chicago. He graduated with a degree in Film & Video, with a concentration in Computer Animation, in 2010. Though he's only worked professionally in the art field for roughly four years, Kevin's been drawing since before he could talk, and to this day, still draws on a daily basis. He met David Arieti through his Animal Behavior class at Columbia, and when David mentioned needing an illustrator for his new book, Kevin readily accepted. Though this is Kevin's first time illustrating for a book, he hopes it won't be the last. Despite not being actively involved with environmental matters, with the exception of this book, he's been exposed to plenty of information and issues presented at Columbia, and believes it doesn't take a genius to understand that something must be done about them.

Miranda Olsen

Miranda Olsen is currently a design student at Columbia College Chicago with a concentration in Illustration. Miranda has been practicing drawing since a young age but didn't ascertain illustrating as a career option up until two years ago. Miranda first met David while enrolled in his Environmental Science class. David saw her drawing animals in her sketchbook and asked if she would be willing to create illustrations for the reissue of his book. She happily accepted the offer! This is the first book she has illustrated in her professional career. With art and environmental cautiousness being huge parts of her day-to-day life, she believes that this book is a great way to start out her professional book illustrating journey.

PRAYER FOR CHILDREN

We pray for children
who sneak popsicles before supper,
who erase holes in math workbooks
who can never find their shoes.
And we pray for those
who stare at photographers from behind barbed wire,
who can't bound down the street in a new pair of sneakers,
who never "counted potatoes",
who are born in places we wouldn't be caught dead,
who never go to the circus,
who live in an X-rated world.
We pray for children
who bring us sticky kisses and fistfuls of dandelions,
who hug us in a hurry and forget their lunch money.
And we pray for those
who never get dessert,
who don't have a safety blanket to drag behind them,
who watch their parents watch them die,
who can't find any bread to steal,
who don't have any rooms to clean up,
whose pictures aren't on any body's dresser,
whose monsters are real. We pray for children.
who spend all their allowance before Tuesday,
who throw tantrums in the grocery store and pick at their food,
who like ghost stories,

who shove dirty clothes under the bed and never rinse out the tub,

who gets visits from the tooth fairy,

who don't like to be kissed in front of the carpool,

who squirm in church or temple and scream in the phone,

whose tears we sometimes laugh at and

whose smiles can make us cry.

And we pray for those.

whose nightmares come in the daytime,

who will eat anything,

who have never seen a dentist,

who aren't spoiled by anybody,

who go to bed hungry and cry themselves to sleep,

who live and move but have no being.

We pray for children who want to be carried and for those who must, for those who we never give up on and for those who don't get a second chance . Take care of all these, Your children, O Loving God and guide us to reaching out to them. AMEN!

Author unknown

INTRODUCTION

WHY WE WROTE THIS BOOK

A lot has changed since we wrote the first edition of this book in 2011. The world's environment appears to be deteriorating, human stupidity has increased astronomically and the climate has become more unstable. Because of these phenomena it occurred to me that I should update this book with additional information. I became horrified at the polarization of America due to the horrors of the Trump Administration. Not only has there been deterioration of civility, we have become more militant and dangerous with shootings and ridiculous and immature statements from our elected representatives in Congress mainly from one political party.(Hint-it begins with the letter R)

There is a complex relationship between land use, climate phenomena, species diversity (biodiversity), disease transmission and human involvement in the environment which has worsened over the past few years. This book is intended to educate people as to how human activities help disrupt the balance of nature and how nature fights back which is the increase in disease levels. Environmental changes such as climate change are occurring at a much faster pace than they used to. Over the past 20 years the number of recorded natural disasters increased from 200 a year to over 400 a year. Seventy percent (70%) of these disasters are climate related.[1] Over 211 million people are directly affected each year.

I was shocked reading two articles with the ominous headings: *"Rising Fears of mass extinction after alarming scientific discovery"*[2] and *"Warning signs of mass extinction event on earth are growing, scientists say.*[3] Another headline says it all which is the most frightening: *Is HumanityDoomed? Science Says We Are In Middle Of*

A 'Mass Extinction Event'.[4] "I was also shocked to see that the author mentioned toxic algal BLOOMS which seem to be proliferating in the world's waterways thus threatening aquatic animals and other species of algae. According to the authors there was a great dying 251 million years ago aided by toxic algae blooms. So I decided to add a whole chapter to the second edition dealing with algal blooms. By the way the study of algae has it's own name, **Phycology**. I know that it sounds like psychology but it is not.

Something that should be pointed out and which is not stressed in the mainstream media is that the majority of the world's carbon dioxide emissions since 1850 have come from a minority of the world's population. The world now contains over 7.5 billion people with the richest half billion, or roughly 7% of the world's population, being responsible for 50% of the world's CO_2 emissions while the poorest 50% of the population are responsible for 7% of CO_2 emissions.

According to Mary Robinson, who is the United Nations High Commissioner for Human Rights, "The poorest have the least role in causing climate change yet they are being hit first, hardest, and worst.[5] This is especially true in poor countries.

I want to emphasize that the reason we wrote this book is because we see everyday the ignorance and stupidity of the human race when it comes to environmental and political issues which have life and death implications. I really don't want to bring this up but it appears that during the Trump administration (2017-2021) we were faced with the COVID 19 epidemic and due to Trump's failure to inform the public of it's virulence, it is possible that 600,000 Americans died because Trump ignored the virus by not telling the public of it's seriousness. It is believed that he could have informed the public and made mask wearing mandatory. Another major threat that faces the world is the burning of the Amazon Rain Forest. During the dry season (August-November) farmers set ablaze the trees in order to clear the forest for their use such as cattle raising and growing food crops.

Harlan Ellison (an award winning author of short stories, screenplays and essays) said *"the two most abundant elements in the universe are hydrogen and stupidity."* .

Einstein came up with a brilliant statement" *The difference between Genius and Stupidity is that Genius has its Limits.*

The following Table lists the current problems facing the world at the present time. However, the main premise of this book deals with diseases caused by climate change. In this edition we will also highlight diseases caused by algae blooms such as paralytic shellfish poisoning and Palm Island Mystery Disease.

TABLE I-1

Some major problems caused by Humans

Climate change (Global Warming)	Deforestation	Increased Toxic Algae blooms
Mass starvation	Fossil fuel dependence	Poverty
Warfare	Greedy corporations	Droughts
Earthquakes	Extinctions	Sea level rise
Disappearing islands	Nuclear Waste	Coral reef bleaching
Air pollution	Land pollution	Water pollution
Ozone depletion	Bioinvasions	Soil erosion
Animal and plant extinction	Garbage increase	Loss of natural pollinators
Radiation contamination	Methane production from domesticated animals*	Oil spills
Environmental refugees	Light pollution	Disease

*The reader can see my lecture called

"Low Methane Cows" given on Feb 6, 2022. Just click on the link below.

https://www.youtube.com/watch?v=T7ZafklcdKA

Carbon Dioxide (CO_2) and methane gas are being released in incredible amounts; specifically, from industry and domesticated animals such as cows and pigs. This release is causing Global Warming. Despite the scientific evidence that seems to prove that global warming is real, many members of the US Congress (mainly Republicans) and leaders of corporations are denying that it even exists and calling it one of the biggest hoaxes perpetrated on humanity[6]. Our book assumes that global warming is real and discusses the consequences, especially on human health.

Much of this information on global warming is not making its way to the mainstream media because of distractions such as the Tiger Woods affairs (golf player), the sex lives of American politicians, whether movie stars are divorcing, and other nonsense that has absolutely nothing to do with the reality that we are all facing **DISASTER IF WE DON'T ACT NOW.**

Why the concern? I have come to the conclusion that if we don't do something now, and I mean now, it will all be over. No more people; yes, no more people. Why??? Because they will all be dead due to new devastating diseases and other insults influenced by globalization and global warming.

We feel that now is the time to really be serious about what humanity is doing to the planet as well for their own future, even though it should have been done many years ago. We humans have wasted precious time by concerning ourselves with profits (the money kind) without regard for anything else.

The majority of this book is about diseases which affect people, yes, **Homo sapiens,** but it should be pointed out that many diseases discussed in this book also affect animals. Viruses, fungi, nematodes, insects, protists and other disease-causing organisms also affect plants, but this book is mainly concerned with the animal kingdom to which we humans belong.

Why call the book, **Prognosis: Disaster?** Simple. Because of humanity's pursuit of resources fueled by greed (where the rich want more and more at the expense of the world's poor), stupidity[7], warfare, ignoring poverty, etc. and ecosystem deterioration. This deterioration is leading to outbreaks of deadly diseases that didn't have to infect the millions of the world's citizens as they have in the past and continue to do so today.

As you, the reader keeps on reading, especially Chapter 4, you will understand what humanity has done to ecosystems kept in balance by Mother Nature and destroyed by Human Nature, **Homo sapiens** are currently destroying their very own home (the Earth's ecosystems).

The subtitle of the first edition is "The Environment, Climate Change, Human Influences, Vectors, Disease and the Possible end of Humanity?" explains how human intervention is mainly responsible for the possibility that humanity may destroy itself as well as the planet's other organisms.

- The Environment--without it there is no life. Damage done may be unsustainable.

- Climate Change--the earth's climate is changing in a negative way. Heat waves, floods, and storms are on the rise with the potential of displacing millions of people, thus making them environmental refugees.

- Human Influences--People are destroying most of the world's ecosystems as well as globalizing the planet, which is the process where countries interact with each other in positive ways and negative ways. The positive aspects are trading and cooperating on treaties, and the negative aspects are the exploitation of people and natural resources, and of course the greatest outrage of all, **warfare**.

- Vectors-- These are organisms, the majority of which are arthropods such as mosquitoes, flies and ticks, that spread disease.

- Disease--The threat of disease is magnified by climate change. The largest section in this book, Section 2, deals with diseases that are most likely to increase due to climate change.

- The Possible End of Humanity? --If we only care about money without regard for the biota on the planet (All Life) we are all doomed because all of us depend on life to sustain ourselves.

The subtitle of the second edition is: How climate change and Disease will ruin your life. UNLESS. Unless we do something now, and I mean now (we should have done it years ago) we are doomed. Humanity is plagued with climate change which is causing severe food insecurity in many countries of the world. These countries have had little impact on the causation of climate change. We also added a large section on the effects of toxic Algal Blooms- A little discussed result of climate change is a large increase in toxic algae blooms. When algae (microscopic plants) bloom (grow in tremendous quantities) they become a hazard not only to humans but to other aquatic organisms such as fish, birds, and other wildlife. Many contain potent toxins. Insects as well as other organisms such as ticks live longer and can spread disease. A large part of this book explains diseases that are influenced by climate change; but there are things that we can and must do now by changing our ways of doing things. WE MUST ACT IMMEDIATELY

---

[1] UNFPA.State Of the World Population 2009: Facing a changing world: women ,population and climate, p.30.

[2] Stanfield, L. 2021. Rising fears of mass extinction after alarming scientific discovery. Express.

https://www.express.co.uk/news/science/1495503/extinction-fears-scientific-discovery-algal-blooms. Viewed 26 Sept 2021.

[3] Sinay, D,2021.Warning signs of 'mass extinction event' on Earth are growing, scientists say. https://www.indy100.com/science-tech/study-mass-extention-human-caused-environment-b1925020

Viewed 26 Sept. 2021.

[4] Sharma, B. 2021. Is Humanity Doomed? Science Says We Are In Middle Of A 'Mass Extinction Event'. Indiatimes (on line) 26 Sept 2021. https://www.indiatimes.com/technology/science-and-future/humans-middle-of-mass-extinction-event-550114.html Viewed 25 Dec 2021.

[5] International Organization for Migration (IOM) Policy Brief. Migration, Climate Change and the Environment. Geneva: May 2009, pp.1-9.

[6] See the book, "The Greatest Hoax by James Inhofe, A former Senator from Oklahoma. A climate change denier.

[7] Einstein said."The difference between genius and stupidity is that genius has it's limits.

Section 1

Chapter 1

The Environment

Chapter 2

Biology

Chapter 3

Who or what is responsible for Climate Change?

Chapter 4

Human Activities Cause Disease. Here's How...

Chapter 5

Vectors

Chapter 1

THE ENVIRONMENT

I t is important to realize that the Earth and its life support systems are in balance, or should I say *were* in balance before humans placed their footprints on it by destroying many of its systems. Picture the earth just like a human, with all of its 11 or so systems which include a respiratory system, circulatory system, urinary system, reproductive system, lymphatic system, and many others. If just one of these systems fails, humans will be in trouble and perhaps die.

Well, think of the earth as a human. When a human is well, all is OK. When the human gets sick he/she is uncomfortable and may die.

The planet Earth is the same way. All of its systems, whether they are the atmosphere or the land and oceans with their plants, must be healthy all the time or trouble may ensue.

When I first went to college in the 1960s, there were very few environmental and ecology courses. Now, almost 50 years later, there are many more books on the subject, and many more environmental science courses taught at colleges and universities throughout the United States and in other countries.

Why are there so many courses being taught about the environment? Because these colleges realize that the environment is in peril and if people are educated about the environment, then they can do something to protect it. By protecting the environment, they are protecting their own HOME. That's right, the environment is our home! And don't you forget it.

THE ENVIRONMENT

To study diseases caused by human involvement in the environment, it is best to define the earth's environment and discuss how it works. What exactly is the environment composed of? How does it function? Why is it so fragile and open to destruction by humankind's lack of understanding? I will try to answer these questions in the next few pages.

Once the reader understands what a normal environment looks like, then he/she can understand what is happening to it. The best way to do this is to start with the Biosphere, also known as the Ecosphere.

THE BIOSPHERE-ECOSPHERE

What exactly is the Biosphere? The Biosphere is also known as the Ecosphere. The Biosphere is the area of the planet that has life. It is divided into three parts:

The Atmosphere (the air portion), The Lithosphere (the land mass), and the Hydrosphere (the water-containing part). Each part of the Biosphere is unique in its components. All three areas have life as well as non-living components. The living components, which include animals, plants, protists, fungi and bacteria, are called the *biota* or biotic factors. Non-living components are called the *abiota* or abiotic factors. See Table 1-1 for biotic and abiotic factors.

ECOSYSTEMS

Ecosystems are areas on the planet that contain living (biotic) and non-living (abiotic) components. Ecosystems can be aquatic, terrestrial, on tops of trees, in a person's intestinal tract, inside a cave, or just about anywhere where life exists. A good example of a unique ecosystem is the arctic Tundra. We know it is cold up there, at least for now, but maybe not much longer because of global warming. This unique ecosystem has a tremendous amount of aquatic life as well as land life.

TABLE 1-1

BIOTIC AND ABIOTIC FACTORS

BIOTIC FACTORS-living	ABIOTIC FACTORS-non-living
Plants-trees	Oxygen, Nitrogen, Carbon dioxide
Animals-zebras, insects	Wind
fungi-mushrooms, molds	Moisture
Protists- below is a list of some members of the protista	Atmospheric pressure
Amoebas	Light-sunshine
Algae	Electromagnetic radiation
Seaweeds	Soil
Malaria parasite	Gravity
Paramecium	Temperature
Red tide organisms (dinoflagellates)	Chemicals
Bacteria(Prokaryotic organisms)	Gamma rays

THE ATMOSPHERE.

This is the part of the biosphere that contains the air. Almost all (99%) of the atmosphere is up to thirty miles from the Earth's surface. The atmosphere consists of many gases, the first four being Nitrogen (N_2), Oxygen (O_2), Argon (Ar)and Carbon Dioxide (CO_2). Nitrogen and Oxygen make up about 99% of all the gases and the rest make up less than 1% of an unpolluted atmosphere. See Table 1-2 for a more complete list of atmospheric gases. Notice that Carbon Dioxide is rising, which means that percentages of the other gases are falling.

TABLE 1-2

COMPOSITION OF A NORMAL ATMOSPHERE

COMPOSITION OF AIR	PERCENT PER VOLUME	PPM
NITROGEN (N_2)	78.09	780,900.00
OXYGEN (O_2)	20.94	209,400.00
ARGON (Ar)	.93	9,300.00
CARBON DIOXIDE (CO_2)	.0420 (May 2022 and rising)[1]	461.00
NEON (Ne)	.0018	18.00
HELIUM (He)	.00052	5.2
METHANE (CH_4)	.00015	1.5
KRYPTON (Kr)	.0001	1.0
HYDROGEN (H_2)	.00005	0.5
NITROUS OXIDE (N_2O)	.000025	0.25
CARBON MONOXIDE (CO)	.00001	0.1
XENON (Xe)	.000008	0.08
OZONE (O_3)	.000002	0.02
AMMONIA (NH_3)	.000001	0.01
NITROGEN DIOXIDE (NO_2)	.0000001	0.001
SULFUR DIOXIDE (SO_2)	.00000002	0.0002

It should be pointed out that that since the Industrial Revolution, which started around 1870, the CO_2 concentration has been increasing. If the reader wants to know the CO2 concentration in the atmosphere for a specific day go to this web site: **CO2.Earth**. It is believed that the concentration of CO_2 prior to the Industrial Revolution was around 0.0270%. Now it is much higher, .0416% and rising. This amount may not seem like a lot, but it is for the Earth's systems to function properly. Cities like Los Angeles, California; Bangkok, Thailand; and Mexico City, Mexico are very polluted and I doubt that the mentioned O_2 and Nitrogen are as high as is listed in Table 1-2.

The atmosphere itself is divided into five major regions depending on temperature changes. See Fig.1-1 for parts of the atmosphere. The main parts of the atmosphere starting with the parts that we are standing in are as follows:

1. Troposphere

2. Stratosphere

3. Mesosphere

4. Thermosphere

5. Exosphere-Area above the thermosphere. It does not appear in Fig.1-1.

Exosphere

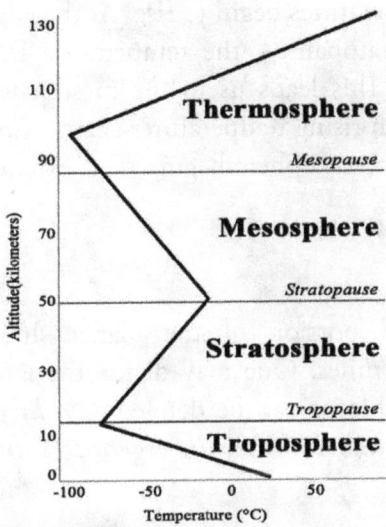

Fig. 1-1. A chart that shows the divisions of the atmosphere. (Illustration by Kevin Bley)

TABLE 1-3

THE ATMOSPHERE

PART OF THE ATMOSPHERE	DISTANCE ABOVE THE GROUND
Exosphere	640-64,000 km (400-40,000 miles)
Thermosphere	80-640km (50-400 miles)
Mesosphere	50-80km (31-50 miles)
Stratosphere (good ozone here)	16-50km (10-31Miles)
Troposphere	16km (10miles)

* A Km is equal to 1000 meters or .62 of a mile. 1.6 Km is equal to a mile.

Each area is divided up into the specific regions by something called a *lapse curve*. Lapse curves are areas with temperature differences. The troposphere, the first area, has falling temperatures as we go higher in altitude. From the edge of the troposphere-- called the tropopause-- begins the stratosphere, where temperatures begin to rise. At the edge of the stratosphere-- called the stratopause-- the temperature falls again, leading to the mesosphere. This leads us to the mesopause, where the thermosphere begins, with rising temperatures again. After the thermosphere we can say that the exosphere begins. This extends up to 40,000 Km above the earth.

THE LITHOSPHERE

The Lithosphere is the land portion of the planet. It is approximately 57.5 million square miles. One may divide the land masses into units called *Biomes*. A biome can be defined as *a large geographical area containing similar climate, soil, vegetation and animal life*.

Table 1-4[2]

A list of the major world's biomes

BIOME	HIGH AVERAGE TEMPERATURE IN DEGREES CELSIUS	AVERAGE RAINFALL IN CM 2.54 CM IS EQUAL TO 1 INCH	BIOTA TYPES
Tundra	5	23	Herbs, birds shrubs
Boreal forest and temperate evergreen forest(Taiga)	15	31	Trees, shrubs, birds
Temperate deciduous forest	24	81	Trees, shrubs, birds
Temperate grasslands	24	31	Perennial grasses and shrubs
Cold desert	22	38	Shrubs, herbaceous plants
Hot desert	38	15	Plant species
Chaparral	17	42	Shrubs, herbaceous plants
Thorn forest and tropical savanna	36	74	Small trees, shrubs
Tropical decidous forest	27	163	Deciduous trees
Tropical evergreen forest	21	262	Trees and vines

In theory, we can also say that the ocean consists of various biomes. There are areas with large reefs, such as the Great Barrier Reef off the coast of Australia, or even areas of the Red Sea with its higher salinity than other oceans. These areas of the ocean have their own endemic (found only in one place) aquatic life.

It should be pointed out that Mother Nature (God) in his/her infinite wisdom made all the biomes of the Earth suitable for specific organisms. The Earth was kept in balance until humans came along to disrupt it. These biomes play a significant role in maintaining the Earth's systems, which to put it bluntly, include the life support systems that we all need to survive.

Later in this section I will discuss the wide biodiversity found in the biomes. (See Table 2-4).

THE HYDROSPHERE-1.36 billion cubic kms of water on the Earth.

Water makes up about 70% of the Earth's surface, with oceans being the major aquatic biome, accounting for the wide variety of life as well as supplying oxygen through photosynthesis-- the oxygen that we need to breathe. About 97.6% of all the water on the Earth is saltwater found in oceans. The rest of the water, which is considered freshwater, makes up the rest. See Table 1-5.

TABLE 1-5

WORLD'S WATER RESOURCES[3]

WATER SOURCE	% TOTAL
Ocean	97.6
Ice and snow	2.07
Groundwater down to 1 km	0.28
Lakes and reservoirs	0.009
Saline lakes	0.007
Soil moisture	0.005
Biological moisture and plants andanimals	0.005
Atmosphere	0.001
Swamps and marshes	0.003
Rivers and streams	0.0001

It should be pointed out that we are losing groundwater due to extensive agriculture usage, which is quite wasteful because much of the water is lost to evaporation and doesn't make it to the crops. It is also evident that future wars will be fought over this precious resource.

Misuse of the world's water supplies will lead to more disease as is evidenced with increased dams. Stagnant water sitting in the reservoirs has the potential to be breeding grounds for mosquitoes which may carry malaria and other diseases.

As mentioned previously, the biosphere is our life support system of which we humans are all a part. When we destroy various parts of the biosphere, as in warfare or just for profit by taking available trees and making paper and furniture out of them, we are damaging our life support, perhaps forever.

IMPORTANCE OF BIOTA ON THE ENVIRONMENT

What is biota? Biota is the term used to describe all the living components of the planet.

Everybody who is alive at the present time depends on the biota for their well-being. How is this so? Simple!!!!. Everybody eats. All the food we eat comes from living organisms. Yes, that's right, living organisms. Vegetables are living or once were. Meat, which comes from various animals, also depends on plants for their' survival. Even those animals that eat other animals depend on plants, because, like us, meat eaters depend on animals that mainly eat plants. These plant eaters are called *herbivores*. Those that eat meat are called *carnivores*. Organisms such as humans eat both plants and animals; these are called *omnivores*. All this leads us to the next environmental topic: nutrition types and food chains.

Nutrition of organisms falls into the following categories:

1. AUTOTROPHIC
2. HETEROTROPHIC
3. SAPROTROPHIC
4. CHEMOSYNTHETIC

1. **AUTOTROPHIC**

Autotrophic organisms are generally plants which get their nutrition from inorganic chemicals. These are chemicals that do not contain carbon in *covalent linkage*. For those of you who don't know what that means, covalent linkage means that chemicals share electrons. Examples of organic chemicals are sugars, alcohol, proteins, starch etc. All of these contain carbon. Examples of inorganic chemicals include water (H2O), table salt (NaCl) and potassium phosphate (K3PO4), and many others.

Plants utilize the chemicals and make food by the process of photosynthesis, which can be expressed in this simple equation.

$$6H_2O + 6CO_2 + \textbf{sunlight} \rightarrow C_6H_{12}O_6 + 6O_2$$

This process takes place in the chloroplast of a plant cell.

Essentially this equations means that water (H2O) and carbon dioxide (CO2) plus sunlight makes glucose (a sugar) and oxygen. Figure 1-2 shows a typical chloroplast where photosynthesis takes place.

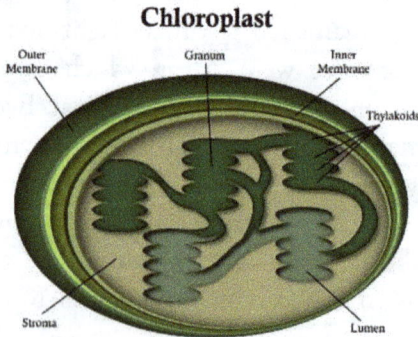

Chloroplast

Fig. 1-2. A diagram of a chloroplast. This is the organelle mainly involved in photosynthesis. (Illustration by Kevin Bley)

2. **HETEROTROPHIC**

Heterotrophic is the term that refers to those organisms that eat preformed organic chemicals. This includes humans, cockroaches, ants, certain bacteria and fungi. It's hard to imagine that when someone orders a steak with potatoes in a restaurant he/she is ordering preformed organic chemicals in the form of protein and starch.

For those of us who eat, we know that we eat food containing nutrients in the form of fats, proteins and carbohydrates, all organic.

3. SAPROTROPHIC

Saprotrophic (formerly called saprophytic) is the term that mainly refers to fungi and bacteria which obtain their nutrients from dead organisms. If you have ever been in the woods and seen a log rotting, you have seen that it is rotting primarily by fungus. This is an excellent example of a saprotrophic organism. It is these organisms that break down dead plants and animals. Imagine if this couldn't happen and a dead cow (for example) would not decompose—soon there would be thousands of dead cows all over the place!

4. CHEMOSYNTHETIC

These are organisms that obtain their energy from the oxidation of hydrogen sulfide gas (H_2S). This results in the synthesis of carbohydrates such as occur during photosynthesis in the CalvinCycle.[4] See equation below.

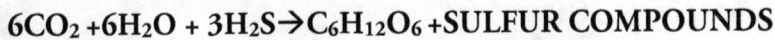

$$6CO_2 + 6H_2O + 3H_2S \rightarrow C_6H_{12}O_6 + SULFUR\ COMPOUNDS$$

Examples of organisms that obtain their energy this way are the deep sea tube worms and mussels found on the thermal vents which are located in areas where tectonic plates merge. (Tectonic plates are masses where parts of the planet move.)Water may reach temperatures of 400°C (472°F). But for these organisms that live there it is a paradise. These organisms have no stomachs and don't eat the way that we are familiar with such as having a mouth and digestive system. They have bacteria which live inside their bodies and are responsible for chemosynthetic activity by using the energy stored in the chemical bonds of H_2S as shown above. These vents were discovered by Robert Ballard of Woods Hole. Ballard was also the scientist who discovered the location of the *Titanic*.

5

Fig 1-3. Tube worms, Riftia sp living at the hydrothermal vents in the deep Pacific Ocean.

FOOD CHAINS

A food chain is the term used to describe levels of food eaten by all organisms. Put simply, it states "who eats whom." A food chain starts with a primary producer, a PLANT, being eaten by a primary consumer, an ANIMAL. This primary consumer is in turn eaten by a secondary consumer, another ANIMAL. The secondary consumer, too, may in turn be eaten by a tertiary consumer, another ANIMAL.

Fig. 1-4: A TYPICAL AQUATIC FOOD CHAIN WITH A HUMAN ON TOP.

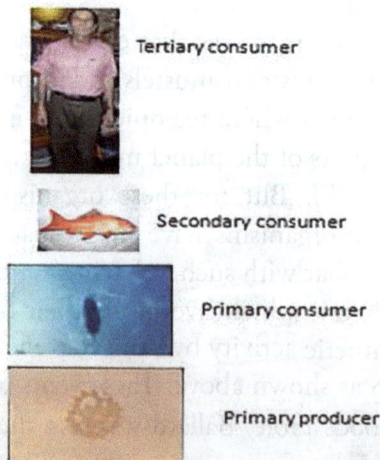

Tertiary consumer

Secondary consumer

Primary consumer

Primary producer

All photos above by David Arieti

THE IMPORTANCE OF ALL ORGANISMS

The importance of all, and I mean *all*, the organisms and ecosystems on the planet cannot be underestimated. Now let me explain. Let me give you, the reader, an analogy. We have a wood block game called JENGA. The object of this game is to remove as many pieces as possible without the tower falling down. Let's pretend that each block represents some aspect of the Earth's biosphere. The blocks could be the air, water, tree species, animal species, grass species, insect species, etc. If we take out each block, the tower gets weaker and weaker until it collapses. Now, if we would say that this tower represents The Planet Earth, we see disaster. In other words, the planet's life support systems would collapse, thus causing most, if not all, life to disappear.

When one is asked, "Does every species have a function on the planet?" the answer is **yes.**

What if we don't know the function of a species? Then we can still say that it has an ecological function. That is why we have to make every effort to prevent species from disappearing, no matter how useless or annoying they are to the human race. A good example of this is the mosquito. We know that on a hot summer day if there are plants and water nearby, humans get bitten by mosquitoes. Obviously, many people consider mosquitoes to be pests. True, but they have an ecological function. What is their function? For one thing, the blood that they get from humans and other organisms that they bite is used to feed their eggs. This creates more mosquitoes. Now, what organisms depend on mosquitoes for their survival?

Simple. Birds, dragonflies and bat species are among the many organisms that use mosquitoes as a food source.

One thing that everyone forgets, or doesn't even notice is the fact we humans think that we own the Earth and that we can do whatever we want to it. If we look at warfare and other activities that damage the Earth, we see the effects. (For a very good discussion of warfare on the Earth's ecosystems see the book *The Ecology of Warfare,* by Susan D. Lanier-Graham, 1993.)

Among the major effects of warfare or other destruction of habitat is depletion of certain organisms such as fish or trees. These depletions have serious consequences that warmongers and developers don't consider. Depletion of fish stocks, whether by overfishing or warfare, has disastrous consequences that cannot be foreseen, such as animal and human hunger. One effect of overfishing are the tremendous blooms of jellyfish all over the world because the fish were eating the jellyfish larvae. (See fig 1-5 below.) If people over-fish there is no fish to eat the jellyfish larvae, hence more jellyfish.

Human hunger is caused by depleting protein, such as killing plants and animals. By doing this, new diseases due to natural selection-- which is a form of evolution-- may occur.

It is evident that without a life-sustaining environment, life as we know it will cease to exist.

Fig 1-5 A bloom of thousands of jellyfish.[6]

NATURE'S SERVICES

One of the most important aspects of our environment is its biodiversity. This is the fact that we have a tremendous variety of organisms (any living thing) on this planet. There are roughly two million organisms identified thus far which include bacteria, fungi, single celled organisms called protists, animals and plants.[7]

Biodiversity isn't just an abstract idea that's important in biology class. Each and every organism has one or more functions that contribute to the successful operation of Planet Earth. These functions can be called "**Natural Services,**" **And** include the following:

ECOSYSTEMS SERVICES AND FUNCTIONS

1. Gas regulation
2. Climate regulation
3. Disturbance regulation
4. Water regulation
5. Water supply
6. Erosion control and sediment retention
7. Soil formation
8. Nutrient cycling
9. Waste treatment
10. Pollination
11. Biological control
12. Refugia (refuge for animals)
13. Food production
14. Raw materials
15. Genetic resources
16. Recreation
17. Cultural

According to many investigators the value of the Earth's natural resources and services amount to approximately $18-54 trillion per year (1997).[8]

Notice that these are the services that we humans and every single organism on The planet Earth needs. So lets discuss each service separately and realize that as I write, various countries and humans in the name of profit are destroying some of these services.

1. GAS REGULATION

What exactly is gas regulation? Notice that everyone who is reading this is breathing. If you look at Table 1-2 you will see the percentages of most of the gases in an unpolluted atmosphere. These gases have remained constant for thousands of years if not millions. Nature, in his or her infinite wisdom allows the percentages of gases to remain constant. However, with the arrival of the human race some gases such as CO_2, methane, and Nitrogen oxides have risen due to their activity such as hunting for oil and burning it and loading up the soil with nitrogen fertilizers.

Hence polluted air.

2. CLIMATE REGULATION[9]

Notice that we have four seasons. Most of the time we have had a stable climate due to The Gulf Stream and the jet stream, rainy seasons and dry seasons which were normal. We knew when the rains will come and go. One might say that the Planet Earth has a built-in thermostat. According to an article in "The New Scientist" there are eight parts of the Earth's thermostat:

1. Volcanoes spew out CO_2.

2. CO_2 enhances the Greenhouse Effect

3. This warmth helps to evaporate sweater which brings rain.

4. CO_2 mixes with rain to form a slightly acidic rain (pH 5.6) which dissolves minerals from the rocks.

5. The rain dissolves carbon-containing minerals which wash into the Earth's waterways.

6. Minerals precipitate out to form carbon-containing rocks.

7. The rocks are subducted (where one tectonic plate goes below another tectonic plate- Geologic event) and CO_2 is released.

8. CO_2 returns to the atmosphere through volcanoes.

Hurricanes help balance the atmosphere and climate by radiating heat out of the tropics which go into space. They act as giant engines forming high winds.

3.DISTURBANCE REGULATION

This provides storm protection, flood control, drought recovery etc.

In March of 1980 The Volcano, Mount St Helens (Skamania County, Washington State, USA) erupted. Notice that the land regenerates itself. Volcanic disruption of land and its recuperation are an example of disturbance regulation.

Fig1-6. Mount St Helens fractured.[10]

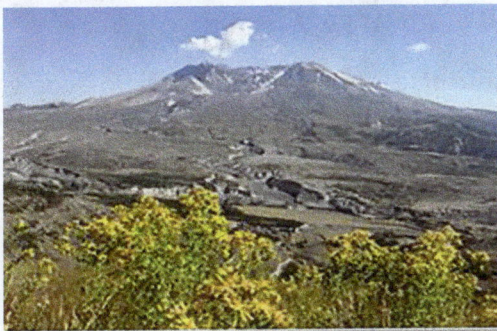

Fig 1-7. Mount St Helens 38 years after eruption.[11]

Fig 1-8 Rejuvination of burnt land.[12]

Here we see a South African mountain range where fire destroyed some land but now it is rejuvenating.

4 AND 5 .WATER REGULATION
AND WATER SUPPLY

This makes sure that we have water for agriculture, milling and transportation as well as provisioning of water by watersheds, reservoirs and aquifers.

I lived on the eigth floor of an apartment house in New York City. Every time I would turn on a faucet I would get water. Other countries are not so lucky but in most of the developed world water is available for crops and personal needs. How come it doesn't seem to run out?

Notice that Niagra Falls has been pouring water over the falls for years. Every second 3160 tons of water flows over Niagra Falls. The water comes from five out of the six great lakes.

At least for now most of us have enough water to supply our needs. This of course is due to natures services. At least until the year 2000 water seemed abundant. But now we may wonder if we will have enough water in the future because of our practice of wasteful irrigation.

Fig 1-9 Here we see wasteful irrigation practices in California. Much of the water gets evaporated.[13] Photo by Paul Hames

6. EROSION CONTROL AND SEDIMENT RETENTION

Prevention of soil loss by wind, runoff and other processes is also a process that nature performs. See Fig 1-10 below showing seagrass holding the soil together and thus preventing erosion.

Fig 1-10 Seagrass: *Zostera marina*[14]

7. SOIL FORMATION
(Pedogenesis is the technical term for soil formation)[15]

Soil is formed by rock weathering and accumulation of organic material.[16] Soil forms continuously by the gradual breakdown of rocks from physical weathering, chemical weathering and biological weathering as well as the accumulation of material through the actions of water, wind and gravity. Soil formation also involves the interaction between parent material, living organisms, climate, topography and time. Over a period of time organisms such as lichens, bacteria and algae help decompose rocks to make soil.[17]

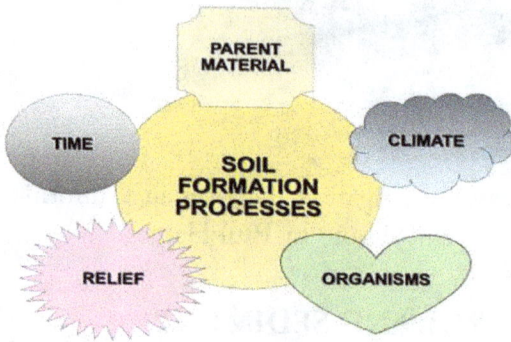

Fig. 1-11 Soil formation[18]

8. NUTRIENT RECYCLING

This includes nitrogen fixation (assimilated into other compounds) of nitrogen, phosphorus and many other elements used as nutrients. When organisms such as animals and plants die they decompose into nutrients.

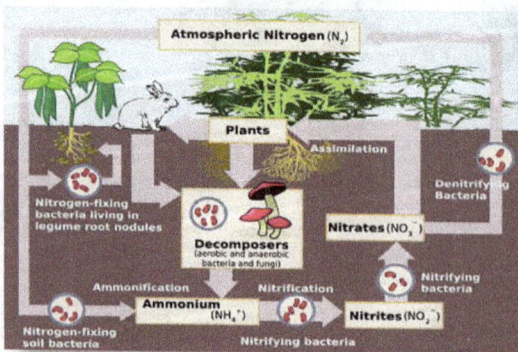

Fig 1-12 Nitrogen cycle.[19]

9. WASTE TREATMENT

Sewage treatment is basically a biological process. Many microorganisms are involved.

Fig 1-13 Waste water treatment plant in Belgium. Photo by Annabel[20]

10. POLLINATION

Pollination is accomplished by many organisms such as bees, bats, birds, mammals and many other animals as well as by wind and water.

Fig 1- 14 Bee pollinating a dandelion.. Notice that his thorax is covered with pollin. Photo byJessie Eastland.[21]

11. BIOLOGICAL CONTROL

Biological control means that nature automatically controls populations of some organisms so that that they don't overwhelm ecosystems. Good examples of this is the fact that most organisms have predators (Trophic-Dynamic relationship) which means transference of energy via food chains. Biological control also means controlling pests by use of viruses, other organisms, bacteria and parasites.

Fig 1-15 Vedalia Beetle eating cottony cushion scale insect.[22] Photo by Katja Schultz

Fig 1-16 *Syrphus hoverfly* larva feeding on aphids (an insect)[23] Photo by Beatriz Moisset

12. REFUGIA

Refugia are habitats for transient (migratory) populations such as geese and other animals. We see this here especially in the Chicago area where we see geese returning from overwintering.

Fig 1-17 Geese that returned to Beck Lake in Des Plaines, Il May 2021. Photo by David Arieti

13. FOOD PRODUCTION

For those of you who have ever eaten, you know that food is available. Did you ever think that this is a natural function of nature? You bet it is.

Fig 1-18 A supermarket with its copious amounts of food.[24]

14. RAW MATERIALS

Here raw materials include extractables like wood, metal ores such as iron, copper and nickel; oil, grain, natural gas etc.

Fig 1-19 Banded iron Ore.[25]

15. GENETIC RESOURCES

Nature gives us unique biological materials. Many medicines for our diseases such as diabetes and cancer can be found in tropical rain forests around the world.

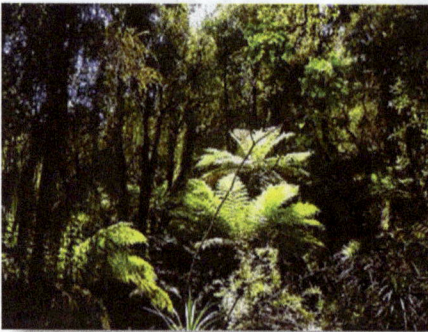
Fig 1-20 Tropical Rain Forest .Photo by Arenda Veenhuizen[26]

16. RECREATION

This includes ecotourism and other outdoor activities.

Fig 1-21 Recreation ground .[27]

17. CULTURAL

Cultural activities include those activities that have no commercial uses. These activities are for educational and spiritual reasons.

Fig 1-22 Cultural activities-rafting.[28]

[1] https://www.CO2.earth/

[2] Life: The Science of Biology, 8th edition, by David Sadava, H.C. Heller, G.H. Orians, W. Purves, D.M. Hillis; Sinauer Associates and W.H. Freeman and Company 2008. p.1117,

[3] Cunningham, W, A. Cunningham and B. Saigo, *Environmental Science: A Global Concern.*, McGraw Hill, 2007, p.374.

[4] The Calvin cycle takes place in the stroma, the part of the chloroplast where CO_2 is converted to sugars.

[5] https://commons.wikimedia.org/wiki/File:Riftia_tube_worm_colony_Galapagos_2011.jpg Photo from NOAA Okeanos Explorer program. Viewed 30 March 2021.

[6] https://upload.wikimedia.org/wikipedia/commons/7/72/Water-jellyfish.jpg Viewed 18 April 2021.

[7] Wilson, E. O. "Vanishing before our eyes." Time Magazine April-May 2000, pp. 29-30.

[8] R.Costanza, R. d'Arge, R.de Groot, S. Farber, M.Grasso, B. Hannon, K.Limburg, S.Naeem, R.V.O'Neil, J. Paruelo, R. Raskin, P. Sutton and M. van den Belt, "The Value of the World's Ecosystem Services and Natural Capital," Nature 385 (May 1997): 253-262.

[9] Lovett, Richard. 2008.Unknown Earth: Why is Earth's climate so stable? New Scientist 24 September 2008.

[10] https://commons.wikimedia.org/wiki/File:Mount_St._Helens_fractured_tree.png. Photo by Reywas92

[11] https://en.wikipedia.org/wiki/Disturbance(ecology)#:~:text=In%20ecology%2C%20a%20disturbance%20is,of%20biotic%20and%20abiotic%20elements

[12] https://commons.wikimedia.org/wiki/File:Contrasts_-_fire.jpg Viewed 19 March 2022.

[13] https://commons.wikimedia.org/wiki/File:Crop_sprinklers_Rio_Vista_California_15_Jul_2004-002.jpg Viewed 19 March 2022.

[14] https://commons.wikimedia.org/wiki/File:Seagrass_Zostera_marina_(Dzharylhach_island).jpg Viewed 16 May 2021.

[15] https://en.wikipedia.org/wiki/Pedogenesis#/media/File:Soil-formation-factors-en.jpg Viewed 28 Dec 2021

[16] https://www.pmfias.com/soil-formation-indian-conditions-factors-that-influence-soil-formation/

[17] https://www.qld.gov.au/environment/land/management/soil/soil-explained/forms Viewed 28 Dec 2021.

[18] https://commons.wikimedia.org/wiki/File:Soil-formation-factors-en.jpg

[19] https://commons.wikimedia.org/wiki/File:Nitrogen_Cycle.svg

[20] https://commons.wikimedia.org/wiki/File:WWTP_Antwerpen-Zuid.jpg Viewed 27 May 2021.

[21] https://commons.wikimedia.org/wiki/File:Diadasia_Bee_Straddles_Cactus_Flower_Carpels_close-up.jpg Viewed 28 May 2021.

[22] https://commons.wikimedia.org/wiki/File:Vedalia_Beetle_(15959056801).jpg Viewed 28 May 2021.

[23] https://commons.wikimedia.org/wiki/File:Syrphid.maggot3554.5.13.08cw.jpg Viewed 28 May 2021.

[24] https://commons.wikimedia.org/wiki/File:EmpressWalkLoblaws-Vivid.jpg Viewed 28 May 2021.

[25] https://commons.wikimedia.org/wiki/File:Banded_iron_formation_Dales_Gorge.jpg Viewed 19 March 2022. Photo by Graema Churchard

[26] https://commons.wikimedia.org/wiki/File:Rain_forest_near_Monroe_Beach_-_panoramio_(3).jpg Viewed 19 March 2022. \

[27] https://commons.wikimedia.org/wiki/File:Long_Lane_Recreation_Ground_2016(4).jpg Viewed 10 June 2021.

[28] https://commons.wikimedia.org/wiki/File:Rafting_em_Brotas.jpg Viewed 22 Jan 2022.

Chapter 2

BIOLOGY

In order to begin a discussion about diseases it would be best to tell the reader about life and the organisms associated with life. What exactly is life? Life as we know it has many characteristics. Most organisms have at least some of the characteristics below. However, viruses (for example) are not considered life by many scientists because they don't meet the requirements listed below. I will discuss viruses after we discuss life.

CHARACTERISTICS OF LIFE

First of all, let's define life. Biologists attribute the following characteristics to life. It should be pointed out that most organisms do not have *all* the characteristics mentioned below, but *most* of them. In order to be considered life an organism must meet the following criteria:

1.) **CELLULAR STRUCTURE**- They must have some **CELLULAR** basis. They all must be composed of cells or a single cell.

2.) **CELLULAR RESPIRATION**-They must have **RESPIRATION**. This is the process in which food is broken down in the cell's body for energy. Different organisms utilize different sources of food and therefore have different types of respiration. For example, animals actually eat with a mouth and food goes into a stomach. Plants don't have stomachs, but they utilize chemicals like nitrogen-containing compounds as a food source, and these are absorbed by the plant body.

3.) HOMEOSTASIS- This means maintaining the internal environment such as blood pressure, temperature, osmotic pressure etc.

4.) ENZYMATIC REACTIONS-All organisms have enzymes. These are mainly made of protein. They work by lowering the energy of activation. Enzymes speed up reactions. In plain English, this means that with enzymes you need much less energy to affect a reaction.

5.) NUTRIENTS-All organisms need food of one sort or another. Plants need inorganic nutrients such as nitrogen, phosphorus and potassium. Bacteria need either organic or inorganic substances, depending on the species and where they live. Animals need preformed organic substances such as plants, other animals and sometimes both.

6.) RESPONSE TO STIMULI- All organisms respond to various types of stimuli such as sound, light, moisture, atmospheric pressure, electromagnetic radiation, gravity, etc.

7.) INTERACTION WITH THEIR ENVIRONMENT- All organisms have to interact with their environment. Otherwise, they wouldn't survive. It is also important that if their environment gets destroyed, catastrophic events may occur which will undoubtedly cause severe effects such as an unlivable environment.

8.) REPRODUCTION- All organisms must reproduce. Every organism must reproduce or they won't be around much longer. Speaking of reproduction I asked my class on a test what is the difference between Eukaryotic organisms and Prokaryotic organisms. One of the answers was, "Prokaryots don't reproduce" Interesting answer but obviously the students didn't understand characteristics of life too well. When you look at trees during springtime trees produce millions of seeds. It's obvious that most will not become fertile.

It should also be pointed out that fish lay thousands of eggs, however, less than one percent make it to adulthood because they are being eaten by other aquatic organisms.

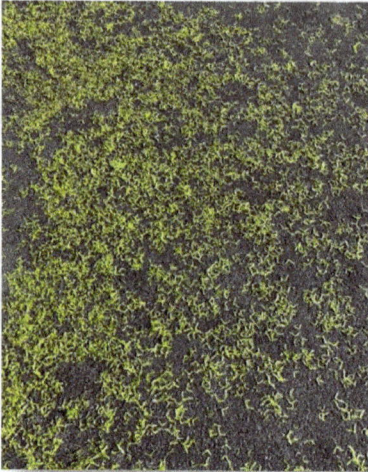

Fig.2-1 Seeds from a tree are released by the thousands. Photo by David Arieti

At present there are around 1.8 million known species of organisms identified but it is predicted that there may be 8.7 million species on the our planet.[1]

Biologists have devised three major groups of living organisms based on their cell structure. They are called domains or superkingdoms. These are called the BACTERIA, ARCHAEA and EUKARYA. The BACTERIA include the bacteria that we are familiar with, the ARCHAEA includes a different type of bacteria and the EUKARYA includes everything else.

The main differences between the Prokaryotic and Eukaryotic can be found in Table 2-1.

TABLE 2-1
DIFFERENCES BETWEEN EUKARYOTIC AND PROKARYOTIC CELLS

	EUKARYOTIC	PROKARYOTIC
Sexual reproduction	YES	NO
Membranebound nucleus	YES	NO
Membranebound organelles	YES	NO

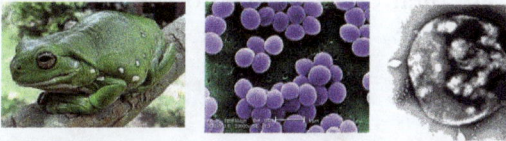

Examples of the three domains: **Eukarya, Bacteria and an Archaea**

(Three-domain system - Wikipedia)

The Eukaryotic kingdoms include:

Plants

Animals

Fungi

Protista

CLASSIFICATION OF ORGANISMS

Before delving more into the biology of these organisms it would be wise to explain how biologists classify organisms. A typical hierarchy of the classification of organisms includes the following taxons:

Domain (also known as Superkingdom)

Kingdom

Phylum

Class

Order

Family

Genus

Species

Biologists classify these organisms based on similar characteristics. For each of these taxons there may be the prefix "super" or "sub." Examples are superkingdom or subkingdom.

As mentioned above the two largest groups are the superkingdoms which are sometimes called Domains. They are followed by the six kingdoms which are then categorized into the various phyla (plural for phylum).

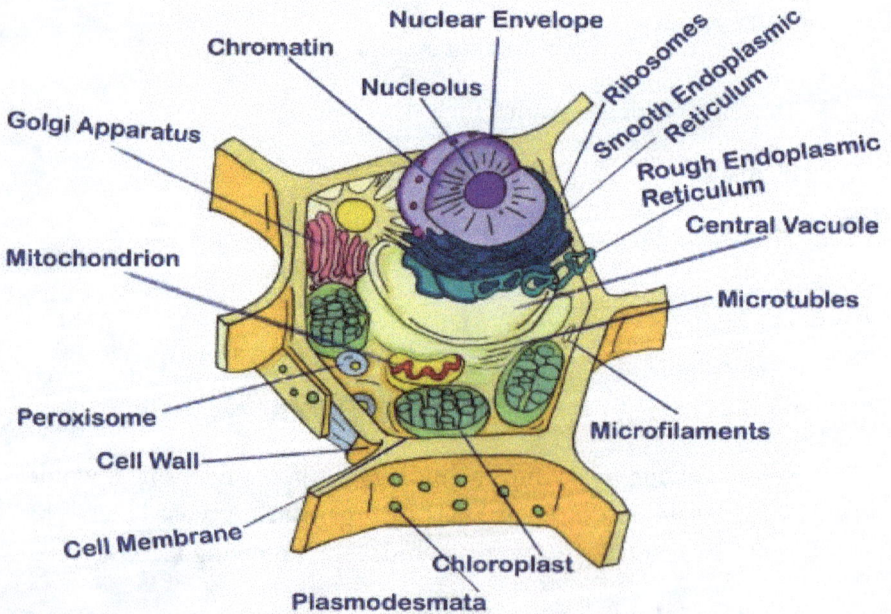

Fig. 2-2. A Plant Cell.[2]
Drawing by Miranda Olsen.

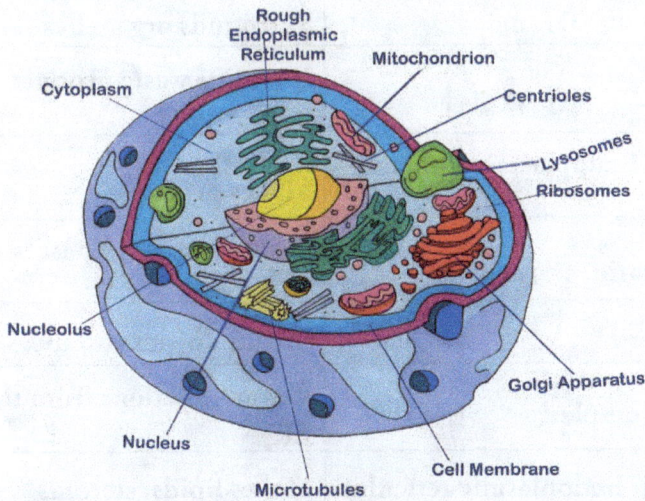

Fig 2-3 An Animal Cell.

Drawing by Miranda Olsen. [3]

I want to point out that plant and animal cells are similar with a few exceptions Plant cells have a large central vacuole, chloroplasts and a cellulose cell wall. Animal cells lack those three organelles but have a lysosome (breaks down substances), centrioles (involved in development of spindle fibers) and some animal cells have a flagella , used to propel cells.

TABLE 2-2

CELL ORGANELLES	FUNCTION
Nucleus	Control center of the cell
Nucleolus	Contains RNA, makes ribosomes
Mitochondria	Energy producers of cells

Plasma membrane	Surrounds organelles
Vesicle	Contains waste products and enzymes
Rough Endoplasmic reticulum –ER	Makes proteins
Chromatin	Genetic material-passes on genes
Lysosome	Breaks down substances
Golgi complex	Wraps secretions from the ER
Smooth Endoplasmic reticulum	Makes lipids, steroids
Peroxisome	Metabolism, detoxification
Chloroplasts	Photosynthesis
Cell Wall (Plant cells only)	Cellulose wall found in plant cells
Large Central Vacuole (In plants cells)	Storge vacuole found in plant cells

The six kingdoms are the following:

THE PROKARYA

1. **ARCHAE** -These include bacteria which have been identified as having different types of cellular components such as cell walls, different types of metabolism, and live in extreme environments (see Table 2-3 for examples of extremeophiles) such as in temperatures which approach that of boiling water, and above; very high and low pH's and in high salinities (high salt concentrations). Some examples are: *Halobacterium sp. and Methanobacterium sp.*

TABLE 2-3

TABLE OF VARIOUS TYPES OF EXTREMEOPHILES[4]

EXTREMEOPHILES	TOLERANCES
Halophiles	high salt concentrations
Barophiles	high pressure
Thermophiles	high temperatures
Acidophiles	pH below 3-very acidic[5]
Psychrophiles	low temperatures of 15°C or lower
Xerophiles	dry conditions
Piezophile	high hydrostatic pressure
Osmophile	high sugar concentration
Oligotroph	nutritionally limited environments
Metalotolerant	heavy and toxic metals
Lithoautotrophic	carbon source is only carbon dioxide (CO_2)
Hypolith	inside rocks in cold deserts
Hyperthermophile	temps of 80-122°C
Endolith	microscopic spaces within rocks
Alkaliphile	high pH concentration of 9 and above

2. **BACTERIA**-This is the category where most of the 5000 species of bacteria belong. These include most of the bacteria that we are familiar with. These include the coliform bacteria such as those found in intestinal tracts including the most famous one of all, ***Escherichia coli,*** and ***Mycobacterium tuberculosis,*** the causative agent of Tuberculosis. Bacteria are classified by the way they look, metabolize certain substances, and interact with other organisms.

Bacteria come in three basic shapes: Rod - shaped called **bacilli**, round ones called **cocci**, and spiral-shaped called **spirilli**. Examples of diseases caused by bacilli are Tuberculosis, Anthrax and *E.coli* infections. Examples of diseases caused by cocci are streptococcus diphtheria, staphylococcus infections and Strep throat. Examples of spirilli-caused diseases are Lymes disease and Syphilis.

Ricketssia and **Chlamydia** are bacteria which are obligate intracellular parasites.

Mycoplasmas are bacteria without cell walls.

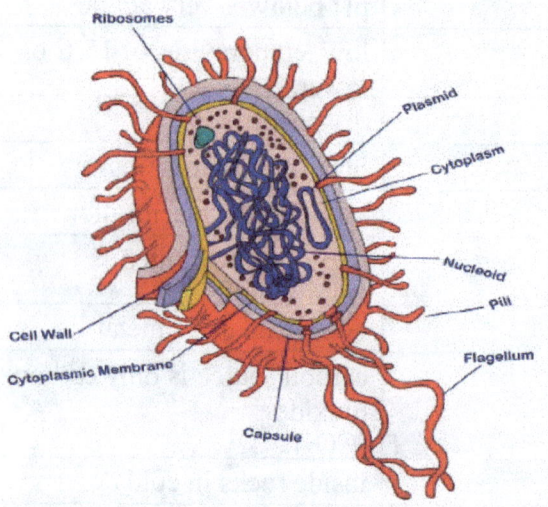

Fig. 2-4. Typical Bacterial Cell
Drawing by Miranda Olsen

THE EUKARYA

1. **ANIMALS-**These include organisms such as humans, worms and the much-maligned cockroach.

   Examples of animals which should be included in this book include mosquitoes, which are vectors of many diseases such as dengue fever and malaria. Nematodes, which are unsegmented roundworms, cause both human and animal diseases such as river blindness, Guinea worm, ascariosis and elephantiasis.

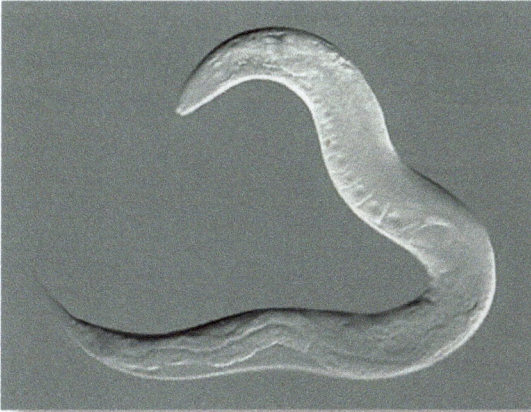

Fig. 2-5. *Caenorhabditis elegans* **Photo by Rob Goldstein-See footnote 1 of Fig 9-1 in beginning of Chapter 9**

2. **FUNGI** - These are organisms which don't contain chlorophyll and therefore can't photosynthesize, have a substance called chitin in their cell walls, and reproduce by spores. Common examples include the common morel, ***Morchella esculenta***, the penicillin-making fungus, ***Penicillium notatum***, and yeast-- the stuff that makes bread rise-- ***Saccharomyces cerevisiae***.

Fig. 2-6. Examples of fungi[7]

3. **PROTISTA**-These include many of the single-celled organisms that we have become familiar with in high school biology classes. Examples include the amoeba, ***Amoeba proteus***;

an algae such as ***Pediastrum simplex,*** and the organism that causes Malaria, ***Plasmodium falciparum.*** This group also includes seaweeds which are multicellular (many cells). A good example is sea lettuce, ***Ulva lactuca.***

Fig. 2-7. An example of a green algae, *Pediastrum simplex*. Photo by David Arieti.

4. **PLANTS-**This group includes those organisms that photosynthesize. We would not be alive without this group, because they produce the food and oxygen that we all eat and breathe.

Fig. 2-8. A plant--magnolia tree in bloom. Photo by David Arieti.

VIRUSES

Viruses are very interesting organisms. They don't meet any of the characteristics mentioned above, with the exception that they reproduce. However, they must utilize the DNA and other cellular components of their hosts to reproduce. They also must live as a parasite inside a living organism.

Viruses have the following characteristics:

1. They are acellular, which means they do not have a cellular organization.

2. They have either DNA (Deoxyribonucleic acid) or RNA (Ribonucleic acid), the genetic material which are the nucleic acids.

3. They have a protein coat called a capsid surrounding the nucleic acid core.

4. They can only multiply inside other cells by using the cell's DNA.

5. They range in size from 20 to 450nm (nanometers)

When dealing with human-induced climate changes, it is mainly these viruses that may cause the human race the most harm. By now we are all familiar with diseases caused by viruses such as the Ebola virus, HIV AIDS virus and the newly emerging disease, Avian flu virus and the COVID19 Virus.

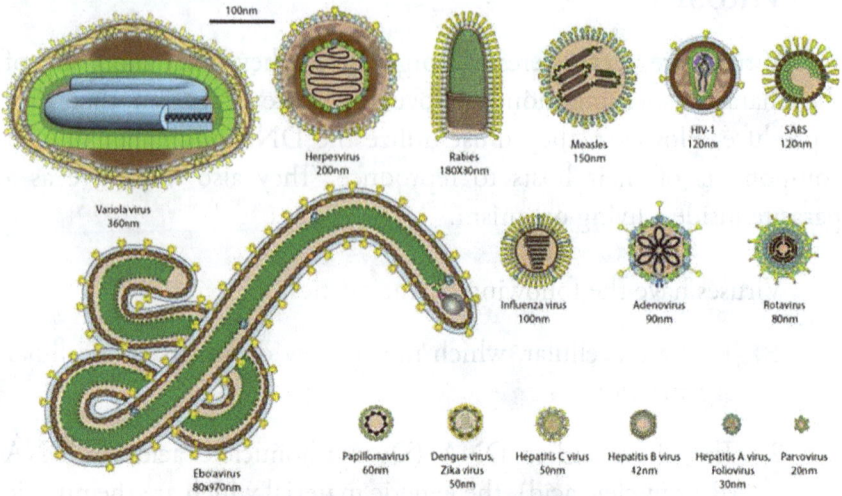

Fig. 2-9 Diagrams of various viruses.[8]

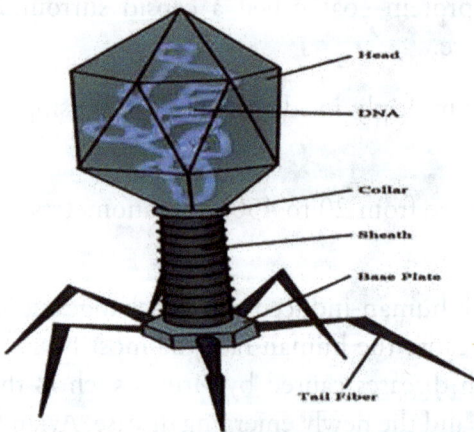

Fig. 2-10. A bacteriophage- A virus that infects bacteria. (Illustration by Kevin Bley)

OTHER AGENTS THAT CAUSE DISEASE

Recently two other entities of interest have been discovered. They are called prions and viroids.

PRIONS

Prions are proteinaceous infectious particles which are Infectious agents believed to be made of protein. The cause of disease is believed to be that there is a change in shape of these particles that somehow enter the brain and fold or bend, thus causing the brain to be damaged. The most common diseases believed to be caused by these agents are Scrapie, which is a disease found in sheep and goats; Mad Cow Disease of cows; and human forms such as Creutzfelt-Jakob disease, Kuru, and fatal familial insomnia.

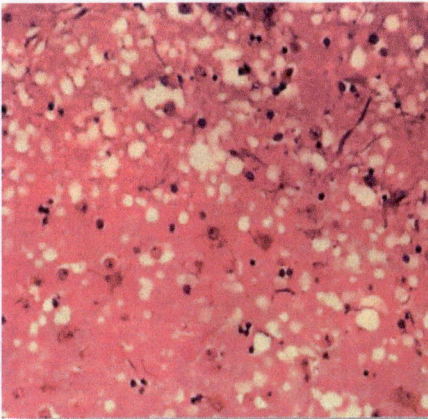

Fig 2-11 Brain tissue infected with prions.[9]

VIROIDS

Viroids are single-stranded, circular, RNA molecules without a protein coat and are capable of autonomous replication. They do not code for proteins as DNA does. They differ from viruses in that they don't have a protein coat. Most viroids consist of between 300-400 nucleotides.

They have been implicated in causing plant diseases in avocados, peaches, potatoes, tomatoes, chrysanthemums, palms and many other plant species. More than 30 viroid diseases have been reported. It is believed that transmission of viroids is mainly achieved by human involvement, hence their inclusion in this book.[10]

Viroids replicate in the nucleus or chloroplasts of plant cells. Some examples of viroid diseases are:

Dapple fruit –Hob Stunt-Viroid

Potato Tuber Spindle Viroid

Coconut cadang cadang viroid (CCV)

Hop latent viroid (HLV)

Hop stunt viroid (HSV)

Apple scar skin viroid (ASSV)

Avocado sunblotch viroid

Fig 2-12 Potato spindle tuber viroid[11]

ANIMAL PHYLA

There are approximately 36 phyla of animal species described at the present time. A list of the most important ones based on number of known species is listed in Table 2-4[12].

TABLE 2-4 KNOWN TYPES OF ORGANISMS ON THE PLANET (BIODIVERSITY)[13]

ORGANISM-ANIMALS	APPROXIMATE NUMBER OF SPECIES
Vertebrates	70, 000
Birds	9,990
Reptiles (snakes, turtles etc.)	8734
Fish	31,153
Mammals	5,500 (25% ARE BAT SPECIES)
Invertebrate species	OVER 1.3 MILLION
Insects	Millions
Crabs and relatives	47,000
Mollusks	85,000
Annelid worms	16,763
Nematode (Unsegmented roundworms)	25,000 ?
Flatworms	20,000
Cnidaria (Jellyfish and relatives)	9795
Porifera (sponges)	6,000
ORGANISM-PLANTS	
Bryophyta-Non-vascular plants (Mosses)	16,236
Vascular Plants (Trees, roses etc)	281,621
ORGANISM-FUNGI	98,998
Lichens-(Fungi and algae living together)	17,000
BACTERIA	7643

PROTISTA -Single cells eukaryots and algae	**60,000**
VIRUSES	**Thousands**

It is very important to realize that the phylum Arthropoda contains the largest group of organisms, with more than one million known at the present time.[14] This is the phylum of insects. It shouldn't be surprising that these organisms are the major vectors of disease. Below are examples of arthropod vectors. It should also be pointed out that arthropods that spread viruses are called Arboviruses. It is short for Arthropod-borne viruses.

All Arthropods have three major characteristics. They are the following:

A. JOINTED LEGS

B. HIGHLY DEVELOPED SENSE ORGANS

C. A CHITINOUS EXOSKELETON (Chiton is a glucose amine. This is a chemical similar in structure to glucose).

Examples of arthropods are insects, millipedes, centipedes, copepods, crabs and spiders. As for humans and the planet, Arthropods are very important. Those of us who like to eat should appreciate them because many, such as bees, help pollinate plants. The majority of vectors belong to this phylum.

EXAMPLES OF ARTHROPOD VECTORS - See Chapter 5

1. **MOSQUITOES**
2. **BLACKFLIES**
3. **SANDFLIES**
4. **TSETSE FLIES**
5. **FLIES**
6. **FLEAS**

7. **LICE**
8. **BITING MIDGES**
9. **TRIATOMINE BUGS**
10. **BEDBUGS**

DNA

In order to understand how organisms can be made to cause disease, it is very important to study DNA, the genetic material of all life on the planet. The little structures that carry the information and of course the genetic makeup of organisms are called *Chromosomes.*

Chromosomes are composed of DNA and little protein units called *histones.* Chromosomes are found in the nuclei (plural of nucleus) of all cells. DNA is also found in the *mitochondria,* the energy producers of all cells; and in *chloroplasts,* the oxygen producers of plant cells.

DNA consists of building blocks called *nucleotides.* Each nucleotide consists of one nitrogen- containing base, a molecule of deoxyribose (a type of sugar), and a phosphate group. Each nucleotide is represented by the first letter of one of the four bases that make up the nucleotide: Adenine, Guanine, Thymine and Cytosine. Thus we use the letters A,G,T,C.[15]

DNA is a molecule that is in the form of a double helix. This structure is probably the most famous biological molecule in the world and for good reason. See Fig. 2-13.

Since its discovery, the science of genetics has increased exponentially. Since the discovery of DNA we know more about disease etiology (causes of disease), mutations, treatments for disease, organ transplantation without immunosuppressive drugs, how to manufacture substances such as insulin in bacteria, how to create human organs outside humans, and the list goes on and on. Now with DNA we also can determine how evolution occurs. Organisms such as bacteria and insects have rapid reproductive rates which lead to new genetic discoveries. It's obvious that if one studies genetics it is better to get organisms that reproduce rapidly, such as bacteria and insects. Can you imagine doing genetic studies on elephants? [16]

Changes in DNA structure are now believed to cause mutations which lead to evolution of new organisms. In the section on Evolution (below) we will discuss how organisms can change. It should be pointed out that not all mutations are necessarily bad. In fact, we would probably not be able to exist in our (human) form if it weren't for mutations. However, on the other hand, some mutations result in diseases such as cancer and other serious illnesses.

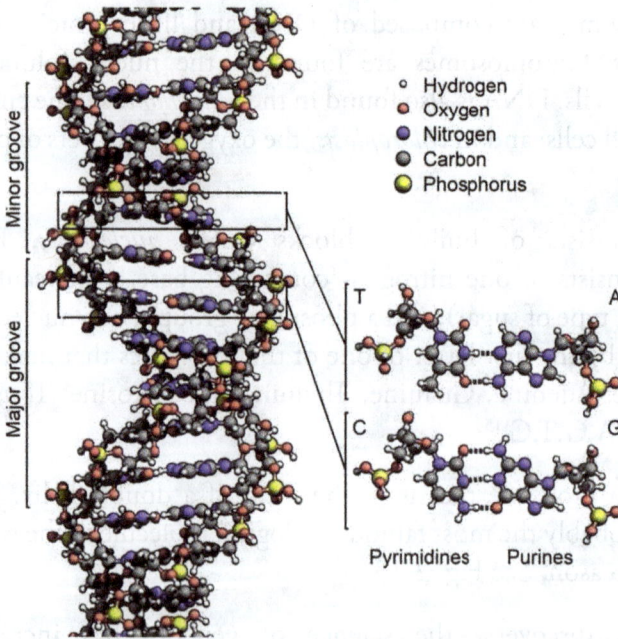

Fig. 2-13 Picture of a typical DNA molecule[17]

EVOLUTION

Evolution is the accumulation of genetic changes in species or in populations. It is an ongoing process. Unfortunately, here in the United States many people in certain States like Kansas don't believe in evolution. However, it is my intention to mention it anyway.[18]

In fact, as you read this sentence something is evolving somewhere on this planet. It could even be a newly dreaded disease of humans that is being created. For those of you who study biology, genetics is generally part of the curriculum. We realize that organisms and traits can change dramatically by what we do to influence them.

A good example of this is the carcinogenic effects of smoking. Most people who don't smoke don't get lung cancer. Many who do smoke do get lung cancer. Why? Because the substances in cigarette smoke trigger the cancer by interfering with the genes of the cell, and thus the proper functioning of the cells. The genes mutate.

This section is an introduction to evolution and how human activities may lead to new diseases.

There are four major mechanisms that give rise to evolution:

1. **Mutation**
2. **Genetic drift**
3. **Migration**
4. **Natural selection**

MUTATION

Mutation is the direct change in the DNA of an individual. Mutations can be beneficial, neutral or deleterious. Mutations can be caused by a mutagen, something that causes a mutation.

Mutagens can be physical, chemical, or biological. An example of a physical mutagen is UV light. A chemical mutagen can be something like Bennzo a-pyrine (found in cigarettes). An example of a biological mutagen can be a virus, which when infecting an organism may cause cancer.

Let me give you an example of how we describe mutation. From this you will be able to understand how a gene change can cause a problem.

TABLE 2-5
SOME TYPES OF MUTATIONS

TYPE	DEFINITION
Point mutation	A change in a single base
Base substitution	Replacing one base with another base
Transition	A point mutation where a C is replaced by a G or an A is replaced by a T
Frameshift mutation	A change of bases not in multiples of threes
Nonsense mutation	A change that results in a stop to the making of a protein
Missense mutation	Change in the base sequence that leads to a change in the amino acid sequence
Silent mutation	A mutation that does not alter the aminoacid sequence

The proper sequence of bases encode for specific proteins. Proteins are composed of amino acids. In humans there are 20 amino acids. A sequence of three nucleotides encode for one amino acid. If one base is either removed or replaced by a mutagen (something that causes a mutation), a whole sequence can be changed, thus causing a mutation which may be neutral, serious, or harmless.

To understand how mutations occur, let me show you in an easily understandable way. As mentioned above, DNA is composed of nucleotides, each containing one of the four bases. The sequence of the bases makes up the genes. Therefore, if there is a change in the sequence, a mutation can occur. Let me show how this works by giving you a hypothetical sequence of bases on a chromosome. I will just use 15 bases as an example. Chromosomes have millions of bases.

1) A-T-C-C-C-C-G-A-T-G-C-A-T-T-A

Notice that I divided the above sequence into three bases. In strand #1, the first sequence is A-T-C. If A is replaced by a G, you may get a different amino acid. If, as in strand #2, you add a base G after the first C, the frame will be totally changed.

2) A-T-C-G-C-C-C-G-A-T-G-C-A-T-T-

This type of change is called a frameshift mutation because the whole frame of the molecule will be changed. Notice that instead of the sequence as in strand #1, the whole sequence of bases has been changed in strand #2. Instead of ATC CCC GAT GCA TTA you will get ATC GCC CGA TGC ATT-A etc. By adding or deleting or changing a base. a disaster can occur. Thus organisms can mutate and those that were once harmless can become more virulent (dangerous) or even vice versa.

PLASMIDS

Plasmids are circular pieces of DNA that contain genes. They are found in bacteria and yeast cells. Plasmids containing genes can be transferred to other organisms, thus imparting traits such as drug resistance, and beneficial genes are used in gene therapy. Below is a diagram of how plasmids (small circles) are transferred to bacterial cells.

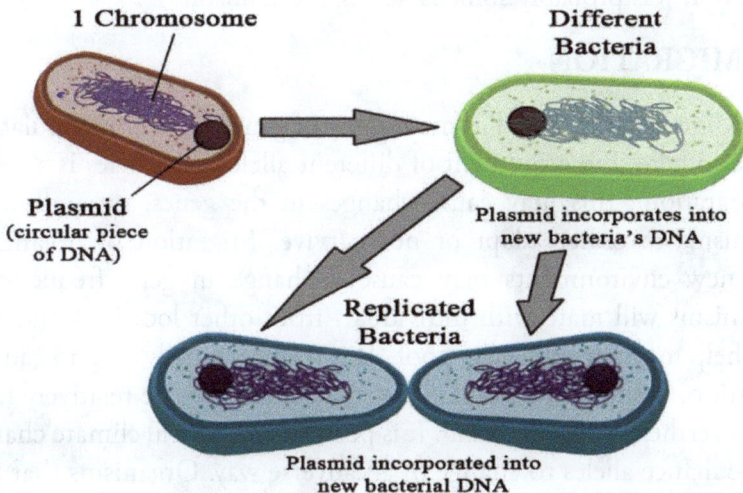

1 Chromosome

Different Bacteria

Plasmid
(circular piece of DNA)

Plasmid incorporates into new bacteria's DNA

Replicated Bacteria

Plasmid incorporated into new bacterial DNA

Fig. 2-14 Drawing of plasmids being transferred to cells. The small circles, called plasmids, are also called R factors. (Illustration by Kevin Bley)

GENETIC DRIFT

Genetic drift is a change in allele[19] frequencies mainly by chance from one generation to another. This is more likely to occur in small populations. A good example of this phenomenon is the Bottleneck effect. In this case, a natural population may be reduced in size by earthquakes, floods, droughts, lava flows, and even human habitat destruction. In these cases the gene pool (available genes in the present population) may be weeded out or totally eliminated. This may cause mating organisms to lose all their genetic variability. Put in plain English this means that if you marry and mate with a close relative you can keep "bad" genes, thus causing some defect.

This is why people or some animals that are close genetically, such as brothers and sisters, may have defective children if they reproduce. These defects can be physical or mental.

The quintessential example of this occurring is with the African Cheetah. These animals are known to have low genetic variability, probably due to an event that occurred around 10,000 years ago.[20] The event was probably some isolating mechanism.

MIGRATION

Migration is defined as movement of genes from one population into another.. The movement of different allele frequencies is a result of migration. This may cause changes in the genes, thus allowing organisms to either adapt or not survive. Migration of organisms into new environments may cause a change in gene frequencies. Organisms will mate with individuals from other locations and this may help to change the gene pool. This may eventually help to cause a mutation. It should be pointed out that mutations are relatively rare, but nevertheless they do occur. It is possible that global climate change may enhance alleles to change in an adverse way. Organisms that live in a stable environment probably don't change much.

I believe this is the reason why the Ebola virus suddenly appeared on the scene. Ebola is a deadly member of the deadliest virus family, called filoviruses.

Ebola made itself famous by appearing in Zaire, a country in Africa. It was made famous by the book, *The Hot Zone*, [21] written by Richard Preston, who has to be one of the most eloquent writers of this day and age.

Scientists are not 100% sure where it originally came from, but there is suspicion that it originated in monkeys. Remember that Africa has a large rainforest. Africa, by the way, is the least densely populated continent on the planet.

After learning about Ebola, and of course teaching environmental science, I began to put two and two together. Remember what I said about Africa being the least densely populated. Now add the fact that there are not enough jobs to go around for everyone, especially in countries in Africa where there are lots of political corruption and poverty. This poverty will cause unemployed humans to earn money any way they can, even if it means cutting, chopping and poaching everything they can. This also means that they (humans) will venture out anywhere.

Many of the areas in the rainforest are far from human habitations. Organisms such as monkeys, bats, insects, etc. were living in harmony with each other. It is possible that viruses such as Ebola live naturally with these animals or plants. But when a new organism, such as humans, encroaches on their territory, the viruses may jump species-- such as from monkeys to humans-- and thus cause disease. They may even mutate to be more virulent.

Currently there are four known strains of Ebola: Ebola Zaire, Ebola Ivory Coast, Ebola Sudan and Ebola Reston (this is the strain made famous in Richard Preston's book, *The Hot Zone).*

It may even be possible that as you read this another strain is evolving.

So here we have migration of the Ebola virus when it infects a human. This may just be the tip of the iceberg. We may be in store for something even worse than the Ebola that we now know. Remember, the symptoms of this virus are relatively horrible. One symptom of being infected with Ebola is blood coming out of any orifice in the body, just for starters. Can anything be worse? I leave it up to you to imagine it.

NATURAL SELECTION

This is a term used to describe a force that changes gene frequencies. There are three main types of natural selection:

1.)DIRECTIONAL

2.)DISRUPTIVE (DIVERSIFYING)

3.)STABILIZING

To make this easier to understand, let's mention two terms:

Phenotypes and *Genotypes.* Phenotype is the term used to describe the way an organism looks. We can say "that person has blond hair." Being blond is an example of phenotype. Genotype is the term used to describe the genetic makeup of an organism. We can say that XX is the genotype of a female human and XY is the male genotype. X and Y refer to one of the 46 chromosomes that humans have.

DIRECTIONAL SELECTION

In directional selection one extreme persists. Let's say that you have black, white and grey butterflies, where grey is the average color. In directional selection black or white will predominate, with the other colored butterflies being almost eliminated or greatly reduced.

DISRUPTIVE SELECTION

In disruptive selection the two extreme phenotypes will predominate. Thus black and white butterflies will predominate while the grey colored ones are reduced in number.

STABILIZING SELECTION

In stabilizing selection the average phenotype will predominate. In this case the grey butterflies will predominate, while black and white butterflies are reduced.

It should be pointed out that there are many examples of natural selection.

Now if evolution occurs, it is possible that organisms can adapt to a changing environment either benignly (harmlessly) or by turning into a virulent (bad) organism.

It could be that in order to survive they may have to cause a disease.

When we see and hear about deadly viruses like Ebola, we can see what may happen when organisms and non-organisms like viruses go wild. Again, I am not saying that this *will* happen, but *may* happen. The fact is that with the knowledge gained through the study of genetics and breakthroughs in biotechnology we can see that virtually anything can happen.

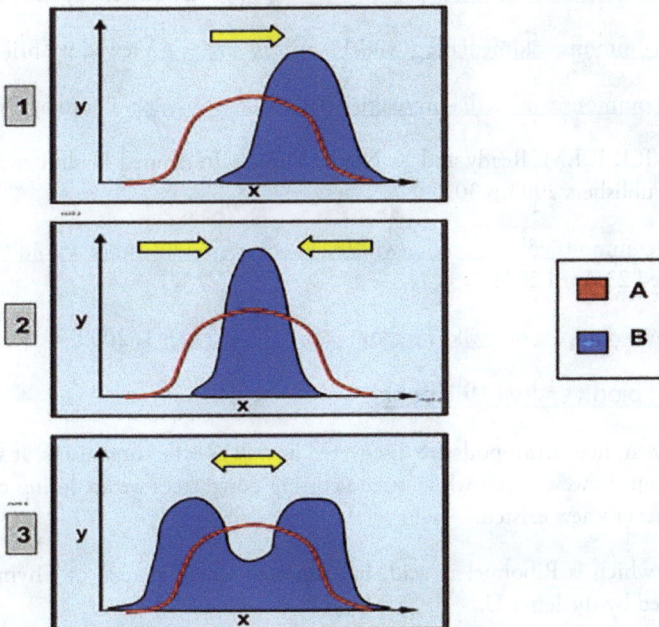

Fig. 2-15 Charts showing the differences between directional, stabilizing and disruptive selection. The colored sections show which groups are prevalent in each graph. #1 Is directional selection; #2 is stabilizing selection and #3 is disruptive selection.[22] The red curve is the average in all three parts of the figure.

---

[1] **How many species haven't we found yet?National Geographic.** https://www.nationalgeographic.com/newsletters/animals/article/how-many-species-have-not-found-december-   26#:~:text=A%20study%20in%202011%20 predicted,maybe%201.6%20million%20of%20them.

[2] Drawing by Miranda Olsen

[3] Drawing by Miranda Olsen

[4] https://commons.wikimedia.org/wiki/File:Banded_iron_formation.png Viewed 28 May 2021.

[5] Wikipedia-Viewed 1/17/09.

[6] File:CelegansGoldsteinLabUNC.jpg - Wikimedia Commons Viewed 11 Dec 2021.

[7] http://en.wikipedia.org/wiki/File:Fungi_collage.jpg Viewed Feb. 20, 2010.

[8] https://commons.wikimedia.org/wiki/File:Virus_size.png Viewed 9 April 2021.

[9] https://commons.wikimedia.org/wiki/File:Histology_bse.jpg. Photo by Dr Al Jenny

[10] Singh, R.P, K.F.M. Ready and X. Nie in Viroids. In Ahmed Hadidi et al editors. Science Publishers 2003,p.30.

[11] https://commons.wikimedia.org/wiki/File:Potato_spindle_tuber_viroid_5356693. jpg Viewed 23 April 2021.

[12] http://en.wikipedia.org/wiki/Phylum   Viewed Oct 1, 2009.

[13] Global biodiversity - Wikipedia

[14] Every year, new arthropods are discovered as well as other organisms. It should be pointed out, however, that while we are cutting down trees we are losing organisms that we never knew existed.

[15] RNA, which is Ribonucleic acid, has the base Uracil instead of Thymine. It is symbolized by the letter U.

In addition to having U it is also single stranded and contains the sugar Ribose instead of deoxyribose.

[16] Elephants have life spans of 60 years and are pregnant 22 months.

[17] https://en.wikipedia.org/wiki/File:DNA_Structure%2BKey%2BLabelled.pn_NoBB.png

Viewed 12 Dec 2021.

[18] It is possible that when this book gets published, states that don't believe in evolution may delete the section on evolution. Cynical you say?

[19] An *allele* is a term used to describe an alternative form of a gene.

[20] Brookner, R.J. *Genetics: Analysis and Principles*. McGraw Hill. 2005, p.705.

[21] Preston, R. *The Hot Zone*. Anchor Books, 1994.

[22] https://commons.wikimedia.org/wiki/File:Genetic_Distribution.svg Viewed 5 Jan 2022.

Chapter 3

WHO OR WHAT IS RESPONSIBLE FOR CLIMATE CHANGE?

The purpose of this chapter is to discuss the human factors that are responsible for Climate Change. Primarily, most of the damage to our planet stems from what I believe to be explicable in light of a brilliant observation by Einstein: "The difference between stupidity and genius is that genius has its limits." So, here we are, a planet struggling to survive for current and future generations.

For the past fifty years, based on indisputable empirical data, scientists have made it widely known to the public and the immensely wealthy fuel companies causing the problems, that fossil fuel burning is creating a warming world. However, most fuel company Chief Executive Officers (CEOs), fully cognizant of these time-sensitive issues, are reluctant to change anything.

Additionally, they downplay the effects to their shareholders. Why? Because they are more interested in their company's profits than the health of the *Biota*. This is the animal and plant life in different geographical regions. Has anyone asked the CEOs: What good is money if the planet reaches temperatures too hot to exist?

In fact, there are many reasons for Climate Change. However, before discussing the factors responsible for the effects of oil induced

Climate Change, there was someone raising the alarm. Wallace Broecker, a Geochemist, was one of the first scientists to predict Climate Change by using the word: "Global Warming." This was during the 1970's, when he collaborated with Exxon.[1] He was known by his peers as: "The Grandfather of Climate Science and Dean of Climate Scientists."[2] In 1975 he published a paper in Science (08 Aug 1975) entitled *Climate Change: Are we on the Brink of a Pronounced Global Warming?*[3]

In 1996 in recognition of his work, Wallace Broecker received the National Medal of Science from President Bill Clinton. Throughout most of his career, Broecker studied world ocean circulation patterns.

Fig 3-1 Wallace Broecker-(See footnote 2 for source)

Unfortunately, oil companies, including Exxon Mobile, continued to ignore his warnings about Global Warming. Even the 41st president of the United States, George Herbert Walker Bush (G.W. Bush's father), refused to sign the Convention on Biological Biodiversity 1992 held in Rio de Janeiro, Brazil in May of 1992.[4,5] Although, all other global leaders at this Earth Summit did sign this vital agreement.

PROOF OF CLIMATE CHANGE IN TODAY'S WORLD

Now in the summer of 2021 we keep hearing on the news the catastrophic events of Climate Change in dying reefs, and hurricanes registering numbers 4 and 5 on the Saffir-Simpson Scale (See Table 3-1 on page 62)

The reality is that Climate Change has arrived. We are experiencing it every day. It is not, as former president (#45) called it: "a Chinese hoax."

The news headlines that I gathered during August 2021 corroborate the reality of Climate Change. Specifically:

- **RED ALERT: IT RAINED IN GREENLAND'S ICE SHEET FOR THE FIRST TIME IN RECORDED HISTORY**[6]

- The climate apocalypse is real, and it is coming.[7]

INTERGOVERNMENTAL PANEL ON CLIMATE CHANGE (IPCC) REPORT

Additionally, according to a landmark report by a UN (United Nations) Climate Panel, the Earth is already feeling the irreversible impacts of Climate Change. The report has determined that this **is** a 'reality check' and a grim warning. Yet, according to the report there's still time to take action to combat the climate crisis. However, the United States needs to join other countries that are taking care of Global Warming now, or it could be too late; the IPCC Report reveals an ominous warning about the fate of the human population.[8]

Below are further recent headlines from online news reports. They are all frightening.

- Five chilling predictions from doomsday climate report predicting 'devastating' future.'[9]

- StanChart CEO says companies must act on Climate Change; we cannot count on our governments.[10]

- Dixie Fire: Firefighters tackle historic California wildfire.[11]

- A harrowing New UN Report Finds Humans Are The 'Unequivocal' Cause of Climate Change.[12]

- Climate Change: UN to reveal landmark IPCC report findings.[13]

- Gulf of Mexico 'dead zone' has grown larger than Connecticut.[14]

- Climate Change disruption to oceans could freeze parts of North America.[15]

- IPCC report shows 'possible loss of entire countries within the century.'[16]

RECENT MAJOR STORM SYSTEMS

On August 26, 2021, Hurricane Ida arrived on the Northern shores of South America and proceeded to ravage the East Coast of the United States. At one point, it developed into a Category Four (4) storm with a one-minute sustained wind of 150 mph (240 km/hr).[17]

As a result, there were seventy fatalities. Damage to homes and other property was ninety-five billion dollars. The countries affected included: Venezuela, Columbia, Jamaica, Cuba, the Cayman Islands, and the Gulf Coast of The United States. Once you realize the current number of hurricanes, which are becoming increasingly frequent, you will see the correlation between these dangerous, damaging storms and Climate Change.

On December 11, 2021, a giant tornado struck six states: Illinois, Arkansas, Kentucky, Missouri, Mississippi, and Tennessee. It left a trail of death and destruction for over 250 miles, which makes it the longest path of a tornado in history. Is this a result of climate change? It was based on unseasonably warm temperatures in the Midwest before this tornado touched down.

Fig 3-2 Tracking Hurricane Ida. August 2021.[18]

Hurricanes are categorized according to wind speeds, utilizing the Saffir- Simpson Scale.

TABLE 3-1-Saffir-Simpson Scale[19]

Category	Wind speed-mph	Damage
1	74-95	Extremely dangerous winds
2	96-110	Extremely dangerous
3	111-129	Devastating damage
4	130-156	Catastrophic damage
5	156+	Catastrophic damage

This data raises an important question: Who or what is responsible for Climate Change? Based on my fifty years of professional experience as a science educator, and after conducting an extensive Review of Literature on Climate Change, I have narrowed down its main causes.[20]

THE REASONS FOR CLIMATE CHANGE (Greenhouse Effect)

The earth's temperature has allowed millions of organisms to inhabit the Earth. Due to natural gases in the atmosphere, including Carbon dioxide (CO_2) and water vapor, the Earth has a Greenhouse Effect. Light reaches the Earth's atmosphere, and it passes through clouds and air. Then it radiates upward as Infrared Heat. However, Greenhouse Gases stop heat from escaping into space. Instead, they remain on the planet, and create a warming environment. Please see Figure 3-3.

Picture of the GREENHOUSE EFFECT.

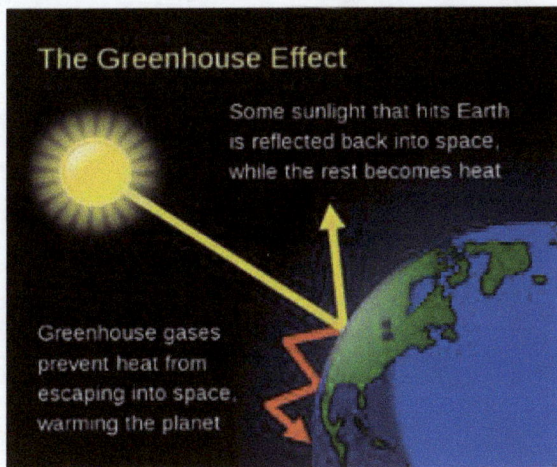

Fig 3-3 Schematic of Greenhouse effect- Efbrazil-author

Additionally, there are many different causes of Climate Change. The following lists include Natural Causes and Human Causes. The latter is a result of greed, recklessness, and a willful ignorance of the severity and urgency of this global crisis.

NATURAL CAUSES OF CARBON DIOXIDE (CO_2) RELEASE

- Volcanos
- Methane leaks
- Decomposition of organisms [22]

- Ocean release

- Respiration

- Others

HUMAN CAUSES OF CLIMATE CHANGE GASES:

- Oil Drilling
- Cement manufacturers

- Burning Fossil Fuels

- Power Plants (Especially Coal)

- Transportation

- Farming

- Deforestation

- Fertilizers

- Natural gas drilling

- Melting Permafrost

- Garbage

- Bitcoins

- Human overpopulation

- CEOs of Fossil fuel, Mining, Plastics, and Agribusiness Companies

- Politicians

Specifically, the people that I believe are most responsible for Global Warming have held the most powerful positions in the world. They could have taken steps to save our planet. However, they have failed us. As a direct result, I contend that all are truly despicable or incredibly obtuse about the way ecosystems operate. My list includes three former U.S. Presidents, and the former President of Brazil and two oil executives.

- George Herbert Walker Bush
- George W. Bush
- Donald Trump
- Jair Bolsonaro
- Lee Raymond
- Rex Tillerson

POLITICIANS

There are many kinds of politicians in the world. Some care deeply about the environment. On the other hand, there are several elected officials who do not value it at all. It is as if they cannot fathom that an Ecosystem sustains all life. These pseudo leaders only care about the economies of their countries. Don't they realize that a good economy is worthless in an environment that will cease to exist? Hello? There are countless politicians that fit this reprehensible category. But we will only discuss these four: George Herbert Walker Bush, George W. Bush, Donald Trump, and Jair Bolsonaro, currently President of Brazil.

George H. W. Bush President #41

Fig 3-4 George H.W. Bush[23]

When it comes to common sense and intelligence, George H.W. Bush is not your man. The earlier referenced 1992 convention on Biodiversity in Rio de Janeiro, Brazil, known as the **Convention on Biodiversity (CBD),** was designed to address international conservation concerns, including mass extinctions and ecosystem degradation.[24]

There were 196 countries present. A total of 168 nations signed the agreement. However, the leader of the Free World, the U.S., President Bush Senior did not sign it. Why? He had concerns that it would not protect the patents of American Biotechnology Firms. These are Pharmaceuticals, prescription drug manufacturers. His Mantra was: The Hell with saving Planet Earth. How infuriating.

Initially, I was hopeful that the United States would be on board to save our planet because of the person President Bush sent down to Rio de Janeiro to represent us. This human being wanted to do something positive for the environment. However, it did not happen.

The United States was on the wrong side of history here.

Then in 1993, President Bill Clinton signed the climate agreement. However, he could not get Senate approval because Republicans opposed it. The following sentence comes from an article about the 1992 convention and Bush: "Mr. Bush finds himself in this position because of the trade-off he made by placing economic and political needs, rather than diplomatic or scientific concerns, at the center of Administration's environmental stance."[25]

The article pointed out that negotiators in Rio were nearing a multimillion dollar accord in which industrialized nations, like the United States, would contribute money to help pay for environmental efforts in the developing world. Bush's refusal to commit to this agreement make it painfully obvious that he could care less about the world's environment. This is pure stupidity on his part mixed with entitled indifference. Lives are at stake, Mr. President. Please think of the Greater Good and the hundreds of millions of people that you are supposed to represent, protect, and serve.

G.W. Bush, son of president number 41.

Fig 3-5 George W. Bush[26]

Not surprisingly, when his son, George W. Bush, became president, he, too, lacked intelligence and good judgement. In addition, it should be noted that George W. Bush started an oil and gas exploration company called Arbusto Energy in 1977.[27] How ironic that the head of an oil company has the initials G.W., which just happen to stand for Global Warming. This president's Frame of Reference, based on his life experiences in the oil business, do not bode well for the environment.

To sum up his environmental record, let us consult another article.[28] First, a spokesperson for the Sierra Club points out: "He has undone decades, if not a century, of progress on the environment." Secondly, the same article elaborates: "The Bush administration has introduced this pervasive rot into the federal government which has undermined the rule of law, undermined science, undermined basic competence ..."

Furthermore, Dr. James E. Hansen, the prestigious scientist who raised awareness of Global Warming with his testimony before Congress during the 1980's, also exposed the Bush Administration. Specifically, he accused them of trying to block data showing that global warming is accelerating. Now, thirty years later, we see that Hansen's fears have now become a reality.

Donald Trump[29]

I did not include a picture of Trump here because it is too distressing to look at. Also, the entire world knows what he looks like. It is really a shame to waste ink writing about Donald Trump.[30] He remains a former one term, twice impeached, seditionist in a failed attempt to remain president. Moreover, he is a COVID 19 denier, who told over 30,573 lies when he was in office. His essence is that of a criminal. Below are some of his major acts that have added to destroying the Earth's environment. The latter are not criminal. Perhaps they should be.

- Pulled out of the Paris Agreement on Climate Accord, which was adopted in December 2012.

- Backed out of the Iran Nuclear Deal, agreed to in 2015, by the Obama Administration.

- Withdrew from UNESCO-The UN Educational and Cultural Organization.

- Exited from The UN Human Rights Council.

- Abandoned the Intermediate Range Nuclear Forces Treaty.

These irresponsible actions will have some environmental impact in one way or another. The most devastating of Trump's insults on the environment were his proposed rollbacks of 125 environmental safeguards.[31] [32]

Safeguards rolled back for protecting Planet Earth can be divided up into seven categories:

- Air pollution emissions
- Drilling and Extraction
- Infrastructure and planning
- Animals
- Water Pollution

- Toxic substances and safety
- Other

I could elaborate. However, how much time do you have?

Jair Bolsonaro

Fig 3-6

Jair Bolsonaro became president of Brazil in 2019. Unfortunately, he was the worst choice for the leadership role of a heavily populated country with approximately one third of the world's remaining Rainforests. Furthermore, Brazil has the most people living in the Amazon Rainforest.(The Amazon is in nine South American countries)

Bolsonaro shares several negative similarities with Trump. Their values, or lack of them, are the same when it comes to protecting the Earth. There is a saying about Trump: "If you lie, you get a job. If you tell the truth, you get fired." With Bolsonaro, people who destroy forests feel safe, and those who protect forests get murdered.[34]

Decisions made by politicians have harmed the planet. For example, Jair Bolsonaro, the President of Brazil, abused his power. A recent article (February 1, 2021), by the Mongabay Website revealed that massive budget cuts of up to 27.4% for Brazilian Environmental Monitoring, was undertaken in a brazen attempt to dismantle national environmental policies. In fact, there were over 600 administrative revisions and rule changes invoked by Bolsonaro's Executive Orders. As a direct result, these changes allowed deforestation rates to soar.

Disgusted with Bolsonaro's decisions and Executive Orders, two Brazilian Indian Chiefs have asked the International Criminal Court in the Hague (ICC) to investigate Bolsonaro for Crimes Against Humanity. As a Scientist, it is my professional opinion that these are not only crimes against humanity, but crimes against *all* organisms on the planet.

Specifically, this includes everything from bacteria to mammals, including humans and animals. Furthermore, these are crimes against 100% of the biosphere to include: the Lithosphere (all land mass), Hydrosphere (all water in oceans and lakes), Atmosphere and any other type of ecosystem. Bolsonaro will be charged with "Ecocide" (destruction of the environment), and with his criminal negligence in managing the COVID pandemic.[35] Thousands of people died that would have otherwise lived.

BOLSONARO'S ENVIRONMENTAL IDEAS AND ACTIONS [36]

- Since taking office his administration has severely diminished Brazil's Environmental law enforcement agencies, falsely accused civil society organizations that could commit environmental crimes and who sought to undermine indigenous rights.[37]

- Supports Agribusiness, the entrepreneurial aspect of Agriculture and their profits.

- Quelled the agendas of nonprofit groups who want to protect the Ecosystems in Brazil.

- Backed out of Brazil's offer to host the 2019 UN Climate Change conference.

- Called Global Warming a plot by "Cultural Marxists."

- Supported plans to reopen the Reserva Nacional do Cobre e Associados (Renca), an Amazon Reserve in the states of Para and Amapa, to permit mining.

- Championed industries that want access to protected Amazon areas.[38]

- Destroyed parts of the Rainforest; this increased by 88% since he took office.[39]

- Rejected millions in aid funding, pledged by the Group of Seven Nations.

- Harassed agents of IBAMA.
  (Instituto Brasileiro do Meio Ambiente E Dos Recursos Naturais Renováveis)
  (Brazilian Institute of Environment and Renewable Natural Resources)

Now, we will examine harmful energy in the United States.

FOSSIL FUELS-COAL, OIL AND GAS

Coal is undoubtedly the worst form of energy because it is responsible for 46% of CO_2 emissions worldwide. Coal fired powerplants are the culprit. They account for 72% of total Greenhouse Gas (GHG) emissions from electricity production.[40] Coal was the fastest primary source of energy between 2001-2010. In the early 1970's, we knew that fossil fuel companies were adding tremendous amounts of atmospheric-heating chemicals into Earth's atmosphere.[41] Yet, despite evidence that the climate is heating due to causes from human beings, we just ignored it. The emissions continued to spiral into the atmosphere, as if it did not matter. Horrifically, there are plans to build another 1200 coal-fired power plants worldwide.

As I write this, COP26 (Conference of the Parties), a United Nations Climate Conference for phasing out Coal and other Greenhouse Gases, has just wrapped up. Many Island states and several countries were upset with the final report from the meeting. They were especially angry at India because instead of phasing out coal as an energy source, they decided to commit less to saving the Earth.

Specifically, India decided to cut down usage with this source of energy. This which means that they will still be burning coal, although, less of it.

The result of this decision is the fact that it will make it harder to reach 1.5°C which should be the maximum allowed. According to the report, the world has warmed 1.1° C .

In fact, there are over five hundred fossil fuel lobbyists who are trying to weaken the efforts to eliminate Coal Usage and other Greenhouse Gases. Do lobbyists have even the slightest concern that the world will become uninhabitable from Global Warming? Are they worried about their families surviving? Are highly paid lobbyists so arrogant that they believe that Climate Change will not affect them?

FOLLOW THE MONEY

In the United States and in other countries, the wealthiest people can strongly influence the way industries are run. We see this constantly in our everyday lives, especially during political campaigns. Who donates millions to politicians, particularly Republican Members of Congress who pass laws and regulations concerning Energy? The answer is simple. Fossil fuel companies, including oil. They are some of the worst polluters in our environment.

Something interesting happened as I was drafting this book. As a scientist, I realized that the main cause of Global Warming is the fossil fuel companies and their extremely dangerous CO_2 Emission. This has always been my theory. An online article from the Toronto Star, August 25, 2021, supports this rationale. It is titled: "Humans aren't as stupid as they seem. Something else is blocking climate action." [42] The article summed it up this way: "We're being blocked by the fossil fuel industry, probably the most powerful set of interests on Earth."

It does not take much wisdom to figure this out. However, the monumental challenge we face is getting the fossil fuel companies to reduce CO_2 Emissions.

How do we achieve this if Republican members always block the reduction? We need to elect more Democratic members of Congress who will support reducing CO_2. This should be a major priority for all future elections in the United States, including the upcoming 2022 Midterm Elections. But, for now, let us focus on two of the most infamous CEOs (Chief Executive Officers) of oil companies: Rex Tillerson and Lee Raymond. Both worked for Exxon.

See Table 3-2. It is obvious that Exxon knew the effects of their product on the climate as early as 1968 that CO_2 was rising in our atmosphere.

Table 3-2 shows the direct correlation between increased CO_2 and higher profits for Exxon and its shareholders. The table is made from data obtained from the Greenpeace's article called: "Exxon's denial history." See footnote 51.

CEOs OF OIL COMPANIES

Lee Raymond –CEO of Exxon from 1999-2005.

Fig 3-7 Lee Raymond[43]

Lee Raymond was one of the most outspoken executives against regulation to curtail Global Warming.[44] As an oil company executive, he owed his allegiance to his shareholders and not to the planet Earth. The reality is that all fossil fuel CEOs will follow suit. This is what they were hired to do. And they are rewarded so handsomely for it, that they never have to work again. For example, CEO Lee Raymond, received a retirement package worth $400 Million.[45] This is so revolting that it should be criminal.

Climate Deniers do not want to see things as they are. Rather, their goal is to make as much money as they can, regardless of the damage to the Earth. As of this writing, 27 August, 2021, wildfires are raging all over the world; Hurricane Ida made her debut; flooding is occurring in Europe and the United States. On the other hand, rivers and reservoirs are drying up, and melting permafrost. This leads to methane releases and new disease organisms. In addition, there are many other catastrophes happening at record pace, including tornadoes. In sum, Climate Deniers know that Global Warming is real; however, they do not care enough to help reduce CO_2, because they love money more.

Greenpeace is a nonprofit, nonviolent global organization that works to save the Earth and its ecosystems. They authored an article about Exxon Mobile climate deniers in a detailed timeline about Lee Raymond. It is entitled: "Exxon's climate denial history: a timeline."

A couple of examples are as follows:[46]

1996

In a speech to the Economic Club of Detroit, Lee Raymond denies the scientific consensus on Climate Change. Raymond claims that: "Currently, the scientific evidence is inconclusive as to whether human activities are having a significant effect on the global climate."

October 1997 (Global CO2 level: 364 ppm, Exxon annual profit: $8.5 billion)

Exxon CEO Lee Raymond tells the 15th World Petroleum Congress in Beijing that: "The world's climate isn't changing, and that even if it was, fossil fuels would play no part."

Some people will say and do anything for money.

Rex Tillerson CEO of Exxon from 2006-2016

Fig 3-8 Rex Tillerson[47]

Tillerson became Secretary of State in 2006. During his confirmation hearing for job, he was asked about Exxon's role in Climate Change. He said that he does not deny Climate Change. Yet, he dodges the role of Exxon.[48]

When asked about Climate Change he said the following: "Our ability to predict that effect is very limited," and precisely what actions nations should take "seems to be the largest area of debate existing in the public discourse." He was wrong.

Our ability to predict climate is excellent, as was mentioned in the first paragraph of this chapter. Specifically, Wallace Broecker warned us of global warming more than five decades ago, in the 1970s.

Tillerson was completely wrong, and it is all driven by the greed to sell his climate destroying product, known as oil. It was pointed out that scientist's dispute that we have limited ability to predict Climate Change. That of course is pure bull.

It is obvious that he is like all CEOs. He knows that his product is instrumental in causing Climate Change. But he is too cowardly to admit it. If he ever did, he would have to reduce CO2. And we cannot have that, because it would mean less money for him, for Exxon, and their precious shareholders.

Darren Woods

Darren Woods is the current CEO of Exxon Mobil. Based on several articles, he appears to follow in the footsteps of his predecessors because limiting carbon emissions is not on his radar. Although, according to Cable News Network (CNN) Money, an online journal, May 27, 2017, Woods did urge Trump to back the Paris Climate Agreement.[49] This was in 2015. Of course, Trump ignored him. He made the United States and himself, *our alleged leader,* look foolish by pulling out of the accord with some incredibly lame excuse that made no sense to the average American. Trump said that it would undermine the U.S. economy and put our country at a permanent disadvantage. All of this was a lie. The actual reason is simple and yet, ugly; Trump did not want to agree to American Oil Companies making less money. This is about greed at any cost.

If Woods did care about the environment, why did Exxon Mobil plan to increase annual carbon dioxide by as much as the entire output of Greece? This was made known by Bloomberg Green. Their data, through leaked documents, showed that the emission increase was planned.[50]

Now, let us look at annually increased CO_2 and Exxon's disgustingly higher profits that resulted from it. See Table 3-2[51].

TABLE 3-2

CO_2 AND PROFITS

Year	CO_2 in ppm	Profits in Billions of dollars
1968	323	1.2
1978	335	2.4
1982	341	4.2
1983	343	5

1989	353	3.5
1992	356	4.8
1995	361	6.5
1997	364	8.5
2000	370	17.7
2002	373	11.5
2004	377	25.3
2008	386	45.2
2013	396	32.6

FOSSIL FUEL COMPANIES[52]

According to the Climate Accountability Institute, the following twenty companies contributed 480 billion tons of CO_2 equivalent since 1965:

Saudi Aramco
Chevron
Gazprom

Exxon Mobile*

National Iranian Oil

BP*

Royal Dutch Shell
Coal India
Petroleos de Venezuela
Petrochina
Pemex
Peabody Energy

ConocoPhillips

Abu Dhabi National Oil Co

Kuwait petroleum Corp

Iraq National Oil Company

Total SA

Sonatrach

BHP Billiton

Petrobras

*These companies made the biggest headlines

Saudi Aramco and Chevron produced the CO_2 equivalent of 102.58 billion *tons.* Notice the two companies that I emboldened: Exxon Mobile and British Petroleum (BP). Exxon Mobile had Rex Tillerson and Lee Raymond as CEOs. BP was responsible for the Deepwater Horizon Blowout on April 20, 2010.

Approximately 134 million gallons of oil spilled into the Gulf of Mexico.

As a direct result, this disaster caused $17.2 billion in environmental damage in the Gulf of Mexico. It is considered the largest marine oil spill in U.S. history. This oil spill killed thousands of marine mammals and sea turtles. It took Pemex nine months to clean up this colossal destruction.

Fig 3-9 Deepwater Horizon Blowout. as seen from space in May[53] 2010.

Pemex is Mexico's government-owned oil company. It had an oil spill in the Gulf of Mexico in 1979. The Ixtoc I oil well spilled 3.4,000,000 barrels of oil (142,800,000 gallons).

Fig 3-10 lxtoc, I Oil Well Blowout in 1979[54]

AIRPLANE FLIGHTS

In 2020, before the Covid Pandemic arrived in the U.S., there were 40.3 million flights by all global airlines. In 2019, there were 763.4 billion passenger miles. After Covid, the number of flights was only 16.9 million.[55] This means that Covid 19 was some good news for the environment because it helped reduce air pollution by limiting flights.

A passenger mile is one mile traveled by one passenger. Passengers emit 285g of CO_2 per RPK (Revenue Passenger Kilometer-km). One mile is equal to 1.6 Kilometers. If a passenger travels 1000 Km by plane, then he or she emits 285,000 grams of CO_2. At 453.57 grams per pound that means that he/she is emitting 628 pounds of CO_2 for a 1,000 km trip. At 2000 pounds per short ton that means for every 1,000 km that is traveled by a passenger, approximately 628/2000=.3 ton of CO_2 is emitted in the atmosphere.

Now, prepare for a shock. From 2007-2020 U.S air passengers logged around 306.2 billion passenger miles or 489.9 billion kms. These numbers are frightening. Now multiply 489.9 billion km by 285g (per KM), and you get 307,827,898,671 g or 678.6 million pounds of CO_2.

From this article we see a compilation of emitted CO_2 from various transportation methods. Now it should be pointed out that not everything is clear cut. But it gives some idea as to how much CO2 is being emitted.

TABLE 3-3-CO_2 EMISSIONS PER PASSENGER KM FROM VARIOUS MODES OF TRANSPORTATION[56]

- 14 g of CO_2 / passenger/km for the train

- 42 g CO_2 / passenger/km for a small car

- 55 g of CO_2 / passenger/km for an average car68 g CO_2 / passenger/km for a bus

- 72 g CO_2 /passenger/km for a two-wheel motor

- 285 g CO^2 /passenger/km for a plane

From the information it is obvious that Jet plane travel from a passenger point of view is the most CO_2 intensive.

BITCOINS[57] AND ENERGY USE

Bitcoins are a virtual medium of exchange that exists only electronically. In April of 2011 bitcoins were worth $1. This past April, 2021, it was worth $65,000. Associated with Bitcoin is the term *Bitcoin Mining*. This type of mining has nothing to do with blowing up mountains to get coal or digging up the Earth looking for gold. No, this mining is performed by vary high powered computers that are needed to solve complex problems.[58] Obviously people started to think that this was a brilliant investment.

Bitcoin uses more electricity (energy) than Argentina. The reason for the energy use is to run computers to solve complex problems. According to Cambridge researchers, it consumes about 121.36 Terawatt-hours (TWh) a year. [59]A terawatt is equal to one trillion watts. A trillion looks like this, 1,000,000,000,000. This electricity is used to mine bitcoins.

HUMAN STUPIDITY-Quintessential Example
The demise of the Aral Sea

I cannot write a second edition of this book without mentioning the primary cause of Climate Change, Human Stupidity. The Aral Sea does not have a lot to do with Climate Change, but I am including it here to show what might happen if the stupidity of humanity intensifies. How could the world's fourth largest inland body of water almost disappear with 12% remaining? In 1960 the area of the Aral Sea was 68,000 km² (26,255 mi²) and as of 2021 it is 8300 km² (3205 mi²). The loss of the Aral Sea proves Einstein right by his brilliant observation: "The difference between genius and stupidity is that genius has its limits."

The remaining part of the Aral Sea is located between Kazakhstan in the north and Uzbekistan in the south. It is known as an *Endorheic Lake* because it does not drain into external bodies of water such as an ocean or river. [60]

In 1957, 48,000 tons of fish were caught. However, when the Sea lost most of its water, fishing diminished entirely. As a result, jobs were lost, and the economy was ruined. In fact, in 2005 Kazakhstan had a program to save the Aral Sea with some success. Aralsk, a city in Kazakhstan, restored parts of the Aral Sea.[61] Check the internet for updates on rehabilitation of the Aral Sea.

The sea was fed by two rivers, the Amu Darya and the Syr Darya. The central planners decided to turn the deserts of central Asia into a cotton growing area.[62]

They diverted the water from the two major rivers that fed the Aral Sea from the lake to irrigate the cotton fields. They began in 1960's by channeling water to irrigate the cotton fields of the Soviet Union. Fig 3-11 shows what was left of the sea in 2014.

Because of the heavy use of pesticides and weapons in the region, assessing the water that remained was difficult. There were significant pollutants which resulted in severe illnesses including Cancer, Tuberculosis, digestive disorders, and anemias, liver, kidney, and eye problems. The latter were exacerbated by dust storms.

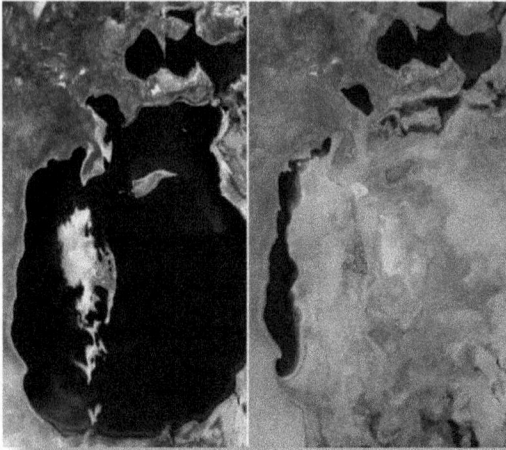

Fig 3-11 The Aral Sea in 1989 (left) and in 2014 (right).[63]

Additionally, we are dealing with other incidences of stupidity, such as never-ending warfare and dictators who do not care about their constituents. This is especially evident as mentioned above with the oil companies. Their executives knew the damage they were causing to the planet's atmosphere.

However, they were more concerned with profits than the lives of the millions of people and other living creatures on the planet, which included members of their own families. They are on the same planet with the atmosphere that everyone needs to survive.

The trouble is that stupidity is never ending and unless we as citizens of the planet do not do anything now to stop Climate Change, we are all doomed by an environmental apocalypse. A sign of hope is that the COP26 is being held in Glasgow, Scotland.

ADDENDUM TO THE COP26 (CONFERENCE OF THE PARTIES) CONFERENCE

Unfortunately, little was accomplished at this conference. It was classified as a monumental failure.[64] This is because the more than five hundred lobbyists from the fossil fuel industry were successful in reducing the number of cuts in fossil fuel usage. These energy reductions are imperative for the survival of humanity and many other forms of life. Think of Einstein's quote about human stupidity. It was discussed at the beginning of the chapter, and it is directly applicable to the outcomes of this conference.

There are other reasons for this negligence. For instance, Australia vowed to keep using coal. India, as mentioned earlier, chose to *Phase Down* coal insteadof *Phase Out* coal completely. Surprisingly, President Biden leased over eighty million acres of public waters in the Gulf of Mexico to fossil fuel companies. To compound these poor actions, a few countries such as Brazil, China. and Saudi Arabia worked actively to further weaken negotiations to the Final COP26 Pact.

To sum up, we must advocate to the leaders who will represent us at next year's COP Meeting, in the hope that they will achieve infinitely better results.

Simultaneously, we must elect strong Democrats in Congress to create laws with a timeline for oil companies to reduce CO2 Emissions. If they do not comply, federal regulators will do it for them. Uncooperative oil company executives will go to a federal prison. Finally, we must all take positive steps to protect our health and do our best to survive and reduce Climate Change. Together, we can make a positive difference. But we need to do it now.

[1] https://news.climate.columbia.edu/2019/02/19/wallace-broecker-early-prophet-of-climate-change/

https://www.npr.org/2019/02/18/695797869/grandfather-of-climate-science-wallace-broecker-dies-at-87

[2] https://commons.wikimedia.org/wiki/File:Wallace_Smith_Broecker.jpg

[3] https://science.sciencemag.org/content/189/4201/460

[4] https://defenders.org/sites/default/files/publications/the_u.s._and_the_convention_on_biological_diversity.pdf

[5] **HOLDING OUT**
The US is the only country that has not signed on to a key international agreement to save the planet https://defenders.org/sites/default/files/publications/the_u.s._and_the_convention_on_biological_diversity.pdf

[6] Red Alert: It Rained in Greenland's Ice Sheet for the First Time in Recorded History (futurism.com)Viewed 20 Aug 2021

[7] https://religionnews.com/2021/08/10/the-climate-apocalypse-is-real-and-it-is-coming/ Viewed 20 Aug 2021

[8] IPCC report shows 'possible loss of entire countries within the century' | Pacific islands | The GuardianViewed 21 Aug 2021

[9] Five chilling predictions from doomsday climate report predicting 'devastating' future - Mirror Online Viewed 20 Aug 2021

[10] https://www.reuters.com/business/sustainable-business/stanchart-ceo-says-companies-must-act-climate-change-cant-bank-governments-2021-08-10/

[11] Dixie Fire: Firefighters tackle historic California wildfire - BBC News Viewed 20 Aug 2021.

[12] https://www.huffpost.com/entry/united-nations-ipcc-climate_n_6110051be4b05f81570b9f50 Viewed 20 Aug 2021

[13] https://www.bbc.com/news/science-environment-58141129 Viewed 20 Aug 2021`

[14] https://www.accuweather.com/en/weather-news/gulf-of-mexico-dead-zone-has-grown-larger-than-connecticut/994180 Viewed 20 Aug 2021.

[15] https://www.audacy.com/wbbm780/news/national/ocean-current-collapse-could-freeze-parts-of-north-america Viewed 20 Aug 2021.

[16] https://www.theguardian.com/world/2021/aug/10/ipcc-report-shows-possible-loss-of-entire-countries-within-the-century Viewed 29 Aug 2021

[17] https://en.wikipedia.org/wiki/Hurricane_Ida Viewed 4 Aug 2021.

[18] https://commons.wikimedia.org/wiki/File:Ida_2021_track.png Viewed 4 August 2021.

[19] https://www.nhc.noaa.gov/aboutsshws.php

[20]https://www.bbc.com/future/article/20200618-climate-change-who-is-to-blame-and-why-does-it-matter Viewed 10 Aug 2021.

[21]https://commons.wikimedia.org/wiki/File:Climate_Change_Schematic.svg Viewed 14 Nov 2021

[22] https://netl.doe.gov/coal/carbon-storage/faqs/carbon-dioxide-101

[23] https://en.wikipedia.org/wiki/George_H._W._Bush#/media/File:George_H._W._Bush_presidential_portrait_(croppe d).jpg Viewed 6 Oct 2021.

[24] Dickie, G.2016. **The US is the only country that has not signed on to a key international agreement to save the planet. Quartz.** https://qz.com/872036/the-us-is-the-only-country-that-hasnt-signed-on-to-a-key- international-agreement-to-save-the-planet/

[25] Wines, M. 1992. President Has an Uncomfortable New Role in Taking Hard Line at the Earth Summit. N.Y. Times (June 11, 1992). https://www.nytimes.com/1992/06/11/world/earth-summitbush-rio-president-has-uncomfortable-new-role-taking-hard-line-earth.html. Viewed 26 Sept. 2021.

[26]https://commons.wikimedia.org/wiki/File:George-W-Bush.jpeg Viewed 6 Oct 2021.

[27]https://en.wikipedia.org/wiki/Arbusto_Energy#:~:text=Arbusto%20Energy%20was%20an%20oil,Bush. Viewedsix oct 2021.

[28]Goldenberg, S. 2009. The worst of times: Bush's environmental legacy examined. The Guardian. Jan 16, 2009.

[29] Wolf, Z and J. Carman. 2019. Here are all the treaties and agreements Trump has abandonedCNN Politics. https://www.cnn.com/2019/02/01/politics/nuclear-treaty-trump/index.html

[30] We all know what he looks like so we do not need a picture.

[31]https://www.nytimes.com/interactive/2020/climate/trump-environment-rollbacks.html

[32] Eilperin, J et al. 2020, Trump rolled back more than 125 environmental safeguards. Here is how. https://www.washingtonpost.com/graphics/2020/climate-environment/trump-climate- environment-protections/ Viewed 6 April 2021.

[33] https://commons.wikimedia.org/wiki/File:Bolsonaro_with_Israeli_Prime_Minister_Benjamin_Netanyahu_at_the_Wailing_Wall.jp Viewed 7 Oct 2021

[34] https://www.reuters.com/article/us-brazil-environment-deforestation/satellite-data-shows-amazon-deforestation-rising-under-brazils-bolsonaro-idUSKCN1T52OQ

[35] https://news.mongabay.com/2021/02/brazil-guts-agencies-sabotaging-environmental-protection-in-amazon-report/Viewed 23 Nov 2021.

[36] https://en.wikipedia.org/wiki/Jair_Bolsonaro#Environment

[37] Chavez. L.T. 2021. Attempt to Greenwash Bolsonaro's Environmental Record Backfires at OECD Human Rights Watch Letter Detailed Disastrous Policies Fueling Brazil's Amazon Crisis. Human Rights Watch. https://www.hrw.org/news/2021/02/11/attempt-greenwash-bolsonaros-environmental-record-backfires-oecd. Viewed 30 Sept. 2021.

[38] Simoes, M. 2019. Brazil's Bolsonaro on the Environment, in His Own Words. The New York Times.27 Aug 2019.

https://www.nytimes.com/2019/08/27/world/americas/bolsonaro-brazil-environment.html

[39] Boadle, A and L Paraguassu, 2029. **Satellite data shows Amazon deforestation rising under Brazil'sBolsonaro | Reuters**

[40] Climate Change. 2001. https://endcoal.org/climate-change/ Viewed 14 Nov 2021.

[41] CO2 is not the only greenhouse gas. Others such as methane, CFCs and sulfur hexafluoride are among the other greenhouse gases.

[42] McQuaig, L. 2021. Humans are not as stupid as they seem — something else is blocking climate action. Toronto Star. Viewed online 27 Aug 2021. https://www.the-star.com/opinion/contributors/2021/08/25/humans-arent-as- stupid-as-they-seem-something-else-is-blocking-climate-action.html

[43] https://commons.wikimedia.org/wiki/File:Premiados_hutchison_y_raymond_(1).jpg

[44] Herrick, T.2001 **Exxon CEO Lee Raymond's Stance on Global Warming Causes a Stir. The Wall StreetJournal. 29 Aug 2001.**

[45] https://en.wikipedia.org/wiki/Lee_Raymond Viewed 28 Aug 2021.

[46] https://insideclimatenews.org/news/22102015/Exxon-Sowed-Doubt-about-Climate-Science-for-Decades-by-Stressing-Uncertainty/
Viewed 21 Nov 2021.

[47] https://commons.wikimedia.org/wiki/File:Rex_Tillerson_official_portrait.jpg

[48] https://www.washingtonpost.com/news/energy-environment/wp/2017/01/11/tillerson-says-u-s-should-maintain- its-seat-at-table-on-fighting-global-climate-change/
Big business wants Trump to stick with Paris climate accord

[49] PETROFF, Alanna. 2017. Big business wants Trump to stick with Paris climate accord.

[50] Crowley, K and A. Rathi. Bloomberg Green October 5, 2020. https://www.bloomberg.com/news/articles/2020-10- 05/exxon-carbon-emissions-and-climate-leaked-plans-reveal-rising-co2-output

[51] **Exxon's Climate Denial History: A Timeline.** **https://www.greenpeace.org/usa/ending-the-climate- crisis/exxon-and-the-oil-industry-knew-about-climate-change/exxons-climate-denial-history-a-timeline/**

[52] Revealed: the 20 firms behind a third of all carbon emissions | Climate change | The Guardian

[53] https://en.wikipedia.org/wiki/Deepwater_Horizon_oil_spill#/media/File:Deepwater_Horizon_oil_spill_-_May_24,_2010_-_with_locator.jpg

[54] https://commons.wikimedia.org/wiki/File:IXTOC_I_oil_well_blowout.jpg

[55] https://www.statista.com/statistics/185744/us-passenger-miles-in-air-traffic-since-1990/
Viewed 23 Aug 2021.

[56] https://www.eea.europa.eu/media/infographics/co2-emissions-from-passenger-transport/view

[57] Cho, R. 2021. Bitcoins Impacts on Climate and the Environment. State of the Planet, Columbia Climate School, Climate, Earth, and Society.

[58] Frankenfield, J. 2021. Bitcoin Mining. Investopedia. https://www.investopedia.com/terms/b/bitcoin-mining.asp.

[59] Criddle, C.2021. Bitcoin consumes 'more electricity than Argentina', BBC News. https://www.bbc.com/news/technology-56012952

[60] https://en.wikipedia.org/wiki/Aral_Sea#Impact_on_environment,_economy,_and_public_health

[61] Chen, Dene-Hern. 2018.Once Written off for Dead, the Aral Sea is Now Full of Life. https://www.nationalgeographic.com/science/article/north-aral-sea-restoration-fish-kazakhstan

[62] Postal, S. 1999.Pillars of Sand.p94. Norton.

[63] https://en.wikipedia.org/wiki/Aral_Sea#/media/File:AralSea1989_2014.jpg

[64] https://truthout.org/articles/climate-diplomacy-failed-again-only-movements-from-below-can-save-the-planet/?eType=EmailBlastContent&eId=49058a06-b6c7-4d14-a775-b88f4c0ce1da

Chapter 4

HUMAN ACTIVITIES CAUSE DISEASE. HERE'S HOW.

Since Al Gore's movie "An Inconvenient Truth," there has been a great interest in the environment and the effects that humans have on it. This movie awakened awareness to the urgency of Global Warming and its effects on biological organisms as well as the physical components of the Earth. The quintessential book about diseases and how humans helped cause outbreaks is *The Coming Plague: Newly Emerging Diseases in a World out of Balance,* by Laurie Garret, published in 1994. It goes into detail as to how humans, while interacting in their environment, help create conditions for the proliferation of various diseases, most caused by microbes.

Much of Garret's book mentions strains of bacteria and viruses which, troublesome as they are, may be considered child's play compared to future diseases more virulent than exist now. Isn't it ironic that *The Coming Plague* is 100% relevant in today's globally warmed world? It is my intention here to delve into other diseases as well as their causes.

Humans can help influence the agents that cause many of the diseases mentioned in Section Two. How does this occur? The answer is quite simple. We help influence the creation of new diseases by aiding evolution (see evolution in Chapter 2 on biology). If we look at the extensive list of damage that human populations cause to the Earth, we can understand why diseases can and will occur in increasing numbers.

The quintessential example of a disease spread by a vector which may have been helped by global warming and Climate Change is Malaria. Malaria is a disease transmitted by the Anopheles mosquito. The organism that causes the disease is a *protist* of the genus **Plasmodium.** This parasite has a large and interesting life cycle. If the climate gets warmer then the mosquitoes which carry Malaria can live longer.[1]

There are enormous numbers of environmental influences that help increase the incidence of Malaria. These include temperature, precipitation, vegetation, deforestation, agricultural activity, and housing construction. The list of natural factors, together with human disruptions, in any combination, can influence the prevalence of most diseases mentioned in this book. Other more disastrous diseases are just waiting to pounce upon us.

Below is a list of the damage that humans do and how this damage can help influence the spread of new diseases. **Table 4-1** shows the types of damage that humans cause. I make many assumptions regarding the phenologies[2] and effects of global warming which are listed in Table 4-1. Based on the knowledge of possible global Climate Change associated with human activity it is possible to predict the reoccurrence of old diseases and the occurrence of new diseases as well.

TABLE 4-1
DAMAGE TO THE ENVIRONMENT THAT HUMANS CAUSE CAN ALSO SPREAD DISEASE.

- **GLOBAL WARMING--threat magnifier and more effects**
- **GLOBALIZATION**
- **LAND USE CHANGES AND AGRICULTURE**
- **DEFORESTATION**
- **LAND DEVELOPMENT**

- AQUACULTURE
- GMO'S-GENETICALLY MODIFIED ORGANISMS
- LOGGING
- POACHING
- DOMESTICATION OF ANIMALS--FACTORY FARMS
- HUMAN OVERPOPULATION
- HUMAN EXPANSION
- URBANIZATION
- INTRODUCTION OF EXOTIC SPECIES-- BIOINVASIONS
- AIRPLANE TRAVEL
- WARFARE
- EMISSION OF GASES
- OVERFISHING
- OVERHUNTING
- OZONE DEPLETION
- OIL SPILLS
- UNEDUCATED COUNTRY LEADERS SUCH AS PRESIDENTS, PRIME MINISTERS, AND KINGS.

GLOBAL WARMING AND CLIMATE CHANGE

In October of 1988, the Office of Policy, Planning and Evaluation and the Office of Research and Development[3] published a draft report to Congress entitled, *"The Potential Effects of Global Climate Change on the United States."* How come, 21 years later, not much has been done about the crisis if we all knew about it? The answer is complex.

However, it is because of the following:

- Stupid politicians who do not want to admit that there is a crisis because they are afraid that if they admitted it, their constituents would not vote for them. A good example of this was demonstrated by the McLaughlin Group, a TV show which discusses politics. On Saturday, Nov. 28, 2009, John McLaughlin asked the panelists, Pat Buchanan and Monica Crowley, syndicated columnists, and Republican sympathizers, if the US Congress will ratify a climate control treaty. President Obama planned to attend the United Nations Conference on Climate Change in Copenhagen, Denmark to be held in December of 2009. Naturally, these two columnists were against the Climate Control Treaty. On the other hand, there were also two Democratic sympathizers like me (Eleanor Clift and Clarence Page). They wanted the climate treaty to be signed. The Republicans claimed that there would be total economic disruption if the United States attempted to limit greenhouse gases, even though the US produces 30% of the world's greenhouse gases. The US is also one of the richest countries in the world, yet it is the poor of the world which suffer the greatest effects of Climate Change. An ironic situation occurred because of the timing of Barak Obama receiving the Nobel Peace Prize and the UN-sponsored conference on Climate Change held in Denmark. Barak Obama flew to Norway to receive the award on December 10, 2009. He then flew back to the United States. Then a week later he flew again to Denmark to attend the Conference on Climate Change. If my geography is correct, Norway and Denmark are close to each other, both being in Scandinavia. It seems to me that the Nobel Committee could have delayed the award ceremony until President Obama was in Denmark, thus saving tons of fuel for Air Force One and not spewing out tons of Carbon Dioxide (CO_2). Fuel capacity of the big jets is 48,445 US gallons or 183,380 liters. 747 burns about one gallon of fuel every second or five gallons[4] per mile (12 liters per kilometer). Put another way, a typical 10-hour flight might burn 36,000 gallons of fuel.

- Greed-- If people were to take Global Warming seriously, then they would not drive as often, and would use less gas. That means that oil companies as well as industries that depend on cars may go broke.

- Many polluting industries worldwide may have to shut down, thus causing a large unemployment problem.

- The science of Global Warming is not completely known, and some people want to wait to see if it real because they think that if they do something about it and the science is false it will affect the economy in a negative way

- People are afraid to heed the old proverb: **"It is better to err on the side of caution,** "even if it means that they will all **die** IF THEY DON'T HEED IT.

The quote in the beginning of this book, **"Homo sapiens, the only deliberate fool that ever evolved, is back tending shop in the good old way!!"** explains it all. To put it in plain English, the quote means that even though we know of the consequences of doing nothing, even though those consequences mean we all die, it still meant nothing to most people because they were afraid of upsetting the status quo. Now things are changing (2009).[5] People are beginning to realize: *"Hey, you know. There may be something to this global warming business?"*

If one looks at the literature on Global Warming and Global Climate Change, he/she will notice that the ten hottest years on record occurred since 2005. Temperature readings began in 1880 (yes, 1880). Below are ten hottest years in order starting with the coolest since 2005.: [6]

- 2005
- 2013
- 2010
- 2014
- 2018
- 2017
- 2015
- 2019

- 2020

- 2016 and 2020[7]

It should be noticed that the hottest years are not sequential but that is the way that the order turned out.

A 1988 report mentioned many events that are happening right now as I author this book. It predicted the following: The southern boundary of the United States could move 700 km, forest compositions of trees could change because of moisture changes (seedlings used to wetter climates will have trouble surviving), agricultural pests may extend northward, and there could be loss of biodiversity and genetic resources that they contain, extinction of species, effects on migratory birds, sea level rise, loss of coastal wetlands, increased salinity of estuaries (areas where fresh water mix with salt water), water resource change, an increase in demand for electricity, air quality deterioration, and other events.

Global Warming is attributed to the effects of adding Carbon Dioxide (CO_2) to the atmosphere thus causing extreme weather events. When talking about weather events we can classify them into three major categories. These categories are as follows:

- Climatological[8]
- Hydrological
- Meteorological

TABLE 4-2

CATEGORIES OF WEATHER EVENTS

CLIMATOLOGICAL	HYDROLOGICAL	METEOROLOGICAL
Heat Waves	Floods	Cyclones
Cold waves	Flash floods*	Hurricanes
Droughts	Storm surges	Typhoons

Wildfires	Coastal floods	Tornadoes
Forest fires	Rockfalls+	Winter storms
Bush fires	Landslides+	Thunderstorms
Brush Fires	Avalanches+	Snowstorms
Scrub fires	Subsidence+	Hailstorms
Grassland fires		
Urban fires		

*These occurred in the State of Georgia and in the Philippines in late September of 2009

+ These are considered wet mass movements

We now know that global warming is the greatest threat to the planet's biodiversity. It is assumed that the major greenhouse gas is CO2. There are many sources of this gas as well as other major gases involved in the greenhouse effect. Table 4-3 shows the sources and their percentages.

TABLE 4-3

SOURCES AND PERCENTAGES OF CARBON DIOXIDE AND OTHER GASES ASSOCIATED WITH GLOBAL WARMING[9]

SOURCE	PERCENTAGE %
Oil (CO_2)	24
Coal (CO_2)	22
Land use change CO_2	17
Methane CH_4 (from energy, agriculture, andwaste)	14

Nitrous oxide from agriculture (N_2O)	8
Fluorine gases from industrial processes	1
CO_2 from other sources	3
CO2 (natural gas)	11

There are many sources of CO_2 including natural sources, like that given off by volcanoes and from decomposition of decaying organisms. But by far it is the human induced (anthropogenic) sources that are the major problem. From looking at the above table the majority of CO_2 comes from the burning of fossil fuels such as coal, oil, and natural gas (methane CH_4). Methane also comes from the cutting-down of rainforests in which termites eat the wood and produce methane emanating from their digestive systems. (Termites produce 20,000,000 tons of methane per year)[10] Methane also comes from the digestive system of cows and other animals raised for meat, including pigs.

There are many gases associated with Global Warming. Carbon Dioxide is the most famous; however, there are others. Most of these gases as well as their Global Warming Potential are listed in **Table 4-4**. Global Warming Potential (GWP) is the heat absorbed by Green House gasses, (GHG) in the atmosphere as a multiple of the heat that would be absorbed by the same mass of carbon dioxide (CO2) which has a GWP of one. [11] Sulfur hexafluoride SF_6 (used in electrical insulation) has a GWP of 22,800 which means that it absorbs heat almost 23,000 times that of CO2.

In Table 4-4 are the approximate numbers when the gasses remain in the atmosphere for one hundred years. Those gases which contain fluorine are called **f** gases. These include the last three on the chart. They are the CFCs, perfluorocarbons, and sulfur hexafluoride.

TABLE 4-4

GASES ASSOCIATED WITH GLOBAL WARMING[12]

GREENHOUSE GAS	PERCENT CONTRIBUTION	GWP-GLOBAL WARMING POTENTIAL	SOURCE
Carbon Dioxide (CO2)	64	1	Fossil fuel combustion and clearing, cement manufacture
Methane (CH4)	19	80	Livestock, rice cultivation, landfills, deforestation
Nitrous Oxide (N2O)	6	310	Industrial processes, fertilizer use
CFC's Chlorofluorocarbons	11	1300-12,000	Leakage from refrigerators
Perfluorocarbons[13]		7,390-12,000	Aluminum production, semiconductor industry
Sulfur Hexafluoride SF_6		22,800	Electrical insulation, Magnesium smelting

In the news, we keep hearing about global Climate Change. The main aspect of this phenomenon is in the form of *Global Warming*. Global warming is the increase in temperature due to increasing levels of gases such as Carbon Dioxide (CO2) and Methane (CH4).

Carbon dioxide comes from burning fossil fuels such as coal, oil, and natural gas, which consists of mainly methane; and biofuels including alcohol and wood. These biofuels are produced from living matter including cellulosic material such as corn plants;trees; methane gas (CH_4) from natural sources, landfills, and anaerobic digesters.

However, increases in temperature are not the only problem associated with Global Warming. As of this writing it is 10°F (-12°C) in Chicago. So how come we keep hearing about Global Warming? The answer is simple. Global Warming has many facets to it. A list of problems associated with Global Warming is found in Table 4-5.

TABLE 4-5

EFFECTS OF GLOBAL WARMING ON THE PLANET[14]

Threat magnifier	Hunger	Warming oceans
Winter loses its bite	Glaciers melt	Sea levels rise
Sea ice thins	Permafrost thaws	Wildfires increase
Lakes shrinks	Lakes freeze up later	Ice shelves collapse
Droughts linger	Precipitation increases	Mountain streams dry up
Lakes heat up	Spring arrives earlier	Autumn comes later
Plants flower sooner	Migration times vary	Habitats change
Birds nest earlier	Diseases spread	Coral reefs bleach
Snowpacks decline	Exotic species invade; bio invasions	Amphibians disappear
Coastlines erode	Cloud forests dry	Temperatures spike
Oceans acidify[15]	Lakes heat up	Islands disappear
Poverty increases	Gulf stream is disrupted	Release of sequestered carbon
Weather-related disasters and economic loss	Increases in toxic algae blooms	Increase in Childhood marriages

Now I will explain each item listed in Table 4-5.

1. THREAT MAGNIFIER[16]

A threat magnifier is something that makes a dire situation worse. Such is the case with Global Warming. Global Warming will cause many hardships to people in many of the undeveloped countries, more so than to those in the rich and fully developed countries such as the United States and in the European Union. It is the poor who suffer because the rich do not care about what they are doing if Wall Street does not suffer. The following problems will be magnified:

- **Water scarcity in areas already suffering water shortages**

As we have seen, weather patterns are changing. Areas that were once wet are now dry and areas that were once dry are now wet. This could mean that areas hit by droughts could totally disrupt local human populations. Water is essential for all life.

- **Crop yields may be diminished thus causing starvation on a massive scale**

Diminishing crop yields will lead to mass starvation. In areas already hit by droughts, this will exacerbate the situation. Just imagine how a person would feel if food became drastically limited. Thus, one will see an increase in children crying for nonexistent food, fights for limited food, family disruption and total havoc. It sounds like a plot for a science fiction movie. Doesn't it?

- **It may cause more frequent and severe natural disasters such as floods mentioned in Table 4-2.**

Land will become flooded, thus changing the topography of the area. If these areas were already compromised, as could be the case in certain areas of Africa, then the cycles of wildlife and plants could be changed in ways that are not hard to imagine.

- **Alter the occurrence of infectious diseases (The premise of this book).**

As mentioned in Section 2 these anticipated changes could lead to more vector populations carrying a deadly cargo of disease organisms.

- **Reduction of air quality**

Air quality could be affected. Asthma cases can increase with more pollen. Scientists have discovered that global warming can cause an increase in pollen. It should also be noted that the health effects of poor air quality caused by ground level ozone and temperature inversions[17] can be exacerbated. It does not take much imagination to figure out what air pollution can do to people.

- **Large scale migration of humans causing population disruption**

Populations already on the brink of starvation and misery may have to migrate to areas more environmentally stable. Floods, droughts, and disease may force people to move. Again, it does not take much imagination to figure out the impact on the human psyche. Take the word misery and add an exponent of one hundred. The effects on these populations will be totally devastating.

- **Increase in malnutrition**

Numerous studies have indicated that grain grown in higher concentrations of CO_2 of 550 ppm (now it is 380 ppm) contains 15 to 30 % less protein, zinc and iron grown at the CO2 level of 2009.[18] Three billion people already have zinc and iron deficiencies and one billion have protein deficiencies.

2. HUNGER

Global warming will increase hunger around the world. According to the FAO (Food and Agricultural Organization of the United Nations) food distribution systems and their infrastructure will be disrupted with the major impact on sub–Saharan Africa.[19]

Reasons for hunger are many, most resulting from the effects of global Climate Change brought about by humankind's activities on the planet and poverty. The effects of Climate Change are those that will affect arable land (cropland). These effects are the following:

- More land being classified as too arid with lack of sufficient moisture.

- Changes in temperatures may make arid regions too hot or dry to raise crops.

- Changes in rainfall will occur. Some crop growing areas may become dryer or flooded, thus making crop growing impossible.

- The possibility that plagues of locusts or some other organism might proliferate in conditions that favor growth of crops.

- There may be other effects that we are not even aware of yet. (Not only is global warming responsible for land degradation but warfare is also a major cause.)

If there is Global Warming then it is possible that areas such as the wheat and corn belts of the United States may move north towards Canada, thus changing the food availability not only for the United States but for other countries that depend on our crops as imports. If weather patterns change, then food insecurity will become prevalent in the United States as it is in countries such as Ethiopia. It is obvious what this means.

Apart from the Dust Bowl in the 1930's most Americans have not known real hunger. People in America do not see situations such as those in Africa because they are blessed with a good climate and soils. Twenty percent of the land in the United States is arable. China has 10% of arable land but much of it is being changed and diverted to road building and cash-producing activities such as golf courses. Rich Westerners like Americans go to China to play golf. (A few years ago, a one-night stay in a fancy Chinese hotel cost as much as a yearly salary for many Chinese citizens.)

The changes will not be limited to the United States. If good climates change direction, that may be catastrophic because the soils may not be conducive to the new climates.

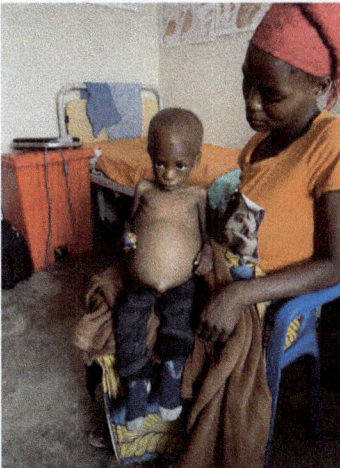

Fig 4-1 A baby with protein malnutrition. Photo by Tim Kubackis[20]

3. WARMING OCEANS

When oceans warm, the world's life support systems may be in jeopardy. This may cause the following disasters: lack of dissolved oxygen for fish to breathe;[21] destruction of the gulf stream;[22] lack of fish for human consumption; disruption of aquatic food chains, and perhaps the introduction of strange diseases that will infect the organisms that live in the oceans (whether they be zooplankton, phytoplankton, fish, mollusks etc.); bio- invasions of strange species (this is the same concept as the appearance of exotic species mentioned in Table 4-1); and most likely some effects that we can't even think of. To put it bluntly, effects on the oceans could be quite frightening, not to mention atmospheric disruptions. The oceans are a major force in climate regulation.

A good example of a strange disease that occurred in the oceans took place in Cordova, Alaska near Prince William Sound in 2004. Yes, the same Prince William Sound where the Exxon Valdez ran aground and dumped barrel loads of that black gooey stuff, we call oil. The temperature had risen above 59°F (very unusual for that time of year), which was perfect for the growth of the bacterium *Vibrio parahaemolyticus,* a relative of the cholera-causing bacterium, *Vibrio cholera.* Oysters were infected with these bacteria. When people ate the infected oysters, they came down with diarrhea, cramping and vomiting. This was an exceedingly rare occurrence in Prince William Sound because the waters were normally frigid.[23]

4. MELTING GLACIERS

Why are glaciers important? One reason is that they supply water to people below the glaciers. Without glaciers the people who depend on them will be left high and dry.

Seventy percent of the tropical glaciers are in the high Andes Cordillera of Peru, Bolivia, and Ecuador. [24] The area of the Andes is home to thirty million people and rich in biodiversity.[25]

Just think what could happen to those people as well as the other organisms that depend on the water from the glaciers if they disappear. Without the water, villages and agriculture will run dry and result in the death of the rich Rainforest biodiversity. The example of the glaciers in the Andes is just one of many examples of what can happen. It also illustrates the value of glaciers, one being the maintenance of life support systems.

In the absence of water, genes can shift, thus allowing plants to become xerophytic (can tolerate low amounts of water) which may in turn cause an evolution of the indigenous organisms through mutations. This **may** result in some strange parasite infecting a plant, animal or even a fungus. See Table 4-6 for a partial list of melting glaciers and ice.

Fig 4-2 Easton Glacier in Washington State (USA) retreated 837 feet from 1990-2005.[26]

Photo by Mauri Pelto

TABLE 4-6

[27]MELTING GLACIERS WORLDWIDE

LOCATION	NAME	NUMBER, PERCENTAGE LOST
Alaska	Columbia	Retreated thirteen kilometers
Montana	Glacier national park	100 of 150 glaciers have melted since 1850
Argentina	Upsula	Retreated 60 km/year
Peru	Quelccaya Ice Cap	Retreated thirty meters per year in the 1990's
Spain	Glaciers	Fourteen of the twenty-seven disappeared
Mt. Kenya	Glacier	Ninety-two percent lost
Central Asia	Tien Shan Mountain	Twenty-two percent of volume since 1960's
China	Duosuogang Peak	Sixty percent since 1970's
Arctic	Ice	Thinned by 40% since 1970's
New Zealand	Tasman Glacier	Thinned by two hundred meters since 1970's

5. SEA LEVEL RISE

One of the problems associated with sea level rise is the fact that millions of people who live in coastal cities will be displaced, thus causing movement inland. This of course will result in the death of land-dwelling organisms and cause salt intrusion in well water. Simply put, this means that ground water will turn salty and unsuitable for domestic use. Sixty percent of the human population lives either on the coasts or within 150 km of the coast.[28] This will result in massive migrations of people to areas which could become overpopulated and disease prone.

Today the human population stands at over 8 billion humans. To make matters worse, countries such as Bangladesh are almost underwater because they are just at one meter above sea level, thus placing 17 million of its 127 million people at risk.[29] Imagine what would happen if this scenario played out many times over throughout the world.

6. ISLANDS DISAPPEAR

Islands may disappear because they will be flooded by rising sea levels. This scenario is being played out as we speak. A few good examples of islands disappearing in the ocean are The Maldives in the Indian ocean, Tuvalu, Kiribati, Majuro Atoll (Marshall Islands) in the Pacific Ocean and Bhola in Bangladesh.[30]

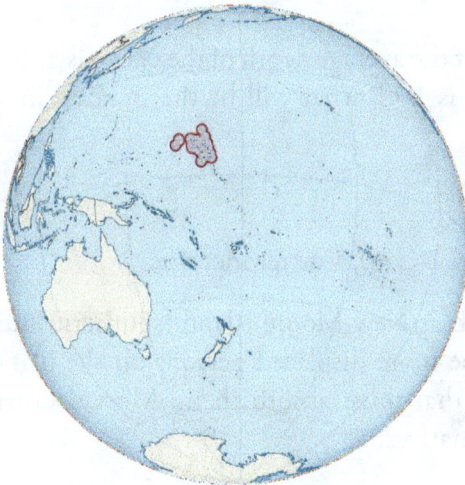

Fig 4-3 Marshall Islands (in red) are in danger of disappearing due to flooding caused by Global warming and thermal expansion of water.[31]

Physicists have developed a formula which can predict the amount of sea level rise based on temperature increase. The formula is as follows:

$\Delta V = \beta Vo \Delta T$ where ΔV equals the volume increase of the oceans; β is the coefficient of volume expansion of water which is 2.07×10^{-4} and ΔT equals the change in temperature. Let us do a sample calculation.

What would the volume of the ocean be if the increase in Temperature is $1^\circ C$? When set up the equation looks like this: $(\Delta V/Vo) = \beta \Delta T$.

The formula which determines the increase in depth is as follows:

$\Delta L = \alpha Lo \Delta T$ where ΔL is the depth increase; α is equal to $\beta/3$; ΔL is the depth increase; Lo is the average depth of the ocean and ΔT is the change in temperature. We divide by three because the number is the cube root.

Let us give an example. If the average depth of the ocean is 4000m and the temperature change is $1^\circ C$, what will be the increase in sea level?

$\alpha = 2.07 \times 10^{-4}/3 = 6.9 \times 10^{-5}$

$\Delta L = 6.9 \times 10^{-5} \times 4000m \times 1^\circ C = 0.3$-meter increase.

If islands disappear, such as New Moore Island (southern India) you can bet that there will be more displaced persons on the planet, thus stressing out countries that must absorb them. Most countries may not want to house non-natives.

7. SEA ICE THINS

Thinning sea ice can prove catastrophic because this means that the earth will heat up more, because the sunlight will not be reflected into space. If sea ice disappears due to excessive melting, the reflectivity (the earth's Albedo) of the ice will be disrupted, thus causing an increased heat buildup. This physical process will just lead to more warming and hence the cycle will keep spiraling out of control. It is obvious that if the water heats up then *stenothermal* organisms (those that can live in a narrow temperature range) may die out. This may disrupt the food chain by causing starvation of the organisms that prey upon them. It is possible that there may even be a change in the biota of the area of the sea ice thinning, as well as in their ecosystem. New organisms might thrive and may become invasive (bio invasions) by multiplying out of control.

8. PERMAFROST THAWS

Permafrost is the soil that remains frozen and is below the freezing point of water (0°C or 32°F) for at least two years. It is found in the northern parts of the planet, specifically in the Tundra or near the Arctic region. If the permafrost thaws, then more carbon dioxide will be liberated, thus worsening the CO_2 problem. When the permafrost is intact it contains roots and other organic matter that decomposes very slowly because of the climate where it exists. However, if the soil begins to melt, the bacteria and fungi which are the main decomposers on the planet will break down the organic matter, thus releasing more Carbon Dioxide (CO_2) and Methane (CH_4), the two main greenhouse gases.[32] This is the last thing that we need. Sometimes bottoms of northern lakes will release methane in great quantities as they warm up, because dead plants and animals will decompose quicker, thus releasing methane. It should also be pointed out that methane is far more potent in absorbing heat than CO_2.[33] I would like to point out that this situation even occurs in Alaska, yes, the 49th State.

CAN METHANE BUBBLES SINK SHIPS? Methane bubbles in large bodies of water can also cause ships to sink. That is right, sink ships. The reason this can happen is due to the fact the density of water will diminish. Ships float according to Archimedes' principle. The principle states the following: "*Any object, wholly or partly immersed in a fluid, is buoyed up by a force equal to the weight of the fluid displaced by the object.*" Simply put, ships float because they displace enough water to compensate for their weight. In other words, if the density of the water gets lower, they will have to displace more water. If there are enough methane bubbles you can sink a ship.[34] Is it possible that this phenomenon could account for the mysterious disappearance of ships in the Bermuda Triangle?[35]

9. INCREASING WILDFIRES

During the week of this writing (February 11, 2009) there was a terrible wildfire in Australia. It was determined that many of the fires were set by arsonists. The fact that the fires did so much damage by killing more than two hundred people and destroying 1831 homes because of drought conditions (1.1 million hectares or 450,000 acres) should wake people up. It made no difference what started the fires.[36] Whether or not the fires were caused by arson or by a natural phenomenon, they should not have burned on the scale that occurred. They burned so violently because of the climate that was changed due to Global Warming. Arnold Schwarzenegger-- yes, the one and only who was governor of California-- said that normally the California fire season was only a couple of months, but now it is all year round. Is anyone listening? I bet the insurance companies are thrilled that now they can get rid of some of their money to help pay off claims. You can bet your bottom dollar (Australian or US) that the CEOs of the insurance companies are delighted with joy that they can help all those people who were burnt out of their homes (a little cynicism there?).

[37] Fig 4-4 Wildfire in Nevada Viewed 4 Jan 2022.

10. LAKES SHRINK

Why do lakes shrink? Because they are running out of water. Why are they running out of water? There are many reasons, but one of the main ones is that they are drying up due to Climate Change. A good example of this is Lake Kinneret (Sea of Galilee) in Israel.

For those of you who have been to Israel, the lake is called Lake Tiberias by American tourists. From speaking to many Israelis, I have gathered that the rain has not been abundant during the past few years. Fishing the famous St. Peter's fish (*Tilapia* species) is forbidden now. This industry once supplied 120 families with an income. The shoreline has receded from its normal position.

The renowned Aral Sea in Kazakhstan exemplifies why lakes dry up. If anyone wants to see what the Aral Sea looks like now, See Fig 3-11. You will be amazed at what you do not see. This was once the fourth largest freshwater lake in the world. Due to Russia's programs to make money, in this case raising the water- intensive cotton crop, the water that once flowed from the two rivers (the Amu Darya and the Syr Darya) was diverted to raise cotton. The lake is now a desert which is blowing its poisonous, pesticide-laden, cancer-causing dust into the faces of the residents.

In addition, lakes dry up because of a process called Eutrophication where these bodies of water get an overabundance of nutrients. In turn the nutrients cause lots of plant growth which then dies and decomposes. This decomposed material settles to the bottom, eventually turning into dry land.

11. LAKES FREEZE UP LATER

Freezing lakes are part of the natural cycle. However, if they freeze up later than usual, ecosystem disruptions can occur. When lakes freeze up later it may be possible that a minor ecosystem change due to a later temperature decrease will occur. This may disrupt normal life cycles of many organisms.

12. ICE SHELVES COLLAPSE

This could be ***catastrophic.*** Why? Because ice reflects sunlight. If there is nowhere to reflect the sunlight the planet can warm up. Did you ever hear of Global Warming? Yes,ice reflects heat. The reader must realize that God or the creator of the planet made sure that everything should be working properly in perfect harmony. Remember, none of us would be here if the planet were not right for us to begin with. There is a term describing the natural reflectivity of the light from the planet Earth. It is called ***Albedo.***

Remember, lighter areas reflect light such as ice and brown or black areas will absorb light and heat. See Fig 4-5.

One would expect that the Arctic and Antarctic regions of the earth would have the highest Albedo. The fact that ice melts in the Arctic should be cause for alarm.

Remember the Earth has a natural reflectivity. If the sun's heat gets absorbed by the water instead of being reflected, the Earth will heat up quicker.

Fig. 4-5. The large arrow represents sunrays hitting the ice. Notice that smaller rays are being reflected into space. More light should be reflected off the earth to prevent more heating of the planet. (Illustrated by Kevin Bley)

A famous ice Shelf, the Larson Ice shelf, is an area off the northwest part of the Weddell Sea extending off the East Coast of the Antarctic peninsula. A part of it, Larsen B, has recently (in 2009) broken off from the Antarctic Peninsula.[38] The Larsen B collapse was partly due to global warming (see footnote #12). This may have profound consequences in that if the ice melts due to a temperature increase, the salinity of the water may change, thus causing a shift in the normal flora and fauna of that ecosystem. This in turn may cause a gene frequency change in the above-mentioned organisms which may have unexpected results for the entire planet.

13. DROUGHTS LINGER

Drought is a situation where there is unusually dry weather where rainfall was previously more plentiful. People will wonder why droughts occur. The answer is obvious. The causes are many, but global warming is one of them. Global Warming causes changes in weather patterns. This means that areas which were once dry are now wet, and those areas which were wet with rain are now dry. We see this all the time, especially here in the United States. Other factors of increased droughts include El Nino, La Nina, and cutting and burning trees and deforestation on a massive scale.

When humans cut trees, they are destroying one of the Earth's climate regulating mechanisms. Trees are particularly important in absorbing CO_2 through the process of photosynthesis.[39] They also add water to the clouds by a process called transpiration which allows rains to continue.

A good example of how a drought can help influence disease is exemplified by an outbreak of St. Louis Encephalitis (SLE) in people in Southern Florida in 1990. To cause a human epidemic, mosquito infection rates must be high. The virus must also be amplified in avian (bird) hosts.

When drought conditions persisted in the state of Florida in the spring of 1990, *Culex Nigripalpus,* the mosquito vector that transmits SLE, restricted its activity to densely vegetated wet habitats where nesting wild birds made their habitats. This brought the mosquitoes and birds into close contact with each other. This proximity of birds and mosquitoes increased the amplification of the SLE virus in infected birds. With many birds infected, one can see the potential for large human infections from SLE.[40] Mosquitoes bite the birds and infect them. Other mosquitoes bite the infected birds, then bite people, completing the cycle.

Another way that droughts help mosquito-borne disease is demonstrated by the spread of Dengue Fever. This is a complex interaction between human economic activity, Global Warming, human behavior, vector biology and infrastructure development.[41] The example that follows took place in Queensland, Australia a short while ago. There was a reduction in rainfall. Homeowners were encouraged to install domestic water tanks in large cities. These water tanks could help Aedes *aegypti* mosquito's breed. Epidemics of Dengue occurred regularly, believed to be spread by travelers. Hence it is possible that human adaptation to the droughts may have encouraged the spread of Dengue to a population which are not immune to the newly arrived virus (Dengue).[42]

Fig 4-6 Drought as exhibited by dried up soil in the Sonoran Desert, New Mexico Photo by Thomas Castelazo[43]

14. PRECIPITATION INCREASES

When the weather patterns change because of Global Warming, one of the things that gets noticed is the fact that precipitation (rain and snow) increases in some areas of the planet while other areas get droughts. We have seen in North America that some areas are getting record floods due to heavy rains. Places like Toronto, Canada are good examples. Heavy rains reached record proportions there in the summer of 2008. The summer of 2007 experienced record droughts. By November of 2008 Toronto reported its wettest year on record with a total 988.2 mm (3.242 feet) of precipitation.[44]

Toronto is just one example out of many. If this happens worldwide, many crops could be damaged by the floods and if these floods take place in warmer climates, then mosquitoes can carry their deadly cargo of disease. One does not need a very vivid imagination to anticipate what can happen with too much rainfall.

15. MOUNTAIN STREAMS DRY UP

When mountain streams dry up it is obvious what happens. Water which is needed by residents downstream will not reach them, and that means that they will have to leave the dry areas for areas with water. Simply put, this means another group of environmental refugees. Could it be that many of these refugees have infectious diseases? Not only will people be dislocated, but this may have detrimental effects on the biota (life) which live in the areas. This may lead to the extinction of organisms which may include plants, animals, bacteria, and fungi.

16. WINTER LOSES ITS BITE

Winter is usually assumed to be cold; however, during the last few years we have noticed at least in the Chicago area the winters have been mild. Proof of this is the fact that a couple of years ago in November I was raking leaves in my short pants. However, during the last two years it was cold here in the Chicago area.

People would ask me, "If Global Warming exists, why it is so cold here?" I tell those skeptics that Global Warming deals with Climate Change all over the world and not just with temperature increases.

True, we had a few snows, but snows do not last long. It will snow one day and by the next day it will be gone because the temperature reaches above freezing. The only good thing is that snowplows do not have to go to work, thus saving the city lots of money.

17. SPRING ARRIVES EARLY

If spring arrives early, this could prove disastrous for many reasons. One is the fact that some animals depend on the flowering of plants for food. If the plants bud early and the flowers which are used for food fall off, then an animal which depends on the flower will not survive and will die of starvation. Let me give you another example which happened. I live in Skokie, Illinois, and there are many magnolia trees. Two years ago, in the spring of 2007, it warmed up early. The pink petals came out. However, a few days after, the temperature was below 32°F, and the petals turned brown. Then as luck would have it, the temperature increased to what it should have been. But the damage was done. No pink petals for the animals that depend on them. Now this may or may not happen again in Skokie, but it will occur in other parts of the world.

18. AUTUMN COMES LATER

If autumn comes later, then ecosystems can change with similar events as those described if spring comes earlier. To put it bluntly, anything can happen. Again, if autumn comes later, warmth-loving organisms such as mosquitoes and other known vectors of disease may linger longer and transmit their dreaded cargo to an unsuspecting animal, plant, or human.

19. PLANTS FLOWER SOONER

This is the same problem as mentioned when spring arrives early. Some flowers last only a few hours or days. Animal reproductive cycles will not change rapidly. Thus, animals may starve because the animals will arrive after their food arrives and disappears, thus causing mass starvation. This also can cause animal behavior changes. Could this behavior change allow unwanted pathogenic organisms to cause a problem? Maybe.

20. MIGRATION TIMES OF ORGANISMS MAYVARY

Many organisms such as geese, lobsters, butterflies, whales, and many others migrate according to temperature changes. Geese are fond of going to Florida or somewhere south, as many Americans do. If temperatures vary, their migration patterns can be disrupted, thus causing other problems. Remember, animals do migrate to mate. They do it for foraging for foods, habitat selection and to escape from predators. There is the possibility that total disruptions of major ecosystems can occur, not to mention starvation of organisms that depend on migrating organisms.

21. HABITATS CHANGE

This is obvious. Global Warming can change habitats as well as the biota that live in them or with them. If the climate changes with higher temperatures, ranges of organisms can change. A good example is the migration of plants higher up in mountains. Can plants climb by themselves? Not exactly. There was a movie called the *Day of the Triffids* (1962) in which giant plants from space walked the streets and killed people.

However, here on Earth plants do not actually walk, but their seeds get planted by various natural means. They can grow in areas that are now warmer, such as on mountain slopes that never had elevated temperatures before.

It has been reported that vectors such as ticks and mosquitoes are inhabiting areas that were once too cold for them. A good example is the case of a tick-borne encephalitis in 1996 in a village called Borova Lada (elevation three thousand feet), in what is known now as the Czech Republic. The tick, *Ioxides rinicus,* had never been seen above 2600 feet.[45]

22. BIRDS NEST EARLIER

The consequences of birds nesting earlier can be bad, as some predators depend on baby birds for food. They would wait until the birds hatched and then go and eat them. It is obvious this could have dangerous consequences, not only for predators but for the bird populations themselves. If certain birds depend on worms for food and the worms have not come out yet, they, the birds, could starve to death. Do you see the connection?

23. DISEASES SPREAD

This is by far one of the major premises of this book. If temperatures rise, vectors of disease such as ticks and mosquitoes could live longer and thus spread their disease organisms to unwilling victims. The most famous such disease is Malaria. This is a disease caused by an organism called Plasmodium. There are four major species of **Plasmodium**. They are the following: **falciparum, vivax, ovale and malariae.**[46]

To use the old proverb, this is just the beginning. There are many insect-borne diseases as well as other vector-borne diseases. Most of those, if not all, like warm temperatures. If temperatures stay high in places such as the Northeastern part of the United States, then vectors can stay around longer and spread their disease organisms to unsuspecting victims.

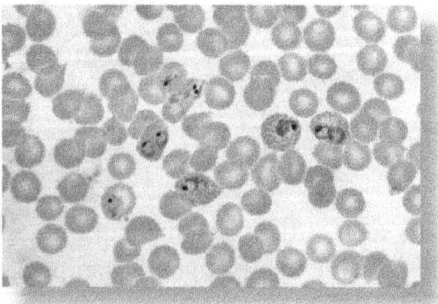

Fig. 4-7 Malaria parasites in human blood.[47]

24. CORAL REEFS BLEACH

There are several diseases of corals, those organisms belonging to the phylum Cnidaria. This is also the group that jellyfish belong to. Corals tend to bleach when temperatures are too high. Corals live with a symbiotic organism called the Zooxanthellae. These are algae. They belong to the group of organisms responsible for many red tides called the Dinoflagellates in the phylum Pyrrophyta, meaning fire algae. Coral is especially important for proper functioning of aquatic ecosystems, or they would not exist. Corals could be compared to Grand Central Station in New York City because that is where the action is. Many aquatic organisms live, eat, play, and reproduce in the vicinity of corals.

Names of some coral diseases are black band and yellow band disease.

Fig 4-8 Coral Bleaching[48]

25. SNOWPACKS DECLINE

We see the tops of mountains as being white because of snow caps and snowpacks. The snow melts and supplies people down the mountains with water. If these packs melt, then there is no water for people, plants and animals that depend on it. It is obvious what that means. This means changed ecosystems which may result in changes in species composition.

26. EXOTIC SPECIES INVADE

What exactly do we mean by exotic species? These are species that do not belong in specific areas. There is another term for this also: Bioinvasions. It sounds like a science fiction movie, doesn't it? The best example of a Bioinvasive organism is *Mnemiopsis leidyi* (see Fig. 4-9). This is a type of organism known as a comb jelly. They belong to the phylum Ctenophora. In the 1980's they invaded the Black Sea inadvertently by a ship's ballast water that was picked up off the east coast of the United States. According to the literature, one cubic meter of the Black Sea contained five hundred comb jellies at the time of their outbreak.[49] This fact proved devastating to the Black Sea because these organisms ate a sizable percentage of the zooplankton which forms the second layer of the food chain. If the zooplankton are gone, then larger organisms such as fish that eat them will also be gone due to starvation.

Just to show the magnitude of this Bioinvasion, it was estimated that if all the comb jellies were hauled out of the Black Sea, they would have weighed more than one billion tons. Yes, you read correctly, one billion tons. Put another way that would have been ten times the world's total fish catch of one hundred million tons.[50]

The above is just one of many organisms that are Bioinvasive. In fact, there are hundreds of species that are invasive. Some Bioinvasive species include other animals such as the zebra mussel, *Dreissena polymorpha*. Some Bioinvasive organisms even include plant species such as Kudzo, Pueraria *Montana var lobata,* and purple loosestrife, *Lythrum salicaria.*[51]

Any group of organisms can become Bioinvasive as we have recently seen. It is important to note that although some Bioinvasive species are pretty, such as purple loose strife, others can be devastating to the local populations of both aquatic and land animals.

Let me give you an example using the comb jelly, *Mnemiopsis leidyi* as mentioned above. These are little jellyfish that eat planktonic organisms such as algae (phytoplankton) and little animals (zooplankton). Now imagine that fishers are fishing these waters for large fish. Well, there may not be any fish left because their food source was destroyed by the Bioinvasive organisms. One does not have to be a genius to figure out what can happen with Bioinvasive organisms.

Land plants can also cause damage by taking over land that is occupied by other organisms.

Fig. 4-9. A comb jelly, *Mnemiopsis leidyi*. Photo by Steven D. Johnson[52]

27. AMPHIBIANS DISAPPEAR

Amphibians include animals such as frogs, toads, salamanders, caecilians (amphibians that resemble earthworms and snakes) and newts. All amphibians lay their gelatinous eggs in water. Lately we have been hearing in the news about the decline of amphibians all over the planet. Many suggestions as to why they are disappearing include global warming, parasitic fungi, and ozone depletion (ozone depletion allows harmful ultraviolet rays to penetrate and kill the babies inside the eggs.) Amphibians function as a food source for birds and other animals. Their loss could mean the extinction of salamander- eating organisms which will have obvious consequences for the Earth's ecosystems.

28. COASTLINES DISAPPEAR

Coastlines may disappear due to flooding. Obviously, that will be disastrous for those populations living on the coasts, which have a combined human population in the millions. If this is the case, there will be millions of environmental refugees. Be aware of what this entails. Every single one will need food, shelter, clothes, and medical care.

29. CLOUD FORESTS DRY UP

Cloud forests are tropical or subtropical evergreen montane moist forests above three thousand feet. All these are characterized by clouds. There are many cloud forests in the world. A good example is the Monteverde Cloud Forest located in Costa Rica. As one would expect, cloud forests have their own special ecosystems and of course their own biodiversity.

30. TEMPERATURES SPIKE UPWARD

Global Warming also means higher temperatures because the globe (the planet Earth) is heating up in various places. Based on current research it has been shown that temperatures have increased. The average temperature of the earth is 59°F or 15°C.[53] However, between 1850-2006 temperatures increased by about 0.8°C.[54] Now, this does not seem like a lot, but it is enough to cause damage. There appears to be a good correlation between CO_2 and temperature. Just imagine what the planet will be like if there is more than a 0.8°C increase. During the week of March 25, 2022 an article appeared in USA Today, the daily newspaper with the headline: **Not a good sign**: **'The temperature was 70 degrees above average near South Pole, a troubling record'**.[55] Certainly not good for the planet.

For specific details on CO_2 there is a web site: **CO2.earth**. This shows the daily CO_2 content in the atmosphere and what it was a year earlier.

When temperatures spike, environments at higher altitudes become more favorable to organisms that carry disease (vectors).[56]

Good examples are mosquitoes that carried Dengue Fever and Yellow Fever. Mosquitoes were limited to altitudes of 1000M (3300 feet), but now they have been found in elevations above 2200m (7200 feet) in Columbia. Malaria has established itself since 1985 at higher altitudes in Rwanda, Zambia, Ethiopia, Swaziland, and Madagascar.

Not only Malaria, Dengue Fever and Yellow Fever are spread, but other diseases as well. In fact, vectors can even live longer at higher longitudes in the United States. A few years ago, when I was teaching at a college in Illinois, (I am still there by the way) I found out that the reason one of my students had to drop out of my class was because she got the West Nile Virus and was hospitalized in a Skokie hospital. (I do not know the outcome in her case.)

To put the problem of vectors such as mosquitoes in perspective, let us discuss optimal temperature. Malaria-carrying mosquitoes need temperatures of 15.5°C (60°F) to survive. To be infective they need temperatures of 17.75°C (64°F).[57] This has ominous consequences for countries that receive immigrants from countries where these disease- carrying mosquitoes (vectors) live. The list of deadly diseases that will increase in warming areas will undoubtedly increase (see "Deforestation" in this section).

31. OCEANS ACIDIFY (See Lecture by David Arieti on Acidification of the Oceans- https://www.youtube.com/watch?v=B4Zix8jiHtk)

Oceans normally have a pH of 8.21 which means that it is alkaline. The pH scales range between 0 and 14 with a pH of seven being neutral. Thus, anything below 7 is acidic and anything above 7 is considered basic or alkaline. Every pH unit is equivalent to ten times the next unit. What this means is the following. Let us say that you have a pH 5. It is ten times more acidic than a pH of six, or one hundred times acidic more than a pH of seven. A pH of four is one thousand times more acidic than a pH of seven.

Now here is the thing. Normally the oceans were considered a buffer. A buffer is a chemical that keeps the pH constant. However, the oceans can only function as a buffer if there is not too much acidity or alkalinity added to them.

Acidification is easy to understand. The reason oceans are acidifying is the fact that when CO_2 mixes with water you get carbonic acid (H_2CO_3) according to the following equation.

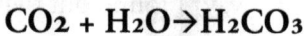

$CO_2 + H_2O \rightarrow H_2CO_3$

This is the reason normal rainwater is slightly acidic with a pH of 5.6. Acid rain has a pH less than 5.6. See Fig.4-10 below.

Fig. 4-10

PH CHART WITH EXAMPLES

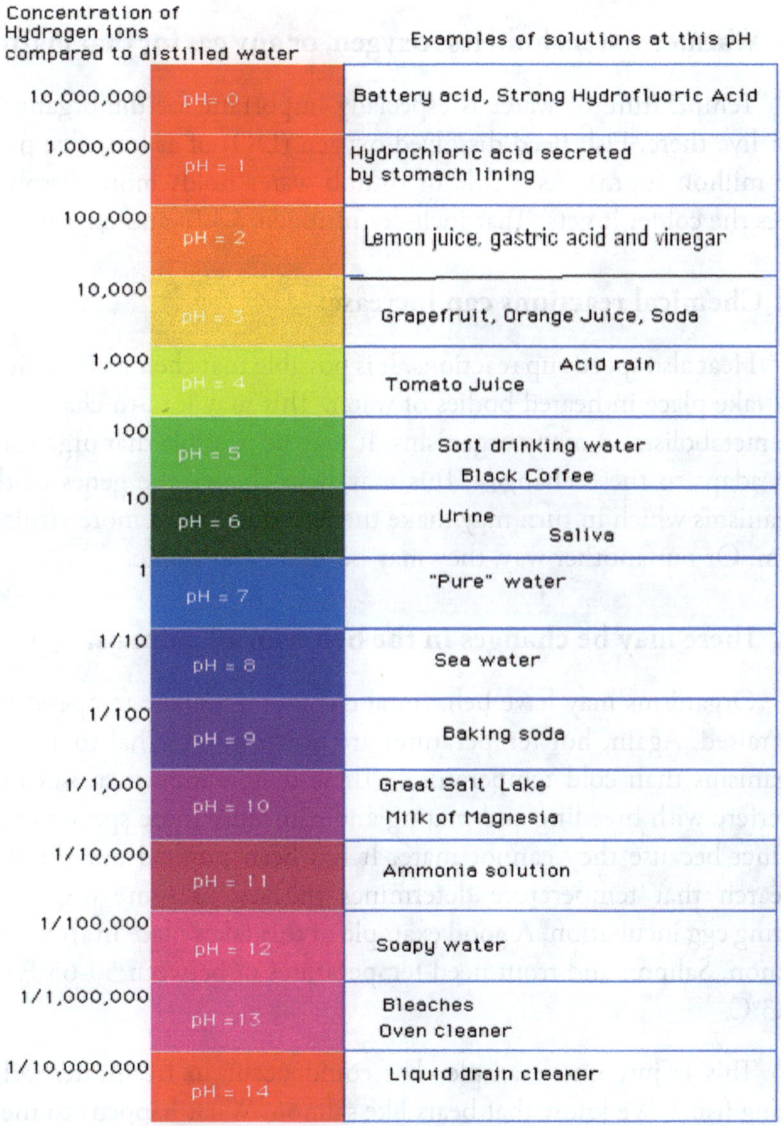

Concentration of Hydrogen ions compared to distilled water		Examples of solutions at this pH
10,000,000	pH = 0	Battery acid, Strong Hydrofluoric Acid
1,000,000	pH = 1	Hydrochloric acid secreted by stomach lining
100,000	pH = 2	Lemon juice, gastric acid and vinegar
10,000	pH = 3	Grapefruit, Orange Juice, Soda
1,000	pH = 4	Acid rain / Tomato Juice
100	pH = 5	Soft drinking water / Black Coffee
10	pH = 6	Urine / Saliva
1	pH = 7	"Pure" water
1/10	pH = 8	Sea water
1/100	pH = 9	Baking soda
1/1,000	pH = 10	Great Salt Lake / Milk of Magnesia
1/10,000	pH = 11	Ammonia solution
1/100,000	pH = 12	Soapy water
1/1,000,000	pH = 13	Bleaches / Oven cleaner
1/10,000,000	pH = 14	Liquid drain cleaner

Public domain-(https://commons.wikimedia.org/wiki/File:Power_of_Hydrogen_%28pH%29_chart.svg

32. LAKES HEAT UP

With the increase in temperature, also known as thermal pollution, four things can happen to lakes: [58]

A. Warmer waters hold less oxygen, or any gas for that matter.

Temperature of water is especially important for the organisms that live there. Fish need dissolved oxygen (DO) of at least five parts per million (ppm). As a rule of thumb water holds more dissolved gases the colder it gets. That includes nitrogen, CO_2, and oxygen

B. Chemical reactions can increase.

Heat also speeds up reactions. It is possible that chemical reactions can take place in heated bodies of water. This may lead to changes in the metabolism of many organisms. It may be possible that organisms can adapt to these changes. This may help change the genes of the organisms which in turn may make them mutate into a more virulent form. Or put another way, they may become pathogenic.

C. There may be changes in the behavior of animals.

Organisms may have behavioral changes if aquatic temperatures are raised. Again, hot temperatures are much more lethal to aquatic organisms than cold temperatures. These temperature increases may interfere with breeding and mating and can cause some species to go extinct because they cannot mate. It has been pointed out through research that temperature determines the sex of some organisms during egg incubation. A good example of this takes place in trout and salmon. Salmon and trout need temperatures of between 50-65°F.10-18.3°C.

This is just one example that could occur in freshwater cold-loving fish.[59] We know that bears like salmon. What happens to them if their food source disappears because of Global Warming? Extinction of them in certain geographical areas. We cannot even imagine what might happen to other organisms. Virtually anything is possible if the local environment changes.

D. Long-term damage such as *Eutrophication* can occur.

Eutrophication is the process where a lake will get too many nutrients which will cause immense growth of aquatic plants such as algae. These algae and plants will decompose, thus diminishing the oxygen supply and causing the lake bottom to fill up with debris from the decomposing organisms. The lake will soon become dry land. Is it possible that eutrophication can cause some organism to mutate and wreak havoc on an ecosystem?

This is a possibility, but unlikely.

I would like to point out that eutrophication can also occur naturally, but if humans are involved it could speed up this process. The opposite of a eutrophic lake is an *oligotrophic* lake. Table 4-7 lists the differences between an oligotrophic lake and a eutrophic lake.

TABLE 4-7

A CHART OF THE DIFFERENCES BETWEEN EUTROPHIC AND OLIGOTROPHIC LAKES

EUTROPHIC LAKE	OLIGOTROPHIC LAKE
High nutrient levels	Low nutrient levels
Poor light penetration	Good light penetration
Low dissolved oxygen	High dissolved oxygen
Shallow water	Deep water
High algal growth	Low algal growth

Most lakes are mesotrophic, which means that they have a moderate level of nutrients. (This chart also appears on page 170 in TABLE 4-14)

33. POVERTY INCREASES

With Global Warming and all of its effects, we can expect millions of people to become impoverished and malnourished, due to loss of jobs because of destruction of their homes and lack of suitable land to raise crops to feed them. As I write these words, California is having tremendous rainfall, resulting in the predicted mud slides and other damage. From watching the news and seeing pictures of the devastation, we can see that many Californians may become homeless, and jobless because their places of business may have been destroyed.

Isn't it ironic that one month prior to the devastating rain California suffered forest fires? It would have been great if the rains had come to put out those devastating forest fires and brush fires.

Why do I mention California? Simple. We in the United States think that we are immune to devastation like that in other countries, but it can and does happen here right this moment. (With California having earthquakes, forest and brush fires, and torrential rains, one may think that the telephone area code for the whole state should be changed to 911.)

One of the hardest hit areas to be affected by Global Warming is Africa. Heavy rains and hotter temperatures will result in more mosquitoes carrying malaria, dengue fever, yellow fever, encephalitis, and hemorrhagic fevers.[60] All this is according to the IPCC (Intergovernmental Panel on Climate Change).

One does not have to be a genius to see how this will result in poverty. Simply put, if many people succumb to diseases like Dengue fever, malaria, they cannot work to feed their families. How will their families cope with this devastation?

Devastation caused by Global Warming also results in water shortages. Many people reading these words live in homes where water is only a few feet away in faucets in bathrooms and in kitchens. Many people in Africa walk a few *miles* a day to get potable water. Now with global warming the situation will just get worse.

Fig. 4-11 Poverty in India as demonstrated by extremely malnourished people. - A very depressing photo taken circa 1876 in Madras. Photo by Willoughby Wallace Hooper (1837-1912)[61]

34. GULF STREAM DISRUPTION

What exactly is the gulf stream? The gulf stream is a warm ocean current and part of the Ocean Conveyer belt that moves from the Gulf of Mexico up along the eastern seaboard of the United States, where it splits. One stream leads to Canada and the other travels to Northwestern Europe and Greenland. It keeps the weather in Europe warm. This is part of a global thermohaline circulation (saltwater movement that depends on temperature). If the salinity of the current is changed due the intrusion of freshwater from Greenland, then the circulation could stop.[62]

If Greenland's ice fields melt, then the warmer water will disrupt the conveyer belt (Gulf stream) and plunge Western Europe in a deep freeze.

If one looks at a map of New York City and Pisa, Italy, he/she will realize that that they are both on the same latitude: 40°N.65" and 43° 43" N, respectively. New York gets very cold in winter while Pisa remains warm. I have been to Pisa, and there is a palm tree in the middle of town. I can assure you that New York City does not have a palm tree in the middle of town unless it is made of plastic.

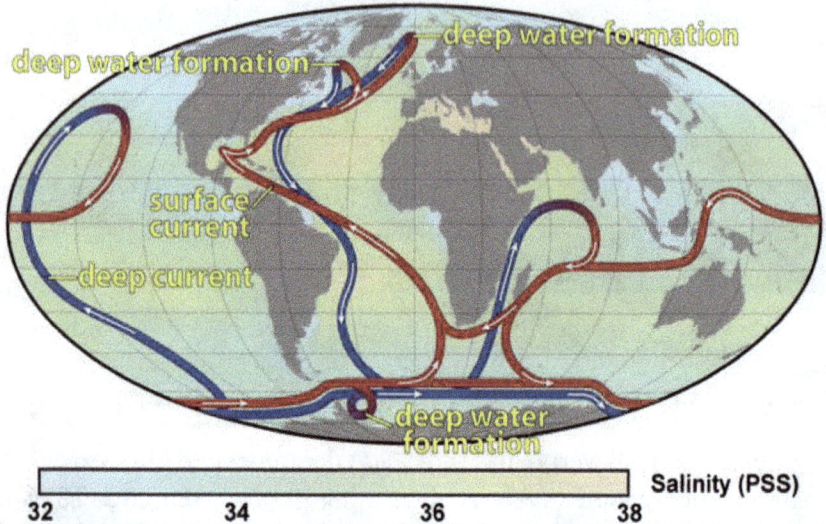

Fig. 4-12. The gulf stream and thermohaline circulation pattern.[63] Blue represents deep-water currents and the red represent surface currents.

35. RELEASE OF SEQUESTERED CARBON AND METHANE(CH₄)

Sequestered carbon is carbon that has been removed from the atmosphere and has been incorporated into plants or soil. The plant or soil can be called a carbon sink.

Unfortunately, the biggest carbon sink is the oceans, which are getting acidified. A good example of what can happen to contained carbon is exemplified in peat.

Peat is partially decomposed vegetable matter saturated with water[64]. Northern peatland soils which are not frozen contain substantial amounts of carbon. When the peat dries out due to a warmer climate there will be a massive loss of organic carbon. It is estimated that in Manitoba, Canada there would be a release of 86% of the carbon that is sequestered in the peat.[65]

Another concern of scientists as it relates to the release of massive amounts of carbon is that of methane hydrates. Methane hydrates are icelike structures which contain frozen methane (CH_4). It is feared that if the permafrost melts (soils in the taiga and tundra regions), billions of tons of carbon and methane could be released into the atmosphere by the year 2100.[66] As of the end of 2020 it is estimated that the world contains around 473.3 trillion cubic feet of gas reserves.[67] Methane hydrates are found in the polar regions and in the oceans and in some soils.

Readers are encouraged to go to the internet and look for more news reports regarding melting ice and the release of methane.

36. WEATHER-RELATED DISASTERS

In 2007 there were 874 weather-related disasters worldwide. Table 4-2 shows the main types of events that are reported. However, during the 1980s there were three hundred events per year; in the 1990s there were 480 events per year and 620 per year during the succeeding 10 years.[68]

These weather-related events include the increased number of named storms. The year 2007 had fifteen storms, which was higher than the average of 10.6 named storms between 1950-2006.

Associated with these weather-related events comes an increase in economic and insured losses. 2005 had the highest losses with Katrina and other storms. The damage was estimated to be around $214 billion. Yes, $214 billion looks like this-----

$214,000,000,000, and it is 21.4% of a trillion dollars. This is a lot of money. In 2007 economic losses were around $69 billion which is lower than in 2005. Dollar loss amounts are not necessarily progressive.

North America and Europe were hit extremely hard in 2007. The United States had heat waves and fires, especially in California. Central and south/southeastern parts of the United were hit with heat waves. Mexico was hit with devastating floods that took place in Chiapas and Tabasco where a million people were made homeless. Europe was also hit hard by a winter cyclone, named Kyrill, and floods as well as temperatures of 45°C (113°F). Many human deaths occurred in these catastrophes.

Damage and deaths are more severe on the people who live in poorer countries rather than those living in rich countries. The wealthier countries are more developed and therefore suffer less damage. Isn't it ironic that the rich and their oil-based economies are responsible for the damage inflicted upon the poor? The poor do not have cars which addexcess CO_2 to the atmosphere which causes warming of the planet. They are just innocent victims of being born in poverty or near poverty.

Not only have there been billions of dollars in damage but also thousands of deaths. These deaths did not have to occur. If the people on this planet had listened to right minded people who preached the fact that we needed more fuel-efficient cars as well as fewer of them (cars) then there would probably be less deaths.

37. INCREASES IN TOXIC ALGAE BLOOMS

Toxic algal blooms, where there may be millions of cells per liter of water, have been on the increase. The tremendous increase in blooms is mostly due to increased water temperatures caused by Climate Change as well as increases in plant nutrients such as phosphates and nitrates. There are lots of debates why algae are so toxic not just to humans but to other organisms as well.

The toxins accumulate in humans and other organisms when they are eaten in popular foods such as clams, oysters, and fish. This is because the algae were food for those organisms. Horrible diseases like **paralytic shellfish poisoning (PSP), amnesiac shellfish poisoning (ASP), diarrheic shellfish poisoning (DSP), Neurotoxic shellfish poisoning (NSP), and Ciguatera Fish Poisoning (CFP) are examples. (Nice names, aren't they?)**

With Climate Change there will be increased toxin production in these blooms. Algae love nitrates and phosphates as a food source and thus will flourish in exceptionally large numbers such as millions of cells per liter of water thus causing havoc on fish and other organisms not only by ingestion but by clogging the gills of the fish which interfere with obtaining oxygen.[69]

There are many documented diseases associated with these algae of which many will be discussed in chapter 12 of this book. Fig 4-13 below shows what an extensive algae bloom can look like.

Fig. 4-13- Algal bloom in Lake Erie, Kelly's Island.[70]

38. INCREASES IN CHILDHOOD MARRIAGE[71]
(See lecture by David Arieti at: https://www.youtube.com/watch?v=gojNnkR4k2k)

You may ask, what does Climate Change have to do with childhood marriage? The answer. Plenty. This fact is one of the least known effects of Climate Change. Here is the issue. Climate Change not only affects countries all over the world, but it affects people in poorer countries more than people in the wealthier countries. This is especially evident in countries particularly vulnerable to Climate Change which results in crop failures thus causing malnutrition in families.[72]

When girls are brought up in countries where Climate Change has drastic effects, girls are encouraged to marry and leave the household because they are considered less valuable than boys. According to the literature there are fifteen million girls that marry before the age of eighteen because they are considered a burden on their families.

Natural disasters, which are caused by Climate Change such as those that occur in poor countries, exacerbates poverty, increases food insecurity, and interferes with their education. This situation occurs particularly in Bangladesh where 52% of girls are married before age 18. Marrying off daughters frees up the parents from another mouth to feed.

As you, the reader guessed, this is cruel, but it occurs in many of these countries such as those in sub-Saharan Africa, Mozambique, and others. Unfortunately, this topic is not discussed in the media. I presented a lecture on this topic on my Youtube channel. (See link above).

OTHER ASPECTS OF CLIMATE CHANGE

GLOBALIZATION

What exactly is globalization? There are three types of globalization: Economic, Political and Cultural. It is the ongoing process where the world's countries are integrated into a global system to include commerce, technology, trade, rapid travel (in airplanes), trade, tourism, urbanization, and the Olympics. Most of globalization deals with economies. Because of Globalization the world is beginning to look the same all over. A quintessential example of globalization is how most countries are beginning to resemble those in the West (developed countries). Look what is going on in China. Have you seen Beijing and Shanghai lately? They look like typical metropolitan cities in Europe or even in the United States. Signs are written in Chinese, however.

Since items such as clothes, jewelry, toys, food, TV's, cars, and a myriad of other items are exported to or imported from countries all over the world, the possibility exists that some dreaded disease-carrying vector (tick, mosquito, or some other organism) may also be imported or exported by accident. This is exemplified and proven by the Dengue Fever-carrying mosquito. Other diseases as well can follow in the footsteps of Dengue.

GLOBALIZATION AND DENGUE FEVER

According to E.Ool and D.Gubler[73] globalization and human movements are responsible for dengue/fever hemorrhagic fever (DF/DHF) which has emerged to become the most important mosquito-borne viral disease in the last 40 years. It is estimated that 2.5 billion of the world's human population is at risk for contacting DF/DHF.

It is obvious that humans (people) just do things on a global scale without thinking of the consequences. (A good example of this is the application of chemical pesticides.

Pesticide companies want to make money by poisoning the world's lands with pesticides. They do not consider the dire consequences such as the poisoning of non-target organisms, the health of the workers in the factories that manufacture the pesticides, or the health of the pesticide sprayers and general contamination of land and water resources.)

Below is a list of problems caused by Globalization.

PROBLEMS CAUSED BY GLOBALIZATION (WHICH LED TO DENGUE FEVER)

- **Economic growth leading to globalization and urbanization**

One of the biggest economic changes that took place during the last ten years is in The People's Republic of China. China now has become westernized, like the United States. Just think of how many countries have been westernized. When countries are westernized, people become more mobile and travel. International movements and trade increase; thus, diseases associated with movements and trade can spread. It is obvious that when the economies of countries improve, the world's environment suffers. An example of evidence is the increase in Dengue fever. Below is a compact list of how globalization can help spread disease. To put it bluntly, one does not need to have a PhD in logic to figure this out.

- **Movement of people and commodities**
- **Increased movement of viruses**
- **Increased vector populations**
- **Increased density of human hosts**
- **Lack of vector control programs**

From the list above it is evident that unless we, meaning those of us who can afford it, do something to improve the health of those in DF/DHF endemic countries, world events like the Olympics and other multinational events may help spread diseases such as Dengue to the host countries. It is highly likely that most of the citizens in the visited countries do not have any immunity to viruses or other diseases, thus rendering them immunologically naive.

MOVEMENT OF PEOPLE AND COMMODITIES

In this day of quick air travel, the spread of major diseases by human movements is on the rise.

A good example of this is the spread of the Dengue fever vector. The major mosquito vector, **Aedes aegypti**, of DF/DHF has spread not only because of increased global temperatures (due to global warming) but also because of human movements such as for business and commerce. The global epidemic of DF/DHF has coincided with air travel. [74]

INCREASED MOVEMENTS OF VIRUSES

Not much more has to be said about this. It is obvious that if people or animals are infected with various diseases and move to other countries or locales, they will carry their cargo with them. In this case the cargo is unwanted organisms such as viruses and bacteria. Vectors are waiting in the wings just to bite an infected person or animal and spread the disease.

INCREASED VECTOR POPULATIONS

A particularly good example of a vector-carrying item brought to many countries are tons of used car tires. Car tires, due to their design, make excellent breeding grounds for mosquitoes because water tends to stay in the tires. A good example of this occurred in 1985 in Houston, Texas, when a shipment of used tires from Japan arrived with the eggs of the mosquito, *Aedes albopictus.*[75] This is the vector of St. Louis Encephalitis and West Nile Virus.

INCREASES DENSITY OF HUMAN HOSTS

As of this writing (February of 2010) Haiti, Chile and Japan had some large natural disasters such as large earthquakes resulting in lots of damage and death. People in those countries are trying to rebuild their countries. Ninety-seven percent of all causalities related to natural disasters take place in developing countries which are not known for their wealth.[76] What would happen if hundreds of thousands of people move to tent cities or to other cities in other countries to escape mounting environmental, ecological, or natural disaster damage? Is it possible that they could unwittingly be hosts?

And what if there is a virus or bacterium going around that can be spread by some vector? Here is a perfect example how a high human density can be infected by a disease. Table 4-8 shows substantial numbers of people leaving due to environmental disruptions. With increased human movements one can see that more potential hosts for vectors will be on the increase.

TABLE 4-8

POPULATION MOVEMENTS DUE TO ENVIRONMENTAL FACTORS [77]

COUNTRY	NUMBER OF PEOPLE MOVING	REASON
Myanmar (formerly called Burma)	800,000	Cyclone Nargis (May 2008)
Mexico	600,000-700,000	Desertification in dry land regions
Brazil	Sixty million	Drought (1970-2005)
Papua New Guinea	2,600 resettled to Bougainville Island	Due to submergence of their home island due to Climate Change

LACK OF VECTOR CONTROL PROGRAMS

It is especially important to have vector control programs. This is especially true for mosquitoes. It is extremely important that time and effort be made in implementing new control measures if the spread of vector-borne diseases is to be halted. This is especially true now, more than ever, because of Climate Change.

There are many types of vectors. It is immensely important to understand the life cycles of all vectors because many can wind up in previous areas due to many reasons. A good example is that of the mosquito, *Aedes aegypti*. (See Figs.4-14 and 4-15.) Even though it is not native to Arizona it has been spreading there. This is partly true because they thrive in urban and in suburban neighborhoods due the fact that there is lots of garbage such as used tires, cans, and other clutter. This mosquito can spread diseases like yellow fever and Dengue.[78]

To understand vectors, it is wise to understand the life cycle of the mosquito.

Considering that the mosquito is such a small member of the animal kingdom (in size only) it is amazing how it has changed the course of history due to its infectious bite.

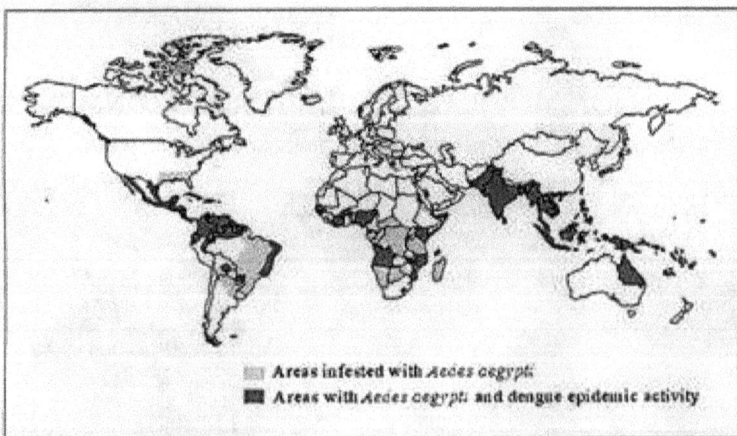

Fig. 4-14. Global distribution of Aedes aegypti and of epidemic dengue, 1980-1998.

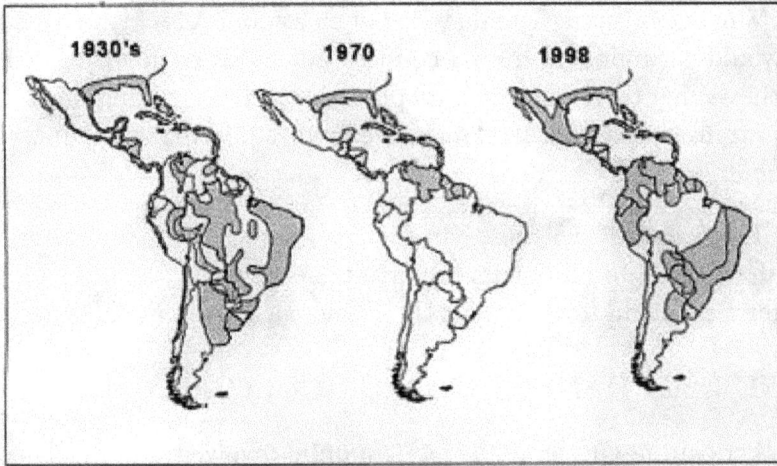

Fig. 4-15. Geographic distribution of Aedes aegypti in the Americas, 1930s, 1970 and 1998.[79] (With permission from Duane Gubler)

GLOBALIZATION AND EBOLA

In 1989 a group of monkeys (to be used for medical research) captured in the Philippines were brought to the primate Holding Facility in Reston, Virginia. A book, *The Hot Zone* by Richard Preston, describes in detail an outbreak of a deadly strain of the *Ebola Virus* in the monkey holding facility. To make a long story short, many of the monkeys came down with Ebola and got extremely sick. The veterinarian became frightened and alarmed when he noticed that the monkeys were sick. If the monkeys were sick then it was possible that the virus could spread to humans. Fortunately, this was not the case. Another factor which saved humanity or residents of Reston, which is less than 50 miles from Washington D.C., is the fact that the strain was species specific. This meant that it would only harm the monkeys and not humans. Unlike human strains, the monkey strain could be spread by air. Those strains that currently infect humans are spread by direct contact and not by air. Those strains that currently infect humans are Ebola Zaire, Ebola Sudan, Ebola Côte d'Ivoire, and Ebola Bundibugyo (Uganda). Direct contact spreads all of these. If there were an airborne virus that affected humans, the consequences would be appalling.

What does this have to do with Globalization? A lot. The fact that we would get monkeys from a country thousands of miles from the US shows that there literally is no place untouched by human hands. This means simply that if humans are involved, virtually anything is possible.

The spread of the Ebola virus is just another example of disease transfer by globalization. I am sure that there are many examples. Does the term swine flu (caused by H1N1 virus) mean anything?

LAND USE CHANGES AND AGRICULTURE

Agriculture and land changes are quite involved on our planet, the main reason being the need to house, feed and clothe all the 8.0 billion people and counting.

One of the most intense uses of land is for agriculture to raise the crops that we need to feed the growing populations of both humans and animals. The main crops in order of tonnage are wheat, rice, and corn. Humans are also involved in the raising of animals such as cows, sheep, and other livestock. These practices can lead to exposure to waterborne disease through exposure to contaminated water supplies. Protistan parasites such as *Cryptosporidium parvum* and *Giardia lamblia* are spread in the feces of domesticated livestock. During heavy rains, the parasites are washed into waterways and drinking water supplies. In 1993 in Milwaukee, Wisconsin, 400,000 people came down with symptoms of cryptosporidiosis and fifty-four people died.[80]

In my book, *The Earth is My Patient* [81] I mention the following: I show that more people equal more pollution; more land degradation; continued loss of living organisms; more water pollution; more air pollution; more land pollution; more unemployment; and more wars! In other words, the planet Earth has more **environmental degradation**. With this environmental degradation we see more imbalances in the world's ecosystems, thus allowing conditions to be set for more disease.

You must remember that every human on the planet needs food, clothes housing, and other forms of life-sustaining items. Because of our habits and diets, we degrade land for pleasure and for food. A good example of pleasure is land conversion for golf courses.

You can imagine what goes into a golf course, including the tremendous amounts of water needed to irrigate them, as well as pesticides.

Households often maintain lawns, and according to one source[82] if all the lawns in the United States were placed together, they would cover twenty-five million acres or ten million hectares. (One hectare is equivalent to 10,000m^2.) Or put another way, ten million hectares would cover the combined area of Vermont, New Hampshire, Massachusetts, Connecticut, Rhode Island, and Delaware.

With more humans we see more overhunting, more fishing and more cattle raising on once- rain forested land, (the Amazon Forest is a good example). We may be depleting natural hosts of unseen or unknown organisms, thus allowing them to jump species and cause disease.

One can see that as humans continue to overpopulate the world there will not be much land left to function as life support systems. A good example is the loss of cropland for food. In 1950 grain land per person amounted to .23hectare (.57 acres) just over half of a football field. But since then, it was cut by half to .12 hectares.[83] More people mean more mouths to feed.

TABLE 4-9

EMERGENT AND RESURGENT VECTOR-BORNE DISEASES
INFLUENCED BY AGRICULTURAL PRACTICES[84]

Malaria	Japanese encephalitis	St. Louis encephalitis
West Nile virus	Oropouche Virus	Western equine encephalitis
Venezuelan equine virus		

DEFORESTATION

Deforestation is one of the most disastrous forms of destruction caused by humankind. Deforestation occurs in South America, Africa, and Asia. Deforestation is the cutting- down of trees in forests. People do not realize that trees and forests are the lungs of the Earth. Trees have many ecological functions such as photosynthesis, stabilizing the soil (preventing soil erosion), providing food, acting as homes for animals, producing medicinal drugs for humans and animals, absorbing CO_2 (carbon dioxide), producing oxygen, and providing shade, just to name a few. There are many other uses of trees of which we do not even know. Some of the countries that have had lots of their rainforest destroyed are the following: Indonesia, Thailand, Malaysia, Bangladesh, China, Sri Lanka, Laos, Nigeria, Liberia, Guinea, Ghana and the Cote d'Ivoire (Ivory Coast). All have lost large areas of their rainforest.

The trouble with this is that it is ongoing as I write this sentence. When I was born 64 years ago (1945) the planet had had around 2.5 million humans. Since then, we have acquired over three times that amount. This means that we have more people dependent on land that was once forested to get the necessary things such as land, homes, food, and clothes that they need to live on.

With the increase in human population, more land is becoming uninhabitable for the residents such as animals and plants. This leads to exposure of new organisms such as the **Ebola** virus which has wreaked havoc on many Africans in Zaire, Côte d'Ivoire, and the Sudan as well as in Philippine monkeys brought to the United States in Reston, Virginia.[85] You may ask what these Philippine monkeys were doing in the US. (They were here because scientists wanted to do experiments on them. The US uses many animals in studies that may benefit humans such as drug efficacy tests and bioassays. I donot want to get into the ethics here, but there is a warning: using jungle organisms such as monkeys may help bring in a very deadly pathogen whether it be a virus, fungus, or bacterium).

Vector ecology changes when deforestation occurs. Vectors, those organisms that help spread disease, tend to increase, or cause human illnesses when deforestation occurs.

Why? Because by cutting down the forests, human activities like logging, agricultural activities, hydroelectric power, mining, cash crops, expose people to these organisms because they are provided with more breeding areas. When trees are cut and then it rains, there will be more puddles for mosquito larvae to develop. Some of these organisms, arthropod species, can adapt to human blood as a new source of food. Some examples of diseases spread this way are arboviruses (arthropod borne viruses such as yellow or dengue fever), malaria, leishmaniasis, and Chagas disease.[86]

A good example of how deforestation can help disease spread is demonstrated by the increase of one mosquito, **Anopheles darlingi.** This is one of the nineteen anopheline mosquitoes (those of the *Anopheles'* species). When areas of the Peruvian Amazon were cut down, the mosquito biting rate increased more than 278 times higher than in forested areas.[87] It is evident that a change in land use such as cutting down forests leads to more mosquitoes. When forest density is greatest there are fewer mosquitoes. This is also true in the Rondônia State of Brazil where *A. darlingi* is the most common.[88] It should also be pointed out that this mosquito is the major vector of **Plasmodium vivax** and *P. falciparum* in the Peruvian Amazon.[89]

When land is reforested, just the opposite situation occurs, not with mosquitoes but with ticks. Over the last several decades when reforestation occurs there will be more deer. More deer mean more ticks. This was and is especially true in the United States where Lyme disease became the most common vector-borne disease.[90]

It should become obvious that as we humans encroach on areas that should be left alone, we will encounter organisms and viruses that lived happily amongst the residents (plants, animals, fungi bacteria) with disastrous results. We just cannot keep cutting, chopping, and digging without considering the consequences from Mother Nature.

Many of these exotic species tend to jump species, as Ebola has, with terrible results. Is it any wonder that now we face newly acquired horrible diseases?

In an experiment in Kenya, investigators found that by reducing shading by deforestation the average temperature rose by 1.8°C, and in aquatic habitats the temperature rose by 4.6-6.1°C. [91] *Anopheles (vector of Malaria)* Gonotrophic cycles (interval between egg laying episodes) are 60% shorter. The mosquito also has a more rapid developmental period. It also has a longer survival. Thus, common sense tells you that the longer the survival time, the more people the mosquito can infect.

There are countless viruses living in the rain forests of the world as well as in other biomes and ecosystems that we are not aware of, that are just waiting to cause new diseases that we never knew existed.

Fig. 4-16 Deforestation in New Zealand Photo by Martin Wegmann[92]

DEFORESTATION AND METHANE PRODUCTION

When trees are felled, termites eat up the wood. Termites have protists living in their digestive tracts. These protists (*Trichonympha*) help digest food (wood) and this process is believed to produce methane gas; yes, that's right, cooking gas.[93] Do not forget that when you chop trees down you are producing little pieces of wood which function as food for those little termites. A forest floor with lots of wood chips may be described as a gourmet restaurant for those little buggers. Each termite produces around .425 µg (microgram)[94] of methane per day.[95]

Fig 4-17 A termite[96] Photo by Sanjay Acharya

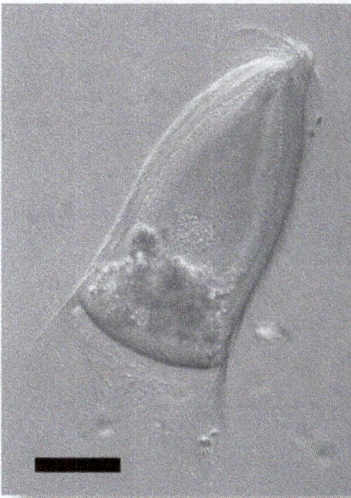

Fig 4-18, *Trichonympha campanula* A wood-digesting protist that lives in the Gut of termites[97] Photo by Tai V James and S, J Perlman, and P.J. Keeling

DEFORESTATION AND CO2 (CARBON DIOXIDE) PRODUCTION

Trees are plants which need water (H_2O) and Carbon Dioxide (CO_2) for photosynthesis. The famous equation for this process is the following:

$$6H_2O + 6CO_2 \rightarrow C_6H_{12}O_6 + 6O_2$$

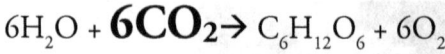

In plain English this means the following: water plus Carbon Dioxide produce sugar and atmospheric oxygen during the process of photosynthesis. Notice that I increased the size of CO_2. This means that Carbon Dioxide is needed for the process of photosynthesis.

Simply put, no photosynthesis, no carbon dioxide removal to make sugar. Cut down the trees and you have no leaves to absorb the CO_2. If you do not have any absorption of CO_2, more of it builds up in the atmosphere and in the oceans, hence the greenhouse effect. As the atmosphere has more CO_2, the hotter it gets. The hotter it gets, the more potential for vectors such as mosquitoes to proliferate and bite people, passing on microbe-caused diseases such as Malaria and Dengue fever. Below is a list of diseases enhanced by deforestation.

TABLE 4-10

VECTOR-BORNE DISEASES INFLUENCED BY
DEFORESTATION[98]

Loaisis	Onchocerciasis	Malaria
Leishmaniasis	Yellow fever	Kyasanur forest disease
Lacrosse encephalitis	Eastern equine encephalitis	Lyme disease

LAND DEVELOPMENT

Those of us who are over 50 years old remember when places where we used to live had more farmland and more open spaces. However, during the last fifty or so years we have seen this farmland converted to other uses such as golf courses, factories, and more roads. Land should be converted to farmland instead of vice versa.

With the growing world's population and affluence, we see more of the planet's life support systems being degraded. Destruction of land, a main support system, is helping to change the ecosystems in such a way that newly emerging diseases are now able to proliferate. The following is a list of changes:[99]

- Expansion of croplands, pastures, plantations, and urban areas
- Biodiversity loss
- Increases in energy use, water, and fertilizer consumption
- Pesticide use increase

This conversion undoubtedly destroys the wildlife present before the land's conversion. The most sinister event that is associated with land use change is the undermining of ecosystem capacity to sustain food production, maintain freshwater resources and regulate climate. The most important function of unchanged ecosystems is the prevention of infectious diseases.

With land conversion, just as in other human activities, ecosystems are being changed. These changes may be for the worse, thus changing the normal flora and fauna and creating areas of biota (life) that were not originally there. Organisms such as mice or other rodents, and insects may help carry disease. The spread of the Junín Virus, (the causative agent of Argentine Hemorrhagic fever) is a good example of how a virus can spread by thoughtless land conversions. A cascade of events which led to the spread of this virus in Argentina is described as follows:

1. WWII caused changes in local agricultural practices.

2. Farmers had trouble growing profitable crops because of short broad-leafed weeds.

3. After WWII pesticides were used, eliminating short grasses, and crop yields increased.

4. At harvest time, tall grasses were prevalent because they were not affected by pesticides.

5. Mice proliferated because they fed on the seeds of the tall grasses.

6. These mice carried the Junín Virus.[100]

Another instance of how land conversions help cause disease is how the epidemic of the Machupo Virus came about in the same way in Bolivia during the 1950's. A cascade of events like the epidemic of the Argentine Junín Virus is described:

1. A social revolution of San Joaquin occurred in 1952.

2. People were left jobless and without a steady source of food.

3. They cut down dense jungles to grow corn and vegetables.

4. By cutting down the jungle they created a wonderful environment for the *Calomys Field Mouse* by providing the rodent with a superior food source, corn.

5. The mice proliferated.

6. Urine from the mice contained the virus.

7. Bolivian Hemorrhagic Fever became prevalent due to casual contact with the urine.[101]

Fig. 4-19. Photo of farmland that is rapidly being developed into expensive homes. See homes in back and left of the white barn.

Photo by David Arieti.

Two of the most important tropical diseases, Malaria and Schistosomiasis are influenced by land use change. Examples of this are found in Tables 4-10 and 4-11. Millions of people are infected worldwide.

TABLE 4-11

Examples of Land Use Change and Increased Malaria Transmission[102]

LAND USE CHANGE	EXAMPLES
Deforestation	In deforested areas of the Peruvian Amazon, biting rates of Anopheles darlingi are almost three hundred times higher than inintact forest, controlling for differences in human density across varied landscapes. Similar associations have been observed in sub-Saharan Africa. In Asia, deforestation favors some vectorsover others but frequently leads to increased transmission. Micro-Dams in Northern Ethiopia increase the concentration ofthe local Anopheles vector and are associated with a sevenfold increase in malaria in nearby villages.
Dams	In India, irrigation projects in the 1990s improved breeding sites for **Anopheles culicifacies** and led to endemic "irrigation" malaria amongtwo hundred million people.
Irrigation projects	In Trinidad in the 1940s, the development of cacao plantations caused a major malaria epidemic. Nurse trees shading the cacao provided ideal habitat for epiphytic bromeliads that, in turn, created excellent breeding sites for A. Bellator, the principal local vector. The epidemic was not controlled until the nurse trees were reduced and plantation techniques changed.

Agricultural development	In Thailand, both cassava and sugarcane cultivation reduced the density of Anopheles dirus but created widespread breeding grounds for a virus with a resulting surge in malaria.
	In Uganda, the drainage and cultivation of papyrus swamps caused higher ambient temperatures and more A. gambiae individuals per household than found in villages surrounding theundisturbed swamps.
Wetland drainage	

Examples of Land Use Change and increased Schistosomiasis Incidence [103]

LAND USE CHANGE	EXAMPLES
Deforestation	In Cameroon, deforestation led to an upsurge in Schistosomiasis as one type of freshwater snail (*Bulinus forskalii*) was displaced by another (*B. truncatus*) that was better suited to cleared habitats. *B truncatus* was an effective host for *Schistosoma haematobium*, a primary cause of urinary tract schistosomiasis.
	Construction of Egypt's Aswan dam in 1965 created extensive new habitat for *B. truncatus*. As a result, *S. haematobium* infection in Upper and Middle Egypt rose from about 6 percent before construction of the dam to 20 percent in the 1980s. In Lower Egypt, intestinal schistosomiasis rose to an even greater extent.
Dams	In Kenya's Tana River region, the Hola irrigation project led to an introduction of snail vectors where they had never been before. Between 1956 when the project began, and 1966, the prevalence of urinary schistosomiasis in children in the region went from 0 to 70%. By 1982, it was 90 percent.

Irrigation projects	In Lake Malawi (Mozambique), evidence suggests that overfishing contributed to the recent surge of Schistosomiasis around the Nankumba Peninsula. Studies found that as populations of snail-eating fish declined, the *B. globosus* snail proliferated in areas that used to be free of it. This sudden surge in host density has been associated with a spike in schistosomiasis in an area historically free of the disease.
Overfishing	In China's Yunnan Province, an economic development project attempted to raise local incomes by giving villagers cows-which also happen to be a key reservoir of *Schistosoma. japonicum* in the region. As the cows spread, they shed schistosomiasis eggs into the waterway, where the parasites could infect local snails. Disease rates surged, infecting up to 30 percent of some villages and correlating directly with cattle ownership.
Livestock Development	

LOGGING

Logging for wood products takes place all over the world. When old growth forests are destroyed, we lose the original biota (all life forms) of the local ecosystems. Logging not only destroys trees but destroys birds and other organisms that live in them. I remember the fuss about the spotted owl. Owls like the spotted owl (*Strix occidentalis caurina*) need lots of trees to eke out a living in the forest. If you cut down the trees for wood products where the spotted owl lives (Pacific Northwest), it can lead to the extinction of these wonderful animals in addition to causing other problems which may LEAD TO AN INCREASE IN DISEASES. Is it possible that the spotted owl has an ecological function that we do not know about? Is it possible that it helps keep a certain virus, bacteria, or other organism in check so that it does not get out of hand?

POACHING

Poaching is the illegal hunting, killing, or capturing of plants and animals, primarily for profit, and for food. It may be possible that many of these poached organisms harbor deadly viruses or may even help to control them by keeping them in check by some unknown mechanism that we have never even considered.

By eliminating animals or plants, it is obvious that whole ecosystems may be damaged, and the total balance of nature can be disrupted, at least in the area where the organisms are poached from. Again, from studying biology and environmental science for over 30 years, this scenario is not out of the realm of plausibility.

In Africa, poaching is rampant, especially of elephants because of their ivory. Bush meat (no relation to the former resident of the White House), which consists of monkeys and other mammals of the forest, comes from poaching dead animals for food. Could it be that the beginnings of the deadly Ebola Virus came from a monkey or even a bat which caused havoc to many villages in Africa?

GMO-GENETICALLY MODIFIED ORGANISMS

These are organisms, specifically food items, which have been genetically modified to meet various specifications such as increased taste, resistance to frost, and other characteristics. We humans are eating many of these foods without knowing that they are genetically modified. It is possible that some of these plants can kill or harm other organisms.

There is one type of genetically modified organism that should bring fear into a normal thinking human being, and that is the use of plants modified with BT (*Bacillus thuringiensis*), a bacterium that is used as a form of biological control of insects. *BT* is the shortened name, and its genes are placed in plants. This bacterium is quite toxic to many insects and kills them by perforating their midgut.

It is used on corn, cotton, tobacco and in other plants as well. One of the main concerns is that if a nontarget organism eats plants which have been genetically modified with BT, it may die, and the dead organism could be a beneficial organism. This could be devastating for food chains.

Remember that this is just one of many scenarios. With GMOs we can be creating Frankenfoods (an expression to explain strange looking and tasting foods that are unnatural.) The word of course comes from the wonderful horror classic, *Frankenstein*.

There are lots of data on GMOs. The reader is encouraged to research more on this topic, but he/she should be aware that GMOs are not necessarily the panacea for feeding the world that we would expect or hope from the descriptions by their advocates.

DOMESTICATION OF ANIMALS

Humans have been domesticating animals for thousands of years. It was not until the 20th century that we were able to produce "factory farms" where animals are produced by the thousands. When used to raise pigs, chickens and other organisms, these factory farms sometimes give antibiotics to the animals to prevent various diseases. Any microbiologist knows that this can cause problems for both the animals getting the antibiotics and those people who eat the animals. The main problem is that of resistance. The microbes that were once susceptible to the antibiotics are now having a field day, or "celebrating," as a friend used to say. Thus, humans can get a certain bacterial disease and the drugs normally given may become ineffective.

In addition to microbial diseases, other diseases not caused by microbes (prions) such as "Mad Cow Disease," also known as "Bovine Spongiform Encephalitis BSE," can be passed onto humans in the form of the human variant, Creutzfeldt-Jakob Disease or Kuru. Humans can get this disease by eating infected cows. You may wonder how cows become infected. The answer boils down to human greed. When a cow is sick and cannot stand up on its own, it is called a "Downer." In order not to lose money, the owners of these cows will kill them and grind them up. They then mix them into the feed, so they are fed to other cows even though cows are vegetarians, or technically "Herbivores" (see chapter on biology).

Another problem associated with animals is the use of CAFOs, short for concentrated animal feeding operations. These are areas where thousands of animals are raised and fed in crowded conditions. CAFOs are common in the raising of pigs, chickens, and cows. It is obvious that not only are antibiotics being fed or injected into the animals, but their waste products (manure) accumulate, thus contaminating the soil with bacteria and nitrogen. Nitrogen clouds appear in the soil, thus affecting soil fertility. These nitrogen clouds are areas containing lots of animal waste. This is no good for the soil because it may over-fertilize it, thus making it unsuitable for crops, or for anything else, for that matter.

As of this writing, May 1, 2009, the world is in a panic over a few hundred cases of a new type of swine flu. Why do we call it swine flu? Is it because we know that it is a virus that was once associated with swine (pigs)?[104] It is believed that this virus came from the pig farms in Perote, Mexico. It is estimated that there are close to 950,000 pigs raised there. It is no wonder that the flu and other viruses are associated with animal raising. The Nipah Virus may also be associated with pig raising. This fact was discovered in Malaysia where pigs were raised in exceedingly high densities. The Nipah virus jumped from bats (one of its known hosts) to pigs, after which it jumped to humans.[105] See Section Two on diseases.

What does one expect when you raise pigs in cramped inhuman conditions? Pigs are raised in so called factory farms where thousands are housed in unsanitary smelly conditions. Ammonia from the pig's waste saturates the air and irritates their lungs. The poor pigs must breathe the polluted air while their immune systems are being overworked. It would appear to any geneticist or microbiologist that organisms can mutate in this environment.

Most industrialized animals-for-food-raising is done in cramped quarters, using antibiotics to compensate for the increased danger of diseases caused by overcrowding. Isn't it ironic that due to the added antibiotics to prevent disease we are creating disease by mutations?

We see this situation with other diseases as well.

Does the term MRSA ring a bell? MRSA Stands for Methicillin-Resistant- **Staphylococcus aureus.** These are bacteria. (See section under bacteria in Biology (Chapter 2). The term MRSA instills fear in any doctor trying to treat patients with a bacterial disease. The reason this strain of bacteria is feared is because it could indicate that there may be other bacterial infections which cannot be treated with the drugs we have today. These bacteria become resistant to the drugs because of something called plasmids, formerly called R-factors. These are circular pieces of DNA which can pass from one bacterium to another through the process of conjugation, a form of sexual contact between bacteria.

(Most bacteria reproduce by splitting into two by a process called Binary Fission.)

Why do humans raise animals in inhumane conditions? Here it comes due to that Magic "word.G" Yes, G for greed. To produce cheap meat, unethical growers would do anything for the proverbial "buck."

In North Carolina there was an outbreak of **Pfiesteria piscicida**, a dinoflagellate which was responsible for the disease, Pfiesteriosis. These organisms cause fish lesions which penetrate deeply into their flesh and organs. When humans are exposed to these organisms, they can get brain damage as well as other problems.[106] How does the *Pfiesteria* get into the water? Simple. The waste is stored in lagoons. Sometimes the lagoons break, dumping the waste into the Neuse River, a major river in North Carolina. Pig waste is loaded with nutrients and is equivalent to a gourmet restaurant for the tiny *Pfiesteria*. *Pfiesteria* will proliferate, thus causing their damage.

ANIMALS' CONTRIBUTION OF GREEN HOUSE GASES

According to various estimates there were between 21.7 to 50 billion head of livestock between 2003 and 2005. The tremendous variation in numbers is attributed to government reports vs. nongovernmental organization reports (NGO). The NGO reports list higher numbers and are more accurate. I am a little cynical about government reports because they want to downplay issues which are not popular with the public, for obvious reasons. A good example of this discrepancy is in a report, *Livestock's Long Shadow.*

The report stated that in 2002, 33.0 million tons of poultry were produced worldwide while the FAO's report (Food and Agriculture Organization of the United Nations, primary office is in Rome, Italy) showed that 72.9 million tons of poultry were produced.[107]

Whatever the case, the contribution of greenhouse gases is not mentioned. People forget that two major gases are produced from animals, including humans. They are methane (CH_4) and Carbon dioxide (CO_2). Methane comes from their digestive tracts. CO_2 comes from cellular respiration. For those of you who are interested, the equation of cellular respiration is the following:

$$C_6H_{12}O_6 + 6O_2 \rightarrow 6CO_2 + 6H_2O + ATP.$$

gives you carbon dioxide and water from the breakdown of the sugar. Does this equation remind you of something? Well, it should. It is the reverse of photosynthesis. See section on deforestation in this chapter.

According to the FAO, livestock produced 7,516 million metric tons of CO_2 equivalent (CO_2eq).[108] Animals which include pigs, goats, sheep, horses, cattle, buffalo, and poultry emitted 5,047 million tons (5,047,000,000) of methane, which accounted for 7.9% of total greenhouse gas emissions. In addition, when measurements of GHG were taken, animals' contributions were uncounted, overlooked, or misallocated.[109] Do not forget that methane accounts for 19% of total greenhouse gases while CO_2 represents 65% of total greenhouse gases.

It was recently discovered that adding, *Asparagopsis*, a seaweed, to cow feed will reduce methane emissions. See my lecture on low methane cows at https://www.youtube.com/@DavidsEarth/videos

EMISSIONS OF GASES OTHER THAN CO$_2$

When oil, coal or gas is burned, sulfur dioxide (SO_2) is emitted into the atmosphere. This of course can cause bronchitis in humans and may cause the death of other air- breathing organisms. When mixed with rain, SO_2 forms sulfuric acid (H_2SO_4) as is demonstrated in the following equation:

$SO_2 + H_2O \rightarrow H_2SO_4$ where SO_2 plus water forms first sulfurous acid (H_2SO_3) then sulfuric acid (H_2SO_4).

In this equation sulfuric acid is a *secondary pollutant*. When a pollutant such as SO_2 mixes with water it turns into something else. That something else has a general name called a secondary pollutant. Secondary pollutants are those that are formed by either air or water. Another example of secondary pollutants is ground level ozone, O_3. O_3 can be a lung irritant. When sulfuric acid lands on the ground or a lake it lowers the pH (pH is the term used to denote whether something is alkaline or acidic). When the pH is low, especially on soil, then metals and other nutrients that plants need are leached from the soil and may poison other animals. This in turn can change the whole local ecosystem by changing the normal fauna and flora present. Fauna and flora refer to animals and plants and other local organisms present. If plants and trees are affected it can lead to their deaths, which will undoubtedly change the makeup of the local biota.

HUMAN OVERPOPULATION

Most environmental scientists will agree that the greatest problem facing humanity is humanity itself. Prior to the Industrial Revolution (around 1870) there were less than one billion people on the planet. Since then, the population has increased over 7.9 times, to 7.9 billion as of 2022. Unfortunately, most of this increase is seen in the least developed countries (LDCs) while in the most developed countries (MDCs) there is a slower growth rate of humans.

I am in favor of having everybody being able to have clean water, good nutritious food, clean air, good shelters (homes) and a good war-free life. A nice dream? Not. Anyway, it is obvious that if we keep producing more people nonstop, we are heading toward a global train wreck.

According to a certain web site[110] there are approximately 150 people born each minute; nine thousand per hour; 216,000 per day; 6,480,000 per month or 77,760,000 per year. This is quite a bit of humanity.

Table 4-12 below show the populations of the ten most populated countries in the world.

TABLE 4-12[111]

POPULATION OF THE TEN MOST POPULOUS COUNTRIES

#	COUNTRY	POPULATION
1	China	1.402 billion
2	India	1.400 billion
3	United states	329.9 million
4	Indonesia	271.7 million
5	Brazil	211.8 million
6	Pakistan	220.9 million
7	Nigeria	206.1 million
8	Bangladesh	169.8 million
9	Russia	146.7 million
10	Mexico	127.8 million

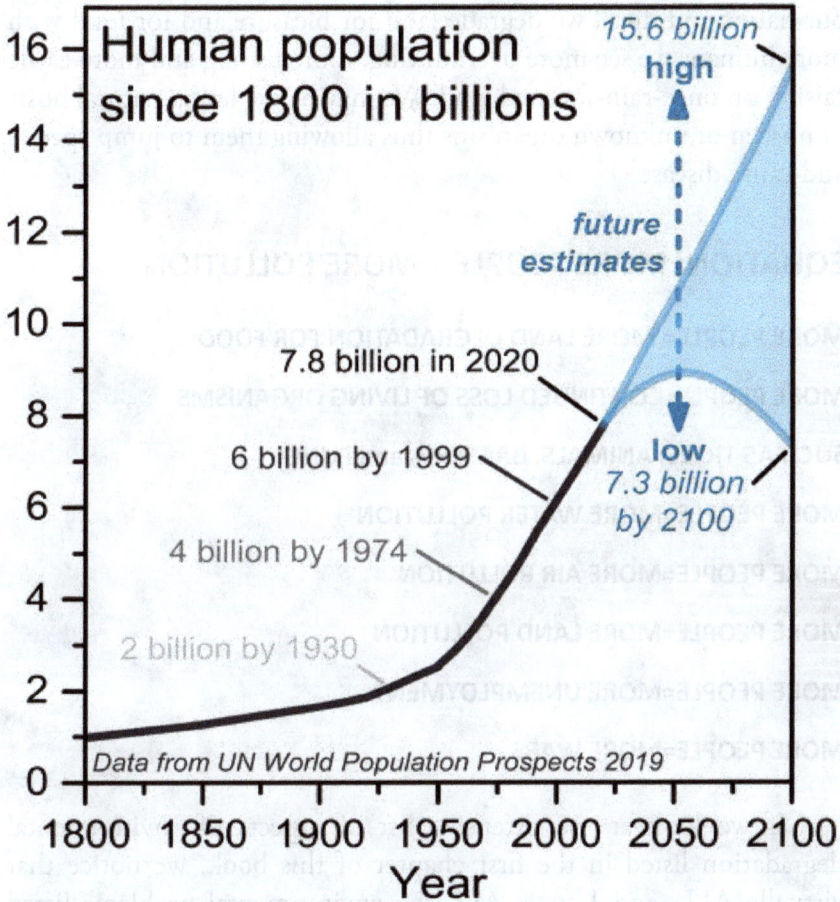

Human population since 1800 in billions

15.6 billion high

future estimates

7.8 billion in 2020

6 billion by 1999

low 7.3 billion by 2100

4 billion by 1974

2 billion by 1930

Data from UN World Population Prospects 2019

Year

Fig. 4-20[112] Population growth chart.

In my book, *The Earth is My Patient* [113] I mention the following: more people will mean more pollution; increased land degradation; continued loss of living organisms; more water pollution; more air pollution; more land pollution; more unemployment; and more wars! In other words, the planet Earth has more environmental degradation. With this environmental degradation we see more imbalances in the world's ecosystems, thus allowing conditions to be set for more disease.

You must remember that every human on the planet needs food, clothes, housing, and other forms of life-sustaining items. Because of our habits and diets, we degrade land for pleasure and for food with more humans we see more overhunting, more fishing and more cattle raising on once-rain-forested land. We may be depleting natural hosts of unseen or unknown organisms thus allowing them to jump species and cause disease.

EQUATION: MORE PEOPLE = MORE POLLUTION

MORE PEOPLE= MORE LAND DEGRADATION FOR FOOD

MORE PEOPLE=CONTINUED LOSS OF LIVING ORGANISMS

SUCH AS TREES, ANIMALS, BACTERIA and FUNGI

MORE PEOPLE=MORE WATER POLLUTION

MORE PEOPLE=MORE AIR POLLUTION

MORE PEOPLE=MORE LAND POLLUTION

MORE PEOPLE=MORE UNEMPLOYMENT

MORE PEOPLE=MORE WARS

If we look at the extensive list of aspects of environmental degradation listed in the first chapter of this book, we notice that virtually ALL, and I stress ALL the environmental problems listed are a direct result of human overpopulation. As of 2010, there are approximately 6.8 billion people on the planet, one third of which live in China and India. The first billion was reached around 1870. The second billion was reached around 1930. In the mid 1950s there were around 2.5 billion. The average growth rate is around 1.3% worldwide. Every second, five people are born and two people die, thus, a net gain of three people,[114] with approximately 10,000 being born per hour. At this current rate, the world's population will double in about forty years to about twelve billion. [115]

Added to the burgeoning human population there are more than 16.1 billion domesticated animals. This fact is significant because such animals require a great deal of food that instead could be used to feed humans. Cash crops such as coffee, bananas and cotton also take their toll on land by utilizing tremendous amounts of water, like the Aral Sea[116] in the former Soviet Union. These crops do not add food value to the residents, but instead, profit multinational food corporations. What do humans have to do with pollution? Simple. We must feed, house, clothe, educate, and take care of all humans as well as the domesticated animals on the planet. To do this, we need food which comes from farmland.

A few years ago, troops from the United States entered Somalia, a country in East Africa, to save millions of starving people, due to much of their food having been looted by bandits. The devastation in Somalia was caused by greed, under the misguided leadership of President Said Barre. It became clear that Somalia was a nightmare, which may have been the precursor to other similar catastrophic events, such as, the genocide in Rwanda, and the Gulf Wars (caused by the presidents named Bush). If we keep producing more humans, the entire world may look like Somalia.

We are already witnessing another catastrophe taking place right now, in the former Yugoslavia. Like Somalia, we see plenty of guns, but not much food. It never ceases to amaze me that there always appears to be a shortage of food, but plenty and plenty of guns. It sure makes you wish that guns were edible. Doesn't it?.

Here in America the average number of children per family is about 1.6. In Somalia it is around 6.9.[117] We must first realize that to curb environmental degradation, we must limit the human population to maintain the ability of the planet to support them. There is a special word for this ability and it is called *The Carrying Capacity.* Simply put, The Carrying Capacity is defined as the maximum population that an area can sustain without bringing in outside resources. If we exceed the carrying capacity of an area, that area will eventually stop supporting its life. The best way to describe this is the analogy about a small house. Think of the planet as a small house. Let us say you normally have three people living in the house with only a $50,000 yearly salary. Now think! How would it be if you had ten people living in it, still with the same $50,000 salary? You could not even buy enough food unless you went to stores like Aldi, located here in Chicago, where people stock up for cheap. You can buy a few bags of Ramen noodle soup for only $.89. You might be able to feed those ten people with only that, but what about clothing, medical care, education, and special needs? It would be tough!

Now, what about space or lack of space? Pretend that this house is the planet earth. We need this space to grow our crops, maintain our institutions, clean up waste, and provide places for recreation. If we continue in our present ways the quality of life would be undoubtedly compromised, making life miserable for everyone. We would no longer be able to sustain ourselves. This is exactly what would happen to the earth if population growth were not halted.

Sure, we talk about human life, but what about animal and plant life (non-domesticated of course)?

Can one imagine a planet without the lovely song of birds, or the tapping and pitter-patter of animal feet? If people keep having children, which will result in overpopulation of this small planet, we will only hear the pitter-patter of children's feet. We will not hear the pitter-patter of other animals.

One of the reasons why we have overpopulation is the influence of the far right, here in America, as well as beliefs of various religions. Under Ronald Reagan, who was president in the 80's, the world's population increased, due in part, to his paying attention to the far right by not aiding countries in family planning programs.[118] I know of families in Israel and in America where there are over ten children per family from the same parents. Sure, it is nice to have many children, but the world cannot tolerate or afford that many per family. The world seems gigantic in scope, but too many people, having too many children, slowly adds to the depletion of the planet's life support systems.

There is a proverb, one I thought up myself. It goes like this: The more people you have, the more pollution (brilliant, isn't it?). This is especially true in the United States because we tend to destroy the environment at a greater rate than other countries, because we have the every- thing-we-see-we-want-now attitude. According to Paul

Ehrlich, each child born in the United States does about two hundred times the damage that a child born in Bangladesh does. The reason for this is because we need more of everything, or so we perceive. Our society makes us believe we need more things.

Because of the media we are being brainwashed into believing that the more you consume the better off you are. The condition of our planet has proved the old saying, ***"The one who dies with the most toys wins."[119]***

Somehow when we watch the evening news on TV in the comfort of our own home here in America, it is as if we are watching a science fiction movie take place in a faraway planet. The media packages significant stories in 30-second sound bites, while more frivolous stories such as the affairs of sports and entertainment celebrities such as Michael Jordan and Tom Cruise are deemed more important and thus given more airtime. Events taking place in Somalia, Ethiopia, Rwanda, and in other turbulent places around the globe are mentioned (briefly) in the media.

These images show us that devastation like death, disease, and turmoil are everyday realities in these places. I do not want to be a pessimist, but we here in the Western world may suffer the same fate if we do not face reality.

What is the best way to determine the amount of damage a person has done over 65 years? Well, let us just count the number of grandchildren. A person who has more than four grandchildren has helped to do some environmental damage. Why four? Because we have just replaced the grandparents with their grandchildren. Once, while I watched one of the daily, morning TV shows, the host was discussing a baseball team made up of players over 80 years old. The host mentioned that one of the players was a grandfather with sixty-three grandchildren (or fifty-nine too many). I was outraged when I heard this. But the host's reaction indicated that this was something to aspire to. However, you do not need to be a great mathematician to realize that if everybody had sixty-three grandchildren, we (all the world's citizens) would really be in trouble.

One of the most egregious abuses of nature due to overpopulation is a river in Indonesia called the Citarum in West Java. The river is loaded with plastic debris, bags, sewage, and other industrial effluents caused by humans.[120] If the reader wants to see what the river looks like I suggest that they google it. There are many pictures of the river as it appears today. In many of the pictures many of the residents of the cities on the river are looking through the pollution looking for plastics that they can recycle and thus earn a small income.

Let us look at carrying capacity that the reader can associate with. Below is a picture of the Space Shuttle, Discovery. Let us say that the space shuttle can orbit the Earth many times. Let us say that there are seven people on board. How many days people can stay on board without getting added resources such as oxygen, water, and food?

You see the shuttle has a carrying capacity just for a brief period (Fig 4-21). The shuttle cannot stop at a local 7/11 on a nearby asteroid or a Walmart Store containing oxygen, water, food. You get the point.

This is the same idea that we have on the planet Earth. How many people can the earth sustain without bringing in outside resources?

Fig 4-21. Photo of the Space Shuttle Discovery.[121]
Photo by Sanjay Acharya.

HUMAN EXPANSION

URBAN SPRAWL AND URBANIZATION

One of the things that humans do quite a bit is to move to areas that were once pristine. We see this all the time. With the increase in human populations, cities get overwhelmed, thus causing people to move out to the suburbs, which extends the populated city limits.

The definition of Urban Sprawl:[122]

1. unlimited outward extension

2. low-density residential and commercial settlements

3. leapfrog development

4. fragmentation of powers over land use among many small localities

5. dominance of transportation by private automotive vehicles

6. no centralized planning or control of land-uses

7. widespread strip commercial development

8. great fiscal disparities among localities

9. segregation of types of land uses in different zones

10. reliance on the trickle-down or filtering process to provide housing to low-income households

4-22. Los Angeles Urban sprawl.[123]

Look at what has happened in the United States. Wealthier people decided to move out to the suburbs and thus spread their presence. Los Angeles is a good example. Here we see tremendous sprawl. This is indicative of many cities such as Mexico City as well. (There was an environmental book that we used once that facetiously said that they wanted to change the name of the city to Make Sicko City because of the air pollution.)

It should be pointed out that sprawl, especially in the United States, consists of the following:

Strip malls, housing subdivisions, fast food chains, shopping malls, and increased automobile use. The last effect, increased automobile use, has caused a tremendous increase in CO_2 in the atmosphere. I have seen this myself in the various places that I lived here in the United States.

It is obvious that people do not understand the environmental consequences that this movement to the suburbs causes. Traffic increases tremendously. People spend a great deal of time on the road, thus depriving them of time in their homes and with their families.

Remember that when building homes, malls, and roads out in the suburbs, the environment will be degraded, thus causing disruptions in the ecosystems, which may cause an increase in disease due to Bioinvasive species and other damage that we cannot even imagine.

Road building will eventually cause flooding by not allowing water to percolate in the soil as it should. As we have constantly seen, flooding under overpasses occurs and roads must be closed because cars cannot make it through. Do not forget, standing water is an excellent breeding ground for disease-carrying mosquitoes. (See Section Two on diseases.)

I call this expansion "the beer-belly effect" because sprawl is exactly like a guy who has an extended stomach because of too much beer. Developers build up the suburbs and people keep on coming out to them. This expansion leads to destruction of the normal biota (plants, animals, bacteria, fungi, and protista) there, thus changing the ecosystems. The best example of this was displayed in a photograph in a weekly news magazine where the desert in Nevada turns into a boom town where all we see are thousands of private homes, one-story homes. Can you imagine what can happen if they run out of water?

It should also be pointed out that urbanization in poorer countries, such as Asia and Africa, can help spread disease. When millions of people move to cities specifically for economic reasons, cities swell way above their carrying capacity, which results in slums, with fewer municipal services, clean water supplies, sewage disposal and waste management. This leads to pools of contaminated water, such as those in tires which make perfect breeding sites for disease- carrying mosquitoes, other vectors, and rodents. Diseases like dengue fever, malaria, filariasis, chagas disease, plague, leptospirosis, and typhus are various diseases that can be spread by these vectors.

TABLE 4-13

LIST OF DISEASES INFLUENCED BY URBANIZATION[124]

Dengue fever	Malaria	Yellow Fever
Chikungunya	West Nile fever	St. Louis Encephalitis
Lyme disease	Ehrlichiosis	Plague

OVERFISHING

Overfishing is becoming quite common in today's world because of sophisticated fishing methods which include satellites and airplanes to spot schools of fish. It seems plausible to assume that if we are overfishing specific species of fish, we may be causing disruptions of the aquatic ecosystems by eliminating fish that are food for other organisms as well as humans.

They may even have a beneficial effect on the ecosystems by supplying necessary nutrients that we do not even know exist.

A good example of overfishing is the reduction of the Orange Roughy, *Hoplostethus Atlanticus*. This is a deep-water fish found around 1500 Meters (a mile is 1,600 meters) below the surface waters near New Zealand. They reach a maximum size of around 50 cm (20 inches) and may weigh as much as 3.9 Kilograms (7.9) pounds.[125] What really makes them so interesting is the fact that they can live up to 149 years; yes, that is right, over 1.5 centuries. In fact, many of the species living today were living at the time of Abraham Lincoln, our 16th president. They reach sexual maturity around 25 years, double that of a human. Another interesting fact is that these fish were not discovered until 1979, only 43 years ago as of this writing. In 1991 around 35,000 tons were harvested, but in 1999-2000 only 15,000 tons were harvested. They were already in decline, so the government of New Zealand had to try to lower the catch so that the stocks can rebuild themselves without going extinct. For those of you who are interested in eating them, you should know that they are quite expensive, although incredibly good. It is because of their good taste that they may well become extinct.

One may wonder how we know that they live to be 150 years old. The scientists use radiometric dating (radioactive dating) to determine age. They also look at their otoliths (ear bones: structures that determine gravity and linear acceleration).

Their diet consists of prawn, fish, and squid. If we allow the Orange Roughy to go extinct, then the organisms that they feed upon may multiply out of control and cause some now-unknown catastrophe which may even lead to some unheard-of-as-yet disease. Fig 4-23 shows how lots of fish are caught rather quickly

Fig 4-23 Chilean purse seine catching lots of fish.[126]
Photo by C Ortiz Rojas

Bushmeat, disease and overfishing

A little-known aspect of overfishing is the good possibility that overfishing can lead to more killing of bushmeat in Africa and increase the spread of disease. Bushmeat is meat from primates (monkeys) and other terrestrial wild animals that roam free in Africa and other countries. In Ghana, a country in West Africa, there appears to be a strong correlation with declining fish due to European Union ships and an increase in bushmeat hunting.

More than half of Ghana's 20 million humans live within one hundred kms of the coast. Many of these people make their living and derive their protein from the sea. With lack of employment and food (fish) the local population had to resort to get bushmeat for their diet. Hunting for bushmeat may lead to infections of dreaded infectious diseases.[127] Diseases like Ebola may be spread by contact or exposure to bushmeat.

Overhunting in Ghana has also resulted in a 76% decline in the biomass of forty-one species of mammals in its nature preserves, as well as local extinction of some species. This overexploitation can lead to disaster, not only in increased human disease, but to ecosystem imbalance as well, because some of the exploited animals may have some unknown ecological function that we do not even know exists.

Overhunting of organisms means hunting organisms to near extinction so that they may have a tough time recovering their natural populations. These organisms may be aquatic such as fish, or land-based animals such as rhinoceroses and primates. It is possible that as we eliminate some species of organisms, we may inadvertently be destroying other species which depend on the overhunted species.

A good example of this was the relationship between the famed dodo bird (*Rhaphus cucullatus*) and the calvaria tree. Dodo birds eat the seeds of this tree. For the seeds of this tree to take root, the outer coat must be eliminated. When dodo birds ate the seeds, substances in their intestinal tracts dissolved the seed coat. The seeds were excreted out and wound up planted in the ground where calvaria trees grow. Now there are still old calvaria trees, but no dodos, so no new calvaria trees. Remember, this is just one example of many.

There may be thousands of such relationships that we do not even know about.

AQUACULTURE AND MARICULTURE

What exactly is aquaculture? Aquaculture is the raising of fish and other organisms by use of human-made ponds or even parts of the ocean. Mariculture is the use of the oceans to raise organisms such as shrimp and seaweeds.

Sometimes when fish are raised in human-made ponds, disease organisms can spread. When his happens, non-farmed fish species can be infected if some of the pond-raised fish escape into the wild and pass on disease organisms.

Raising shrimp is also bad because mangrove trees must be destroyed to make areas for the shrimp. Mangroves are the trees with roots that extend into the waterways. They are prevalent in tropical and subtropical areas within areas of high salinity. It is amazing that every ecosystem has its own biota. Remember, if the environment is not perfect for any organism, it will not survive.

FERTILIZERS AND EUTROPHICATION

Fertilizers come in many forms. They are applied to enhance the growth of crops. However, many fertilizers wind up in bodies of water and in other non targeted areas. This is especially true in aquatic areas. This leads to the growth of algae and other aquatic organisms. By allowing algae to grow, the environment is being changed. There is a special term used to describe this process. It is called Eutrophication. What happens is that, with all the algae growth, the body of water affected will eventually turn into dry land. I know that this sounds strange but that is exactly what happens. With this algae growth, organisms that depend on the water may die out. The table below compares a Eutrophic Lake with that of a lake that does not have many nutrients, called an Oligotrophic Lake. Most lakes are mesotrophic; they have a moderate amount of nutrients. Notice that the root, *trophic,* is in these three words. Trophic means feeding or food related.

TABLE 4-14

CHARACTERISTICS OF EUTROPHIC AND OLIGOTROPHIC LAKES

EUTROPHIC CHARACTERISTICS	OLIGOTROPHIC CHARACTERISTICS
Poor light penetration, Turbid (cloudy)	High light penetration
High nutrient levels	Low amount of nutrients
Shallow waters	Deep water
Low dissolved oxygen	High Dissolved Oxygen
High algal growth	Low algal growth

AIRPLANE TRAVEL

As we have seen, when temperatures spike there is the possibility of more disease. Many of these diseases will travel from one country to another in less than 24 hours thanks to the worldwide use of airplanes or air travel. It does not take a genius to figure this out. When diseases have long non-infectious stages or incubation periods of at least a few days or weeks, then the possibility exists for a viral plague to infect entire cities as was demonstrated in the movie, "Outbreak" (1995). This can be caused just by a simple plane trip to anywhere. We now live in a world united by globalization. If you have the money, you can go virtually anywhere in the world by plane, whether you are infected with a disease or not.

OZONE, ITS IMPORTANCE, AND ITS LOSS

Another aspect of Climate Change involves OZONE depletion. Ozone (O_3) is a form of oxygen in a triatomic state, which means that it is a form of oxygen containing three oxygen atoms. Ozone is especially important in that its function is to filter out ultraviolet

(UV) light rays from the sun. We keep hearing about the carcinogenic effects of UV light. Now you may ask, what exactly is UV light and how does it cause cancer?

UV light is part of the electromagnetic spectrum. According to various sources The electromagnetic spectrum is the entire distribution of electromagnetic radiation. It consists of things that we can see, cannot see but can hear, and cannot hear such as gamma rays, x-rays, UV light, visible light, infrared light, microwaves, radio waves, and electric waves. Below is a chart of electromagnetic radiation. On the chart in Fig. 4-24, one sees wavlengths. These are waves, and the numbers on the left indicate distances. See Table 4-15 which explains meanings of the distances:

TABLE 4-15

EXPLANATIONS OF SYMBOLS

SYMBOL	Number per meter
nm-nanometer	1,000,000,000
μm-micrometer	1,000,000
mm-millimeter	1,000
cm-centimeter	100
m-meter	1
Km-kilometer	1000m=1km

For reference purposes there are 91.44 centimeters in one yard.

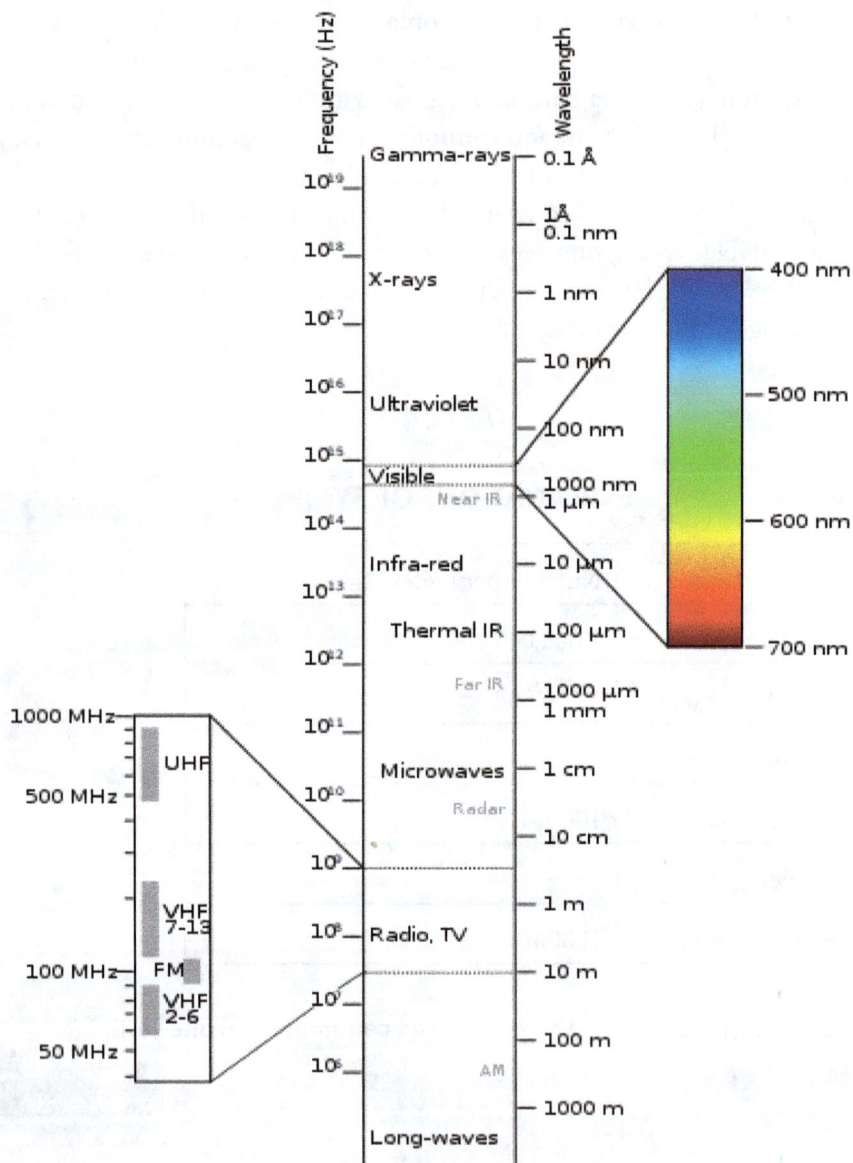

Fig 4-24. Electromagnetic spectrum.[128]

UV light is divided up into three groups: UV A, UV B and UV C. [129]

TABLE 4-16

WAVELENGTHS OF ULTRAVIOLET LIGHT

UV LIGHT TYPE	WAVELENGTH IN NANOMETERS[130]
UV A	315-400
UV B	280-315
UV C	100-280

UV A is the least dangerous whereas UV C is the most dangerous for humans. If it were not for ozone, most life on The Planet Earth would be extinct, except for those organisms living deep in the ocean or deep underground.

The primary mechanism of carcinogens by UV light is the process whereby UV light rays, especially those of UV B, damage the DNA of cells, specifically skin cells since they are the closest to the sun's rays. It is estimated that every 1% decrease in Ozone will cause an increase of 2% UV B radiation. This will cause a 4% increase in basal cell carcinomas and a 6% increase in squamous cell carcinomas.[131]

Ozone is measured with a satellite called TOMS. TOMS stands for Total Ozone Monitoring Spectrophotometer. Every October, when ozone is at its lowest, TOMS takes pictures from space. Ozone levels are measured in units called Dobson Units (DU)[132] . A DU is equal to .01mm of ozone at one atmosphere pressure. The higher the number of Dobson Units the more ozone there is.

If UV light is not filtered out by ozone, then the possibility exists that cancer can form, especially in the skin. The reason this happens is because the hydrogen bonds that hold the double helix of the DNA molecule together break up. This can cause problems to the genes and thus cancer can form. See Chapter 2 on Biology.

In 1995, Mario Molina, Sherwood Rowland and Paul Crutzen won the Nobel Prize in Chemistry for figuring out how CFCs destroy Ozone. CFCs are the refrigerants used in air conditioners and in other cooling systems. There are three major categories of refrigerants based on their chemical structure.

TABLES OF REFRIGERANTS AND EXAMPLES

TABLE 4-17A

CFCs – CHLOROFLUOROCARBONS-These contain carbon, chlorine, and fluorine

EXAMPLES	NAME	FORMULA
R-11	Trichlorofluoromethane	CCl_3F
R-12	Dichlorodifluoromethane	CCl_2F_2
R-113	Trichlorotrifluoroethane	CCl_2FCClF_2
R-114	Dichlorotetrafluroethane	$CClF_2CClF_2$
R-115	Chloropentafluroethane	$CClF_2CF_3$

TABLE 4-17B

HCFCs-Hydrochlorofluorocarbon

These contain Hydrogen, fluorine, chlorine, and carbon

EXAMPLES	NAME	FORMULA
R-22	Chlorodifluoromethane	$CClHF_2$
R123	Dichlorotrifluoroethane	$CHCl_2CF_3$

TABLE 4-17C

HYDROFLUOROCARBONS

These contain Hydrogen, fluorine, and carbon

EXAMPLES	NAME	FORMULA
R-125	Pentafluoroethane	CHF_2CF_3
R-134A	Tetrafluoroethane	CH_2FCF_3

The discovery of how the Ozone gets destroyed by CFCs, was made by Mario Molina, Paul Crutzen and F. Sherwood Rowland. They realized that ozone is destroyed in the stratosphere where most of the normal ozone resides. The little black dot after the Cl means that the chemical is a free radical. Free radicals do not exist long in nature. To make it simple, the mechanism of ozone destruction is as follows:

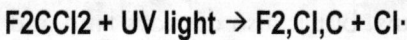

F2CCl2 + UV light → F2,Cl,C + Cl·

Cl· + O3 → ClO + O2

Let us start with dichlorodifluoromethane which is represented by R-12. R-12 gets broken down with UV light. UV light dislodges a chlorine atom, which in turns takes an oxygen atom from ozone (O_3) in the presence of polar stratospheric clouds (PSC) which are clouds in the winter polar stratosphere at altitudes of 15,000-25,000 meters. One type is composed of supercooled droplets of water. ClO which is called chlorine monoxide, and atmospheric oxygen (O_2) are the result. Chlorine monoxide, when prevalent in the atmosphere, is indicative of ozone depletion. Atmospheric oxygen is not good at filtering out UV light.

The above is the basis of the chemical reaction that Molina and his colleagues figured out.

There are other theories as to how ozone gets broken down. The SSTs (supersonic transports, planes going faster than sound, which are no longer in use) also are believed to help break down ozone by producing nitric oxide (NO) which helps to remove an oxygen atom from ozone in an equivalent manner, shown in the following equation.

$$NO + O_3 \rightarrow NO_2 + O_2$$

What is happening here is the fact that NO (nitric oxide) is taking an oxygen atom from ozone.

Ozone depletion is not only bad for animals and humans, but it can inhibit photosynthesis, yes, the process whereby plants produce all the food on the planet that animals eat, including humans. Without photosynthesis no one will eat.

WARFARE

One of the cruelest endeavors that humans engage in is that of warfare. War not only kills humans, but it also destroys land and many of its organisms, such as the simplest bacterium or a big water buffalo. Could it be that these organisms have an ecological function in their own ecosystem? By destroying the land, a catastrophe may be in the making. The plants which get destroyed or the animals which die may harbor some deadly virus, with which they live within perfect harmony. By destroying the ecosystems where these larger organisms reside, a new ecosystem can evolve, thus causing an imbalance which may lead to a deadly epidemic.

We know that lots of land gets destroyed without a thought from the politicians, who start these wars without thinking of the consequences. We should all try to make friends out of our enemies instead.

A good example of a continent rich in biodiversity and ravaged by war is Africa. Look at the following: Ethiopia, Sudan, Somalia, Democratic Republic of Congo, Uganda, Zimbabwe, Rwanda, Sierra Leon, and Liberia, just to name a few. Are these countries icons of peace, prosperity, and longevity? Absolutely not. Under what conditions do we hear of these countries in the news? Yes, that is right, because of war. These countries either have been tearing themselves apart by war (Idi Amin in Uganda) or are doing it as I write these words, such as Sudan (think of Darfur).

One thing that most of these warring African countries have in common is the fact that they have a rich biodiversity, which includes countless species of plants, animals, fungi, protists, bacteria, and viruses. Every single one of these organisms has an ecological function on the planet. If they did not have an ecological function they would not exist. The leaders of these countries never took a course in Environmental Science or Biology.

WATER POLLUTION

There are eight categories of pollutants. They are the following:

1. sediment,

2. sewage. (Sewage contains many disease organisms such as bacteria, viruses, and worms.)

3. disease- causing organisms,

4. organic plant nutrients

5. inorganic nutrients

6. thermal pollution (heat addition)

7. radioactive particles

8. organic nutrients

Again, as with land, we may be destroying the environments of friendly organisms. When we kill the friendly organisms, the not-so-friendly ones may take over. A good example will suffice. Why do people get a sore throat? They get a sore throat because their immune system is not working properly. Therefore, bad organisms take over. Now think of the planet the same way. When we destroy the "good guys" the "bad guys" will take over. As with warfare, with thoughtless human interventions like pollution because of overpopulation, and a lack of government controls, we are creating conditions that can lead to disease

Fig 4-25 Water pollution caused by dumping fecal sludge into a river.[133]

ADDITION OF CHEMICALS SUCH AS PESTICIDES AND OTHER DEADLY SUBSTANCES

Of course, the addition of chemicals, many of which are deadly in minute amounts, can change ecosystems, whether they be on land or water or even in the air. Unfortunately, after WWII pesticides became fashionable to use against insects and other perceived pests. Since then, millions of pounds of these toxic substances have been used.

In 1999, 1 million short tons (two thousand pounds)[134] of herbicides were used worldwide: 0.6 million short tons of insecticides and about 250,000 short tons of fungicides.[135]

Notice the fact that when we add pesticides, we not only kill target organisms, but we may also kill non-target beneficial organisms. By doing this we can also change the natural ecosystems of the areas, thus leading to a new dynamic between the resident organisms and newly acquired organisms because of the ecosystem change. This is a well-known phenomenon called *directional selection*. (See section on evolution in the chapter on Biology.)

RIVER DIVERSIONS

River diversions can help cause disease in unexpected ways. The best example of this was demonstrated with the Aral Sea, which must be one of the worst human-induced environmental tragedies since the creation of humanity.(See figure 3-11) The Aral Sea is in Kazakhstan and Uzbekistan, both republics now independent from the former Soviet Union. It was the fourth largest lake in the world. Because of the diversions of the two major rivers, The Amu Darya and the Syr Darya, the water volume has been diminished to less than one third of its original volume since 1960. This was because the area was a good cotton-growing area and the water from the two rivers was diverted to grow cotton, which is very water intensive.

Not only did the region lose water, but now the fishing economy of the region has virtually disappeared and cases of cancers, respiratory ailments and other respiratory diseases are prevalent in the[136] local population.

But one of the most amazing sites that one will ever see is to visit the region, and you will see fishing boats semi buried in the sand. Anyone can see these pictures by doing a Google search on a computer. Search for the Aral Sea. You will be amazed at what you see-- or better, what you do not see (the water, of course).

With river diversions we are also destabilizing the natural environments and ecosystems. By doing this, we can disrupt the normal fauna and flora (animals and plants) of the area, thus allowing changes that may help influence disease.

Environmental changes can influence disease transmission by altering exposure to soil and water.

OIL SPILLS

As we write these words (May 2010) there is a tremendous oil spill in the Gulf of Mexico. This has the possibility of destroying all life on the coasts of Louisiana, Alabama, and Florida. This catastrophe has the potential to alter the ecosystems of the Gulf forever. Is it possible that this alteration may give an advantage to opportunistic organisms to cause disease? If we destroy natural organisms that normally live in the Gulf, we may create conditions for unnatural organisms. (See Bioinvasions.)

There may even be some relationship between the Gulf organisms and the rest of the world's oceans. Remember, many organisms migrate to the Gulf for several reasons including to spawn. Many go back to the other oceans where they live. If they do not return to their main habitats, we may see disaster globally and there may be an ecological melt down.

Fig 4-26. Oil spill[137]

Birds and other land-dwelling organisms are also affected. Oil coats the feathers of pelicans and other aquatic birds thus allowing birds to die of hypothermia (cold). The oil also depletes the dissolved oxygen (DO) content of the water. Some fish need a minimum of 5 ppm (parts per million) dissolved oxygen.

This is not the first time that offshore spills have occurred but we should all strive to make it the last. Since 1967 there have been thirty-six giant oil spills in the oceans. Each one dumped over one million gallons of oil.[138] There are forty-two gallons per barrel (bbl) of oil.

Corporations and governments worldwide must take responsibility in assuring a fossil fuel free world.

[1] Wilson, Mark, "Ecology and Infectious Disease." In *Ecosystem Change and Public Health: A GlobalPerspective*, Joan L. Aron, and Jonathan A. Patz Editors; Johns Hopkins University Press: 2001, p.301.

[2] Phenologies- This is the science of the relationship between science and periodic biological phenomena such as pollination and bird migrations.

[3] Smith, Joel, D.A. Tirpak, Dennis, The Potential Effects of Global Climate Change on the United States: Draft Report to Congress: Executive Summary. U.S. EPA, Office of Policy Planning, and Evaluation.

October 1988.

[4] http://www.howstuffworks.com/question192.htm viewed 20 December, 2009.

[5] This quote reminds me of a bumper sticker that says, "The one that dies with the most toys wins." Put another way, some people are so brainwashed when it comes to the economy that they would rather die with a lot of money rather than live with a little money.

[6] NOAH 29 Jan 2021. https://www.ncei.noaa.gov/news/projected- ranks#:~:text=-Global%20temperature%20data%20document%20a,2005%20(tied)%2C%20re-spectivelyViewed 6 April 2021.

[7] https://www.nasa.gov/press-release/2020-tied-for-warmest-year-on-record-nasa-analysis-shows

[8] Low, Petra. "Weather-related Disasters Dominate." In *Vital Signs 2009*. Worldwatch Institute, pp. 62- 64.

[9] Russell, J. Carbon Emissions on the rise but policies growing too. In *Vital Signs 2009*; Worldwatch Institute. P. 59-61.

[10] Holland, Dr Daryl .2018. How a termite's mound filters methane-and what it means for greenhouse gases.Phys.org. 27 Nov 2018.

[11] https://en.wikipedia.org/wiki/Global_warming_potential#:~:text=Global%20warming%20potential%20(GWP)%20i s,gas%20and%20the%20time%20frame.

[12] Source: Hinrichs, Roger, M. Kleinbach. *Energy: Its Use and the Environment*. Brooks/Cole 2002,p.288.

[13] These are chemicals that have all their hydrogen atoms replaced by fluorine. An example isoctafluoropropane, C_4F_8.

[14] Source: *National Geographic*, September 2004. p.3.

[15] http://blog.wired.com/wiredscience/2009/02/rising-ocean-ac.html   v i e w e d Feb 1, 2009.

[16] Meyers, Samuel," Adapting to the Health Impacts of Climate Change," Worldwatch Report 181 In Global Environmental Change: The Threat to Human Health, 2009, p.30.

[17] Ground level ozone is a product of photochemical smog. Temperature inversions result when there is not enough wind to blow pollutants away. A normal stratification of air consists of warm air, followed by a layer of cool air and that followed by cooler air. In an inversion the following exists: Cool air, warm air, and cool air.

[18] Op. cit. Meyers, p.25.

[19] Pellella, Philip. "Global Warming will Increase World Hunger"-UN May 27, 2005. http://www.planetark.com/dailynewsstory.cfm/newsid/31000/story.htm   Viewed Sept. 26, 2009.

[20]   https://commons.wikimedia.org/wiki/File:Protein_malnutrition.jpg Viewed 16 Jan 2022.

[21] Cold water can hold more dissolved gases than warm water. These gases may be Oxygen, Carbon Dioxide, Nitrogen, or any other gas. Fish need at least 5ppm oxygen to survive.

[22] If the Gulf stream is disrupted, the event can place Europe in a deep freeze. Pisa, which is a city in Northern Italy, has a palm tree in the center of town. If you look on a map, Pisa is on latitude like that of New York, yet it is relatively warm compared to New York.

[23] Chong, Jia-Rui, "Global Warming: Enough to make you Sick." The Los Angeles Times. Feb 25, 2007. http://www.truthout_org/docs_2006/022607N.shtml; viewed 2/27/07.

[24]   http://www.ens-newswire.com/ens/apr2008/2008-04-28-01.asp Viewed Feb 7, 2009.

[25] See glossary for definition of biodiversity.

[26] https://en.wikipedia.org/wiki/Retreat_of_glaciers_since_1850#/media/ File:Eastonterm.jpg
Viewed 27 Dec 2021

[27] Worldwatch Magazine. Nov-Dec 2000. Vol 13 #6 pp5-7.

[28] http://wps.prenhall.com/esm_abel_issuesocean_2/0,6649,228188-,00.html viewed Feb 10, 2009.

[29] http://wps.prenhall.com/esm_abel_issuesocean_2/0,6649,228188-,00.html viewed Feb 10, 2009.

[30] http://www.futureatlas.com/index.php/2009/02/28/eight-disappearing-islands/ viewed June 1, 2009.

[31] File: Marshall Islands on the globe (small islands magnified) (Polynesia centered). svg - Wikimedia Commons

[32] http://www.sciencedaily.com/releases/2008/09/080903134309.htm Viewed Feb. 10, 2009.

[33] http://www.sciencedaily.com/releases/2007/09/070911092139.htm Viewed Feb 10, 2009.

[34] http://stupac2.blogspot.com/2006/09/how-to-sink-ship-with-bubbles.html Viewed Feb. 10, 2009.

[35] http://images.google.com/imgres?imgurl=http://www.bermuda- triangle.org/assets/images/Methaneart.gif&imgrefurl=http://www.bermuda- triangle.org/html/methane_hydrates.html&usg=__fxvDDrN1783TzLe7m4nQ- crUcWI=&h=222&w=300&sz=32&hl=en&start=6&um=1&itbs=1&tbnid=lw3ua665c1eliM:&tbnh=86&t bnw=116&prev=/images%3Fq%3Dmethane%2Bhydrates%26um%3D1%26hl%3Den%26sa%3DN%26t bs%3Disch:1

[36] http://www.google.com/hostednews/afp/article/ALeqM5j4YVkl5OcsYMs1zKnx-qPmc4OGggAviewed Feb 13, 2009.

[37] https://commons.wikimedia.org/wiki/File:2011-08-

04_20_00_00_Susie_Fire_in_the_Adobe_Range_west_of_Elko_Nevada.jpg Viewed 4 Jan 2021.

[38] http://www.sciencedaily.com/releases/2008/02/080210100441.htm viewed Feb. 16, 2009.

[39] Photosynthesis is represented by the following equation: $CO_2 + H_2O \rightarrow C_6H_{12}O_6 + O_2$. Simply put, the equation states that carbon dioxide (CO_2) and Water (H_2O) are transformed by plants into carbohydrates (sugars such as $C_6H_{12}O_6$) and oxygen (O_2).

[40] J. Shaman, J.F. Day, and M. Stieglitz, "St Louis Encephalitis Virus in Wild Birds During the 1990 South Florida Epidemic: The Importance of Drought, Wetting Conditions, and the Emergence of *Culix nigripalpus* (Diptera: Culicidae) to Arboviral Amplification and Transmission," *Journal of Medical Entomology*, July 2003, pp. 547-54.

[41] Op cit., Gubler p.576.

[42] Ibid.

[43] https://commons.wikimedia.org/wiki/File:Drought.jpg
Viewed 29 Dec 2021

[44] http://toronto.ctv.ca/servlet/an/local/CTVNews/20081230/to_wx
_2008_081230?hub=TorontoNewHome viewed Feb 23, 2009.

[45] http://colli239.fts.educ.msu.edu/1996/06/15/borova-1996/ viewed Feb.5, 2010.

[46] Marcus, B.A. Malaria. Chelsea House Publishers, p.17.

[47] https://pixnio.com/science/microscopy-images/malaria-plasmodium/malarial-para-
sites-undergo-asexual- multiplication-in-the-erythrocytes-ie-erythrocytic-schizogony
Viewed 4 June 2021.

[48] https://commons.wikimedia.org/wiki/File:EL18p-R%C3%A9union.jpg
Viewed 4 Jan 2022.

[49] Bright, Chris. *Life Out of Bounds: Bioinvasion in a borderless world.* Norton Press,
1998, p. 157.

[50] Ibid.

[51] http://www.issg.org/database/species/search.asp?st=100ss. Viewed March 13, 2009.

[52] http://en.wikipedia.org/wiki/File:Sea_walnut,_Boston_Aquarium.jpg) viewed
Feb.5, 2010.

[53] http://www.ucsusa.org/global_warming/science_and_impacts/science/glob-
al-warming-faq.html#2
viewed March 21, 2009.

[54] http://images.google.com/imgres?imgurl=http://www.caring- planet.com/im-
ages/ibrowser/350pxinstrumental_temperature_record.png&imgrefurl=http://
www.caring- planet.com/Page/title/About_Global_Warming/&usg= lRWeDEg-
MqRQLEmFIF7oVCbhwPjM=&h=26 1&w=350&sz=64&hl=en&start=16&tb-
nid=L0Itit3RTmiQcM:&tbnh=89&tbnw=120&prev=/images%3F
q%3Dtemperature%2Bincreases%2Bdue%2Bto%2Bglobal%2Bwarm-
ing%26gbv%3D2%26hl%3Den%2 6sa%3DG viewed March 21, 2009.

[55] Moya,. J 2022. Not a good sign': The temperature was 70 degrees above average near South Pole, atroubling record. https://www.msn.com/en-us/weather/topstories/not-a-good-sign-the-temperature-was-70-degrees-above-average-near-south-pole-a-troubling-record/ar-AAVgPY7?ocid=msedgntp#image=1 Viewed 25 March 2022.

[56] Brookesmith, Peter. *Biohazard: The Hot Zone and Beyond*. 1997, p.132.

[57] Ibid. p.133.

[58] Henrichs, Roger, M. Kleinbach. *Energy: Its Use and the Environment*. Brooks/Cole 2002, p.305.

[59] "Global Warming Threatens Cold-Water Fish."
As temperatures rise, salmon and trout are likely to disappear from streams across the United States -- unless global warming pollution is reduced. http://www.nrdc.org/globalwarming/ntrout.asp= viewed Feb.6, 2010.

[60] http://www.afrol.com/Categories/Environment/env056_warming_poverty.htm viewed October 14, 2009.

[61] https://commons.wikimedia.org/wiki/File:1876_1877_1878_1879_Famine_Genocide_in_India_Madras_under_Briti sh_colonial_rule_2.jpg Viewed 5 Dec 2022.

[62] http://environment.about.com/od/globalwarmingandweather/a/gulf_stream.htm Viewed October 24,2009.

[63] http://en.wikipedia.org/wiki/File:Thermohaline_Circulation_2.png viewed March 19, 2010.

[64] http://wordnetweb.princeton.edu/perl/webwn?s=peat viewed March 3, 2010.

[65] http://www.oeb.harvard.edu/faculty/moorcroft/publications/publications/ise_etal_08.pdf viewed 3 March 2010. Ise, T et al. High sensitivity of peat decomposition to Climate Change through water-tablefeedback. Published online 12 October 2008; doi:10.1038/ngeo331.

[66] United Nations Population Fund. 2009. State of the World Population 2009. P.15. [67] EIA. (U.S. Energy Information Administration) Natural gas explained. https://www.eia.gov/energyexplained/natural-gas/how-much-gas-is-left.php#:~:text=According%20to%20U.S.%20Crude%20Oil,trillion%20cubic%20feet%20(Tcf). Viewed 26 March2020.

[68] Low, Petra, "Weather-Related Disasters Dominate" in *Vital Signs 2009*. The Worldwatch Institute, pp.62-64.

[69] *The New York Times*, August 7, 2014. https://carlzimmer.com/cyanobacteria-are-far-from-just-toledos-problem-157/

[70] https://commons.wikimedia.org/wiki/File:Algal_bloom_in_Lake_Erie,_Kelley%27s_Island_%288740853803%29.jpg\
Viewed 4 June 2021.

[71] How Climate Change is Driving Child Marriage. Girls Not Brides. 2017. https://www.girlsnotbrides.org/articles/hidden-connections-climate-change-child-marriage-bangladesh/ Viewed 2 Sep 2021.

[72] Sundaram, L. 2017.How Climate Change drives child marriage. The New Humanitarian. Viewed 2 Sep 2021. https://deeply.thenewhumanitarian.org/womenandgirls/community/2017/11/06/how-climate-change-drives-child- marriage

[73] Ool, E and D. Gubler: Global spread of epidemic dengue: the influence of environmental change.
Future Virology 4 (6) pp.571-580 (2009).

[74] Ibid. Gubler p. 573.

[75] http://www.epa.state.il.us/land/tires/mosquito-borne-illnesses.html viewed 20 December, 2009.

[76] International Organization for Migration (IOM) Policy Brief. Migration, Climate Change, and the Environment. Geneva: May 2009, pp.1-9.

[77] Ibid. IOM 2009

[78] http://www.maricopa.gov/EnvSvc/VectorControl/Mosquitos/MosqInfo.aspx viewed March 16, 2010. [79] Gubler, D.1998 Resurgent vector-borne diseases as a global health problem. Emerging infectious diseases 4(3) pp.442-450.

[80] W.R. Mac Kenzie et al., "A Massive Outbreak in Milwaukee of Cryptosporidium Infection Transmitted Through the Public Water Supply," *New England Journal of Medicine*, 21 July 1994, pp.161-67.

[81] Arieti, David F., *The Earth is My Patient*. Authorhouse Books, 2005, p.127.

[82] M. McKinney and Schoch, R.M. *Environmental Science: Systems and Solutions*. Jones and Bartlett,1998, p.425.

[83] Gardner, Gary. "Asia is Losing Ground." *World Watch* magazine, November/December 1996, p.23. [84] Gubler, D. Resurgent vector-borne diseases as a global health problem. Emerging infectious diseases4(3) pp.442-450.

[85] See the book, *The Hot Zone* by Richard Preston. Random House, 1994. [86] http://www.ncbi.nlm.nih.gov/pubmed/8488073 viewed July 12, 2009. [87] Lemon, S. M. et al. In *Vector-Borne Diseases*, 2008, p.18.

[88] Walsh, J.F., D.H. Molyneux and M.H. Birley. "Deforestation: effects on vector-borne disease."
Parasitology (1993). 106: S55-S75

[89] Limited Diversity of *Anopheles Darlingi* in The Peruvian Amazon Region of Iquitos Viviana Pinedo-Cancino, Patricia Pheen, Eduardo Ttarazona-Santos, William E. Oswald, Cesar Jeri, AmyYomiko Vittor, Jonathan A. Patz, and Robert H. Gilman* http://www.pubmedcentral.nih.gov/articlerender.fcgi?artid=1559519: viewed July 21, 2009.

[90] Op cit. Lemon, p.19.

[91] S.S. Meyers, Global Environmental Change: The Threat to Human Health. Worldwatch Report #181,2009, p.21.

[92] https://commons.wikimedia.org/wiki/File:Deforestation_NZ_TasmanWest-Coast_2_MWegmann.jpg
Viewed 27 Dec 2021.

[93] http://www.globalwarming.org.in/causes-of-global-warming.php Viewed Feb. 9, 2008.

[94] One microgram is equal to one millionth of a gram. 1,000,000 micrograms make up one gram.

[95] http://www.iitap.iastate.edu/gcp/studentpapers/1996/atmoschem/brockberg.html: viewed March 28, 2008.

[96] https://commons.wikimedia.org/wiki/File:Isoptera.jpg Viewed 9 Jan 2022.

[97] https://commons.wikimedia.org/wiki/File:Trichonympha_campanula.png Viewed 9 Jan 2022.

[98] Gubler, Duane. Resurgent vector-borne diseases as a global health problem. *Emerging InfectiousDiseases*. 1998.(3) pp.442-450.

[99] Foley, J, A et al. Global Consequences of Land Use. *Science* 209 (5734) pp.570-574.

[100] Garrett, Laurie, 1994. *The Coming Plague: Newly emerging diseases in a world out of balance.*
Penguin Books, pp.27-28.

[101] Ibid. Garrett

[102] Myers, S. "Table 1. Examples of Land Use Change and Increased Malaria Transmission, Global Environmental Change, Threats to Human Health," pp.16-17. Worldwatch Institute. Used with permission

[103] Myers, S. "Table 2. Examples of Land Use Change and increased Schistosomiasis Incidence," Global Environmental Change, The Threat to Human Health. pp.16-17 Worldwatch Institute: Worldwatch report#181.www.worldwatch.org. Used with permission

[104] http://www.independent.co.uk/opinion/commentators/johann-hari/johann-hari-lifethreatening-disease-is-the-price-we-pay-for-cheap-meat-1677067.html viewed May 3, 2009.

[105] See footnote 47 from S. Meyers, p.20.

[106] Arieti, David. *The Earth is my Patient*. Authorhouse Books. 2005, p.251.

[107] Goodland, R., and J, Anhang. 2009. Livestock and Climate Change. Worldwatch Vol.2, #6November/December 2009, p.14.

[108] CO_2 equivalent is a universal expression to describe the effect of all greenhouse gases in their relation to Carbon dioxide. Therefore methane, CFCs and other gases are expressed in their heating potential with GWP of CO_2 being 1. See Table 3 for the GWP of the various GHGs.

[109] Op cit. Goodland, p. 11.

[110] http://www.medindia.net/patients/calculators/worldPopulation.asp: viewed March 28, 2008.

[111] Population Reference Bureau, 2020 World Population Data Sheet.

[112] https://commons.wikimedia.org/wiki/File:Human_population_since_1800.png

Author Bdm 25 Viewed 14 Dec 2021.

[113] Arieti, David F., The Earth is My Patient. Authorhouse Books 2005, p.127.

[114] http://www-popexpo.ined.fr/eMain.html viewed April, 27,2004.

[115] Doubling times of populations are calculated according to the following formula: 70 is divided by the percent growth rate. For example: If a country has a growth rate of 2%, the doubling time is calculated as 70/2, which equals 35 years. For those of you who are interested in investing money, you can figure out the amount of time that it would take to double by using the same formula. It is obvious that you would want a growth rate which is very high on your income when it comes to investing, but when it comes to population increase, you would want the smallest percent growth rate as possible.

[116] This is probably one of the most quintessential examples of human-induced disaster. The Aral Sea was once the fourth largest inland lake in the world, located in Kazakstan (it used to be part of the former Soviet Union), but due to the former Soviet Union's desire to acquire money by raising cotton, they took water from the two major rivers, the Amu Darya and Syr Darya, which caused the lake to virtually dry up.

[117] Population Reference Bureau 2021. Data Sheet.PRB.ORG

[118] Paul R. Ehrlich, Anne H. Ehrlich, *The Population Explosion*. New York: Simon and Schuster, 1990, pp.190-195.

[119] Seen on a bumper sticker in Chicago.

[120] http://www.guardian.co.uk/environment/gallery/2008/dec/05/water-pollution-citarum-river viewed Jan 30,2009. and http://www.treehugger.com/files/2007/06/up_the_citarum.php

[121] https://commons.wikimedia.org/wiki/File:Space_Shuttle_Discovery_at_Udvar-Hazy_Center.jpg
Viewed 14 Dec 2021.

[122] http://www.plannersweb.com/sprawl/define.html viewed May 24, 2010.

[123] https://commons.wikimedia.org/wiki/File:Los_Angeles_-_Echangeur_autoroute_110_105.JPG Viewed 15 Dec2021.

[124] Op cit. Gubler 1998.

[125] http://www.starfish.govt.nz/science/facts/fact-orange-roughy.htm#the; viewed May 10, 2008.

[126] https://commons.wikimedia.org/wiki/File:Chilean_purse_seine.jpg
Viewed 16 Jan 2022

[127] J.S. Brashares et al, "Bushmeat Hunting, Wildlife Declines and Fish Supply in West Africa," *Science,* 12 November 2004, pp.1180-1183.

[128] https://commons.wikimedia.org/wiki/File:Electromagnetic-Spectrum.svg Viewed 14 Dec 2021.

[129] Electromagnetic radiation is organized by wavelength and frequency of which UV light is one aspect.

[130] A nanometer is equal to one billionth of a meter (10^{-9} of a meter).

[131] http://www.nas.nasa.gov/About/Education/Ozone/radiation.html. Viewed April 7, 2009.

[132] DU also stands for depleted Uranium and The University of Denver.

[133] https://commons.wikimedia.org/wiki/File:Dumping_of_faecal_sludge_into_the_river.jpg Viewed 16 Jan 2022

[134] A metric ton equals 2206 pounds or 1000 kilograms.

[135] Cunningham, W., and M. Cunningham. *Environmental Science: A Global Concern.* 2008, p. 210.

[136] Postel, Sandra. *Pillars of Sand.* Norton Press, 1999, pp. 93-95.

[137] https://commons.wikimedia.org/wiki/File:Oil-spill.jpg Viewed 15 Dec 2021.

[138] http://www.infoplease.com/ipa/A0001451.html viewed May 24, 2010.

Chapter 5

VECTORS

Vectors are organisms that spread disease. Most vectors belong to the phylum *Arthropoda*. Many common organisms, not only arthropods, can be considered vectors, including domesticated animals such as pigs, horses, humans, and even aquatic organisms such as clams and oysters.

MAP 5-1 below shows deaths from vector-borne diseases throughout the world. It is apparent that vectors are particularly important in disease transmission. Notice that Africa has the highest number of deaths from vector-borne diseases. [1]

MAP 5-1

Deaths from vector-borne disease

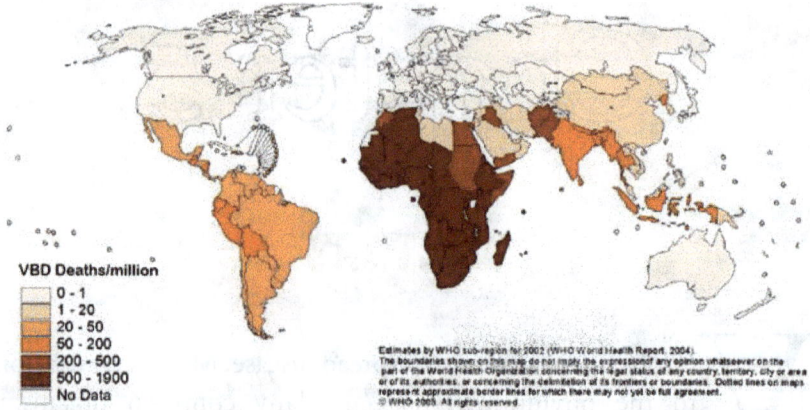

VBD Deaths/million
- 0 - 1
- 1 - 20
- 20 - 50
- 50 - 200
- 200 - 500
- 500 - 1900
- No Data

Estimates by WHO sub-region for 2002 (WHO World Health Report, 2004). The boundaries shown on this map do not imply the expression of any opinion whatsoever on the part of the World Health Organization concerning the legal status of any country, territory, city or area or of its authorities, or concerning the delimitation of its frontiers or boundaries. Dotted lines on maps represent approximate border lines for which there may not yet be full agreement.
© WHO 2005. All rights reserved.

The following is a list of vectors and examples of diseases that they spread.

TABLE 5-1

TABLE OF VECTORS

VECTOR	PHYLUM	Example of a disease spread by the vector
Mosquito	Arthropoda	Malaria
Ticks	Arthropoda	Typhus
Sand flies	Arthropoda	Leishmaniasis
Lice	Arthropoda	Typhus
Triatomine bugs	Arthropoda	Chagas Disease
Tsetse flies	Arthropoda	Sleeping Sickness

Fleas	Arthropoda	Bubonic plague
Humans	Chordata	Flu
Horses	Chordata	Hendra Virus

PHOTOS OF TYPICAL VECTORS
AND LIFE CYCLES

M.OSQUITOES

Phylum Arthropoda
Class Insecta

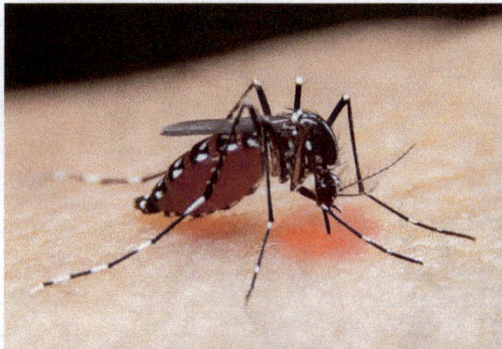

Fig. 5-1. Aedes aegypti biting a human, US Dept. of Agriculture [2]
Photo from US Department of Agriculture

There are approximately three thousand known species of mosquitoes. Only the female mosquitoes bite to feed their eggs. It appears that the CO_2 that we exhale is their first cue to the location of a human to bite.

The mosquito's life cycle

The life cycle of the *Aedes aegypti* mosquito:

1. **Oviposition**-egg laying. This takes place in the afternoon in dark colored water, rich in organic matter such as decaying leaves. Females lay eggs in batches between 30-50.

2. For eggs to hatch they need exposure to high humidity at the water line for two-three days. If eggs dry out during this period they will die; however, if conditions are right and they remain unhatched, the eggs can stay in a cured position, which means they can live up to six months.

3. There are four larval stages which take between five-ten days for development.

4. The transformation from the pupal stage to adult takes two-three days. It is possible that the transformation from egg to adult can take as little as 10 days.

5. Their life span is between two weeks to a month.

6. Their flight range is between 50-100 meters (164 -328 feet).

TICKS

Phylum Arthropoda

Class Arachnida

Fig. 5-2 *Amblyomma americanum* **Photo by James Gathany**[3]

The life cycle of a tick

There are two general types of ticks, soft ticks, and hard ticks.

There are approximately 650 species of hard ticks and 170 of soft ticks.[4]

The life cycle of ticks:

1. Female lays between 1,000 to 8,000 eggs. Mother dies and father dies after mating.

2. A six-legged larva emerges (it resembles an adult, except it has six legs instead of eight). It climbs onto grass.

3. A warm-blooded mammal (host) comes by and the larva attaches itself to it.

4. The larval tick will be called a nymph after digesting its first meal. Those ticks that live only on one host will live under the host's fur.

5. The larva becomes an adult after eating its second meal. After the final meal, it will lay its eggs and then die.[5]

SANDFLIES
Phylum Arthropoda
Class Insecta

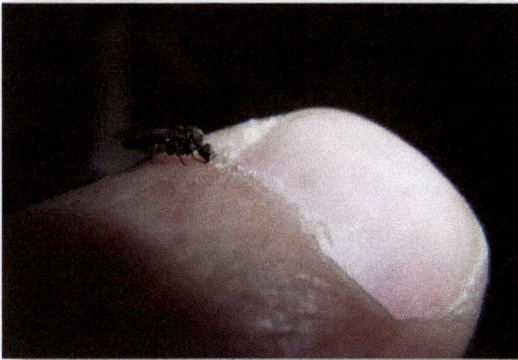

Fig. 5-3.The sand fly. Photo by Emilio .[6]

There are around seven hundred species of sandflies, but only 10% are known to transmit disease. Most are found in tropical locations, but some are found in temperate regions.

Sandflies develop in terrestrial environments rather than aquatic environments.[7]

The life cycle of sandflies

1. Females lay between 30-70 elongated, oval shaped eggs.

2. Eggs will hatch within one to two weeks unless it is very cold. If it is cold eggs will enter a phase called *diapauses*, which is defined as a type of sleep for insects. Eggs do not grow during this stage; they grow only after the temperature heats up.

3. Larvae emerge (they have a dark head) from the eggs. They begin to feed on dead organic matter.

4. After the larval stage they enter a pupal stage (golden brown) which will stay on debris floating in water. This stage takes five to ten days.

5. Adults come out before dawn. Males emerge 24 hours before the females.

6. Males and females feed on plants. Females need to suck blood to have egg production.

7. Sandflies will mate near a host and the cycle will begin again.

LICE

Phylum Arthropoda

Class Insecta

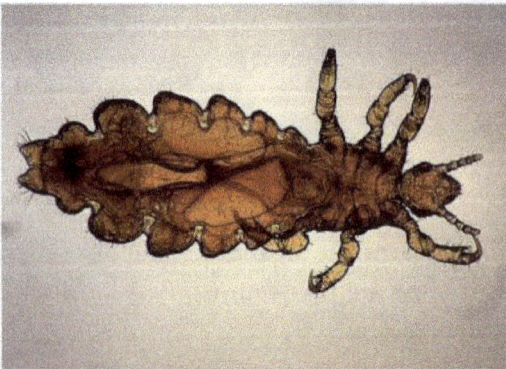

Fig.5-4. *Pediculus humanus.* **From** PD-USGOV-HHS-CDC.[8]

Lice are six-legged insects. There are three major types of lice that parasitize humans: head lice, crab lice and body lice. Body lice cause diseases like Rickettsia and Spirochetes (spiral shaped bacteria). There are approximately 5000 species worldwide.[9]

The life cycle of body lice

1. After feeding on blood, body lice deposit eggs in clothing that are in contact with a host's body (humans). A female lays between 50-150 eggs in her lifetime, but some may lay more (275-300).

2. Incubation depends on temperature of the host which is five-seven days. They hatch at temperatures between 23°C-38°C (73°F-100.4°F).

3. Nymphs look like their parents when they hatch. They immediately begin to feed. The Nymph stage lasts between 16-18 days.

4. After one-two days of maturity, females lay eggs.

5. Body lice live 30-40 days. A four to five degree rise in temperature can be fatal for them.

6. Body lice are transferred between people when they come in close personal contact.

TRIATOMINE BUGS

Phylum Arthropoda

Class Insecta

Fig.5-5. Triatomine bug *Panstrongylus geniculatus*-Vector of Chagas Disease[10] Author-Fernando Otálora Luna

Triatomine bugs are bloodsucking insects that are found in the southern United States and in Latin America. There are 130 known species.[11]

Life cycle of triatomine bugs

1. Egg hatches, releasing a nymph called an instar.

2. The instar passes through instars 2,3,4,and 5.

3. The Fifth instar turns into an adult.

TSETSE FLIES

Phylum Arthropoda

Class Insecta

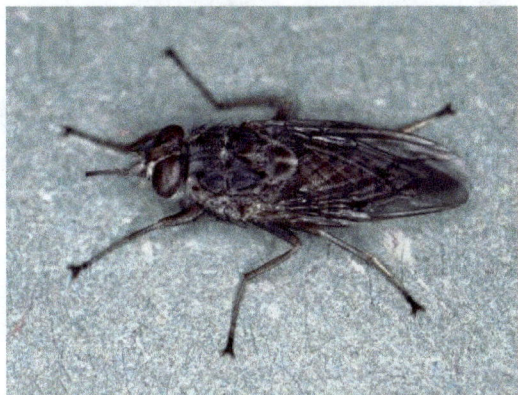

Fig.5-6. Tsetse fly- *Glossina morsitans* **Vector of African Sleeping Sickness**[12]

There are approximately twenty-two species of tsetse flies worldwide. They are responsible for 250,000-300,000 deaths per year worldwide due to the disease they help spread.

The life cycle of tsetse flies [13]

1. Females do not lay eggs. The larva develops in the uterus of the female tsetse fly for 10 days.

2. The larva then is deposited fully grown in soil, under bushes, under stones and roots.

3. It will bury itself and turn into a pupa.

4. Depending on the temperature, the fly emerges 22-60 days later.

5. Females mate only once in a lifetime but can produce a larva every ten days.

Tsetse flies like shaded areas. They rest near food resources where they look for food foronly short periods. People can be bitten on forest trails, water collection points and near vegetation near water collection sites.

FLEAS

Phylum Arthropoda

Class Insecta

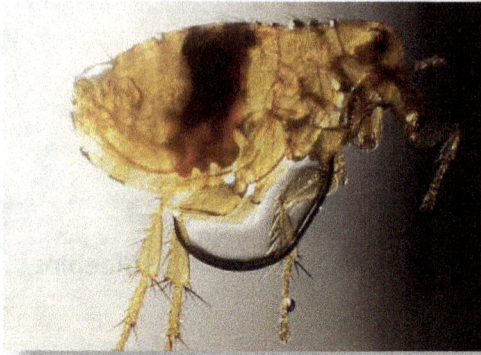

Fig. 5-7. Rat Flea—Xenopsylla cheopis -vector of the plague[14]**Photo by Dr Pratt from the CDC**

Life Cycle of the rat flea [15]

1. Females lay eggs in debris in groups of between 3-25 a day. (During her life span a female can lay 300-1000 eggs.)

2. Eggs hatch between 2-14 days, depending on conditions.

3. The larva emerges and avoids light.

4. The larva feeds on humans or animals.

5. After about two hundred days the larva pupates.

6. The adult emerges after 14 days.

7. Fleas can delay emergence if they do not detect a host.

8. Both sexes take several blood meals a day. If the host dies, the flea will move to a new host.

HUMANS

Phylum Chordata
Class Mammalia

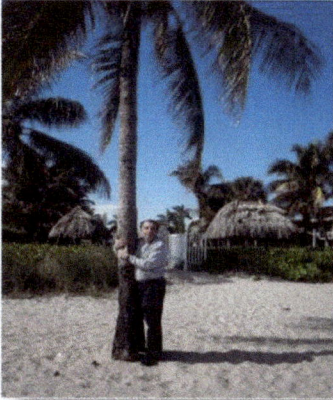

Fig.5-8. A human vector of the flu—*Homo sapiens* Photo by Randolph Swiller.For the life cycle of humans and horses consult any major biology book.

HORSES

Phylum Chordata
Class Mammalia

Fig.5-9. Horses—*Equus* Vector of the Hendra Virus. Photo by Diane Jedlicka.

OYSTERS AND CLAMS

Phylum Mollusca

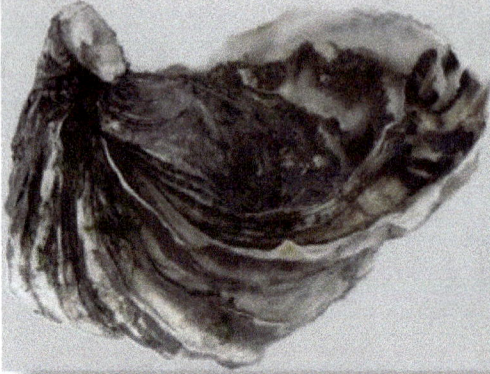

Fig 5-10 Pacific oyster *Crassostrea gigas*[16]

Fig 5-11 Mixed seafood with oyster in Dubai. Typical shellfish used as human food.[17]

TABLE 5-2

BACTERIAL INFECTIONS AND THEIR VECTORS

VECTOR	DISEASE	KINGDOM OF PATHOGEN	GENUS AND SPECIES
Uncooked meat	E.coli	Bacteria	*Escherichia Coli*
Humans	Typhoid Fever	Bacteria	*Salmonella Typhi*
Tick *Ixodes scapularis* *Pacificus*	Lyme Disease	Bacteria	*Borrelia burgdorferi* (USA) *Borrelia afzelin* and *Borrelia garinii*, in Europe
Humans	Tuberculosis	Bacteria	*Mycobacterium Tuberculosis* (Worldwide)
Humans	Shigellosis	Bacteria	*Shigella Sonnei, S. Flexneri; S. Boydii*
Contaminated water (not really a vector)	Legionnaire's Disease	Bacteria	*Legionella Pneumophilia*
Sandfly *Phlebotomus Peruensis*	Carrion's disease Fever, Verruga Peruana, Carrion Fever	Bacteria	*Bartonella Bacillformis*
Cat flea *Ctenocephalides sp*	Cat scratch disease	Bacteria	*Bartonella henselae*
Body louse *Pediculosis Humanus corporis*	Trench Fever Bacillary Angiomatosis	Bacteria	*Bartonella Quintana*

Ticks- *Dermacentor andersoni, D. Variabilis, Amblyomma americanum)*	Tularemia	Bacteria	*Francisella Tularensis (formerly Pasteurella Tularensis)* *Francisella Holarctica,* (found in voles, beaver and muskrats in the US and rodents and hares in Europe)
Rat Flea	Disease	Bacteria	***Yersinia pestis*[18]**
Ticks *Amblyomma americanum* (lone star tick) *Ixodes Sacpularis* (Black-legged tick) *Ixodes pacificus* (Western black-legged tick) in California *Ixodes Ricinus* in Europe *Ixodes Persulcatus*	Ehrlichiosis/Anaplasmosis	Bacteria	*Ehrlichia Chaffeensis* – *Ehrlichia Ewingii Anaplasma Phagocytophilum*
Rodents, pigs, dogs, cattle, deer, sheep, and wild animals	Leptospirosis	Bacteria	**Leptospira** Icterohaemorrhagiae Canicola Pamona **Grippo Typhosa Bratislava**
Algae, copepods	Cholera	Bacteria	***Vibrio cholera***
Humans	Meningitis	Bacteria	***Neisseria meningitides***

TABLE 5-3

VIRAL INFECTIONS AND THEIR VECTORS

VECTOR	DISEASE	TYPE OF PATHOGEN	TYPE OF VIRUS
Humans	Ebola Hemorrhagic Fever	Virus	Filaviridae
Humans	Marburg Hemorrhagic Fever	Virus	Filaviridae
Natal multimammate mouse *(Mastomys natalensis)*	Lassa Fever	Lassa Virus	Family Arenaviridae
Rodent Calomys Callosus	Bolivian hemorrhagic Fever	Machupo Virus	Family Arenaviridae
Mosquitoes Aedes aegypti, Aedes albopictus	dengue Fever / dengue hemorrhagic Fever	Dengue virus of four different serotypes (DEN-1, DEN-2, DEN-3 DEN-4) *Genus Flavivirus*	Family Flaviviridae
Mosquitoes Culex Pipiens C. Tarsalis Culex Guinquefasciatus	West Nile Virus	RNA virus-single stranded	Family Flaviviridae
Humans	Hepatitis	RNA virus-single stranded	Family Picornavirus
Humans	Hoof and Mouth Disease	Virus-genus Aphthovirus	Family Picornaviridae
Mosquito: Aedes aegypti, A. Albopictus	Chikungunya Fever	Genus Alphavirus	Family Togaviridae

Mosquito: Anopheles Funestus	O'nyong-nyoung Fever	Genus Alphavirus	Family Togaviridae
Mosquito: Aedes aegypti: A. Simpsaloni, A.africanus	Yellow Fever	Virus	Family Flaviviridae
Mosquito: Ochlerotatus Vigilax	Ross River	Ross River Virus	Family alphavirus
Birds	Avian Flu		Sub type H5N1
Pigs	Porcine Reproductive and Respiratory Syndrome- Blue Ear Virus	Arterivirus	Family Arterivirus
Midge: Culicoides paraenesis Mosquito: Ochlerotatus serratus	Oropouche Fever	Oropouche virus	Family Bunyaviridae
Bat, horses, pigs	Henipavirus	Virus	Family Paramyxoviridae

For detailed descriptions of viruses consult the many microbiological textbooks that are out there.

TABLE 5- 4

PARASITIC INFECTIONS AND THEIR VECTORS

VECTOR	DISEASE NAME	PATHOGEN	GENUS AND SPECIES OF PATHOGEN
Copepod-Cyclops	Dracunculiasis	Nematode Worm	Dracunculus medinensis

Blackfly Simulium	Onchocerciasis River Blindness	Nematode Worm	Onchocerca volvulus
Mosquitoes: Aedes, Anopheles, Culex, species	Filariasis Lymphatic filariasis – Subcutaneous filariasis Elephantiasis	Nematode Worm	*Wuchereria bancrofti, Brugia malayi, Mansonella streptocerca Dracunculus medinensis Mansonella perstans, Mansonella ozzardi*
Cattle Ticks-Amblyomma and Rhipicephalus sp.	African tick bite Fever	Rickettsia (Bacteria)	Rickettsia africae (primarily) Rickettsia conorii (rarely)
Chiggers: Trombiculid Mites (Leptotrombidium deliense)	Scrub Typhus Orientia tsutsugamushi	Rickettsia (Bacteria)	tsutsugamushi
Rat fleas, cat fleas	Murine Typhus	Rickettsia (Bacteria)	Rickettsia felis
Body Louse Pediculus humanus, Pediculus capitis Phthirus pubis	Epidemic Typhus	Rickettsia (Bacteria)	Rickettsia prowazekii
Dog Tick *Dermacentor variabilis* Brown Dog Tick *(Rhipicephalus sanguineous)* Other ticks *(Amblyomma cajannense)*	Rocky Mountain Spotted Fever	Rickettsia	*Rickettsia rickettsii*

Brown dog tick *(Rhipicephalus sanguineus)*	Mediterranean Spotted Fever or Boutonneuse Fever (MSF)	Rickettsia	Rickettsia conorii
Mosquitoes: Female anopheles mosquitoes	Malaria	Protista	Plasmodium falciparum Plasmodium vivax - Plasmodium Ovale, Plasmodium Malariae
Tsetse Fly Glossina Palpalis Glossina Morsitans	African trypanosomiasis	Protista	Protista Trypanosoma Brucei Gambiense and Trypanosoma Brucei Rhodesiense
Tsetse flies	Nagana Pest Animal African *Trypanosomiasis Souma* or *Soumaya* in Sudan	Protista	

TABLE 5-5

ALGAL DISEASES AND THEIR TOXINS

VECTOR AND TOXIN	DISEASE	ALGAL GROUP	GENUS OF ALGA
Shellfish Domoic Acid	Amnesiac shellfish poisoning-	Diatom	Pseudo-nitzschia Chondria Armata
Mussels Azaspiracid	Azaspiracid Shellfish Poisoning	Dinoflagellates	Azadinium

Cycad seed flour containing BMAA	ALS and Parkinson's	Cyasnobacteria	Nostoc
Bivalves Such as clams oysters Okadaic acid	Diarrhoeic Shellfish Poisoning	Dinoflagellates	Dinophysis Prorocentrum
Water	Microcystin Poisoning	Cyanobacteria	Microcystis
Clams and oysters Saxitoxin	Paralytic shellfish poisoning	Dinoflagellates	Gymnodinium, Alexandrium and others
Water	Palm Island Mystery Disease	Cylindrosperm-opsin	Cylindrospermopsis, Aphanizomenon, Umezakia, Anabaena (now Dolichospermum), Lyngbya, and Raphidiopsis
Cylindrospermopsin			
Water Pfiesteria	Pfiesteria Disease	Dinoflagellates	Pfiesteria

[1] Reprinted with permission from WHO. http://www.who.int/heli/risks/vectors/en/vbdmap.pdf. December2009.

[2] http://pt.wikipedia.org/wiki/ficheiro:aedes_aegypti_biting_human.jpg, viewed March 20, 2010.

[3] https://en.wikipedia.org/wiki/File:Amblyomma_americanum_tick.jpg Viewed 30 Dec 2021.

[4] http://www.advantix.info/tick-species.959.0.html, viewed March 19, 2010.

[5] http://images.google.com/imgres?imgurl=http://static.howstuffworks.com/gif/tick-10.gif&imgrefurl=http://animals.howstuffworks.com/arachnids/tick2.htm&usg=t_5mslgcpgbhamgzelccw4pbm5s=&h=500&w=400&sz=32&hl=en&start=2&sig2=ekfdb1qp6_9sshalt14-kg&um=1&itbs=1&tbnid=tjyewb-sqernum:&tbnh=130&tbnw=104&prev=/images%3fq%3dtick%2blife%2bcycles%26um%3d1%26hl%3d en%26sa%3dn%26tbs%3disch:1&ei=nlols4fujyuutg-fxl-n2cq , viewed March 19, 2010.

[6] File:Biting sandfly.jpg - Wikimedia Commons Viewed 8 Dec 2021

[7] http://pcwww.liv.ac.uk/leishmania/life_cycle_habitats.htm, viewed March 20, 2010.

[8] http://en.wikipedia.org/wiki/file:pediculus_humanus_var_capitis.jpg

[9] http://www.discoverlife.org/mp/20o?search=phthiraptera, viewed March 20, 2010.

[10] Vector of chagas disease- http://en.wikipedia.org/wiki/file:pgeniculatus2.jpg

[11] http://www.kaieteurnewsonline.com/2009/03/01/the-triatomine-or-kissing-bug/

[12] https://commons.wikimedia.org/wiki/File:Glossina-morsitans.jpg Viewed 8 Dec 2021.

[13] http://www.who.int/water_sanitation_health/resources/en/vector178to192.pdf viewed March 20, 2010. [14]https://commons.wikimedia.org/wiki/File:Xenopsylla_cheopis_flea_PHIL_2069_lores.jpg Viewed 8 Dec 2021

[15] http://www.cbwinfo.com/biological/vectors/fleas.html, viewed March 22, 2010.

[16] https://en.wikipedia.org/wiki/Oyster Viewed 30 Dec 2021.

[17] https://en.wikipedia.org/wiki/Oyster#/media/File:Oyster_dubai.jpg

Viewed 30 Dec 2021.

[18] Black rat (*Rattus rattus*), Brown rat (*Rattus norvegicus)* *are considered secondary vectors.*

Section 2

Chapter 6

Bacterial Infections

Chapter 7

Rickettsia Infections

Chapter 8

Viral Infections

Chapter 9

Parasitic Infections

Chapter 10

Fungal and Protista Infections

Chapter 11

Other Arthropod-borne and Non-Microorganismal Conditions

Chapter 12

Harmful algal blooms (HABs) and their toxins

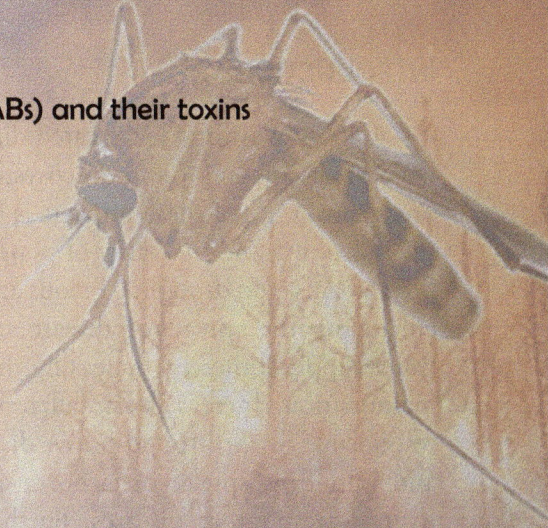

Section Two

DISEASES CAUSED AS A RESULT OF HUMAN ACTIVITY

INTRODUCTION:

Global Warming has created dire climate changes. These changes have great impacts on the ecosystem and their populations of all organisms. One of these impacts is health. Unstable climates increase the morbidity and mortality of populations due to floods, heat waves, fires, storms, and droughts. Floods and heavy rainfall become frequent, enhancing the spread of disease from contaminated and polluted water supplies, thereby increasing the incidence of gastroenteritis and malnutrition. Increased rainfall results in more habitats for vectors responsible for arthropod-borne diseases. Due to increased warming and humidity, spread of airborne pathogens has increased, leading to more respiratory diseases. Furthermore, unstable climates would change the distribution of some infectious diseases.

The following table illustrates the types of transmission of diseases based on four types of transmission systems.

TABLE 1

SYSTEMS OF DISEASE TRANSMISSION [1]

SYSTEM	AGENT*	ENVIRONMENTAL CAUSES	EXAMPLES
Direct Anthroponosis	Human	Proximity, Sexual Contact	Syphilis, Leprosy
Indirect Anthroponosis	Human Vector	Agriculture Growth, Hygiene, Irrigation, Urbanization	Malaria, Schistosomiasis

Direct zoonoses	Reservoir, Human	Deforestation, Water Impoundment	Cryptosporidiosis, Tetanus
Indirect zoonoses	Reservoir, Human, Vector	Deforestation, Reforestation, Climate Change	St. Louis Encephalitis, The Plague

*Agent is the vector that may contain and spread the disease.

The chapters in this section discuss diseases according to the kingdoms of their causative organisms.

LIST OF DISEASES ASSOCIATED WITH CLIMATE CHANGE

Chapter 6: Bacterial Diseases

1. Escherichia Coli Infection

2. Typhoid Fever

3. Lyme Disease

4. Tuberculosis

5. Shigellosis

6. Legionnaires Disease

7. Carrion's Disease

8. Oroya Fever

9. Verruga Peruana

10. Cat Scratch Fever

11. Trench Fever

12. Tularemia

13. Plague

14. Leptospirosis

8. Ross River Virus Infection

9. Avian Influenza

10. Human Papilloma Virus Infection

11. Porcine Reproductive and Respiratory Syndrome (PRRS)

12. Oropouche Fever

13. Nipah Virus Infection

14. Rift valley Fever

15. Murray Valley Encephalitis

16. Marayo Fever

17. Bluetongue disease

Chapter 9: Parasitic Infections

1. Dracunculiasis

2. Onchocerciasis

3. Subcutaneous Filariasis

4. Serous Filariasis

5. Lymphatic Filariasis

6. Schistosomiasis

7. Cryptosporidiosis

Chapter 10: Fungal and Protozoan (Protista) Diseases

Fungal infection: 1. Sporotrichosis

Protozoa parasite:

1. Malaria

2. African Trypanosomiasis

3. American Trypanosomasis or Chagas Disease

4. Animal Trypanosomiasis

    a. Nagana

    b. Dourine

    c. Surra

    d. Baleri

    e. Taliagos

    f. Peste-bola

    g. Galzieka

5. Leishmaniasis

6. Human babesiosis

7. Giardiasis

Chapter 11: Arthropod-spread and Non-microorganismal conditions

Arthropods

Microorganismal conditions:

1. Tungiasis

2. Tick Paralysis

Non-microorganismal Conditions:

1. Asthma

2. Sunburn

3. Heat Wave

4. Heat Exhaustion

5. Heat Cramps

6. Heat Stroke

7. Tropical Sprue

8. Malignant Melanoma

9. Basal Cell Carcinoma

10. Squamous Cell Carcinoma

11. Merkel Cell Carcinoma

12. Kaposi Sarcoma

13. Photodermatoses

Chapter 12: Toxic Algae Diseases

1. Amnesiac shellfish poisoning

2. Paralytic shellfish poisoning

3. Diarrhoeic shellfish poisoning

4. Neurological shellfish poisoning

5. Azaspiracid shellfish poisoning

6. Palytoxin poisoning

7. Microcystin poisoning

8. Palm island mystery disease

9. Pfiesteriosis

10. Ciguatera food poisoning

11. BMAA (beta-methylamino-L-alanine)

[1]Wilson, Mark. Ecology and Infectious Disease in Ecosystem Change and Public Health, J. L. Aron, J. Patz. 2001, p.292.

METRIC UNITS OF TOXINS WHICH ARE FOUND IN AQUATIC ANIMALS AND HUMANS.

ug or μg in this book means microgram. One million micrograms equal one gram. 100ug/100 grams mean there are 100μg per 100 grams of the body Weight. Mg means milligrams. 1,000 milligrams equals a gram. MU means mouse units. The amount of toxin needed to kill a 15-gram mouse in 15 minutes.

Chapter 6

BACTERIAL INFECTIONS

All bacteria are prokaryotic and contain only one chromosome known as a Genophore. There are around five thousand known species with millions undiscovered. The majority are beneficial but as with everything else we only hear about bad ones. Many bacteria live in symbiotic relationships with animals and plants such as those which occupy a human throat and intestinal tract. Most bacteria are good but some cause diseases such as Lyme Disease and Tuberculosis.

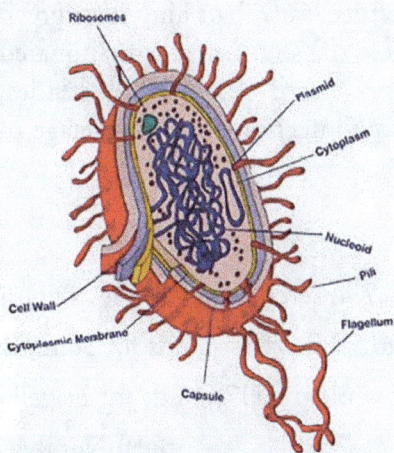

Fig 6.1 A Typical Bacterial Cell.[1]

1. ESCHERICHIA COLI (E. coli) INFECTION[2]

Increased flooding, as well as increased rainfall, would cause the rise of waterborne infections. One of them is the *Escherichia coli* infection. *Escherichia coli* are a large group of bacteria and most of them are harmless inhabitants of the human body. Some are beneficial like the *E. coli* that thrive in the Intestine of healthy people and animals. They assist in the digestion of food and protect against intestinal colonization of harmful microbes. Some *E.coli* are even used as a marker in water contamination. The presence of *E. coli* in drinking water is a sign of contamination, but this does not mean that the *E. coli* is harmful. There are some "bad apples" though, and one of them is the *E.coli* O157H7. This type of *E. coli* causes bloody diarrhea with severe abdominal pain. It can also lead to Renal Failure and Hemolytic Disease called Hemolytic-Uremic Syndrome (HUS).

E.coli O157H7 was first recognized in 1982 after gastrointestinal outbreaks traced from contaminated hamburgers in the United States and Japan. It was, however, a year later in 1983 that the microorganism was identified. Several sporadic outbreaks followed. In 1993, there were 477 people who suffered from this type of infection after eating undercooked hamburger in Washington State. In 1994, 15 cases were reported of having *E.coli* infection after eating salami, and 11 people were reported sick after eating ground pork. Later it was discovered that other sources, aside from meat, are lettuce, salad bars and other green vegetables. There were even cases where the source was contaminated in swimming water and petting farms. Based on the 1999 studies, there were around 73,000 cases of *E.coli* infections with an average of 60 deaths here in the United States.

CAUSATIVE MICROORGANISM:

Escherichia coli serotype O157:H7, a gram Negative Bacterium. The O157 stands for Somatic Antigen Number while the H7 stands for Flagella Antigen Number.

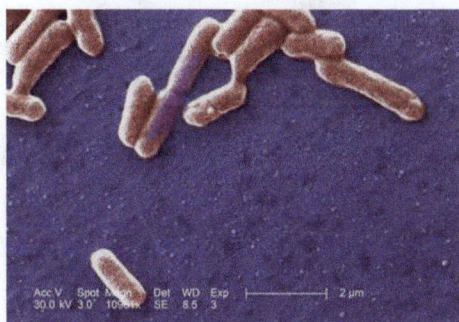

Fig. 6-2. Colorized SEM of *E. Coli*

Photo by CDC /Janice Haney Carr[3]

Diagnosis and Laboratories Pathogenesis

Escherichia Coli O157:H7 lives in the intestines of some mammals like cattle, deer, and swine without causing any symptoms. The bacterial toxin requires specific receptors to have its effects. These receptors are not present in these mammals. That is why they do not get ill. These microorganisms are continuously shed off with the feces to infect humans. In the process of slaughtering, these bacteria can be mixed with the meat while they are ground into hamburger. The meat looks and smells normal so there is no suspicion of it being contaminated.

This *Escherichia coli* infection is sometimes called "Hamburger Disease" because it was commonly acquired from eating hamburger. Now the fast-food chains have devised a cooking method to eliminate this bacterium. They must cook the meat up to 160 F (72 C). Cases of hamburger disease nowadays are usually from homemade meat that may not have been cooked at 160 F.

Since the bacteria may be present in cattle's udders, drinking unpasteurized milk is another way of acquiring the infection. Other ways are eating raw vegetables, sprouts, and lettuce that have been in contact with manure of infected mammals.

Since the microorganism is continuously shed off through the stool by infected humans even for two weeks after improvement, these bacteria can easily spread, especially if the person is not very hygienic. That is why this infection can cause an outbreak in places with crowded people, like prison cells and dormitories. In day care centers, there is a greater chance for the bacteria to spread easily because children are not that concerned about self hygiene.

Symptoms

Some infected persons are asymptomatic while others may have mild or moderate symptoms. Serious cases may develop into life-threatening conditions. There is a saying that goes "***E. coli*** in your spinach salad or hamburger, makes a memorable meal." Yes, your meal is memorable, because afterwards you experience severe abdominal pain and bloody diarrhea. The usual symptoms are abdominal cramps, diarrhea, and vomiting. The diarrhea is grossly bloody. Some would describe it as "all blood, no stool" which gives rise to the word hemorrhagic colitis. If there is fever, it is typically low to moderate grade (no more than 105 F or 38.5 C). The disease usually lasts for a week.

In 15% of cases, complications set in, such as Hemolytic Anemia and Acute Renal Failure. This is the so called Hemolytic-Uremic Syndrome (HUS). This is more common on the extremes of life, in children under five and the elderly.

Presented with bloody diarrhea, there are many infectious conditions to think of. Typhoid fever and Shigellosis are part of the differential diagnosis. History may give some help, but it is really the laboratory tests that would confirm the diagnosis. A sample of the stool is taken for culture. It uses the sorbitol-MacConkey agar (SMAC). This culture, however, takes several days to have the results. Other fast diagnostic methods are the Polymerase Chain Reaction or the Fluorescent and Antibody detections.

Treatment

Usually, symptoms would go away even without antibiotics. There is no evidence that antibiotics provide beneficial effects against this infection. Some would even say that the antibiotics may precipitate kidney failure. Antidiarrheal medications are not advised because they decrease the peristalsis of the intestine, thereby slowing down the elimination of the bacteria from the intestine. Infected persons are generally given supportive treatment only, like complete bed rest and increased fluid intake to prevent dehydration. Antipyretic medications may be given for moderate fever.

Those patients whose symptoms last more than a week must be thoroughly checked for kidney functions and bleeding tendencies to prevent the fatal complication called Hemolytic-Uremic Syndrome. If indeed there is a hemolytic condition, a blood transfusion must be initiated. If there is renal failure, dialysis must be performed.

Prevention

Prevention should start from the processing of the meat from slaughtering to the grinding. All meat products must be prepared following the strict guidelines of the federal government. Secondly in the cooking of meat products, especially hamburger or any processed beef (beef steak), they must be cooked no lower than 160 degrees F to kill all the *Escherichia coli*. The federal government has earned a good mark in this aspect of prevention; but has had little progress in securing safety of commercial vegetables. Their safety guidelines for commercial vegetables must be as strict as those for meat.

Other preventive measures are:

- For raw vegetables, wash them properly with water. Do not mix raw vegetables with ready-to-eat foods.

- If cooking a homemade hamburger, cook it to at least 160 F. If possible, use a thermometer to make sure that all the thickness of the meat is at that temperature. If no thermometer is available, make sure that all the meat turns brown (not pink).

- Drink only pasteurized milk or juice.

- Refrain from drinking water coming from an unknown source. It is wiser to drink from

- bottled water or canned sodas. If there are no other drinks, boil the water for at least one minute.

- Avoid drinking from recreational water like swimming pools or lakes.

- Proper hand washing before and after eating, and after coming from the toilet, is essential. For those caring for children in day care centers, make sure to clean hands thoroughly after changing diapers.

There is no vaccine to date for the Escherichia coli infection, though research is on going on for a potential vaccine in the future.

2. TYPHOID FEVER[4]

Typhoid Fever is also called Enteric Fever, Bilious Fever, or Yellow Jack. It is a life- threatening infection of the gastro-intestinal tract. This infection is directly transmitted from humans to other humans without any intermediate host or vector. It is acquired through the fecal-oral route. How is this increased in global warming? One of the effects of increased temperature is drought and a decrease in the availability of water supplies. This will eventually result in poor hygiene and sanitation. People may minimize hand washing or do no hand washing at all, which would increase the fecal-oral borne infection. Moreover, this warming will decrease the availability of drinking water.

Typhoid Fever has been thought to be present since time immemorial. Scientists believed that the devastating plague that killed one third of the population of Athens in 430 – 326 BC was due to Typhoid Fever. They were able to trace DNA sequences like the causative microorganism of Typhoid Fever from the Athenian mass graves. Others, however, are disputing the findings due to some flaws in the study.

One documented outbreak of Typhoid Fever was in Chicago in the late 19th century when there was an average of 65 deaths / 100,000 people. It became worse in 1891 when the rate rose to 174 deaths / 100,000 people.

In 1907, a New York cook by the name of Mary Mallon was believed to be responsible for 47 cases of Typhoid Fever, three of whom died. She was believed to be a carrier of the bacterium that was transmitted to her customers. She quit her job as a cook but worked again using a false name. She caused another outbreak and was quarantined until she died of pneumonia. She was called "Typhoid Mary."

A vaccine for Typhoid Fever was developed by Edward Wright in 1897, which decreased the incidence of the disease in the following years. Coupled with improvements in sewage and sanitation, there was a significant reduction in the morbidity and mortality in the early part of 20th century. In 1942, an antibiotic was developed against Typhoid Fever, further reducing the incidence of the disease.

At present, in the US, approximately four hundred cases of Typhoid Fever are reported each year and 75% of them are acquired abroad, mostly from the Indian subcontinent and Latin America. In developing countries where sanitary conditions are poor, approximately 21.5 million cases are reported each year. According to WHO, there are estimated to be 16-33 million cases of Typhoid Fever annually with 500,000-600,000 deaths, mostly in children between 5-19.

The last reported outbreak of this disease was in Congo (Africa) in 2004-2005 where 42,000 cases with 214 deaths were reported.

CAUSATIVE MICROORGANISM: *Salmonella typhi, Salmonella parathypii* (less virulent); both are gram negative

Pathogenesis

Typhoid Fever is acquired by the fecal-oral route which means you get it by ingesting food or water contaminated with the bacteria. The bacteria originate from the excreta/ refuse of infected persons or from an asymptomatic carrier. They continuously shed the bacteria in their stools and some in the urine. If these individuals are unhygienic (poor toilet habits, improper hand washing), they are potential distributors of the bacteria.

They will leave behind the bacteria on everything they touch. Flies take part, too. Flies love to swirl around and step on these excreta and later do the same on uncovered food and water, transferring the bacteria.

Prolonged wet seasons (increases in rainfall or extensive flooding like the Katrina disaster) due to climate change will affect waste disposal and sanitation. These conditions equate to an increased incidence of Typhoid Fever. On the other hand, a prolonged hot season would make our water supply scarce, affecting sanitation and waste disposal. These conditions also increase the incidence of Typhoid Fever.

Upon ingestion, the bacteria pass through the stomach. These bacteria are resistant to the acid environment of the stomach unless the pH is 1.5 or less when they are destroyed.

This is the reason people taking medications for acidity (antacids, H2 receptor antagonists, proton pump inhibitors) are more prone to Typhoid Fever, because fewer bacteria are destroyed as they pass through the stomach. Then the bacteria reach the small intestine where they are phagocytosed by the intraluminal dendritic cells. These infiltrated dendritic cells stick to the lining cells of the intestine, causing inflammation of the mucosa leading to diarrhea, the usual initial symptom. These dendritic cells further go in reaching the Lamina Propria and the submucosa and concentrate in the Peyer's patches (a mucosa associated lymphoid tissue),where they continuously multiply. They later go with the flow of the lymph through the lymphatic vessels, passing through mesenteric lymph nodes and thoracic duct to seed the reticula-endothelial tissues – liver, spleen, bone marrow, as well as lymph nodes. Here they continuously multiply and enter the bloodstream where they seed other parts of the body. From the blood or from the liver, the bacteria reach the gall bladder. There they further multiply and are released with the bile and re-infect the small intestine. Some are continued with the stool and subsequently excreted to be transmitted again. Others re-infiltrate the mucosa to concentrate and re-infect the Peyer's patches.

There are a few bacteria that are brought to the kidneys by the hematogenous route to be excreted with the urine. The second wave of bacterial invasions is through the bile or hematogenous route to re-infect the Peyer's patches which become hyperplastic and necrotic, creating a bleeding ulcer on the intestinal wall. This causes the lower GI bleeding in Typhoid Fever. In severe cases, the ulcer extends, leading to intestinal perforations causing peritonitis.

Some patients, about 1-4 %, become chronic carriers. They continuously release the bacteria with the excreta for several years.

Symptoms

The clinical manifestations of Typhoid Fever can be divided into four stages with each stage lasting for a week.

Stage I or First week, corresponds to the initial invasion of the bacteria. The symptoms are fever, body malaise, headache, cough. Most of the patients have additional sinus bradycardia (Sphygmo-Thermal Dissociation), and Epistaxis. Complete blood counts at this stage show leukopenia and eosinopenia. Widal tests are still negative currently, but blood culture and bone marrow cultures are most likely positive.

Stage II. The fever turns high grade, at times reaching 104°F. Rose spots are seen on the abdomen and lower chest. These are about 1-4 cm salmon colored blanching maculopapular rashes that resolve after 2-5 days. These rose spots are due to bacterial emboli on the skin. These lesions are considered the classical symptoms of Typhoid Fever. Because of the toxicity and the high-grade fever, patients at this stage usually develop delirium. Patients have diarrhea about 6-8 times a day described as having a pea-soup smell. If the Peyer's patches of the small intestine are so inflamed, they may constrict the intestinal lumen causing an obstruction, thus the patients have constipation.

Usually there is accompanying hepatosplenomegaly due to the seeding of bacteria in the liver and spleen. With liver involvement, the transaminases (liver enzymes) are increased in the blood. At this stage, the Widal test is already positive for O and H antibodies and the blood is still positive for bacterial culture.

Stage III. Patients turn more toxic with loss of weight. Complications usually arise at this stage. The common complications are:

1. Intestinal hemorrhage due to congestion of the Peyer's patches that ulcerate the intestinal linings, causing bleeding. The gastrointestinal bleeding may become severe.

2. Intestinal perforation which occurred if the Peyer's patches are much congested, affecting all layers of the intestinal wall. This complication will spread the bacteria to the peritoneal cavity, causing peritonitis.

3. Metastatic abscesses. The bacteria having spread throughout the body can concentrate on the different body organs causing abscesses. These lead to Cholecystitis (gallbladder), Endocarditis (heart), Osteitis (bone), Arthritis (joints).

4. Encephalitis / Meningitis is due to the presence of bacteria in the brain. The patient may manifest with delirium to coma.

Diagnosis and Laboratories

Stage IV- With treatment people begin to recover. The high fever comes down and other symptoms subside.

Based on the medical history, it is hard to be certain if Typhoid Fever is the correct diagnosis, because the presenting manifestations are the usual constitutional symptoms. The patient is more toxic, though, compared to other infectious conditions. The appearance of the rose spots on the skin is the classical symptom, but they are not always present. The typical bradycardia may be absent as well.

The basis of diagnosis is more from the laboratory tests. A fourfold rise in the antibodies against the antigen of Typhoid Bacilli is indicative (but not confirmatory) of Typhoid Fever.

The best way to be certain of the diagnosis is to identify the typhoid bacilli in the cultures. In the first week (Stage I), the typhoid bacilli can be isolated from the blood and bone marrow. On the third to fifth week, the bacteria can be isolated from the stool.

Treatment

If the evidence is so strong for Typhoid Fever, empirical antibiotic therapy is started right away, after blood has been drawn for the laboratory tests. There are many antibiotics that are effective against Typhoid bacilli. The drug of choice is Chloramphenicol. If the patient has a good bowel status, the drug is taken orally rather than intravenously because the oral route is as effective as the intravenous route. Other antibiotics of choice are the cephalosporins (Ceftriaxone, Cefoperazone) and the Quinolones. Supportive treatments such as fluids, electrolytes and nutrition must be given. It is also important to monitor the status of other body organs to make sure that they are functioning well despite the invasion of the bacteria.

For those with intestinal perforation, surgical intervention must be done immediately. Surgeons prefer simple surgical closure of the perforation because of the edema and inflammation of the tissues. For multiple perforation, however, intestinal resection is necessary. This management, however, has poor prognosis.

Prevention

Since Typhoid bacilli are transmitted through the feces and urine, sanitation, proper waste disposal and hygiene are the mainstays of prevention. Proper hand washing is the most important form of sanitation. Wash hands thoroughly after using the toilet, and before and after eating.

Avoid risky food and drinks. Avoid drinking water from an unknown source. It is wiser to drink bottled mineral water or canned carbonated drinks. Avoid taking ice coming from an unknown source. It must come from boiled water or from bottled water. Avoid uncooked food, raw vegetables, and fruits. If you want to eat fruit, peel it yourself.

Avoid food from street vendors because it is difficult for the food and water to be kept clean on the street.

If travelling in areas endemic for Typhoid Fever, bring with you enough bottled water or canned carbonated drinks. Ask a doctor for the prophylactic vaccine for Typhoid Fever. There are two types of vaccines recommended by WHO. The oral Ty21a (brand named Vivotif Berna) and the injectable Vi (brand named Typhim Vi). They are given one week before travel, and they are effective in 50-80 % of cases.

3. LYME DISEASE OR BORRELIOSIS[5]

Another infectious disease linked to climate change is Borreliosis. The vectors of this disease thrive in the woods and usually come out during summer and early fall. With global warming, the longer the summer season, the more time the vectors are present in the woods to transmit the infections.

As early as 1883, in Europe, a physician by the name of Alfred Buchwald described a degenerative skin lesion present in a man for almost 16 years. He called this lesion Acrodermatitis *Chronica Atrophicans*. In 1909, a Swedish doctor, Arid Afzelius, described a ring-like skin lesion on a woman bitten by a tick. He called the lesion an *Erythema Migrans*.

Several physicians reported similar skin lesions as the years went by. It was, however, in 1975, when medical authorities were puzzled by the increased incidence of arthritis among children in and around a town in Connecticut named Lyme. This was where the name Lyme Disease originated.

In 1982, a researcher of The National Institute of Health, Willy Burgdorfer, isolated a spiral bacterium from the mid gut of a deer tick. These were Spirochetes of the genus *Borrelia*. This microorganism was later named *Borrelia burgdorferi* in his honor.

This infection is more common in summer and early fall because the ticks which are the vectors are more apparent at these times. They stay in the shrubs, grasses in the woods or forest, and sometimes on the ground, ready to stick to passersby. People who are fond of camping or mountaineering are vulnerable to this infection. Also prone are children who are fond of playing or walking among grasses or shrubs.

Lyme disease is the most common tick-borne infection in the United States.

From 1991 to 2006, the average number of cases per year of Lyme Disease was 17,000 but by 2002 it rose to 20,000 per year. As reported by the CDC, from 2003-2005 there were 64,384 cases of Lyme Disease in the US, of which 59,770 (93%) were from the nine states of Connecticut, Delaware, Maryland, Minnesota, New Jersey, New York, Pennsylvania, Rhode Island, and Wisconsin. The most recent report in 2006 showed almost the same incidence rate and same distribution.

CAUSATIVE MICROORGANISM: three species of Borrelia

Borrelia burgdorferi- most common cause of Lyme disease in US

Borrelia afzelii and *Borrelia garinii*, causes of Lyme disease in Europe

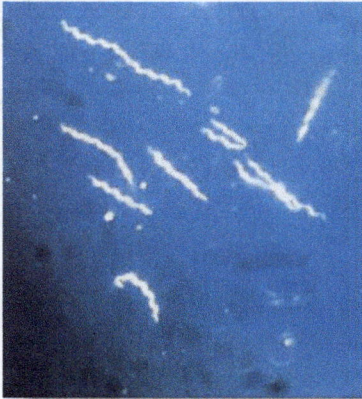

Fig. 6-3. *Borrelia burgdorferi*-causative agent of Lyme Disease[6]

Vectors

Black-legged ticks (*Ixodes scapularis*) usually found in the Northeastern and North Central United States; Western black-legged ticks (*Ixodes pacificus*) usually found on the Pacific coast.

(These Ixodes ticks are smaller in size, no bigger than the size of a pinhead.)

Pathogenesis

The natural reservoir of **Borrelia burgdorferi** is the white footed mouse. For whatever reason, the bacterium does not infect these rodents. The ticks serve as the vector of *Borrelia burgdorferi*. The life cycle of the ticks takes two years to complete and has three distinctive stages. It starts with the adult ticks laying eggs, usually during springtime.

These eggs will develop into the larval stage by summer. By fall and winter the ticks remain as larvae and develop into nymphs by next spring (one year elapsed). The ticks will remain as nymphs throughout the following summer. They become adults by fall and winter and by spring lay eggs again to repeat the life cycle.

The development from larvae to nymphs to adults needs blood meals that they get from animals. The larvae and the nymphs would usually feed from small animals such as the white-footed mouse, while the adult ticks would prefer to feed on white tailed deer. These ticks acquire the causative organisms once they bite an infected animal. These microorganisms remain in the tick until they are injected again to another animal or human during feeding. These ticks remain in the shrubs, grass, leaves of other plants and even on the ground where they will stick to any passersby, be they animals or humans. This is how humans acquire the disease.

Once the ticks bite the skin, the bacteria are injected together with the saliva. The tick's saliva serves as a protective medium protecting the bacteria from the effects of the initial immune response of the person. Then the spirochetes go deeper into the dermis where the inflammatory reaction occurs. resulting in the so called "bulls eye" skin lesion of Lyme Disease.

After several days or weeks in the dermis, the bacteria spread to other parts of the body like the heart, causing palpitations or arrhythmias. In the joints there are joint pains. In the peripheral nervous system, radiculitis and numbness. In the central nervous system meningitis and encephalitis, and in distant skin the so called Borrelial Lymphocytoma (a purplish discoloration) that develops on the ear lobe, nipple, and scrotum.

In chronic cases, erosion of the bone and cartilages in the joints can lead to permanent damage. Those affecting the brain can lead to permanent neuropsychiatric problems.

Symptoms

The symptoms of Borreliosis are divided into three stages:

Stage I (early localized infection). This is the so-called erythema migrans (EM) or erythema chronicum migrans or the so-called bull's eye lesion that appears at the site of tick bite. This skin lesion appears 3-32 days after the tick bites, corresponding to the inflammatory reaction to the bacteria in the dermis of the human host. This is described as a central dark red indurated (localized hardening of soft tissue of the body) area surrounded by a clear area and reddish borders resembling a bull's eye. It can extend up to 6-10 inches in diameter. The whole round lesion is warm on touch but painless. This lesion, however, may not be present in every patient with Borreliosis. It is only present in 60-80% of cases.

Stage II (early disseminated infection). Weeks after the initial infection, the bacteria spread to other body organs through hematogenous route. Another EM may develop in other areas of the skin. Another skin lesion that may appear is the so called *Borrelial Lymphocytoma.*. This is a purplish lump that may develop on the ear lobe, scrotum, or nipple. Those that invade the joints cause joint pains and surrounding muscle pain.

Heart involvement manifested as arrhythmias, palpitations, and heart block. Those involving the nervous system may lead to encephalitis, and/or meningitis manifested as headache, neck rigidity and photophobia.

Stage III (Late persistent infection). After several weeks, months or even years those who are inadequately treated may develop chronic disabling symptoms attributed to the multi-organ involvement. The most common chronic complication is arthritis. Patients manifest joint pains and swelling involving the knee and other large joints. In 11% of cases, bone and cartilage erosions are present, resulting in major disabilities. Other chronic manifestations are arrhythmias, neurological deficits like facial paresis, numbness, weakness, and tingling sensations of the extremities. Frontal lobe symptoms may be present such as loss of memory, impaired intellect, disturbance in mood and sleep patterns.

In general, the symptoms of Borreliosis are very variable. Some would have arthritis as the presenting manifestation without any rashes. Some would present with only neurologic symptoms.

Diagnosis and Laboratories

The diagnosis of Lyme Disease is difficult because the clinical manifestations resemble other disease entities. In fact, Lyme Disease is called *Great Imitator*. The arthritis and the muscle pains can be manifested by other diseases like *Systemic Lupus Erythematosus*, Chronic Fatigue Syndrome, Rheumatoid Arthritis. Fibromyalgia and many others. The neurologic symptoms can be attributed to Multiple Sclerosis, polyneuritis and many other neurodegenerative diseases.

Secondly, there is the variety of the presenting manifestations of the patients. The appearance of the Erythema migrans or the so-called Bulls Eye lesion may be considered typical for Lyme Disease, but this is present only in 60-80 % of cases. Others do not have arthritis while still others have only the neurologic deficits as their presenting symptoms.

Thirdly, the serologic tests (ELISA, Indirect Fluorescent Antibody tests, Western blot tests) are unreliable, especially in the early stage of the disease. There are cases of false negative results (meaning the serologic test results were negative is spite of the presence of the disease). There are also false positive results. These serologic tests have a high degree of cross- reactivity with other conditions like Rocky Mountain Spotted Fever, Mononucleosis, and Syphilis.

The diagnosis is based more on the history of tick bite or history of staying in a grassy or wooded areas in endemic regions. The presence of the Bull's Eye lesion with this history is highly pathognomonic of Lyme Disease. The serologic tests are only helpful in the late part of the disease because more antibodies are present currently.

The serologic tests that may be of help are the Enzyme Linked Immunosorbent Assay (ELISA), Indirect Fluorescent antibody determination (IFA), Western Blot test (WB).

Treatment

A person presenting simply with a history of tick bite or history of staying in grassy or wooded areas does not warrant drug therapy. He would be under observation, however. Those with constitutional symptoms like fever, joint pains, and muscle pain can be given symptomatic treatment. With the diagnosis, however (the presence of a bull's eye skin lesion with a positive history), antibiotics are given right away. The drug of choice is Doxycycline (for adults), Amoxicillin (for children) and Ceftriaxone. Those patients that are resistant to the antibiotic of choice are given hydroxychloroquine or methotrexate.

Patients with chronic symptoms of joint pains are usually given non-steroidal-anti inflammatory drugs (NSAIDS) or Aspirin. Those with debilitating conditions are given Ceftriaxone intravenously at 2 grams / day for 4 weeks. Those with neurologic symptoms are given Minocycline because of its ability to cross the blood-brain barrier, thereby entering the brain.

Prevention

Whenever possible, avoid wooded areas with leaf litter, low-lying vegetation, brushy or overgrown grassy habitat, especially in the summer months of May, June, July. These areas are highly likely to be infested with ticks. If you have grassy or tall grass around your house, have these areas cleared. If it is very necessary for someone to be in an infested area, he should wear light colored clothes so that the ticks can be easily spotted and removed. Cover your body as much as possible by wearing long sleeved shirts with gloves. Tuck pants into the socks or boots to prevent the ticks from approaching the skin.

Before going into the infested area, spray your clothes with insect repellants containing DEET or treat the clothes with permethrin. This will kill the ticks immediately upon contact. The ticks transmit the bacteria (*Borellial burgdorferi*) 36 hours after attachment so there is enough time for someone to check his clothes and remove the ticks.

Take note of the body of the tick. If it is well embedded on the skin, it is not advisable to remove the tick with bare hands because the tick body is usually crushed on the skin. It is recommended to remove ticks using tweezers, grasping the tick close to the skin, and gently pulling it with steady force.

Programs for isolating or managing the deer population, especially in endemic areas, must be implemented because there is a positive correlation with deer populations and the abundance of deer ticks.

A vaccine has been developed by LYMErix in 1998 licensed by the Food and Drug Administration. This vaccine is recommended for those people who frequent tick-infested areas and those who stay longer in these areas. However, it was pulled out of the market after several years because of poor sales.

4. TUBERCULOSIS

Tuberculosis is an airborne contagious disease. It affects some parts of the body like the skin, joints, bones, muscles, urinary system, central nervous system, lymphatic system. The most affected organ is the lungs. This infection affects 8-9 million people every year with a mortality rate of 1.5 million per year. It is estimated at present that one person dies of TB every second. Even with the advent of active screening, and new medications, tuberculosis is still a global menace as of this time. Several factors contribute to this reality, the unimproved socio-economic status of some countries, coupled by increased international travel and the emergence of drug-resistant strains of the causative organism. Environmental changes brought about by humans also contribute to the spread of this disease.

Air pollution from car exhaust, factory exhaust and other CFCs (Chlorofluorocarbons) not only cause allergic reactions or bronchial asthma; but contribute to the spread of the disease. The nasal irritation and throat irritation brought about by air pollution cause frequent coughing or sneezing. The warm temperatures caused by the greenhouse effect, coupled with poor ventilation in crowed spaces like prison cells, shelters for the homeless, and day care centers can further enhance the spread of the disease.

See Table 6-1 below for the global distribution of TB as reported in 2005.

TABLE 6-1

ESTIMATED TB INCIDENCE, PREVALENCE AND MORTALITY, 2005[7]

WHO region	Incidence[a]				Prevalence[a]		TB Mortality	
	All forms		Smear-positive[b]					
	number (thousands) (% of global total)	per 100 000 pop	number (thousands)	per 100 000 pop	number (thousands)	per 100 000 pop	number (thousands)	per 100 000 pop
Africa	2 529 (29)	343	1 088	147	3 773	511	544	74
The Americas	352 (4)	39	157	18	448	50	49	5.5
Eastern Mediterranean	565 (6)	104	253	47	881	163	112	21
Europe	445 (5)	50	199	23	525	60	66	7.4
South-East Asia	2 993 (34)	181	1 339	81	4 809	290	512	31
Western Pacific	1 927 (22)	110	866	49	3 616	206	295	17
Global	8 811 (100)	136	3 902	60	14 052	217	1 577	24

[b]*Smear-positive cases are those confirmed by smear microscopy and are the mostinfectious cases.*

pop indicates population

Causative organisms: ***Mycobacterium tuberculosis***, the most common worldwide

Mycobacterium africanum, found in Africa

Mycobacterium avium, least common among the three, found in the developed countries but limited by pasteurization of milk.

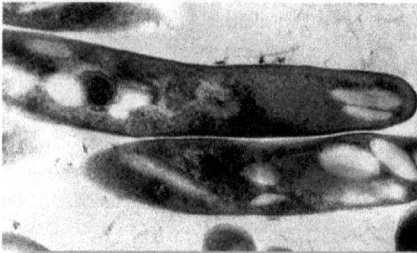

Fig. 6-4. TEM *Mycobacterium tuberculosis*-Causative agent of TB-Photo by Eliazabeth "Libby" White[8]

Transmission

As an airborne infection primarily affecting the lungs, it is acquired by inhaling the bacteria. A contagious person can spread the *Mycobacterium* as far as twenty feet (The distance for an eye exam using the alphabetical chart) by coughing, sneezing, spitting or simply talking or singing. These *Mycobacteria* remain in the air for several hours. They also can be transmitted by direct contact with contaminated personal items like eating utensils. It does not mean, though, that one will get the infection right away. Body reactions to the *Mycobacterium* depend on several body defenses but are due to the immune defenses of the person. Studies show that a person must have a prolonged exposure to acquire the infection. It was also stated in the studies that persons who take the complete medications regularly can be contagion-free as early as 3 weeks after starting medications.

Signs and Symptoms

The presence of *Mycobacterium* in the body stimulates the Immune System to a response.

Depending on all the body defense mechanisms (this includes the Immune System). There are three outcomes:

1. With a good competent Immune System, the *Mycobacteria are* killed, resulting in no infection.

2. With a compromised Immune System, the *Mycobacterium* may stay in the body in an inactive form causing no symptoms. Here the bacteria are surrounded by the phagocytes rendering them inactive. This is the so-called Latent TB or TB Infection. The bacteria remain dormant and will be active only when the Immune System is diminished. The person in this case is not contagious.

3. With an incompetent and inadequate Immune System (as seen in HIV cases), the bacteria multiply and destroy the lung tissues causing cavitary lesions. This is the so called "Active TB or TB Disease." The person in this case is contagious.

The signs and symptoms of the infection are the following:

Cough: Usually chronic type, decolorized sputum. It may be yellowish or blood stained. In severe cases, patients may have hemoptysis (Spitting out blood). The cough may be accompanied by chest pain or dyspnea.

Sweating or Diaphoresis: The person tends to sweat more and his perspiration is not congruent to the outside temperature. This is the so called "Night Sweat." Unexplained weight loss.

Fever, usually slight with bodily malaise.
Other less common symptoms are fatigue, loss of appetite, or chills.

Diagnosis

Presented with typical signs and symptoms of this disease coupled with a previous history, it would not be difficult to consider Tuberculosis. However, in cases of doubt, there are confirmatory diagnostic tests that can be performed.

1. Mantoux test or TB test: This is an intradermal test just like any allergic test.

A PPD tuberculin is injected intradermally and observed after 48-72 hours. The presence of a hard red bump is a positive result indicating that the personhad acquired a *Mycobacterium* before. It has limitations, however. It is non- specific. It does not tell if the subject had an TB Infection or TB Disease, either at present or in the past.

It can be false positive: as in a previous BCG vaccination.

Other mycobacterium infections include **Mycobacteria leprae** which causes Leprosy.

It can be a false negative - as in a recent infection. It takes 8-10 weeks for the body to react after acquiring the infection as in patients with compromised immune systems, such as those with AIDS or cancer, and those having chronic steroid treatment.

Despite these limitations, the Mantoux Test is an important diagnostic tool in many countries.

2. QFT (QuantiFERON-TB Gold): This is a blood test that determines the reactions of the blood defenses against **Mycobacterium tuberculosis**. This is a relatively new diagnostic tool for TB and it is not yet widely used.

3. MODS (Microscopic Observation drug-susceptibility). This test is still under investigation.

However, it is still used in some laboratories. It detects the presence of Mycobacterium *Tuberculosis* in the Sputum Specimen.

4. Chest X-Ray- TB lesions show white spots on X-ray film. This spot is a TB bacterium that has been surrounded by Lymphocytes. In Active TB, they can present as nodular or cavitary lesions.

5. Acid-Fast Smear – This is a more reliable test. Here the sputum or gastric secretion specimen is smeared for the presence of the **Mycobacterium bacilli.** It appears as red rods. For a *Sputum Smear*, make sure that is really Sputum and not Saliva.

Of immense importance is to advise the person to cough out as hard as he can to get the real sputum. One advantage of this test is that the result is ready in a few hours. Culture and Sensitivity Test: This is the best method of diagnosis. The specimen (Sputum or gastric secretion) is placed in a Petri dish with a specialized culture medium. The presence of mycobacterial growth is a confirmation of the disease. The second part of this test is to place several preferred antibiotics in the bacterial growth and check after several days. Those antibiotics which have destroyed the most bacteria are the best choices for that patient's disease.

Treatment

In the past, the main treatment or cornerstone of TB treatment was Isolation, which means confining patients in a sanitarium or hospital. Now medication is given on an out- patient basis. It is particularly important that the patient complete the treatment course which is 6-12 months. In the first few weeks of treatment, improvement may be noticed by the patient (in fact, studies showed that with complete medication, a patient is contagion-free after three weeks of treatment) but it is very necessary to complete the course of treatment to eliminate the bacteria. There are some bacteria that are well embedded in the granulomatous cavitary lesions.

The "walling off" of phagocytes and lymphocytes around the lesions prevent some antibacterial drugs from reaching the embedded bacteria. Stopping the medication too soon will result in the resurgence of the remaining bacteria. Some of the bacteria mutate, thereby producing different strains that resistant to the medication. This will make the treatment harder and more expensive eventually.

In some countries, one must follow the patient with Direct Observed Therapy-Short course (DOTS). Here, medical staff of a health unit give a patient a free supply of complete medication for one month and the patient must come back every month for the next month's medications. In this way the patient is monitored by the medical staff at the same time.

If the patient fails to return for his medication, the medical staff will go to the extent of visiting the patient in his/her residence to check on the patient's progress and give the medications. The moment the treatment course is completed, the sputum must be examined for acid fast bacilli *(Mycobacterium tuberculosis)* and must have three negative exam results before being declared as treated.

Treatment of Tuberculosis depends on whether the person has Latent TB or Active TB.

Active TB: Four drugs (Isoniazid, Rifampicin, Ethambutol, Pyrazinamide) are given for the first 2-3 months of therapy. This is called the extensive phase of treatment and depending on the results after 2-3 months, some drugs may be stopped or continued.

Then the patient will take the medication for the remaining months. This is called the maintenance phase.

Latent TB: One of two drugs may be given to this type of patient, either Isoniazid or Rifampicin. This medication destroys the dormant bacteria. It is, however, important for the medical staff as well as the patient to monitor any signs and symptoms of Hepatitis because this is a common complication of TB treatment. The Medical staff should often request liver enzymes to check for hepatitis.

Prevention

There is not much advice to give a healthy person to prevent acquiring the *Mycobacterium*, except to keep oneself healthy with proper diet and adequate sleep, and to avoid crowded places as much as possible. One must be observant of signs and symptoms of the disease and submit to diagnostic tests right away upon the appearance of the signs and symptoms.

For people at risk, like prisoners in crowded cells, aged people living in nursing homes (They have poor immunity making them more vulnerable to the disease), or children at day care centers, skin testing must be done at least annually. For health care workers, always use a face mask and wear gloves once exposed to the disease. Wash hands as often as possible. Observe all other guidelines in the Universal Precautions of the OSHA (Occupational Safety and Health Administration).

The main thrust of prevention, however, lies with the patient with the infection/disease. These are the important steps:

1. Stay home as much as possible. Take a leave of absence from school or work if you have Active TB and avoid crowded places.

2. Cover your mouth with a mask once you start to cough or sneeze. If it is necessary for you to be around other people, use a mask. Note that one can spread the ***Mycobacterium*** as far as twenty feet in coughing, sneezing, or simple talking and singing.

3. Complete the course of treatment, which is 6-9 months. A person may be germ- free after a few weeks of treatment, or symptom free, but the patient will have to complete the treatment and have a sputum exam to confirm that he or she is cured.

4. Make sure ones place of residence is well ventilated so that the subject always has fresh air.

5. SHIGELLOSIS[9]

Shigellosis is an infectious diarrhea caused by a group of bacteria called *Shigella*. In developing countries, it is also called *Bacillary Dysentery* because the most common causative species in these countries is **Shigella dysenteriae**, manifested by infectious diarrhea with fecal-oral transmission. The incidence of this disease is correlated to climate change. Increased rainfall and floods would enhance the transmission of this infection.

The word "Shigella" originated from Shiga, a Japanese microbiologist who first reported this infection more than one hundred years ago. One thing peculiar with this disease is that only a few bacteria are needed to cause the infection. Ten to two hundred microorganisms are enough to initiate the infection. Therefore, there is a high rate of transmission. This is common in locations where the people are crowded, living in unsanitary conditions like concentration camps in war time, or evacuation centers in case of natural calamities like earthquakes and massive flooding. It is also common in day care centers where the toddlers are not yet toilet trained. This is true also in nursing homes where old people do not care much about personal hygiene and sanitation. Also vulnerable are people with compromised immunity like AIDS patients, cancer patients, transplant patients, and those under long term steroid maintenance.

According to the CDC, in the US around 448,240 cases of Shigellosis (85% of which are caused by **Shigella sonnei**) occur annually with 14,000 of them confirmed in the laboratory. In other countries, the most common causative species are the *Shigella flexneri* and Shigella dysenteriae.

Epidemics due **to** Shigella Dysentery have been reported in Africa and Central America with a mortality rate of 5-15%.

CAUSATIVE MICROORGANISM:

Shigella sonnei, also known as "group D;" this species accounts for more than two-thirds of Shigellosis in US.

Shigella flexneri, also known as "group B," this species accounts for the rest of the Shigellosis.

Shigella boydii is exceedingly rare in US.

Shigella dysenteriae is also rare in US but common in developing countries. These species can cause deadly epidemics.

Pathogenesis

Shigellosis is transmitted by the fecal-oral route. The bacteria are passed from the stool of a person and, by whatever means, to the mouth of somebody else. It is particularly likely to be transmitted when personal hygiene and basic sanitation are not observed. The bacteria are present in the watery stools of the infected people and if they are not so concerned with hand washing after defecation, they would carry the bacteria on their hands and leave them on any object.

Anybody who happens to hold the same objects is vulnerable to acquire the bacteria. Keep in mind that, unlike in other infections, Shigellosis needs only a small number of bacteria to initiate the infection. Other methods could be through water and food. It is essential that food be well cooked and well prepared. One should make sure that the water source is clean. Here enters climate change. With heavy rainfall and extensive flooding in well populated urban areas where waste and sewage disposal are poor, these bacteria can easily spread out. Another means of transmission is through flies. We know that flies breed in human excreta, and they travel and prefer to walk on food, transmitting the bacteria

Once the bacteria enter the body, they pass through the gastrointestinal tract, surviving the acidity of the stomach, and end up in the colon. They stick to the colonic mucosa and penetrate the cells. They multiply inside the cells until they destroy the host cells and their underlying **Lamina propria**, spreading to neighboring cells. Added to the virulence of the Shigella is the enterotoxin secreted by this bacterium. This enterotoxin, called Shigella toxin, enhances further the virulence of the bacteria These results in ulceration of the colon, bleeding, and *desquamation* (shedding of outer layers of skin) of the mucosa with the mucus. Since the main function of the colon is absorption of water, making the stool solid as it reaches the rectum, failure to do so makes the stool watery. A patient suffering from Shigellosis will have frequent scanty diarrhea with blood and mucus. The infected person releases the bacteria with the stool throughout the course of the disease and two weeks after recovery, so the cycle of infection is continuous.

Patients usually recover after 5-7 days, with a shorter course if taking medications. Some patients, however, take several months to normalize their bowel movements. With some Shigellosis caused by *S. Flexneri*, 3% of the cases developed complications like Reiter's Syndrome which is a triad of joint pains, eye pain and difficult or abnormal urination. Those cases due to *S. dysenteriae* can lead to Hemolytic Uremic Syndrome, a disorder with diminished platelets, lysis of red blood cells, and nonfunctioning kidneys. These abnormalities are due to the virulence and the toxin of the bacteria.

Symptoms

The incubation period of Shigellosis is one to four days. The most distinguishing clinical presentation is the blood in stool with some mucus or pus. The stools in Shigella are not as voluminous compared to those with Cholera, so patients seldom develop dehydration. Other accompanying symptoms are abdominal pain ranging from mild abdominal discomfort to full-blown crampy, severe pain, fever, tenesmus, nausea, and vomiting.

Diagnosis and Laboratories

Empiric diagnosis is established if there is high index of suspicion during an epidemic. There are several differential diagnoses to consider presenting with bloody diarrhea like Amebiasis, Salmonellosis, *Escherichia coli* infection, and ulcerative colitis.

Complete blood counts show increased Hemoglobin and Hematocrit (due to Hemoconcentration from volume depletion) and increased white blood cells. Stool examination may show the presence of red blood cells and white blood cells. The bacteria may be cultured from stool specimens. In AIDS patients, they may be cultured from the blood. Proctosigmoidoscopy with biopsy is seldom done unless it is necessary to rule out other causes of bloody diarrhea.

Treatment

Presented with bloody diarrhea with probable dehydration, the goal of treatment initially is to correct the fluid and electrolytes. Antibiotic treatment may be started right away based on the educated guess of the diagnosis. Usually, a broad-spectrum antibiotic is given. If there is high suspicion of Shigellosis, especially in epidemics, the antibiotic of choice must be given. These are the combination of Trimethoprim- Sulfamethoxazole, 800/160 mg, given two times a day for five days. An alternative drug is the Ciprofloxacin 500 mg given two times a day for five days. Anti diarrheal drugs like Loperamide and Diphenoxylate are not given in this case because they prolong the course of the disease and make the illness worse.

Those patients with complications like Hemolytic-Uremic Syndrome must be put in an Intensive Care Unit, and must be seen by Gastroenterologists, Nephrologists, and Infectious Disease Specialists.

There are some patients, however, who do not need antibiotics. They are given only symptomatic treatment for fever, abdominal pain, or possible mild dehydration. These are patients who are young and with a competent immunity.

Prevention

Just like any diarrheal disease which is transmitted from person to person through the fecal-oral route, the main goal of prevention is personal hygiene and basic sanitation.

Hand washing is the most important means of sanitation. It is essential to wash one's hands thoroughly with soap and water after using the toilet. Children who are not yet toilet-trained, especially in day care centers, must be monitored while washing their hands.

Likewise, elderly people, especially those in the nursing homes who are not careful about sanitation, must be checked after washing their hands.

If one is changing diapers of an infected child or infant, it is important to properly dispose of the diapers and wash one's hands thoroughly with soap and water afterwards. Also disinfect the diaper changing areas with the usual bleaching solutions.

Observe basic guidelines in food and water safety. Never take food or drink water if you are not sure of the sources. When water supplies are not satisfactory, make sure you boil the water before drinking. If you are a traveler, never eat food from street vendors, and take along enough bottled water. If eating fruit, it is wiser to peel it oneself.

Once a traveler gets infected, it is most important that he is treated completely, and is not a carrier when returning to his country; otherwise, he will be bringing the bacteria with him. There is no vaccine to date against Shigella.

6. LEGIONNAIRE'S DISEASE[10]

In 1976, the medical communities were stunned by the increase incidence of a severe form of Pneumonia among the attendees of the American Bicentennial Convention of the American Legion. About 230 people got sick and 34 of them died. After six months in perplexity, they were able to identify the bacteria that caused the infection. This was later named as *Legionella pneumophilia* and the disease was called Legionnaire's Disease. They were able to sort out the source of the infection as the water supply in the hotel that the attendees were staying in.

Since then, occasional outbreaks were reported in hospitals, nursing homes, hotels, prisons and even cruise ships around the world. In 1995, The CDC reported 1,241 cases of Legionnaire's disease at present, Legionnaire's disease is considered the most atypical pneumonia among hospitalized patients. It is the second most common community- acquired bacterial pneumonia. It is estimated that there are 8,000-18,000 cases of Legionnaire's disease each year in the U.S.

with fatality rates of 5-30%. The last fatality reported as of this writing was in Orlando, Florida. In March 2008, the Orange County Health Department in Orlando reported two cases of Legionnaire's disease. These two people acquired the bacteria from the pool and spa of the hotel that were not chlorinated.

CAUSATIVE ORGANISM: *Legionella pneumophilia,* a ubiquitous organism that thrives in warm water (25-45 C) with an optimum at 35 degrees.

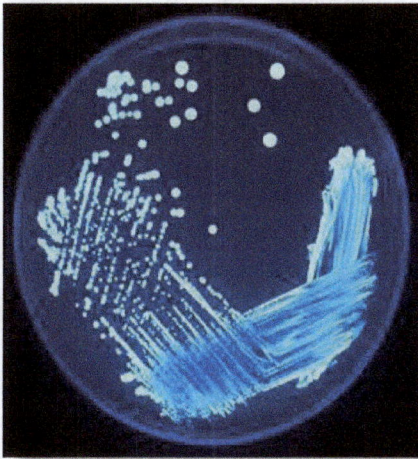

Fig. 6-5. *Legionella sp* on a petri dish. [11]

Pathogenesis

The *Legionella pneumophilia* thrives in bodies of water. They can survive in lakes, rivers, hot springs, or moist soil, but their level in these natural water environments is too low to cause infection. Indoor water systems are more conducive for them to grow in greater numbers, enough to cause infection. These are the whirlpools, spas, air conduction systems, ice cream making machines, and even water misters in grocery stores. The main method of transmission is inhalation of water droplets strung with bacteria. These droplets may be sprayed from faucets, showers, or whirlpools. These droplets are inhaled.

Once inside the lungs, the droplets dry up, leaving the nucleus of the bacterium in the linings of the respiratory tract. Inflammatory process starts with neutrophils invading the area infected, followed by the macrophages which surround the bacteria. Once phagocytized by the macrophages, the bacteria-- rather than being destroyed by the macrophages-- grow and multiply inside the macrophages. When the increasing number of bacteria cannot be contained anymore, the macrophages rupture, releasing the young bacteria to infect other cells, spreading and worsening the infection.

Although the primary organ infected by the *Legionella pneumophilia* is the lung, other body organs are also infected. The bacteria may reach the heart to cause Endocarditis, or pericarditis. They may reach the liver to cause hepatomegaly. The bacteria may also involve the nervous system and the musculoskeletal system. Severe cases may result in Septic Shock, Acute Kidney Failure requiring intensive care. Worse, it can lead to Acute Respiratory Failure, a condition when the lungs fail to provide enough Oxygen to the tissue and the lungs fail to eliminate the CO_2, eventually resulting in death.

Another way of acquiring the bacteria is by direct contact. There were reports of gardeners who got infected after working with contaminated potted soil.

Symptoms

The incubation period for *Legionella pneumophilia* infection is 2-14 days. Initial symptoms include headache, muscle pain, body malaise, and fever as high as 104° F with chills. On the third day, respiratory symptoms appear. The cough is described as dry and non-productive (90% of cases). Shortness of breath, pleuritic chest pain, anorexia, mental confusion, and hemoptysis develop. Gastrointestinal symptoms include nausea, vomiting, and diarrhea. Pertinent physical findings include the presence of rales Septic Shock, Acute Kidney Failure, rhonchi in the lungs with high respiratory rate (tachypnea) and high pulse rate (tachycardia). Neurologic examination may reveal abnormal mental status.

Laboratories

Initial laboratory findings point to an emerging infection. These are the increased number of white blood cells in the blood (leukocytosis) and increases in polymorphonuclear leucocytes and monocytes in the sputum exam with absence of bacteria in the sputum staining test.

The core in the diagnosis is the identification of the bacterium in the sputum culture. This laboratory test, however, will take several days to yield the results. A test that gives an early result is the determination of the Legionella antigen in the urine. Another serologic test that can be done is the Indirect fluorescent antibody test. A chest X-ray is not specific to Legionnaire's pneumonia. The appearance of lower lobe consolidation and the presence of pleural effusion in 50% of cases may also be present in other types of pneumonias. CT scans of the head are done only in those with severe headache and altered mental status. The result of the CT scan in Legionnaire's disease must be normal.

Diagnosis

The diagnosis of Legionnaire's disease is based on the confirmatory laboratory tests. However, presented with a patient with a typical history (middle age or old age, with high grade fever, shortness of breath; smokers, immuno- compromised persons), and typical symptoms of non-productive dry cough, fever as high as 104°F with chills, it is important to have an educated guess on the diagnosis. With a tentative diagnosis, it is necessary to start giving the drug of choice right away and give other appropriate measures. The results of the laboratory tests are just to confirm the diagnosis.

Treatment

The mainstay in the treatment of Legionnaire's disease is drug therapy. The earlier it is given, the fewer the complications. The drug of choice is Erythromycin. This oral drug is highly effective in eliminating the bacteria. This drug, however, may cause stomach upset like diarrhea, nausea, and vomiting. So, patients with gastrointestinal symptoms of Legionnaire's disease may take Tetracycline (Doxycycline). Other alternative drugs are Cotrimoxazole and Ciprofloxacin.

For patients with severe cases, hospital admission is indicated. Those in respiratory distress must be connected to the respirator. All other body systems must be checked to determine if they are infected, too. Appropriate management must be done to prevent organ failure and septic shock.

Prevention

Studies show that 40-60% of the cooling towers evaluated had *Legionella pneumophilia*.

This bacteria can spread airborne as far as six kilometers. In 2003-2004, in Pas-des- Calais, Northern France, there were 86 confirmed cases of Legionnaire's disease and eighteen of them died. They looked for the source and found out what it was. They were able to identify the source of the bacteria, which was a cooling tower of a petrochemical plant about 6-7 kilometers away.

Here is some information as to the fate of the bacterium in different water temperatures:

70-80 C (158-176 °F) – disinfection range

66 C (151° F) –Legionella dies within 2 minutes

60 C (140 °F) – Legionella dies within 32 minutes

55 C (131 °F) – Legionella dies within 5-6 hours

Above 50 C (122°F) – Legionella can survive but not multiply

35-46 C (95-115 °F) – Ideal growth range

20-50 C (68-122 °F) – Legionella growth range

Below 20 C (68° F) – Legionella can survive but are dormant

Based on this table, the recommendation is to keep water tanks at 140° F (60 °C) and tap water for faucet and showers must be kept at 122 F (50 °C) .

On the government side, they have recommended regulations and guidelines in eradicating and preventing the growth of **Legionella** *Pneumophilia* in the water system.

Several manufacturers have already developed contemporary designs in the water- cooling system to prevent the growth of the bacteria. The Health and Safety Executive (HSE) in the UK government recommended weekly microbiological testing, using the "dip slide" to all water-cooling systems. For specific *Legionella pneumophilia* testing, they recommended quarterly and more frequent tests, if the bacteria has been found previously. In Garland, Texas, they assess the cooling tower for apartment buildings yearly for *Legionella pneumophilia*.

Bartonella[12]

The following five diseases 7-11 are all caused by the genus, *Bartonella*.

7. CARRION'S DISEASE

8. OROYA FEVER

9. VERRUGA PERUANA

10. CAT SCRATCH FEVER

11. TRENCH FEVER

This is a group of diseases characterized by high grade fever and progressive anemia caused by bacteria of the genus **Bartonella**. This group of diseases includes the following: Carrion Disease, Oroya Fever, Verugga Peruana, Trench Fever, Cat Scratch Fever, Bacillary Angiomatosis, Peliosis Hepatitis.

This disease is initially thought to be present only at an altitude of about 2,000 – 9,200 feet on Andes Mountain in the country of Peru, Ecuador, and Colombia. Now it is distributed worldwide. In the continental United States, Cat Scratch Fever occurs at a rate of 1/100,000 people. Trench Fever was found to be common in homeless people and alcoholics. In Seattle, Washington, about 20% of people have an antibody titer for Bartonella.

HISTORY

Carrion's Disease

This is an eponym for a medical student named Daniel Alcides Carrion of Cerro de Pasco, Peru. His great interest and dedication to the disease lead him to voluntarily get the disease by requesting that a close friend, Dr. Evaristo Chavez of Dos de Mayo National Hospital, inoculate him with the bacteria coming from pus of an infected patient (Carmen Paredes) in1885. He eventually died without developing a cure, but he was able to prove that *Oroya Fever* and *Verruga Peruana* are the same disease, but at various stages.

It was a microbiologist, Alberto Barton of Peru, who finally discovered the bacteria in 1909, which was later named in his honor. At present there are twenty-three identified species of this bacterium.

At present these three disease names are synonymous. They are found in Peru, Ecuador and Colombia.

Cat Scratch Disease

Cat Scratch Disease was believed to be caused by bacteria called *Afipia felis,* but further studies in 1992 by Dolan and colleagues isolated the bacterium called ***Rochalimae hensalae*** (now called ***Bartonella hensalae***) from a patient with Cat Scratch Disease.

====

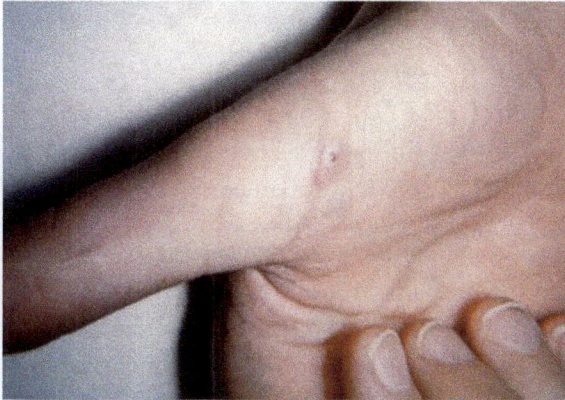

Fig. 6-6

https://commons.wikimedia.org/wiki/File:Cat-scratch_disease_lesion.jpg Photo by Dr Thomas F. Sellers

This bacterium was also associated with the bacillary angiomatosis *Peliosis hepaticus* in HIV patients and in others who are immunocompromised.

At present, Cat Scratch Disease is worldwide in distribution. In the United States, though Cat scratch fever is not a reportable disease, the incidence is 9.3 cases per 100,000 population. This disease is more common in areas with elevated temperature and humidity like the Pacific West coast.

Trench Fever[13]

This fever is also called *Five-Day Fever, Quintan Fever* and *Wolhynia Fever*. This disease was initially described by soldiers in trenches in World War I where the word was coined.

Now it is known as Urban Trench fever because it is common in homeless and alcoholic people, intravenous drug users, and those people with poor hygiene. This disease is also common in regions with famine and malnutrition. The incidence of Trench fever in the United States is unknown, but a clinic in Seattle, Washington that caters to homeless people reported that 20% of their patients have positive antibody titers to *Bartonellosis*. They also found out the positive correlations between alcoholism and the incidence of *Bartonellosis*.

Other names synonymous to *Trench Fever* are *Quintan Fever, Shinbone Fever, Tibialgia Fever, Wolhynia Fever*, and *His-Werner Disease*.

CAUSATIVE MICROORGANISM: Bacteria of genus *Bartonella*

1. *Bartonella bacilliformis* – causes the Oroya Fever, Verruca Vulgaris,, Carrion Fever

2. *Bartonella quintana* – Trench Fever, Bacillary Angiomatosis

3. *Bartonella henselae* – Cat Scratch Fever

4.   Less common species are the- *B. elizabethae*

-*B. vinsonii*

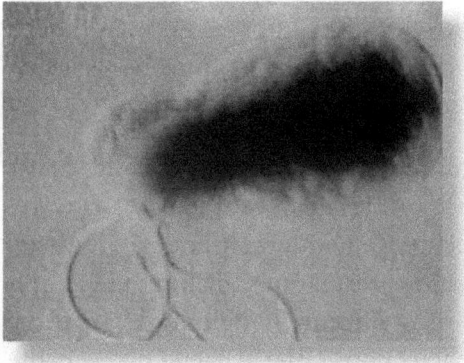

Fig. 6-7. *Bartonella.* Photo by Ceshencam.[14]

Reservoirs and Vectors

Carrion's disease is transmitted by sandflies of the genus *Lutzomyia* (vector) and humans are the reservoirs. Cat scratch disease is transmitted by cat fleas of genus ***Ctenocephalides.***

Felis and cats are the reservoirs. Trench fever is transmitted by the human body louse ***Pediculosis humanus corporis*** and humans are the only known reservoirs.

Pathogenesis

Carrion's disease is transmitted by bites of the sandfly. With their polar flagellum, the bacterium travels within the body and goes with the flow of blood. The bacteria invade the red blood cells, enter and multiply inside. The red blood cells cannot accommodate their increasing numbers; the red blood cells hemolyze and cause transient immunosuppression. This stage represents the acute phase of the disease (Oroya fever). The disease itself is not so serious, but with superimposed infections like Salmonellosis and Toxoplasmosis, the mortality rate is extremely high.

Aside from invading the red blood cells, the bacteria adhere to the endothelial cells and produce endothelial cell-stimulating factors that cause proliferation of both endothelial cells and blood vessels leading to formation of angiomatosis and wart-like lesions (*Verruca Vulgaris*). These wart-like lesions begin as small lesions that subsequently grow, then they ulcerate and bleed. These wart-like lesions grow in groups weeks to months after the episodes of Oroya fever; mostly healed in the form of fibrosis that last for several months.

Cat Scratch Disease is transmitted by cat scratches, cat bites or cat licks, or bites from cat fleas. Once inside the body, the bacteria travel and are trapped by the lymphocytes in the lymph nodes. However, they cause necrotizing granulomatous formations, making the lymph nodes tender and enlarged. The most affected lymph nodes are those in the neck, axilla, and inguinal regions. The bacteria initiate the formation of endothelial cells –stimulating factors. In the feline RBC, the bacteria adhere to the red blood cells and can cause hemolysis, leading to anemia, in cats.

The human body louse transmits Trench Fever. The causative organism is found on the stomach walls of the body louse and injected while sucking blood. Once inside the body, they invade red blood cells and the endothelial cells and they multiply inside. These bacteria are protected from the attack of the inflammatory cells. Instead, the monocytes produce cytokines (interleukin 10) that cause attenuation of the immune responses.

Symptoms

Carrion's Disease has two clinical stages:

1. Acute (Hematologic) stage, also known as Oroya Fever. This stage lasts for two to four weeks. This illness may be mild to severe. In severe cases, since the bacteria are in the blood, signs and symptoms of bacteremia are present. These are high grade fever, chills, headache, sweating, body pains, difficulty breathing, mental status abnormalities and even seizures. Due to hemolysis, patients have a decreased level of RBCs, transient immunosuppression, and thrombocytopenia.

Smears show the bacteria clumping on the red blood cells. Due to the immunosuppression, the patient is very vulnerable to secondary infection. Most common opportunistic infections are Salmonellosis and Toxoplasmosis. The bacteria can spread also to the nervous system causing Neuro Bartonellosis and spinal meningitis.

2. Chronic (Eruptive) stage, also known as Verruga peruana. In this stage, the patient has angiomatous reddish-purple nodules of assorted sizes appearing in crops on the skin. They increase in size and last for several weeks or months. They may erupt, bleed, ulcerate, or heal by Fibrosis. Bacteria can be identified from the eruptive lesion using the silver stain (Warthin-Starry Method).

Cat Scratch Disease

The incubation period for CSD is three days. Most often the patient has a papule or pustule at the site of the bite or scratch a week after exposure. This is self- healing and lasts only for a few weeks, and usually does not bother the patient enough to seek consultation. Some patients have low grade fever and body malaise. The appearance, however, of the enlarged painful and tender lymph nodes on nearby areas (Commonly in the axilla, neck, and groin) is more bothersome to the patients and there are the usual complaints during consultations (commonly on the axilla, neck, and groin) which are painful and tender, and necessitates consultation with the doctor. These lymphadenopathies, however, are self-limited, lasting only for two to three months. Sixteen percent of patients present no lymphadenopathies. In fact, the absence of lymphadenopathies would include the CSD as a differential diagnosis in Fever of Unknown Origin.

Other presenting symptoms are *Parinaud Oculoglandular Syndrome* (POS) in 50 % of patients; encephalopathy in 2.3% of cases; systemic disease in 2% of cases; erythema nodosum in 0.6% of cases, atypical pneumonia in 0.2% of cases; breast tumor in 0.2% of cases and thrombocytopenic purpura in 0.1% of cases.

Trench Fever

An overwhelming majority of patients with Trench fever are asymptomatic or in subclinical state.

The incubation period is usually five to 40 days. Patients present the usual constitutional symptoms of fever, body malaise, headache. Fever is usually relapsing at five-day intervals (range four to eight days), which is why at times it is called five-day fever or quintan fever. The most distinguishing symptom would be pain on moving the eyeballs, and *Paresthesia* of the Shinbone. The pain in the legs would worsen, preventing the person from rising from bed. This pain usually radiates to the loins and the back.

The patient may have some gastrointestinal symptoms like anorexia, nausea, vomiting, diarrhea, constipation, and weight loss.

Diagnosis:

Carrion's Disease

In the acute phase of Carrion's disease, diagnosis may be based on the presenting symptoms of bacteremia bolstered further by the laboratory findings of hemolytic anemia, Thrombocytopenia, and elevated liver function tests. Blood smear may show the bacteria aggregating in the red blood cells.

In the chronic phase, presenting with lymphadenopathies on the axilla, groin, neck, and a patient coming from endemic areas, the first consideration of any clinician is Verruga peruana. The definitive diagnosis, however, lies in the isolation and identification of the bacteria from the discharge of the lesion by using silver stain (Warthin-Starry method).

Cat Scratch Disease

The initial pustule in the scratch site seldom bothers the patient. The usual reason for consultation is the appearance of the painful and tender lymphadenopathies. When asked, however, the patients often say they have a history of cat scratch or cat bite several weeks ago.

Patients with these lymphadenopathies in an endemic area would be considered to display enough signs for experienced clinicians to consider CSD right away and start medications. Confirmatory laboratory tests, however, are still warranted. The only laboratory test that can confirm the diagnosis of CSD is the determination of antibody titer for *Bartonella*. The presence of Immunoglobulin M in the patient is a sign that the CSD is still recent and acute, while the presence of Immunoglobulin G in the patient is a sign of chronic infection. Culturing the bacteria in the skin lesion is difficult; besides, it is less sensitive.

Some physicians opt for the Polymerase chain reaction (PCR) which is more sensitive.

Trench Fever

The presentation of patient with Trench Fever is very non-specific. A clinician may have several differential diagnoses of any insect bites or any infectious disease. The symptoms that would point to Trench fever would be a filthy, homeless individual or an alcoholic, or an intravenous drug user with eye pain and leg pain. Laboratory tests would support this.

Trench fever has a serologic test called Weil-Felix test. (Note: there are diseases that give a positive Weil-Felix test). This test is usually positive one to two weeks after the initial appearance of the symptoms.

Treatment

Carrion's Fever

The acute phase of Carrion's Fever is treated with *Ciprofloxacin* and *Chloramphenicol* for several weeks. If complicated by secondary infections like *Salmonellosis* and *Toxoplasmosis*, the treatments are directed to these opportunistic infections. In the chronic phase the lymphadenopathies are treated with Rifampicin and Macrolides.

Cat Scratch Disease (CSD)

CSD, being a self-limited infection, has no recommended treatment. The clinicians may give a patient symptomatic treatment like analgesic, antipyretic, anti-inflammatory drugs. If secondarily infected, patients may receive an antibiotic for the secondary infection, not for the Bartonellosis.

Trench Fever

Trench Fever is treated by Tetracycline with Gentamycin given seven to ten days. For those contraindicated for tetracycline, *Chloramphenicol* is a good alternate drug.

Prevention

Carrion's Fever

Since Carrion's Disease is primarily in the Andes Mountain regions (Peru, Ecuador, Colombia) travelers to these areas should follow guidelines as to prevention of bites from sandflies. Refrain from going to sandy areas in coastal territories and other areas with substantial amounts of water, like marshland.

Cat Scratch Fever

There are few preventive measures for CSD, except to minimize being scratched by the cats and to clean the cats often. The idea of immunizing the cats was brought out, but as of now there is no study yet on this premise.

Trench Fever

Prevention is geared towards hygiene and sanitation. People should care for their bodies by bathing themselves often to eliminate lice from their bodies.

12. Tularemia[15]

This is also known as "Rabbit fever," "Deer fly fever," "Ohara fever," or "Francis disease." This is a serious bacterial disease affecting humans and animals. For whatever reason, rabbits are more prone to this illness; that is why it is also called Rabbit fever.

This is such a deadly disease that its causative microorganism has been used as an agent for biological warfare.

Tularemia occurs in North America, parts of Europe, Russia, China, and Japan. In the United States the disease is rare with an incidence of one per one million people.

This disease usually happens during hunting season (early winter) and when ticks and deer flies are abundant (summer).

This disease first came to attention in 1911 when there was an outbreak of a disease that killed plenty of ground squirrels in Tulare Lake in California. Now the disease is not limited to the animals but has spread to humans. This illness is more common in agricultural workers, cooks, hunters, and laboratory workers.

CAUSATIVE MICROORGANISM: *Francisella tularensis (formerly Pasteurella tularensis)*

There are two types: Francisella tularensis also known as Jellison type A, virulent for human and domestic rabbits and Francisella holarctica, also known as Jellison type B, occur in beaver, muskrat and voles in North America, rodents and hares in Eurasia.

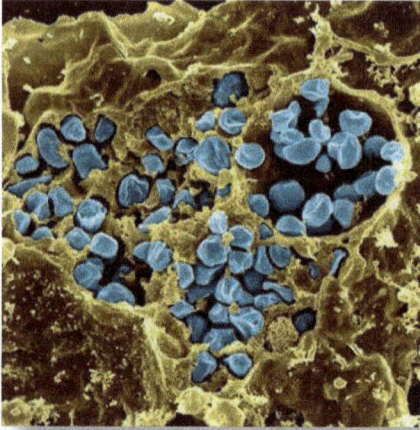

Fig. 6-8. *Francisella tularensis*, Colorized SEM of a macrophage infected with *Tularensis*[16]

Transmission

Tularemia can be transferred in several ways, especially any direct penetration of the skin or mucous membrane like insect bites, inhalations, or ingestions. Most common are bites from arthropod vectors which are the ticks (*Dermacentor andersoni*, *D. variabilis*, *Amblyomma americanum*) and deerflies. The bacteria can survive in the arthropod vector for two weeks, while in ticks they can survive throughout their lifetime.

Tularemia can also be transmitted by eating meat that is not well cooked from infected rabbits. It has been discovered that the organism can survive for three years in frozen rabbit meat at -15 degrees Celsius. Inhalation or ingestion of the organism from water or food is another way of acquiring the disease. The bacteria can survive for a prolonged period in fomites (objects or inanimate materials that carry and spread disease) and can be transmitted by inhalation or direct contact. Direct contact from person to person, however, is not possible. In Russia, especially, mosquitoes often transmit this microorganism.

Symptoms

The clinical picture of Tularemia has six forms depending on the point of entry into the body. These are the following:

Typhoidal: This is manifested by fever with chills and excessive sweating, headache, prostration, nausea, and weight loss. There is no lymphadenopathy. If severe it can become a more serious form called pneumonic. The bacteria in this form are acquired by inhalation, ingestion, or inoculation.

Ulceroglandular

The constitutional symptoms are like typhoidal but in this form there are enlarged lymph nodes that are tender. These lymphadenopathies may ulcerate and drain out fluid. There are usually some skin lesions (papules or ulcerating pustules) at the site of bite or contact.

Oropharyngeal: This is like Ulceroglandular but appears only on the throat. This is usually manifested as exudative pharyngotonsillitis with cervical lymphadenopathies.

Glandular : Here, there are tender lymphadenopathies with fever, but no ulcerations or skin lesions.

Oculoglandular: Involvement of the eye is usually manifested as painful purulent conjunctivitis. There may be minute ulcerations or nodules on the conjunctivae. This is usually accompanied by periorbital edema, ecchymoses and periauricular and cervical lymphadenopathies.

Pneumonic: The pneumonia is usually atypical and fulminant. This is manifested as dry non-productive cough with substernal pain. In severe cases, it may lead to lung consolidation. The bacteria in this form can be acquired by inhalation, leading to a single form of tularemia. There may be a complication of other forms of tularemia like Ulceroglandular or glandular forms that reach the lung by a hematogenous route.

Diagnosis

It is hard to isolate **Francisella tularensis** because they do not grow in standard culture media. It needs a cysteine or sulfhydryl compound in the medium. Some researchers use buffered charcoal and yeast extract medium (BCYE). The specimen may be blood, sputum, gastric washing, or exudates from ulcers. More used are serologic tests like the ELISA. There must be a rising titer of the antibodies in a few weeks to be indicative of Tularemia. Cross-reactions with *Brucella and Yersinia* must be taken into consideration. The polymerase chain reaction (PCR) may be used also, especially to differentiate *F. tularensis* from *F. Tularensis holarctica*.

Treatment/Prevention

The causative microorganisms are sensitive to antibiotics. The drug of choice is Streptomycin or gentamycin given for at least ten days. Other effective drugs are tetracycline (doxycycline) given for two to three weeks, chloramphenicol, fluoroquinolones. Vaccines are available only for high-risk groups.

If one is in an endemic area or in a tick-infested area, it is essential to wear long- sleeved clothes and use insect repellants to prevent the ticks from contacting your skin. Use personal protective equipment like gloves, eye goggles, mask if handling rabbits or lagomorphs (Rabbit- like organisms). Of course, refrain from eating uncooked rabbits or drinking untreated water.

Francisella Tularensis as a Biological Weapon

In 1954 in Pine Bluff Arsenal in Arkansas, a US researcher started studying this organism for biological warfare. It was found out that this is an extremely attractive biological weapon because (1) it is easily aerosolized, (2) highly effective needing only 10-15 bacteria to start an infection, (3) it is highly incapacitating but has low lethality. In three days, the infection starts lasting for one to three weeks if treated and for two to three weeks if untreated. It has low mortality, making it attractive for spreading to enemy soldiers close to non-combatants, (4) it is easily decontaminated. These bacteria have been included in the biological programs of US, USSR, and Japan.

The US uses Agent UL (standardized SCHU S4 strain) for US M143 spherical bomblets.

It was later changed to TT (wet type) and ZZ (dry type) for anonymity. Later new strain (425 strain) was developed and it was standardized as Agent JT. This was better than Agent UL because it is more incapacitating but with less lethality. No vaccine is ready yet for public use currently.

13. PLAGUE

Plague is an infectious disease in humans and animals. This infection is transmitted by fleas that thrive on rodents, squirrels, prairie dogs, wood rats, chipmunks, ground squirrels, deer mice and voles. Other animals that eat theses rodents (carnivores) acquire the disease, too. At home, pet cats and pet dogs transfer these fleas.

Plague is one of the diseases that had a major part in the history of humanity. It is one of the dreadful diseases that caused pandemics in human history. In the 14th century, for four years between 1347 to 1351, 40-50% of the population died in Europe, 40% of the population in Egypt, and half of the 100,000 population of Paris. This era was the so- called "Black Death;" black in the sense that the skin turned black because of the subepidermal hemorrhage, and the tips of the extremities usually turned dark due to gangrene (Acral necrosis). By the year 1350, there was a significant decrease in the mortality rate from Plague but there were still reported outbreaks. For the whole 14th century, it was estimated that there were 75-200 million deaths due to Plague. Succeeding centuries had reported outbreaks, especially in France, Italy, and Norway-- all in Europe.

Between 1924-1925, the last reported human Plague epidemic in the United States was in Los Angeles. Thereafter, there were sporadic cases, mostly in rural areas, with an average of 10-15 persons/year. In the 1980s, there were eighteen reported cases with a death rate of one for every seven persons with Plague. Most of these cases were in the state of California, western Nevada, southern Oregon, northern Arizona, and southern Colorado. The last reported cases of Plague so far in the United States were in 1995 when there were 28 confirmed cases of Animal Plague (13 wild rodents, fifteen carnivores) and fifteen confirmed cases of human Plague.

CAUSATIVE MICROORGANISM: *Yersinia pestis* (previously known as *Pasteurella pestis*)

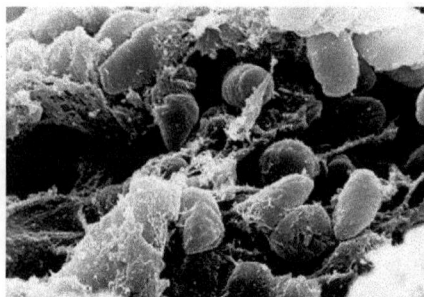

Fig. 6-9. *Yersinia pestis* from Rocky Mountain laboratories. [17]

PRIMARY VECTOR: Rat fleas (mostly), human lice, ticks

SECONDARY VECTORS: Black rat (*Rattus rattus*), Brown rat (*Rattus norvegicus*).

Other wild animals like squirrels, prairie dogs, wood rats, chipmunks, ground squirrels, deer mice and voles.

Transmission and Pathogenesis

Humans acquire the disease from a bite of an infected flea. Other arthropods that can transmit Plague are ticks and human lice. The transmission can be from animals to animals, animals to humans, humans to humans (but not in Bubonic Plague). This disease can also be transmitted by direct contact of the animal handler with the fluid of infected animals, although this is exceedingly rare. Airborne transmission is also possible by inhaling the bacteria expelled from coughing.

There are three types of Plague:

1. **Bubonic plague**, the most common form of Plague. This is the type of Plague that is acquired from flea bites. Once the causative bacteria enter the body, they travel through the lymphatic system and invade the lymph nodes along the way. The unusual thing with **Yersinia** *pestis* is that they resist the phagocytes. Instead of being killed by the phagocytes, they multiply inside them and kill the phagocytes This results in swollen, tender lymph nodes that later become hemorrhagic, necrotic, and ulcerative. Most affected lymph nodes are those on the neck, groin, and armpit. If the causative bacteria spread to the lungs, they can lead to the second type of Plague, called Pneumonic Plague. It is so severe that it can lead to the third type of Plague, called Septicemic Plague. Bubonic Plague cannot be transmitted from person to person.

2. **Pneumonic Plague:** This is the type of Plague acquired from inhaling the bacteria in the air. A cough from an infected person can spread the bacteria as far as twenty feet. Once inhaled, the bacteria cause an inflammatory reaction in the alveoli leading to Pneumonia. The person continuously coughs out bloody sputum or watery sputum. Those with severe forms may develop respiratory failure followed by shock and death. This is the type of Plague that can be transmitted from person to person by airborne means.

3. **Septicemic Plague:** this is the least common among the three types of Plague but the most serious. This is a complication of the previous two types. In this type, the bacteria are in the blood and have been widespread throughout the body, affecting the brain

(Meningitis or Encephalitis) or the heart (Infectious Myocarditis), or all the body organs

(Multiple Organ Failures). The blood of the patients may consume clotting factors due to the presence of the bacteria, resulting in Disseminated Intravascular

Coagulation (DIC). Patients untreated at this stage have 100% mortality, while those treated have 40% mortality.

Symptoms

The general symptoms common to the three types of Plague are fever, chills. headache, body ache, joint pains, prostration, confusion and dizziness, stiff neck, low blood pressure, irregular heartbeats. Others would present with gastrointestinal symptoms like abdominal pain, nausea, vomiting and diarrhea.

Now on specific types, the Pneumonic Plague (aside from the general symptoms) would present with respiratory symptoms like cough with watery sputum or bloody sputum. Patients may have hemoptysis, shortness of breath, or worse, have difficulty breathing.

The Bubonic Plague would present the skin manifestations: the blackish color of the skin due to subepidermal hemorrhage and gangrene. The lymph nodes are enlarged with oozing blood and pus. There is severe pain. These lesions are due to actual decaying and decomposing of the skin.

Septicemic Plague is the result of the other two types. Here the bacteria have spread throughout the body and cause blood poisoning. Seldom would a person have Septicemic Plague alone. It is usually complications of the other two types. Here the patient is more prostrated with high fever and chills, spontaneous bleeding, and signs of disseminated intravascular coagulation.

Diagnosis

A Bubonic Plague would be easy to consider clinically. With all the oozing lymph nodes plus the fever and prostration supported by history of flea bites and occurrence in an endemic area, a clinician is almost certain to be dealing with the Bubonic Plague.

For Pneumonic Plague, since it is a Pneumonia, it hard to be certain clinically, because there are so many microorganisms that can cause Pneumonia. However, in a patient with Bubonic Plague that shows signs of Pneumonia, a clinician is almost certain of the etiologic microorganism of Pneumonia.

In Septicemic Plague, being a blood infection, there are many possible etiologic microorganisms, but a history of Bubonic Plague or Pneumonic Plague would certainly support the diagnosis of Septicemic Plague.

Confirmatory tests are necessary. The best tests are the culture and sensitivity test of the nodal discharges (in Bubonic type), sputum (Pneumonic type) and blood (Septicemic type). For those who are not sure of a possible Pneumonic Plague, a chest x-ray is immensely helpful.

Treatment

Plague is a serious and dreadful disease. It has a noticeably short incubation period and extremely high mortality rate in untreated cases. Once Plague is suspected, the patient should be isolated right away and needs immediate care in a hospital setting. Specimen collection or extractions are done right away for all the laboratory tests. Then antibiotic therapy is started. The drugs of choice are Streptomycin, Gentamycin and Tetracycline. In serious cases, the patient should be treated in the Intensive Care Unit.

Plague must be reported to the local government authorities. Even suspected cases must be reported. Some people may have to take prophylactic antibiotics and other measures to prevent the spread.

Prevention

Preventive measures that were geared to eliminate rodent population have been costly and unsuccessful in the rural areas in Third World Countries. In an urban setting, however, they might prove successful. This is done by spraying insecticide to all shelters of rodents; the insecticide not only kills rodents, but it also kills fleas. There must be a constant surveillance of cases of Plague and increased rodent populations.

Maintain sanitation around dwellings, whether houses or other buildings. Remove all shelters for rodents. Remove food sources for the rodents; also dispose of left- over food properly. Insecticide spray can eliminate the rodent population as well as the fleas. Report to local government authorities any increase in rodent populations. For those with pet animals, make sure that your dogs and cats are free of fleas. Regular examination and treatment must be done to control the fleas.

14. LEPTOSPIROSIS

Leptospirosis is a rare zoonotic disease. This spirochete has world-wide distribution including US. It is more common in tropical countries, sparing the polar regions. In past decades, Leptospirosis has gained much attention in the medical community, due to its increased incidence and its positive correlation with the climate changes. The more floods we have, the more incidence of Leptospirosis.

This infection was first described by Larry in 1812 among the troops of Napoleon that invaded Cairo. Later it was known in Europe as "bilious typhoid." In 1886, Adolf Weil reported in a paper, describing an acute infection with a yellowish eye, spleen enlargement, and kidney involvement. This disease was later called "Weil's Disease." In 1907, the microorganism was isolated from the kidney after a postmortem examination. They initially called this microorganism *Spirochete interrogans*.

CAUSATIVE MICROORGANISM: *Leptospira spp.*, a spirochete bacterium, with five important species: -*L. icterohemorrhagiae*

-L. canicola

-L. ponoma

-L. grippotyphosa

-L. bratislava

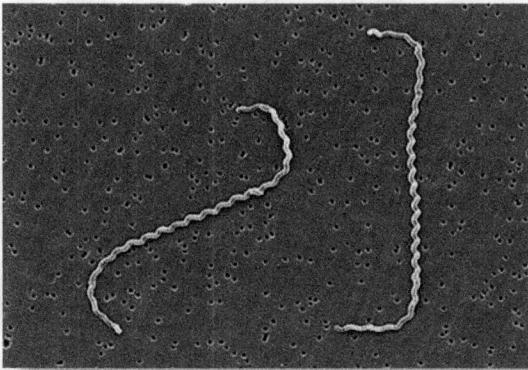

Fig. 6-10. *Leptospira sp.* Photo by Janice Carr CDC.[18]

VECTORS: rodents, pigs, dogs, cattle, deer, sheep, and wild animals

Pathogenesis

For some unknown reason, the *Leptospira* species live in the renal tubules of the vectors for several years. The animals excrete them together with their urine on the soil or in water bodies. Areas most likely to have the bacteria are muddy riverbanks, ditches, gullies, swamps, and muddy areas where these animals usually pass through.

The bacteria stay in the water and are acquired through direct contact of the bacteria with the abraded skin. Other ways of acquiring the microorganism are inadvertent drinking of water, inhalation of aerosolized fluid into the lungs, and through contact with the mucosa of the nose, conjunctivae, and through the placenta during pregnancy. This infection is common in people who wade, stride, or submerge their bodies in water. Also susceptible to this infection are veterinarians, farmers, sewage workers, slaughterhouse workers, workers in abandoned buildings and those people living in areas frequently flooded.

During persistent rainfall and floods where the habitats of these vectors are submerged under water, these bacteria are freely present in the water. So, people who wade or stride are very vulnerable to this type of infection. Once inside the body, the **Leptospira** gain access to the bloodstream and lymphatics and reach the liver where they multiply and later spread to other body organs. This stage corresponds to the appearance of the symptoms. The organs most frequently involved are the liver and the kidney (Hepato-Renal syndrome). The **Leptospira** can cause interstitial nephritis leading to renal failure. In the liver, the bacteria can cause centrilobular necrosis leading to hepatocellular dysfunctions, the heart may develop myocarditis and pericarditis, the presence of the bacteria in the anterior chamber of the eye can cause uveitis leading to the conjunctival suffusion, in the muscles **Leptospira** can cause myofibrils vacuolization, and in the respiratory tract it can cause a mild cough to alveolar hemorrhage.

In severe forms, disseminated intravascular coagulation ensues. This is manifested by spontaneous bleeding in all parts of the body. It can come from the lungs as hemoptysis, from the colon as melena, from the stomach as hematemesis and from other orifices of the body.

Kidney involvement is manifested by dark colored urine, and if it worsens may lead to renal failure. Liver involvement results in hepatitis and may also lead to hepatic (liver) failure. Involvement of the brain leads to meningitis. Lastly, those with severe bleeding may develop shock and death.

Symptoms

The incubation period of Leptospirosis ranges from 3-30 days with an average of 5-14 days. There are two syndromes in Leptospirosis which are:

-Anicteric syndrome – This is a milder, self-limited infection manifested by constitutional symptoms of fever, body malaise, myalgia, anorexia, headache. Its septicemic phase (acute phase) and its immune phase are indistinguishable and last only for few days (one to three days for the acute phase and four to 15 days for the immune phase).

-Icteric syndrome - Here, the septicemic phase and the immune phase are not distinguishable and last longer. The presenting manifestations here are more severe. The fever is high grade (40 °F and above). There are rigors, nausea, and vomiting, and muscle pain mostly on the calf and abdominal muscles. The person may show jaundice (yellowish discoloration of the skin and conjunctivae) with an enlarged liver on palpation. The presence of uveitis causes redness of the conjunctivae without exudates, leading to the so-called "conjunctival suffusion." This is a peculiar manifestation of Leptospirosis.

The cough may be aggravated by difficulty breathing, especially in those patients with alveolar hemorrhage. The urine turns dark-colored due to the nephritis. In severe cases when the microorganisms reach the brain, patients may present with neck rigidity which is a sign of meningitis. A worse complication is the Disseminated Intravascular Coagulation (DIC), a condition usually due to severe infection, consuming all the blood clotting factors, leaving none for subsequent clotting so the person continuously bleeds. This condition is hard to manage. All patients with DIC succumb to death.

Diagnosis and Laboratories

A person who presents with the usual constitutional symptoms of fever, body malaise, myalgia, headache, anorexia, and nausea has several differential diagnoses. It is exactly right for a clinician to consider all possibilities. If the person's history has an episode of wading or submerging the body in water in an endemic area, it would be right to consider Leptospirosis as the primary diagnosis. Once the conjunctival suffusion, jaundice and the deep brown urine appear, it is appropriate to eliminate other differential diagnosis and direct the management to Leptospirosis; more so if there are signs of spontaneous bleeding.

Laboratory tests are used to confirm the diagnosis as well as to know the extent of the infection. Since the bacteria have disseminated throughout the body, all the body fluids are positive for the bacterium. The most practical body fluid to get into is the urine, as this does not require invasive procedures. From the urine specimens, Antileptospiral antibodies are detected using the Microscopic Agglutination Test (MAT).

Other laboratory tests are requested to monitor the functions of the body organs especially the liver and the kidneys (the most affected body organs in Leptospirosis). A complete blood count will show how acute the infection is. Blood platelets and clotting factors levels would show whether DIC is impending or is present. Blood urea nitrogen (BUN) and blood creatinine levels would indicate renal involvement; whether renal failure has set in already or is impending to set in. Transaminase levels indicate the status of the liver.

Treatment

There are two main goals of treatment: to eradicate the causative microorganism and the prevention of complications. The mainstay of the eradication of the microorganisms is drug therapy and here are the chosen drugs:

-Penicillin is the drug of choice for Leptospirosis. The usual dose is 20-24 million units per day given parenterally. Oral penicillin preparation like Amoxicillin can be given to less serious patients. This is given at .5 -1 gram every eight hours PO.

-Erythromycin is the drug of choice for pregnant patients and those with allergy to Penicillin. It is given 500 mg IV or orally every eight hours.

-Doxycycline is usually given for less severe infections and is the choice drug for prophylaxis. 100mg is usually given every 12 hours, orally or IV.

For severe cases, aside from drug therapy and supportive treatment, it is especially important to monitor the functions of the vital organs of the body. A renal failure due to Leptospirosis is reversible in the early stage, so it is necessary to detect the renal status correctly to prevent further complications. Patients with respiratory distress must be placed on mechanical ventilators. Clotting factor levels must be monitored because any impending decrease must be replaced right away to prevent the fatal complication of Disseminated Intravascular coagulation. One should monitor the patient for any arrhythmias because this is a sign of heart involvement. If arrhythmias are present, they must be treated right away to prevent acute heart failure.

Prevention

Much of the preventive measures are directed to the elimination of vectors. Since the common vectors are rats, mice, and other rodents which thrive in the surroundings, it is necessary to strictly adhere to hygiene and sanitation both indoors and outdoors.

Workers who are at risk of infection in the performance of their jobs should wear personal protective equipment. Avoid wading or striding along in floods. If it is necessary to do so, use high cut shoes or waterproof suits to prevent water contact with the skin.

There is a vaccine for Leptospirosis but its effect is only for a few months; that is why it is not popular. For people in endemic areas where people are at risk of having the bacteria, immunization must be given.

15. CHOLERA[19]

One of the diarrheal diseases linked to heavy rainfall and contaminated water is Cholera. This disease is seasonal, suggesting sensitivity to climate change. Studies have shown increased warming of the atmosphere and water which are due to climate change are factors which favor the reproduction of this bacterium. In US and other developed countries, Cholera is exceedingly rare because of advancement in water treatment, (filtration and chlorination) and good sanitation systems. However, in other parts of the world, especially in the Indian subcontinent and sub-Saharan Africa, this condition is still quite common. In January 1991, there was an outbreak of Cholera in South America spreading to several continents. This outbreak did not even spare the US. A few US tourists from this region were reported to acquire the disease. The last major outbreak was in 1998 in Africa where more than 400,000 cases were reported. In US, the last outbreak was reported in 1911.

In history, Cholera was initially reported in the subcontinent of India where the Ganges River was believed to be the reservoir. This disease eventually spread to other continents. To date, there are seven reported pandemics of Cholera across Africa, America, Europe, and Asia killing millions of people:

-The First Cholera pandemic struck between 1811-1826 in India spreading to China

-The Second Cholera pandemic was between 1829-1851 in Europe spreading as far as Russia

-The Third Cholera pandemic was between 1852-1860 affecting Russia, spreading to London

-The Fourth Cholera pandemic was between 1863-1896 in Hamburg, Germany

-The Sixth Cholera pandemic was between 1899-1923 affecting Russia

-The Seventh Cholera pandemic and the last reported Cholera pandemic was during 1961-1970 in Asia. It was also called El Tor based on the specific strain of bacteria. This began in Indonesia and spread as far as India, Bangladesh, Russia, Africa and Europe.

Just to mention a few famous people who succumbed to Cholera: James Polk, the eleventh US President; Elizabeth Jackson, the mother of US President Andrew Jackson; and May Abigail Fillmore, daughter of US President Millard Fillmore.

CAUSATIVE MICROORGANISM: *Vibrio cholera*, a gram-negative bacteria

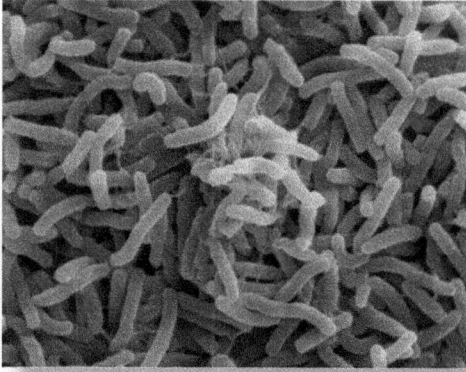

Fig. 6-11. *Vibrio Cholera.* [20]

Pathogenesis

Vibrio cholera bacteria live in the aquatic environment attached to particular algae or copepods (members of a group of arthropods also called zooplankton). They may be toxic or non-toxic. Those non-toxic strains can become toxic by viral mutation. (See the section on plasmids and viruses in the chapter dealing with biology). With poor environmental management such as poor sanitation, improper waste disposal, poor sewage and improper treatment of drinking water, these bacteria are easily spread, leading to an outbreak.

Once these bacteria are ingested through contaminated food and water, they are destroyed by the acidity of the stomach. The organisms are sensitive to acidic environments. However, some bacteria do survive and pass through, reaching the small intestine where they stick to the intestinal walls through the formation of flagella. Along the intestinal wall, they produce a toxin that stimulates the mucosal cells to continuously pump chloride ions, tagging with it sodium. These create an osmotic mixture that attracts more water from the intestinal fluid in a few hours. The person can lose as much as eight liters of fluid per day.

This is due to continuous stimulation of the mucosal cells which pump chloride ions. Eventually the person develops severe dehydration and electrolyte imbalance. In severe cases the patient may develop hypovolemia complicated by acute renal failure and the risk of death. The immediate cause of death is severe hypovolemia and electrolyte imbalance. Individuals who have less acid in the stomach are more prone to Cholera.

These vulnerable groups are infants, elderly, and people with weak immunity like *Immunodeficiency Syndrome* (AIDS) Patients.

Symptoms

A typical symptom of Cholera is the so called "painless diarrhea." The person does not feel any abdominal pain in spite of the frequent watery stools occurring every 15 minutes. Other patients develop vomiting. The enormous loss of water and electrolytes can cause severe dehydration manifested as sunken eyeballs, "washerwoman's hand" (severely wrinkled hands) and leg cramps. If not immediately treated, this may lead to hypovolemic shock manifested as a weak thready pulse and exceptionally low blood pressure. Acute renal failure ensues because not enough blood volume perfuses the kidney to maintain its filtration functions. The immediate causes of death are the hypovolemia and electrolyte imbalances.

Diagnosis and Laboratories

The clinical presentation of a person suffering from Cholera is so very typical that it is seldom misdiagnosed by a clinician. The recurrent bouts of painless diarrhea with signs and symptoms of hypovolemia are pathognomonic and warrant immediate treatment. Rectal swabs are done to obtain specimens for confirmatory diagnosis by serotyping.

Microscopic identification is not reliable.

Treatment

Treatment is initially directed towards the prompt replacement of fluid and electrolytes to maintain adequate fluid volume circulating in the body. In mild cases, patients may be given oral hydration therapy. For those with hypovolemia, intravenous fluid is immediately administered. At times, it is even necessary to put in two to three intravenous lines at the same time. After recovering the normal fluid volume of the body, the treatment is directed to the elimination of the causative organism by giving antibiotics. The drug of choice for Cholera is Tetracycline. For pregnant patients or those with allergies to *Tetracycline*, alternative drugs *are Cotrimoxazole, Erythromycin, Chloramphenicol,* and *Furazolidone.* Since a common complication of Cholera is renal failure, it is of best interest to check for kidney function right away. This can be attained by checking blood levels of urea nitrogen and creatinine.

Prevention

Sanitation is particularly important in the prevention of Cholera. Proper disposal of human excreta and proper sewage disposal are utmost. Since drinking water is a mode of transmission, proper purification and chlorination of drinking water must be done. For those people managing these patients or in direct contact with these patients, proper use of personal protective equipment is necessary. Items that contacted the patients must be disposed of properly. Hospital materials must be sterilized. Bedding must be washed with hot water using chloride bleach. Avoid eating improperly cooked food, fish, or shellfish because they carry the bacteria.

For those traveling to endemic areas, the following precautions are recommended:

-Avoid drinking water if the source is unknown. Better bring with you bottled mineral water or carbonated drinks.

-Avoid drinking tea, coffee with non-boiled water.

-Avoid putting ice in the drinks, especially if the source of the ice is unknown.

-Avoid uncooked salad, raw or improperly cooked vegetables, raw fruits.

-Eat only well-cooked fish and vegetables, preferably still hot when served.

-If you want to eat the fruit, better peel it yourself.

As the saying goes, "Boil it, cook it, peel it, or FORGET IT!"

16. Meningococcal Meningitis

The meninges are three layers lining the surface of the brain and spinal nerves. Several microorganisms can infect these meninges. One of them is the ***Neisseria meningitidis***. The meningitis caused by this type of bacteria is of importance because it can easily spread to epidemic proportions, especially in areas where people are crowded like student dormitories or prison cells.

This disease is included here because it has been correlated to hot and dry seasons. The incidence of this disease can increase to epidemic proportions during the hot season and ends on the start of the rainy season.

Meningococcal Meningitis was first reported in 1805 by Vieusseux in Geneva, Switzerland, but it was in 1887 when the causative bacteria was isolated.

In the start of the 20th century, this disease caused high mortality and the age group commonly involved was the infants and children. In the continent of Africa, there is such thing as the Meningitis belt which extends from Ethiopia in the east to Senegal in the west. In United States, since 1960, the incidence of this disease is .9-1.5 cases/100,000 people. The last epidemic of Meningococcal meningitis, as reported by WHO, was in 1996 with 300,000 cases reported.

CAUSATIVE MICROORGANISMS: *Neisseria meningitides*. There are four serotypes that cause epidemic

(A, B, C,W135) with B and C serotypes causing most of the diseases.

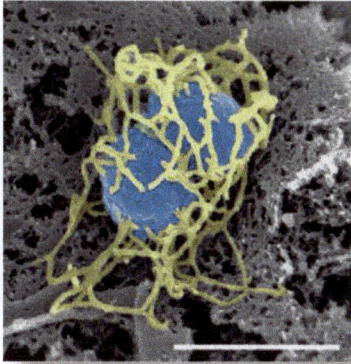

Fig. 6-12. SEM photo of *Neisseria meningitides*. Photo by Arthur Charles-Orszag[21]

Transmission/Pathogenesis

Neisseria Meningitides is transmitted by air or by droplets. People with this infection can spread the causative Microorganism by coughing, sneezing, kissing, and sharing the same utensils. Once inhaled, the bacteria lodge on the mucosal surface of the nasopharynx causing meningococcal pharyngitis. At times, they lodge in the urogenital tract and anal canal. On the mucosa they are carried by membrane-bound phagocytic cells to the deeper submucosa. Colonizing the submucosa may cause mild symptoms or subclinical infection; that is why some patients are asymptomatic.

In five to ten percent of cases the microorganisms enter the blood vessels where they are vulnerable to the immunologic cells. With their rapid replication, however, they may overpower the immune system in the blood to cause meningococcemia and later spread to the brain to cause meningitis. In severe cases this can lead to diffuse endothelial necrosis, thrombosis and perivascular hemorrhage and cause circulatory collapse and DIC (Disseminated intravascular coagulation).

Signs and Symptoms

A person suffering from Meningococcal Meningitis would present with fever, usually moderate to high grade, and severe headache. The severity of headache would cause nausea and vomiting and neck rigidity. Skin rashes may or may not be present. These initial manifestations will be followed by the mental status changes such as lethargy and drowsiness. In severe cases it may lead to stupor or coma. A patient with Meningococcal Meningitis in coma has an extremely poor prognosis.

According to one study, the "Netherland Meningitis Cohort Study" published in 2008, in 70% of cases, patients would manifest with the so-called classical triad of Meningococcal meningitis which are: fever, neck rigidity and mental status changes.

On physical examination, patients may be lethargic, drowsy, or stuporous. They are febrile, with signs of neck rigidity which are the Kernig sign (with patient in supine position, flexing the hip 90 degrees then extending the legs or straightening the knee would elicit pain) and *the Brunzinski Sign* (with patient in supine position, flexing the neck would elicit flexion of the lower extremities). Adults present petechial skin rashes in 62% of cases and children in 81% of cases. If the causative bacteria spread throughout the whole body, *Waterhouse-Friderichsen Syndrome* develops. This is a life-threatening condition with *Petechial Hemorrhages* in the skin and mucous membrane, disseminated intravascular coagulation, circulatory collapse leading to septic shock, and eventually death.

Laboratories

The most important and the best laboratory study is the examination of the *Cerebrospinal Fluid* (CSF). Upon lumbar tap, the pressure of the CSF is increased, the fluid contains less sugar, more protein, with pleocytosis (increased Polymorphonuclear Cells).

The CSF is brought to the microbiology lab for culture and sensitivity test. Polymerase Chain Reaction (PCR) used as complementary test. This test is more a confirmatory test rather than a routine test.

Other tests that may be of use are the CT scan or MRI of the brain. These imaging procedures may show brain lesions such as cerebral edema, cerebral ischemia, dilated ventricles and other meningeal lesions.

Electroencephalography (EEG) can also be of use, especially for those conditions with tendencies to develop seizures.

Treatment

Any type of meningococcal disease, be it pharyngitis, meningitis, or meningococcemia, must be considered as a serious condition that needs immediate treatment. The patient is hospitalized right away and given empiric antibiotics (giving antibiotics based on an educated guess before confirming the diagnosis). The purpose of this method is to prevent delay in the management of the disease. Once the CSF and the blood have been extracted for the laboratory test, the patient is given antibiotics. Once the results of the test are known, you can easily change the antibiotic if the bacteria are resistant to it.

The usual antibiotics that are given are the third generation cephalosporins like ceftriaxone. Other antibiotics used are the quinolones (ciprofloxacin), rifampicin, sulfonamides. Surgical procedures resorted to if the patient develops hydrocephalus, subdural effusion, or empyema.

[1] Drawing by Miranda Olsen

[2] www.cdc.gov/nczved/dfbmd/disease_listing/stec_gi.html

www.doh.wa.gov/EHSPHL/factsheet/ecoli.htm

www.wikipedia.org/wiki/Escherrichia_coli_0157:H7

[3] https://commons.wikimedia.org/wiki/File:Scanning_electron_micrograph_of_an_E._coli_colony.jpg viewed 30
Dec 2021.

[4] www.en.wikipedia.org/wiki/Typhoid_fever

www.cdc.gov/ncidod/dbmd/diseaseinfo/Typhoidfever_g.htm

www.en.wikipedia.org/wiki/Typhoid_fever

[5] www.en.wikipedia.org/wiki/Lyme_disease

www.textbookofbacteriology.net/Lyme.html

www.emedicine.medscape.com/article/1053863-overview

www.textbookbookofbacteriology.net/Lyme.html

[6] https://commons.wikimedia.org/wiki/File:Borrelia_burgdorferi_(CDC-PHIL_-6631)_lores.jpg
Viewed 30 Dec 2021.

[7] http://www.who.int/mediacentre/factsheets/fs104/en/index.html viewed May 23, 2010. Reprinted withpermission from the WHO.

[8] https://commons.wikimedia.org/wiki/File:Mycobacterium_tuberculosis_01.jpg
Viewed 30 Dec 2021.

[9] www.textbookofbacteriology.net/Shigella.html

[10] www.mayoclinic.com/print/Legionnaires-disease

www.en.wikipedia.org/wiki/Legionellosis

[11] https://commons.wikimedia.org/wiki/File:Legionella_Plate_01.png
Viewed 30 Dec 2021

[12] www.interamericainstitute.org/bartonellosis.htm

www.emedicine.medscape.com/article/213169-overview

[13] www.emedicine.medscape.com/article/213169-overview

[14] https://commons.wikimedia.org/wiki/File:Bartonella.jpg
Viewed 30 Dec 2021

[15] www.en.wikipedia.org/wiki/Tularemia

[16] File:Macrophage Infected with Francisella Tularensis Bacteria (5950310835).jpg - Wikimedia CommonsViewed 30 Dec 2021.

[17] https://commons.wikimedia.org/wiki/File:Yersinia_pestis.jpg
Viewed 30 Dec 2021.

[18] https://commons.wikimedia.org/wiki/File:Leptospira_interrogans_strain_RGA_01.png
Viewed 30 Dec 2021.

[19] www.en.wikipedia.org/wiki/Cholera

[20] https://commons.wikimedia.org/wiki/File:Cholera_bacteria_SEM.jpg
Viewed 30 Dec 2021.

[21] https://commons.wikimedia.org/wiki/File:Neisseria_meningitidis_Charles-Orszag_2018.png Viewed 30 Dec 2021.

Chapter 7

Rickettsia are all gram-negative, nonspore-forming obligate intracellular bacteria. They cannot live outside of cells but they get transmitted by chiggers, ticks, fleas, and lice. Many are Rickettsial Diseases, which are cured by antibiotics.

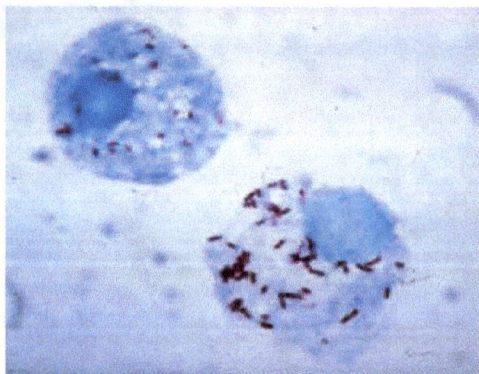

Fig 7-1. Rickettsia rickettsii[1]

1. AFRICAN TICK BITE FEVER (ATBF)[2]

The United States has its Rocky Mountain Spotted Fever, the Middle East has its Mediterranean spotted fever, and Africa has its own

African Tick Bite Fever. This is like Mediterranean spotted fever except for the presentation of skin lesions and the etiologic microorganism. African Tick Bite Fever is usually acquired in wild, rural areas in Africa. People who are hiking or camping are prone to this arthropod borne infection.

In the United States cases of ATBF are usually from returning travelers from Africa.

In May 1998, there were reports in Oregon of nine people who came from a three-week humanitarian construction project in Swaziland in Africa. They developed annular skin lesions with flu-like symptoms. Later it was confirmed that two of them were suffering from African Tick Bite Fever.

CAUSATIVE MICROORGANISM: *Rickettsia africae* (mostly)
Rickettsia conorii (rarely)

Vector

Cattle ticks come from two families: *Amblyomma* and *Rhipicephalus and Amblyomma* stick on humans to feed, while the ***Rhipicephalus*** stays on grasses in the wilderness waiting to bite hikers.

Pathogenesis

Ticks stay in the wilderness and feed on animals for food. These may be wild animals, including wild game, hares, and birds, or domesticated animals like cattle. The causative microorganism stays in these animals. When the ticks bite animals for food, they get the Rickettsial bacteria with the blood. Then these ticks stay in the grasses waiting for the hikers and campers. Once disturbed they stick to humans and bite for food, also introducing their saliva with the bacteria. The microorganisms cause vasculitis leading to epidermal and dermal necrosis. Spread of the virus by the hematogenous route would cause the constitutional symptoms of the patients. *Rickettsia africae* is not so harmful compared to other rickettsiae-caused diseases. The course of African Tick Bite Fever is mild. Rarely, complications or death set in.

Symptoms

It takes five to seven days for the clinical manifestations of the *African tick bite fever* to appear. These are the constitutional symptoms of fever, headache, body malaise, and myalgia, nausea, and vomiting. The site of the tick bite is a large annular skin lesion called *Tache noire* with black crust in the center, and surrounded by a red halo, which also appears concomitantly. These skin lesions may be multiple if there are several bites. Occasionally there is regional lymphadenitis with some exuding lymphangitis. Sometimes there is aphthous stomatitis, arthritis. Later *Maculopapular Rashes* appear, starting from the extremities going centripetally to the trunk. In some cases, this rash is absent.

Diagnosis

Presented with the clinical manifestations, the clinician would inquire if the person has been in contact with ticks, and/or has been in an endemic area a week before. If the person is positive for exposure, the appearance of the rashes and the eschar are strong indications of African Tick Bite Fever.

A Serological laboratory test can be done to confirm the diagnosis, but although this will turn positive in the later part of the disease, it might mislead a clinician during the early part of the disease. So, it is the clinician's discretion based on experience and his/her expertise to consider the diagnosis and to start empirical treatment. These tests are determining the antibodies against Rickettsiae and the detection by immunofluorescence of the microorganisms in the biopsy taken from the eschar.

Treatment

The clinical course of African tick bite fever is milder compared to other Rickettsial diseases. The patients seldom have complications. Some would even improve with simple symptomatic treatment meaning antipyretic for fever, analgesic for headache and bodily malaise, plus increased intake of fluid. The benefits of antibiotic therapy are to shorten the course of the disease and to prevent complications. The drug of choice is Tetracycline (*Doxycycline*), followed by Chloramphenicol and Ciprofloxacin.

Prevention

Just like any other tick-borne disease, the mainstay of prevention is to avoid contact with the ticks. Refrain from going to wilderness and rural areas. If you are camping, hiking, or mountaineering in the wilderness, make sure you have some protective clothing. Have your skin covered as much as possible and spray with insect repellant, not only for ticks but for other insects. *Bayticol*, a product of Bayer, can be sprayed on your clothes. Should you have a tick on your skin, remove it by using tweezers. Grasp it firmly close to the skin to include the head and mouth and then pull it up until the tick lets go. Apply antiseptic solution or even alcohol on the bite site. Never light a match to remove the ticks because it will not work.

2. SCRUB TYPHUS (ORIENTIA TSUTSUGAMUSHI)[3]

Scrub typhus was first described by the Chinese 2000 years ago but it was in 1930 when the causative bacterium was isolated. Initially this bacterium was classified under Rickettsiae but with different cellular structures it was later considered as a different bacterium. This infection is endemic in those regions within the so-called "Tsutsugamushi triangle" which extends from Australia in the south, Pakistan or Afghanistan in the west and Japan in the east. These are the countries of Australia, Japan, Korea, India, Thailand, Vietnam, Laos, Myanmar (Burma), Malaysia, Philippines, Asiatic part of Russia, and China. Cases seen in the US are said to originate from the regions within this triangle.

They are usually travelers, military personnel, and immigrants from these areas.

The term "Scrub" was coined from the geographical area between the forest and the clearings where the vectors were initially isolated. But now it is a misnomer, because the viruses occur in any terrain, be it sandy areas or semi-arid areas.

This disease has renewed interest because of its association with AIDS. There are also reports that there are emerging strains of this bacterium resistant to antibiotics.

At present there are around one billion people exposed to Scrub typhus and one million of these succumbed to death due to this infection.

CAUSATIVE ORGANISM: *Orientia tsutsugamushi*, a gram negative intracellular bacterium with Five major serotypes (Kato, Karp, Gillian, Kawasaki, Boryon)

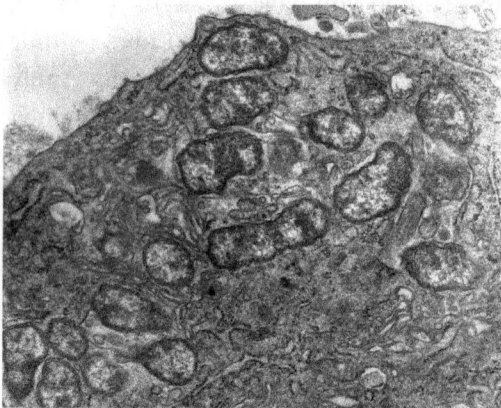

Fig. 7-2. O*rientia tsutsugamushi-* **causative agent of scrub typhus. Photo from CDC.**[4]

VECTOR OR RESERVOIR: *Trombiculid mites (Leptotrombidium deliense)*

Pathogenesis

This infection is transmitted to humans by a bite of an infected chigger, the larval stage of the mite. Once inoculated, they cause an inflammation which later suppurates, ulcerates, and develops eschars at the inoculation sites. This will be followed by regional lymphadenopathies which later become generalized together with hepatosplenomegaly. Just like any other Rickettsial Disease there is *Perivasculitis* of the small vessels. In severe cases patients develop complications, most commonly Pneumonia or Acute Respiratory Distress Syndrome (ARDS), Myocarditis and Disseminated Intravascular Coagulations (DIC). Patients who have been treated with complete medications almost always recover from the disease.

Symptoms

The incubation period is five to ten days. Patients would usually present with high fever of 104-105 Degrees F (100%) with other constitutional symptoms of headache (100%), myalgia (32%), cough (45%), obtundation (reduced mental capacity) (28%), and eschar like cigarette burns (50-100%). Usually patients have localized painful lymphadenopathies (85%) which later become generalized together with hepatosplenomegaly. Some patients may have Central Nervous System symptoms like tremors, nervousness, slurred speech, nuchal rigidity (inability to move the neck muscles) and deafness.

Find

Blood tests would show decreased lymphocytes (*Lymphopenia*) and decreased platelets (thrombocytopenia) and increased liver enzymes (due to liver involvement). There are two confirmatory tests that can be done: the identification of the microorganism by the Polymerase Chain Reaction (PCR) technique and the detection of the antibody titers by the Immunofluorescent Assay . A fourfold increase in titer of antibody is indicative of Scrub Typhus while a onefold increase is a "probable" diagnosis of scrub typhus. In some countries where Immunofluorescent Assay and Polymerase Chain reactions are not available, they rely on the Weil-Felix OX-K tests. This is not so reliable.

Diagnosis

Clinically the most significant finding (not necessarily definitive) pointing to Scrub Typhus is the eschar lesion on the skin (like a cigarette burn). All the rest of the signs and symptoms are non-specific which can be present in other infectious conditions.

These typical skin lesions together with history of travel or staying in an endemic area are enough for a presumptive diagnosis of this infection. The definitive diagnosis, however, will rest on the positive results of the confirmatory tests: the PCR and the Immunofluorescent Assay.

Treatment

Even with a presumptive diagnosis, it is necessary that the patients be started with the medications to prevent complications or mortality. At the same time, all necessary laboratory tests must be done, preferably before the first dose of the medication, so that the results of the tests are not affected by the medications. In the event of negative results from the confirmatory tests, further testing may be done to rule out other similarly presenting infectious conditions. If the patient is improving despite the negative confirmatory test results, the medication must be continued. If the patient is not improving with a negative confirmatory test result, it is at the discretion of the medical practitioner based on his/her expertise as to what medication must be given.

The drug of choice for Scrub Typhus is the Tetracycline brand of Doxycycline. A dose of 1000-2000 mg./day in four divided doses for seven days must be given. In children, tetracycline is contraindicated because of the possibility of teeth discoloration. Instead, the drug Chloramphenicol is the alternate. In case of resistance to both Tetracycline and Chloramphenicol, Rifampicin or Azithromycin can be given. Aside from giving antibiotics, supportive treatment must be given to prevent complications. When patients are seriously ill, it is very necessary to have them hospitalized.

Prevention

There is no vaccine currently for Scrub Typhus. If people such as military personnel are to set foot in a high-risk endemic area, chemoprophylaxis medications of either Tetracycline or Chloramphenicol can be given on a weekly basis to start before exposure and six weeks after exposure.

Aside from these recommendations, precautionary measures must still be observed like wearing protective clothing, shoes, gloves, and covering all your body to prevent the ticks from encountering your skin. One can add repellant sprays on clothing to discourage ticks.

3. MURINE TYPHUS[5]

With climate change making our environment hotter and more humid, insects and rats would infest rotten old buildings, dirty harbors, river lines and other coastal areas. One bacterial infection that is expected to increase is Murine typhus. This is caused by a Rickettsial type in humans and *Rickettsia felis* in cats, dogs and other Peridomestic animals.

Murine typhus is distributed worldwide, chiefly in humid tropical countries, the full year- round. In the United States, Murine typhus dates to the 1930s and 1940s when there were 42,000 cases reported. At present, there are around 80 US cases of Murine typhus, in the states of Texas, California and Hawaii.

CAUSATIVE MICROORGANISM:

Rickettsia typhi, Rickettsia felis.

VECTORS: rat fleas, cat fleas.

Pathogenesis and Symptoms

Murine typhus is transmitted to humans by the bite of infected rat fleas. After the fleas bite and feed, they defecate, shedding the bacteria with their stools at the site of bite. There is a tendency for the person or animal to scratch the bite, thereby allowing the bacteria to enter the body and spread to the bloodstream. Once in the circulation, the bacteria cause rashes which are initially located on the chest and later spread to the sides and back, and later to the arms and legs. They rarely involve the palm, sole or face. These initial symptoms usually appear six to 14 days after the flea bite. Other constitutional symptoms are fever, headache, body malaise, arthralgia, chills and muscle pains, nausea, vomiting and cough.

Mostly, Murine typhus illness is mild. In untreated cases, the infection may last for several months. Only around 10 % of cases have developed into severe form requiring hospitalization with mortality rate of less than 1%. These deaths usually occur in the elderly, debilitated patients, and those with a depressed immune system.

Diagnosis

Initially it is hard to say a person is suffering from *Murine typhus*. The symptoms presented are like other presentations of viral infections. The only probable clue is a patient coming from an area infested with rats. Laboratory procedures can be done. Although of low sensitivity and specificity due to its reaction to some non-Rickettsial antigens, the Weil-Felix Agglutination Reaction is used to diagnose Murine Typhus. The best confirmatory test is the gold standard serological test (Indirect Immunofluorescence Assay). The confirmatory level is a titer of 1:64 or more of Ig M against Rickettsia typhi, or a fourfold or greater increase in Ig G.

Treatment

Just like other Rickettsial diseases, the treatment of choice for Endemic Murine Typhus is Tetracycline. There are several brands of Tetracycline like Sumycin, Doxycycline, Minocycline, and others. These are taken orally or may be given intravenously. Serious cases and older patients usually are treated by the intravenous route. If there are contraindications for Tetracycline, an alternate drug is Chloramphenicol, which is given orally or intravenously. Elderly patients and immuno-compromised patients are usually hospitalized due to the greater chances of increasing severity and complications.

Prevention

The first line in prevention is to eradicate the fleas in your house. There are several commercial products that can be used to control the fleas. If you have pet cats or dogs, make sure they are clean and free from fleas.

Eliminating fleas is not enough. The household surroundings must always be kept clean to deter rats, opossums, and stray cats and dogs from staying. Leftover foods and left- over dog or cat foods must not be left open, for these will attract other animals. Throw the trash away properly, keep grass mowed, and keep firewood on the ground.

4. EPIDEMIC TYPHUS

Epidemic typhus is a prototype of all the typhus diseases. Its disease course has similarities with other typhus diseases. It causes the most severe clinical symptoms among Rickettsial diseases. That is why when someone says, "typhus fever," he is referring to the epidemic typhus.

This is one of the oldest diseases known to humanity. In the fifteenth century it was reported as a louse- borne epidemic. The number of deaths in the Spanish army due to epidemic typhus was (17,000 soldiers) more than the casualties of war (3,000 soldiers). With the remaining Spanish soldiers fleeing from the war returning to Spain and being reassigned to other parts of the world, they helped spread the disease worldwide. In the 17th century, during the Thirty Years War (1618-1648), typhus, together with the plague and starvation, were the cause of ten million deaths; far more than the military casualties which were 350,000. In fact, an impending battle in Nuremburg, Germany was prevented from happening due to typhus killing 18,000 soldiers from opposite camps.

In the twentieth century, Charles Nicolle (1866-1936) was the first person to perform experiments on this disease. His discovery that body lice were responsible for the transmission of the disease eliminated the mystery of typhus fever in the medical community. He was awarded a Nobel Prize in 1928 in medicine. Previously, in 1916, Henrique de Rocha Lima had proved that a bacterium caused Epidemic typhus. He named it **Rickettsia prowazekii** in honor of two zoologists, H.T. Ricketts, and Stanislaus von Prowazek. The latter died in 1915 while investigating a disease in a cell.

In the Second World War, although Typhus fever did not cause havoc, its prevention was a major concern of the armies. The invention of a chick-embryo vaccine by Cox along with the applications of **Lousicides** (kills lice), DDT and MYL, significantly reduced the incidence of the disease.

This disease is associated with wars because soldiers in wars are forced to live in unsanitary conditions with infrequent bathing and changing of clothes and forced to live in crowded conditions in inadequate housing, with greater personal contact that favors the spread of lice. Epidemic typhus is also associated with famine, poverty, and malnutrition. Due to these associated conditions, Epidemic typhus is called by lot of names: War Fever or *Frebris militaris*, Camp fever, Famine fever, Gaol-fever, Jail fever, Ship fever, Sharp fever, and Putrid fever, just to name a few.

At present, there are sporadic cases of Epidemic typhus in Central and South America, Africa, and northern China and the mortality rate is approximately three to four percent, common in elderly and immuno-compromised patients. In the United States, Epidemic typhus was historically unimportant. The last reported epidemic of Epidemic typhus was in 1883 in New York, brought by immigrants from Europe.

Aside from the small outbreaks among the Navajo Indians in 1915 and 1921, the United States has been free from Epidemic typhus.

CAUSATIVE MICROORGANISM: *Rickettsia prowazekii*

VECTOR: *Pediculus humanus corporis,* **a human body louse**
Pediculus capitis, Phthirus pubis

Pathogenesis

The causative microorganisms stay in the gastrointestinal tracts of the louse. Once the louse bite humans for a meal, it defecates in the site of the bite, shedding off the microorganisms. The bite causes a pruritic (itching) reaction enough for the human to scratch the bite site, crushing the louse with its excrement entering the host's skin, serving as a portal of entry for the causative microorganisms.

The microorganisms travel via the bloodstream to spread throughout the body. They are stacked up in the endothelial linings of the smaller blood vessels. They proliferate on these cells causing swelling of the cells leading to *Multiple organ vasculitis*. These Vasculitis can trap leukocytes, platelets, and macrophages creating a thrombosis that could block blood flow, leading to gangrene of parts distal to the blocked vessel. Commonly affected parts are the tips of extremities, nose, ear lobes and genitalia. This vasculitis can also cause perforations in the vascular walls, leading to loss of osmotic proteins in the blood, causing hypovolemia and organ failure.

A milder form of Epidemic typhus can reemerge after several months or years in remission. This is called Brill-Zinsser disease. This usually happens following incomplete antibiotic therapy or malnutrition and in immunocompromised or elderly people. The reason *Rickettsia prowazekii* would linger inside the human body for a long time to re-emerge later is not understood.

Symptoms

Corresponding to the Rickettsemia (spread of the *Rickettsia prowazekii* in the blood), the patient develops fever of abrupt onset, followed by non-remittent headache, myalgia, or body malaise. After five to seven days, the patient develops petechial rashes that start on the trunk and axilla and spreads later to the extremities. CNS involvement may ensue, ranging from mild confusion and mental dullness to a serious state of stupor or coma. Serious conditions may include gangrene in distal body parts.

Diagnosis

Clinical diagnosis of Epidemic typhus may be favored if the patient has the history of the following: coming from an endemic area, military personnel from war, a person coming from a crowded population or living in filthy environments.

Laboratory diagnosis can be attained by serological examinations like complement-fixation reactions, serum agglutination tests and the Weil-Felix test.

Treatment

The core of treatment involves using antibiotics and the drug of choice is Tetracycline. An alternate drug that can be given is Chloramphenicol, orally or intravenously. Patients with severe conditions are usually hospitalized and treated in the intensive care unit.

Prevention

The main goal of prevention is sanitation. There is no better way than simple sanitation to prevent Epidemic typhus and all infections. Another factor is the improvement of living standards of people in endemic areas. Here is the government's role in the prevention of the disease. The government must have preventive programs. Incorporated here is the health education of the people and the use of Lousicides (DDT, MYL) in the eradication of the vectors. There is no commercial vaccine for Epidemic typhus.

5. ROCKY MOUNTAIN SPOTTED FEVER[6]

Among the Rickettsial diseases, Rocky Mountain Spotted Fever is the most common and the most serious. Despite medical advancements, Rocky Mountain Spotted Fever is still a life- threatening infection with a present mortality of three to five percent. Before the advent of Tetracycline and Chloramphenicol, the mortality rate was as high as 30%. At present there are 250-1,200 cases of Rocky Mountain Spotted Fever reported every year.

In 1896, a condition characterized by black rashes was noted in the Snake River Valley in Idaho. This was later labeled as "black measles" to differentiate it from ordinary measles. It eventually spread to the continental USA. In the early 19th century, Dr. Howard Ricketts did an extensive study on this condition. He was the first person who identified the causative organism that was a bacterium. He was also the first person who identified ticks as the vectors of the causative bacterium. In 1910, after his extensive work on this type of disease, he died of a certain disease later identified as typhus, which is another form of Rickettsial disease. Because of his work, this disease organism was named for him.

The phrase "Rocky Mountain" is a misnomer because at present it is not limited to the Rocky Mountain region. Although the disease was originally established in the Rocky Mountain region, it is now widespread to the whole continental United States. In fact, there are more cases of this infection in the South-Atlantic Region of US than the Rocky Mountain area.

This disease has now spread to other countries. There are different names, however. In Brazil it is called *Febre maculosa* or *Sao Paulo fever,* in Mexico it is called *Fiebre manchada;"* in Colombia they call it *Tobia fever*. In other countries, it is simply called *Tick typhus.*

CAUSATIVE ORGANISM: A bacterium named *Rickettsia rickettsii*

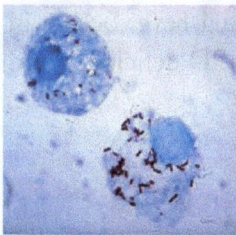

Fig. 7-3. *Rickettsia rickettsii*. Photo by CDC.(See endnote 1 on page 316)

VECTOR AND NATURAL HOST: American dog tick (*Dermacentor variabilis*), Rocky Mountain wood tick (*Dermacentor andersoni*), Brown dog tick (***Rhipicephalus sanguineus***),Other ticks (*Amboyna cajannense)*

Pathogenesis

Rocky Mountain spotted fever is a disease of animals that can be transferred to humans (zoonosis). The ticks in this condition serve as the vector as well as the natural host or reservoir. Humans are just an accidental host. Humans are not a part of the natural cycle of the tick. The ticks acquire the microorganisms by biting infected animals or humans. The ingested virus stays in the tick at any stage of its life cycle (egg, larva, nymph, adult). The bacterium can be transferred by a male tick to a female tick through the fluid coming from the male or from the sperm during mating. The female tick can also transfer it to its offspring by the so called transovarial transmission route (transferred from the mother to eggs).

In humans, the bacterium primarily infects the mucosal linings of the small and medium sized blood vessels. Once inside the mucosal cells, either in the cytoplasm or nucleus, they multiple exponentially until the host cells are damaged and ruptured, disseminating the young bacteria to invade other cells. This goes on and on. Since the damaged cells are lining mucosa, blood leaks from the blood vessels to the surrounding tissues, resulting in the *Characteristic black rashes*.

Blood vessels in body organs are affected too, resulting in organ damage or organ failure. Usually the involvement is systemic, affecting multiple organs requiring hospitalization. Studies show that when severe, it could prove fatal. Rocky Mountain Spotted Fever occurs in patients who are in advanced age, patients with G6PD deficiency and those with chronic alcohol abuse.

Symptoms

The incubation period of this disease is five to ten days. The initial symptoms are nonspecific: headache, bodily malaise, nausea, vomiting, fever, muscle pain and anorexia.

A medical practitioner cannot categorically say that a patient is suffering from the disease based on these initial symptoms because they are common in all infectious conditions. The only clue that might favor Rocky Mountain spotted fever is the history of a tick bite.

Two to five days after the appearance of fever, the initial rashes appear. These rashes are pink macules, non-pruritic, that usually appear first on the wrists, forearms, and ankles. When pressed, the pinkish macules turn pale. After several days, these skin lesions become prominent and appear more as petechiae rather than a rash and are now distributed throughout the body including palms and soles. Not all patients, however, have petechiae. Only about 35-60 % have the typical petechiae. In severe cases, patients have multiple body organs which are affected, mostly those of the respiratory system, central nervous system, gastrointestinal system, and urinary system.

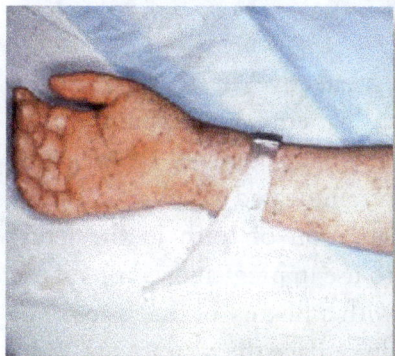

Fig. 7-4. Rocky mountain spotted fever-CDC Laboratory [7]

Patients with Rocky Mountain spotted fever would usually have low platelet count (thrombocytopenia), low serum sodium (hyponatremia), and elevated liver enzymes. The standard confirmatory test is the detection of Ig M, IgG by Immunofluorescent Antibody test (IFA). This is the confirmatory test currently used by the CDC. Other useful serologic tests are ELISA, latex agglutination, and dot immunoassay. Note, however, that these serologic tests are for all Rickettsial diseases. It is a non-specific test for Rocky Mountain Spotted Fever. The test that is specific for the *Rickettsia rickettsii* is the Polymerase Chain reaction (PCR). However, it takes several weeks to get the results of the PCR.

Diagnosis

With very non-specific initial symptoms, it is hard for medical practitioners to diagnose Rocky Mountain Spotted Fever. It is based on the doctor's own discretion, based on clinical experience and expertise. Usually the presence of the so- called *Triads* of tick bite, fever and Petechiae are enough to start antibiotic medications and to request routine laboratory tests. If still needed, the physician may do serologic tests to confirm the diagnosis.

Treatment

Treatment with antibiotics must be started right away, even with mere suspicion of Rocky Mountain Spotted Fever. Empiric treatment must be started at the same time as laboratory tests. At least the patient has ongoing medication while waiting for laboratory results. These results are there for confirmation only.

Tetracycline proves to be highly effective in treating Rocky Mountain Spotted Fever. In fact, a patient not responding to Tetracycline argues against a diagnosis of this disease. In pregnant patients, however, Tetracycline is contraindicated because it causes inhibition of bone growth and discoloration of teeth in the fetus. Chloramphenicol is an alternate drug to use in pregnancy. Prophylaxis treatment for persons who are not ill with a history of tick bite is not recommended because the drug may camouflage the appearance of the symptoms and may just delay the diagnosis of the disease. Seriously ill patients with multiple organ involvement must be hospitalized.

Prevention

Since there is no current vaccine for Rocky Mountain Spotted Fever, the core prevention is avoidance of contact with ticks. Below are recommended criteria. They are not solely applicable to Rocky Mountain Spotted Fever but to all diseases transferred by the tick, such as Lyme disease.

-If going to a known tick-infested area, wear light clothes to spot a crawling tick easily.

-As much as possible, no parts of the body should be exposed to the ticks. Tuck your shirt into your pants, your pants into your socks, use gloves, and wear head protectors.

-Spray repellant (phentermine) onto your boots and clothing to prevent the ticks from attaching.

-For the skin, apply DEET, repeating every few hours. Be cautious in applying the DEET to children because it can cause adverse reactions, especially if applied on a wide area of the body.

-Have a complete body check for ticks after coming from a potentially tick-infested area. Use a mirror to examine the whole body.

-If you have pets with you, examine the pet's hair because ticks often stay on their hair.

The Appropriate Way to Remove Ticks

-Never remove ticks with a bare hand. Use tweezers or a notched tick extractor.

-To make sure all the parts of the ticks are removed, grasp the tick close to the skin and pull it steadily. Never attempt to bend the tick side to side as it will just break the mouth parts and leave a part in the skin which may still cause skin reactions.

-Never squeeze the tick because fluid containing microorganisms can be left on the skin.

-Wash the hands and apply disinfectant (H2O2, Alcohol, Betadine, or detergents) after removing ticks.

- For confirmation, the extracted ticks are placed in a plastic bag, preserved, and sent to the laboratory for identification.

6. MEDITERRANEAN SPOTTED FEVER OR BOUTONNEUSE FEVER (MSF)

A milder form of Rickettsial disease is Mediterranean spotted fever or Boutonneuse Fever, so named because it is the most common

Rickettsial disease in the Mediterranean, especially on the Italian island of Sicily. In other areas of the world, they use the name of the country (Israel spotted fever in Israel, Indian tick typhus in India, Kenya tick typhus in Kenya). It is a mild form of Rickettsial infection which primarily occurs in children. In six to ten percent of cases severe complications set in; those usually happen in the elderly and immunocompromised patients. This disease is found to be more prevalent in males than females, and this may be due to occupational reasons where males have more contact with dogs and other *Peridomestic animals*.

MSF was first recognized in 1910 by Conor and Bruch in Tunisia, but the discovery of the dog tick as the carrier came into being 20 years later. Initially it was considered a mild form of infection that failed to get attention in the medical community. In 1980, however, there was a severe case of Mediterranean spotted fever reported in France that resulted in death. Since then, MSF spread out to other countries. This disease is prevalent in southern Europe, Africa, and central Asia. In 2006, there was a study done in Sicily in which 415 children were diagnosed to have MSF. It was found that there was an average incidence of ninety-eight cases in 1999 and 28 cases in 2004.

MSF is rare in the United States. There have been only fifty confirmed cases (by CDC) of MSF and all of these were imported.

CAUSATIVE MICROORGANISM: *Rickettsia conorii*

VECTOR: Brown dog tick (*Rhipicephalus sanguineus*)

Pathogenesis

The microorganisms stay in the gastrointestinal tract of dog ticks and are transmitted to humans by tick bites. For some reason, these ticks love to stick with dogs and other Peridomestic animals but have less affinity to humans. Once human skin is bitten, the microorganisms enter the body and then later are disseminated via the hematogenous route. In the lining endothelium of the small blood vessels, the bacteria enter the endothelial cells and injure the cells, leading to cell necrosis. This would explain the appearance of the Tache noire or eschars at the bite site.

Thus, with further spread the bacteria eventually invade blood vessels of different body organs.

The bacteria invade the endothelial cells leading to the injury of the blood vessels that initiate the subsequent injury to whole organs. This can cause complications such as renal failure, splenic rupture, and eye involvement. Patients with poor immunity, such as patients who are immunocompromised, the elderly, and those with steroid therapy, are prone to complications.

Symptoms

The incubation period of MSF is four to seven days after the tick bite which is often unnoticed. The initial presentations are flu-like symptoms of fever, body malaise, headache, myalgia, arthralgia, nausea, vomiting and erythematous rashes. These appear together with the Tache noire at the bite site. The fever is high grade. Severe cases can manifest as nervous system involvement (stupor, coma), gastrointestinal involvement (jaundice, hematochezia, hematemesis), and blood involvement like jaundice and hematomas with spontaneous bleeding in other parts of the body.

Diagnosis

With initial constitutional symptoms presented by a patient it is hard to be certain of a diagnosis for MSF. However, items in the patient's history that would put a clinician in a confident situation are history of tick bites, history of contact with dogs, history of living in endemic areas. The triads of high-grade fever, purpura, and the appearance of the Tache noire (eschar) are most characteristic of the MSF.

There are serologic assays that can be done but most of the *Rickettsial Diseases* would turn these tests positive. They cannot identify the specific rickettsiae. The indirect fluorescent antibody test, however, can be used to confirm the diagnosis of MSF. Isolation of the microorganism from the skin lesion biopsies can be done. Other tests that may be of help are the ELISA test to detect antibodies for **R. conorii**. For CNS involvement, MRI can show multifocal disturbances in the white matter.

Treatment

Antibiotic therapy is the mainstay in the treatment of MSF. For mild cases, antibiotics are given for a week while more severe cases are treated for two weeks. The drugs of choice as in other Rickettsial Diseases are Tetracycline (Doxycycline), Chloramphenicol, and Quinolones (Ciprofloxacin). The Doxycycline is given at a dose of 100 mg every 12 hours for seven days.

Chloramphenicol at 50 mg/kg/day orally every 6 hours for four doses. Ciprofloxacin at 750 mg/kg/day every 12 hours.

Prevention

The best preventive measure is to avoid contact with dog ticks. If you are exposed to this vector due to your work or your place of residence, protective clothing must always be worn. The clothing is preferably applied with *Permethrin* and *Pyrethroids*. The skin that is exposed must be sprayed or otherwise covered with insect repellant. Be observant of any itchiness or any crawling insect in your clothes or body once you come from endemic areas. These ticks can be removed manually.

7. EHRLICHIOSIS / ANAPLASMOSIS

Ehrlichiosis is a tick-borne bacterial disease that affects humans and animals. The causative microorganisms are two distinct but related bacteria under the Rickettsial Family namely: *Ehrlichia* and *Anaplasma*. The *Ehrlichia Bacteria* affect the monocytes thereby calling it *Human monocytic ehrlichiosis* (HME), while the *Anaplasma bacteria* affect the granulocytes thereby calling it *Human granulocytic anaplasmosis* (HGA).

Ehrlichiosis may be compared to Lyme disease because the two diseases have the same vector and geographical range. The clinical symptoms are the same in both conditions, except for the rashes which are less common in Ehrlichiosis, and the initial *Ecthyma migrans*, which is absent in Ehrlichiosis. Treatments are different. While Lyme disease is treated by Amoxycillin, Ehrlichiosis is treated by Tetracycline.

The first reported incidence of Ehrlichiosis was in Japan in 1954. In the US, the first documented case was in 1986. In 1991, one military recruit in Fort Chaffee, Arkansas, was reported to have Ehrlichiosis. In 1994, 12 cases were reported in Minnesota and Wisconsin. In 1999, four cases were reported in Missouri.

CAUSATIVE MICROORGANISM:

Ehrlichia chaffeensis – causes Ehrlichiosis

Ehrlichia ewingii – caused the Ehrlichiosis in the four cases in Missouri.

Anaplasma phagocytophilium – causes Anaplasmosis (Both bacteria are members of the Rickettsial family. They are gram negative obligate intracellular coccobacilli.)

VECTORS:

Amblyomma americanum (lone star tick) See photo 4-2 Ixodes scapularis (black-legged tick)

Ixodes pacificus (western black-legged tick) in California

Ixodes ricinus in Europe

Ixodes persulcatus

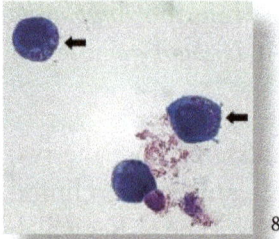

Fig 7-5 *Anaplasma phagocytophilium*

RESERVOIRS OR COMMON NATURAL HOSTS: humans, deer, rodents, cats, and dogs

Pathogenesis

The presence of the causative microorganism in the animals and ticks is a part of the natural cycle. The animals serve as reservoirs while the ticks serve as vectors. The ticks sticking on the skin of animals acquire the microorganisms by taking a blood meal from the animals. Contact of these animals with other animals or humans would serve as ways for the ticks to transfer to other reservoirs. In the secondary reservoirs, when they suck a blood meal for a second time, they transfer the microorganism to the secondary reservoirs in return. These secondary reservoirs may be humans.

Once the microorganisms enter human blood, they are engulfed by the white blood cells (WBCs). For whatever reason, ***Ehrlichia chaffeensis*** (causing Erhlichiosis) are taken by the macrophages while ***Anaplasma phagocytophilium*** (causing Anaplasmosis) are taken by the granulocytes. They are eventually phagocotyzed by the leukocytes forming "morulae" (distinctive membrane-bound intracytoplasmic aggregates). The presence of morulae in the body organs (the liver, lymph nodes, spleen) causes focal necrosis and eventually suppression or deregulation of the immune system. Those organisms trapped in the lungs can cause *Interstitial Pneumonitis* and pulmonary hemorrhage. They may be trapped in the bone marrow causing granulomas as seen in a bone marrow biopsy. Other organs may be invaded causing *Perivascular lymphohistiocytic* infiltrates in these organs. Other patients may have complications, such as renal failure, respiratory failure, intravascular coagulopathy, meningitis, encephalopathies, seizures, and coma. These complicated patients require hospitalization. Mortality rate is two to three percent in HME while in HGA it is 1%.

Symptoms

It does not always follow that once the microorganisms enter the body, symptoms would appear. Infected persons with competent immunity may be asymptomatic (show no symptoms). Some may have mild symptoms. Usually, the symptoms are non-specific like those with other infectious or non-infectious diseases. These are fever (98 Degrees F and above), headache, fatigue, and myalgia. Others may have nausea and vomiting, diarrhea, cough, joint pains, confusion. Skin rashes are more common in Ehrlichiosis than in Anaplasmosis. Those people who are immunocompromised may have severe cases due to the further suppression of immunity in these two conditions (12% of patients in HGE and 6% of patients in HMA). Some of the signs and symptoms of severe Ehrlichiosis and Anaplasmosis are Hepatomegaly (enlarged liver) in liver involvement, nuchal rigidity in Meningitis, Pharyngeal and Tonsillar exudates, Genital, and oral ulcers.

Diagnosis

Since these diseases have non-specific signs and symptoms, they have the potential to become severe and fatal. It is especially important for clinicians to differentiate these conditions from milder diseases. To quote Dr. E. Dale Everett, "The problem is on the diagnosis of ehrlichiosis but it is more on the unawareness of some physicians of these conditions and their failure to recognize the clinical syndrome."

These are some helpful diagnostic tests:

1. Microscopic evidence of the morulae in the cytoplasm of the leukocytes. This is done by blood smears stained with

eosin-azure dyes, such as *Wright-Geimsa stain*. This is positive in 20% of patients with HME and 20-80% of patients with HGA.

2. Positive polymerase chain reaction (PCR) assay or positive serologic tests demonstrating a four-fold or greater change in antibody titer in the blood.

3. Immunofluorescence antibody (IFA) titer or higher together with the presence of morulae.

4. Other non-specific laboratory parameters are the following: abnormal liver enzymes, leukopenia, thrombocytopenia, hyponatremia, elevated C-reactive protein in the first week of illness that is resolved by the end of the second week.

Treatment

With fatal complications, initial treatment is critical. In those persons with a history of being bitten by a tick, Ehrlichiosis must be considered. Medication must be started right away. The drug of choice is Doxycycline for Ehrlichiosis and Anaplasmosis. Since tetracycline is contraindicated in patients less than 8 years old, an alternative drug, Chloramphenicol can be given in this age group. For those patients who are pregnant or have allergic reactions to tetracycline, rifampicin can be given. Complicated patients must be hospitalized with intensive care therapy.

Prevention

1. Refrain from going to tick-infested areas. If it is necessary to be in these infested areas, wear light-colored clothes so that you can easily see the ticks crawling on them. Tuck your shirt into your pants and tuck your pant legs into your socks so that the ticks cannot contact your skin. Apply repellants containing permethrin on your clothing and boots to discourage tick attachment.

2. For exposed skin, apply a repellant containing DEET (n,n-diethyl-meta-toluamide). Caution is necessary for children to avoid adverse reactions.

3. Upon exposure, search your body for attached ticks, especially in your hair. You may use mirrors to see all the parts of your body.

4. If a tick is found, use a tweezer, or use a tissue, paper towel or rubber gloves to shield your fingers and remove the tick. The goal is to remove the point of attachment (Mouthparts of the tick) without destroying the tick by pulling the head of the tick slowly and steadily. It is not recommended to smother the tick with alcohol, petroleum, jelly, twisting or rubbing the tick off because these procedures will destroy the tick, releasing its body fluid (saliva and gut contents) with the microorganism onto the surface of the skin of the person.

5. Once the tick is removed, use any disinfectant to wipe the bite site, and clean hands with soap and water. It is also recommended to preserve the tick by placing it in a small plastic bag and storing it in the refrigerator so that after several weeks of the disease, these ticks can aid in making a correct diagnosis.

Finally, public health must have some form of management strategies to reduce the incidence of these tick-borne infections. This can be done by applications of Acaricides (chemicals that destroy ticks and mites) in those infested areas.

[1] https://en.wikipedia.org/wiki/Rickettsia Viewed 12 April 2021.

[2] www.cdc.gov/mmwr/preview/mmurhtm/00055615.htm

[3] www.emedicine.com/PED/topic2710.htm

[4] https://en.wikipedia.org/wiki/File:Orientia_tsutsugamushi.JPG Viewed 11 Jan 2022.

[5] www.cdph.ca.gov/HealthInfo/discond/Documents

[6] www.en.wikipedia.org/wiki/Rocky_Mountain_Spotted

[7] https://upload.wikimedia.org/wikipedia/commons/a/a1/Rocky_mountain_spotted_fever.jpg
Viewed 6 Jan 2022

[8] https://commons.wikimedia.org/wiki/File:Anaplasma_phagocytophilum_cultured_in_human_promyelocytic_cell_li ne_HL-60.jpg Viewed 24 Feb 2022

Chapter 8

VIRAL INFECTIONS

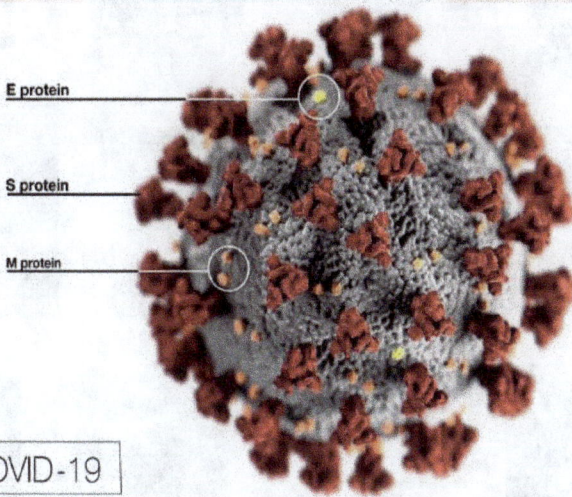

Fig 8-1 Covid-19 Virus[1]

Viruses are infectious agents which are composed of either RNA or DNA and surrounded by a protein coat called a capsomere. They do not meet any criteria of the characteristics of life except reproduction where they can only reproduce inside living cells. They have been implicated infecting all life forms including animals, bacteria, fungi, plants, humans, bacteria, and even other viruses called virophages.

The recent pandemic of the COVID-19 Virus has added a burden of pollution as seen on city streets and parking lots. These are masks left as garbage on the streets. People just scatter the masks all over the place. All photographs by David Arieti.

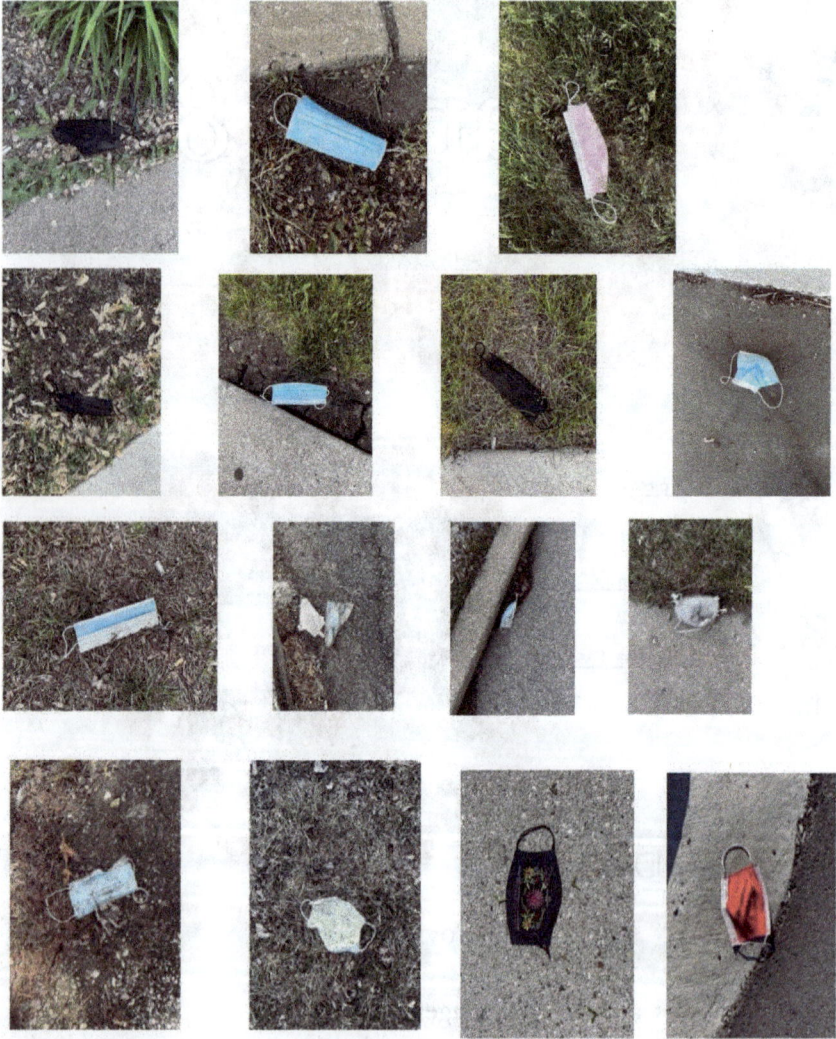

Fig 8-2 Masks found on the streets in Skokie and Chicago, Illinois

1. VIRAL HEMORRHAGIC FEVER

Viral Hemorrhagic Fever is a group of febrile illnesses caused by several families of viruses. These viruses are enveloped and have RNA genomes. Here enumerated are the four types of viruses and the corresponding names of Hemorrhagic fevers they cause.

1. **Filovirus** - Ebola virus (Ebola Hemorrhagic Fever)
   Marburg virus (Marburg Hemorrhagic Fever)

2. **Arenaviurus** – Lassa virus (Lassa Fever)
   Junin virus (Argentine Fever)
   Machupo virus (Bolivian Hemorrhagic Fever)
   Guanarito virus (Venezuelan Hemorrhagic Fever)
   Sabia virus (Brazilian Hemorrhagic Fever)
   Lymphocytic choriomeningitic virus
   Flexal virus

3. **Bunyavirus** – Rift Valley Fever
   Hemorrhagic Fever with Renal Syndrome
   Crimean Congo Hemorrhagic Fever

4. **Flavivirus** – Dengue Hemorrhagic Fever

A. EBOLA HEMORRHAGIC FEVER[2]

In 1976, there was an outbreak of a febrile disease in the Republic of Congo (formerly Zaire) and Sudan. The infection was characterized by fever, headache, myalgia followed by blood diathesis resulting in death. The *Zaire ebolavirus*, more commonly known as *Ebola virus*, outbreak resulted in a 90% fatality rate while the outbreak in Sudan resulted in 50% fatalities.

The Zaire outbreak started on August 26, 1976 in Yambuku, Zaire, when a schoolteacher, Mabalo Lokela, was reported to have high fever, muscle pain, and headache. She was treated as a case of Malaria. After several days, however, she developed blood diathesis (bleeding easy) in the form of bloody diarrhea, hematemesis, rashes, and eventually spontaneous bleeding from her body orifices like nose, eyes, mouth, and anus, leading to death. Soon after, several other people came in with the same signs and symptoms and succumbed like Lokela to death from bleeding, causing the outbreak. There were even two nurses working in the same hospital who became ill with the same signs and symptoms.

The Sudan outbreak started in the same year. It was reported that a cotton factory worker had died with the same symptoms. This was followed by a nightclub owner who presented with the same symptoms. More patients were observed suffering from the same illness leading to the outbreak. Later, doctors were able to identify the Ebola virus as the cause of the infection.

In 1989, in Reston, Virginia, a similar presentation was seen in several rhesus monkeys. These infected monkeys turned out to be positive for the Ebola virus. What made this interesting was that it only affected the rhesus monkeys, not the humans. If you have read the novel *The Hot Zone* by Richard Preston, this was based on the Ebola virus found in the rhesus monkey in Reston, Virginia.

In November 1994, a subtype of Ebola virus was isolated from two dead chimpanzees in Côte d'Ivoire Ivory Coast). The last documented outbreak was on November 24, 2007 in the Bundibugyo district of Uganda. Another strain of Ebola virus was isolated. This new strain caused 149 cases, thirty-seven of which died.

CAUSATIVE ORGANISM: *Ebola virus of Family Filoviridae with several strains as follows:*

Zaire Ebola virus
Reston Ebola virus
Sudan Ebola virus
Ivory Coast Ebola virus
Bundibugyo *Ebola* virus

The name "Ebola" was coined after the Ebola River Valley in

The Democratic Republic of the Congo that was close to the site of the first outbreak.

Natural Reservoir / Vector: Unknown

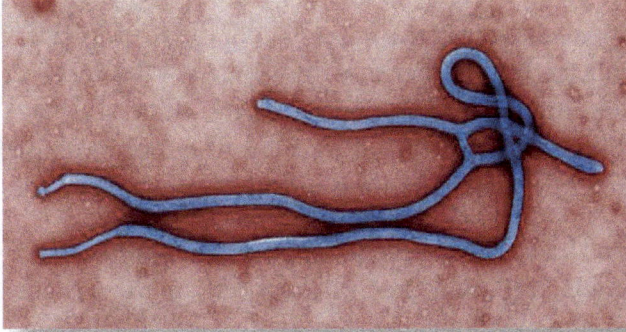

Fig. 8-3. Ebola virus-TEM. Photo by Cynthia Goldsmith (CDC).[3]

Pathogenesis

Until now, investigators have not discovered the natural reservoir of Ebola virus. Several studies are being done in the Republic of Congo and Gabon to identify the reservoir. The researchers believe that somewhere in the rain forest of Africa or Western Pacific resides the reservoir of this virus.

It is believed that these viruses quietly reside in the forest until they are disturbed by animals like gorillas, chimpanzees, and rodents, thereby infecting them in return. Humans acquire the Ebola virus by direct contact with these infected animals. Documents showed that humans got the infections from handling infected gorillas, chimpanzees, or forest antelope, whether dead or alive. Human to human transmissions are also through direct contact with infected human body fluids like blood, saliva, sputum, urine, and feces. Airborne transmission is hardly possible in humans.

Once inside the body, the viruses attach to the G protein on the cell membrane of the cells and eventually enter the cytoplasm where the viral membrane fuses with the vesicular membrane, releasing the nucleoside of the virus that will alter the normal translation of protein in the cytoplasm. More viral proteins are synthesized until they bud out of the host cells, damaging the cells and releasing the virus. This stage is manifested by the constitutional symptoms like fever, myalgia, headache, anorexia. If left untreated, the virus spreads out to other cells of the body.

With the presence of the platelets in the blood, the virus interacts with these, producing a chemical that cuts the capillary walls, forming holes. These result in internal bleeding in the organs of the body and external bleeding like the petechiae and purpura. If severe, multiple organ failure may result, or worse, Hemorrhagic Hypovolemic Shock.

Sporadic outbreaks usually occur in poor, isolated areas where the basic hygiene and sanitation are still luxuries. In these places needle sharing is common without adequate sterilization. There are no basic barrier devices like gloves, masks and aprons, and no autoclave machine.

Ebola virus is less likely to cause an epidemic of substantial proportion because first of all it is not transmitted airborne; secondly, it has a short incubation period. The victim would show right away the symptoms that could easily identify the illness, thereby limiting his/her ability to transmit the disease to others. Thirdly, Ebola is transmitted by direct contact, which is easily prevented by isolating the victim and the use of all personal protective equipment in managing the victim.

Symptoms

It is hard for a medical practitioner to diagnose Ebola virus infection during the initial part of the infection. The presenting symptoms of a patient infected with an Ebola virus is like Malaria, Typhoid Fever, Influenza. These symptoms have a high fever of 101.8 F, severe headache, muscle pain, joint pains, abdominal pain, myalgia, and nausea and vomiting. The time to suspect a Hemorrhagic Fever is when the patients start to show bleeding signs like bloody diarrhea, hematemesis, reddish eye, maculo-papular rashes, and petechiae.

If severe, patients may show signs of hypovolemia like low blood pressure, tachycardia, and pallor. It may lead to disseminated systemic necrosis or disseminated intravascular coagulation in several body organs, resulting in Multiple Organ Failure.

Less than 50 % of cases, however, have only the constitutional symptoms and do not develop hemorrhage.

Diagnosis

Making a definitive diagnosis is difficult, presented with initial symptoms of fever, headache, myalgia, and body malaise because these are the same symptoms common in many other infectious conditions. With these symptoms, a medical practitioner would arrive at a diagnosis of Malaria or Typhoid fever which is more common. However, after several days with the appearance of bleeding signs, one is certain to be dealing with a hemorrhagic fever.

Routine laboratory tests like CBC are of supportive value in the diagnosis, except for the platelet count which shows Thrombocytopenia, favoring the diagnosis of Hemorrhagic fever. There are highly complex tests on blood, urine or saliva specimens that confirm the diagnosis. These are the isolation of the antigen by Elisa Testing, PCR Testing; or determining the Immunoglobulins: M, Ig G antibodies against the Ebola virus. The virus itself may be isolated.

Treatment

The mainstay of treatment for Ebola Hemorrhagic Fever is supportive therapy. There is no standard treatment. Symptomatic treatment is given for fever, pain, and headache. It is important to make sure that the patient has balanced fluid and electrolytes in the body.

If bleeding appears, whole blood transfusion or platelet transfusion of clotting factors is needed, depending on the deficient elements in the blood. If secondary bacterial infections ensue, appropriate antibiotic treatment must be started.

In animals, some experiments using hyper-immune sera were performed, but there was no protection demonstrated against the disease.

Prevention

With unknown identity and location of the reservoir, there are only a few established preventive guidelines. In Africa where endemic areas are common, this disease is a challenge for them to establish preventative measures.

For health care workers, it is especially important for them to be well versed with the pathogenesis, symptoms of the disease, so that they can easily detect and diagnose patients suffering from this type of hemorrhagic fever and later isolate them to prevent spreading the infection. For dead patients, it is necessary that no one have direct contact with the corpse.

Several vaccines have been tried in humans but no success has been reported. Although scientists have developed a vaccine with 99% effectiveness against Ebola and Marburg virus, the vaccine was only for monkeys, not for humans.

B. MARBURG HEMORRHAGIC FEVER[4]

This is a rare but severe form of Hemorrhagic Fever caused by a virus like Ebola virus. In 1967, in the towns of Marburg, Frankfurt and Mainz in Germany and in Belgrade, Yugoslavia, there were reports of thirty-one laboratory workers who became ill with hemorrhagic fever. Seven of them eventually died. After a few weeks, six members of the medical staff who treated the primary infections developed the same illness. This second group was composed of two doctors, a nurse, a postmortem attendant, the wife of a veterinarian, and one additional person. So, this second group acquired the infection through contact with the first group (31 laboratory workers). They were able to source out that the monkey imported from Uganda for the development of polio virus was the carrier of the virus.

In 1975, a man returning from Zimbabwe infected three people in Johannesburg, South Africa, resulting in one death. Similar cases were reported in succeeding years. In 1980 and 1987, European visitors who visited Kitum Cave in Kenya developed the disease and later died. From 1998 to 2000, there were reports in the Republic of Congo of 149 cases with 123 fatalities. The outbreak was caused by miners in Congo.

In 2005, in Angola, three hundred fatal cases were reported and about one hundred of them died. In that outbreak, fourteen nurses and two doctors were also infected. This 2005 Angola outbreak caught much attention that scientists tried to source out for the reservoir of Marburg virus. Teams of scientists went to forest areas of Gabon and Uganda. In 2007, their efforts paid off. The Gabon team found the virus in the African fruit bat. It was the first time they found the virus not to be from humans or other primates. Those that went to Uganda had similar findings. They were able to find Marburg virus genes in the Egyptian fruit bat named " *Rousettus aegyptiacus.*" At the same time, they found Marburg antibodies in the healthy bats. Though they were not able to find live viruses from the bats, they were sufficiently confident to claim that the reservoir of the virus was the bats.

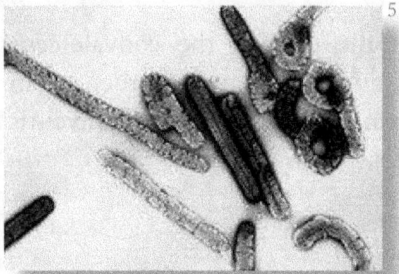

Fig. 8-4. *Marburg virus.* **TEM[6]**
(CDC/ Dr. Erskine Palmer, Russell Regnery, Ph.D.)

Pathogenesis and Symptoms

Just like the Ebola virus, the Marburg virus can be acquired by humans through direct contact with the body fluid of infected humans or infected animals. It may be a close contact or contact with equipment or other items contaminated with the infected body fluid. As mentioned in the 1967 outbreak in Marburg, there were members of the medical staff (two doctors, a nurse, a post-mortem attendant, and the wife of a veterinarian) managing the primary group, who later developed the same illness.

Once the virus enters the body, symptoms appear after a five to ten-day incubation period. The initial symptoms are the usual symptoms common in viral infections, which are fever, chills, myalgia, headache, anorexia. After five to seven days with the symptoms, maculo-papular rashes develop along the trunk. This is then followed by gastrointestinal symptoms like abdominal pain, diarrhea, vomiting, though others develop sore throat and chest pain. These symptoms usually last for two to three weeks.

Severe cases, however, develop multi-organ involvement like jaundice (liver), acute pancreatitis (pancreas), delirium and other neuropsychiatric symptoms (brain) and bleeding or hemorrhage (platelets). Some patients develop prolonged bleeding leading to hypovolemic shock and multiple organ failures. Blood may ooze from any orifice of the body. Fatality rate ranges from 23-90 %, which is considered quite high.

For patients who recover from the disease, the convalescent period may be prolonged with signs and symptoms of other organ involvements like orchitis (testes), hepatitis (liver), transverse myelitis (spinal cord), uveitis (eye) and parotitis (parotid gland).

Diagnosis

Just as with any other hemorrhagic fever, presenting with quite common symptoms, it is hard for a medical practitioner to arrive at a definitive diagnosis of Hemorrhagic fever, unless there is an ongoing epidemic where the incidence is quite common. It is not a mistake initially to consider these symptoms as those of Malaria or Typhoid fever, but it is a big mistake to limit one's consideration to Malaria and Typhoid fever only. Once the bleeding appears, it is becoming clear that the case would be that of a Hemorrhagic fever. All the necessary laboratory tests must be done on the first visit or on admission to the hospital. All other procedures or tests for organ functions must be done to be sure that the body organs are not yet involved.

For confirmation, the following tests can be done; Antigen capture by enzyme linked immunosorbent assay (ELISA), an antibody-capture for *Marburg virus* by ELISA or by Real- time polymerase chain reaction (RT- PCR) or simply by isolating the virus.

Treatment

Like Ebola hemorrhagic fever, there is no specific treatment. The mainstay is supportive therapy which includes symptomatic treatment, monitoring blood volume and electrolytes to make sure they are in balance and will not aggravate the situation. When prolonged bleeding is present with signs and symptoms of hypovolemia, vasopressor drugs can be given to maintain the hemodynamics. This will be followed by whole blood transfusions or fresh frozen plasma, or transfusion of specific clotting factors that are deficient in the patient.

Prevention

Although the Marburg Hemorrhagic fever is a rare human disease, it has a greater potential to spread, especially among the health care staff and the family members. That is why it is particularly important for a medical practitioner to be knowledgeable and observant of the signs and symptoms of the Hemorrhagic fever so that he/she can easily detect the disease and have the appropriate preventions. With advancement in technologies and easy transportation, better diagnostic tests can be done with the results released at the earliest time. This will be a significant factor in curtailing the spread of infection.

For the medical staff overseeing this type of hemorrhagic fever, make sure you are well protected before managing the patients. Always use your personal protective equipment like your gloves, impermeable laboratory gown, face mask, eye goggles and shoes that cover the whole foot.

As mentioned in the section on Ebola hemorrhagic fever, there is an already developed vaccine (presently used) with an extremely high efficacy, but for whatever reason, it is good only for other primates, but not humans.

C. LASSA FEVER[7]

Lassa Fever is an acute viral hemorrhagic fever commonly occurring in West Africa.

This condition was first recognized in 1969, when two missionary nurses in the town of Lassa, Borno State, Nigeria, died from a hemorrhagic fever type of infection. Researchers subsequently found out that this infection was caused by a virus they later called Lassa virus in reference to the town where it was first reported. It has been estimated that there are 300,000 – 500,000 cases of Lassa Fever every year with a mortality of 5,000. At present, this hemorrhagic fever is endemic in the countries of Nigeria, Liberia, Sierra Leone, Guinea, Congo, Mali, and Senegal.

CAUSATIVE ORGANISM: *Lassa virus,* a member of the Family Arenaviridae

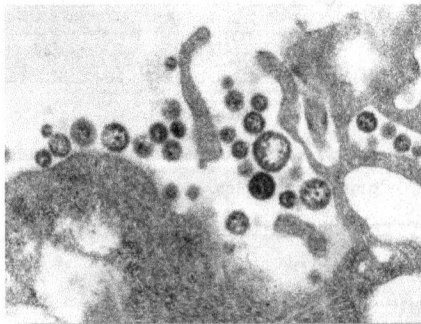

Fig. 8-5. Lassa Fever. TEM Photo by C.S. Goldsmith from CDC[8]

VECTOR: Natal multimammate mouse(*Mastomys natalensis*)

Pathogenesis

The viruses have the rodents as their reservoir. They cause no disease in the rodents. These types of rodents are so numerous because they breed more often and produce more offspring. These rodents would often colonize houses. Therefore, they can easily spread the infection, leading to an epidemic.

In rodents, the viruses are continuously shed off through the urine or through droppings. The most common means of acquiring the virus is by inhalation of the aerosolized droppings. Other means are direct contact with the virus, or contact with materials contaminated by the virus, eating food contaminated by materials with the virus, or direct contact with a cut or abrasion on the skin. In some areas, rodents are a food source, so there is a direct contact with the skin in the preparation of food; more so if the food is half cooked or is not cooked properly.

Human to human transmission, especially among health care workers, is possible through contact with infected patients or with contaminated equipment like syringes and needles. This is in those areas with poor sterilization methods and poor sanitation. There are no reports of transmission by sexual contact.

The incubation period for Lassa Fever is 8-18 days. Once the virus is in significant quantity (viremia), patients will manifest the generalized symptoms. Later the virus will react with the body platelets, causing minute holes in the vascular wall of the capillaries leading to bleeding diathesis. This bleeding can affect all body organs. This is a critical stage of the disease because the patient may develop hypovolemic shock. If not immediately treated, the patient may lose vast amounts of blood, leading to organ failure which has a worse prognosis and is difficult to treat.

In severe forms it may even lead to hypovolemic shock and eventually death.

In experimental monkeys, Lassa Fever leads to pulmonary congestion, pleural effusion, pericardial edema, and other organ hemorrhages.

Symptoms

The initial symptoms are just like any other hemorrhagic fever. They are non-specific like fever, headache, body malaise, myalgia, anorexia. These initial symptoms correspond to the viremic stage of the disease. After several days, bleeding ensues which is manifested as petechiae in the skin, melena, hematochezia or hematemesis from the gastrointestinal tract, hematuria from the urinary tract, hemoptysis from the respiratory tract or simply bleeding from orifices of the body. If the patient shows seizures or coma, this is indicative of signs of intracranial bleeding.

Diagnosis

Initial diagnosis is presumptive considering the commonness of the symptoms. It is more appropriate to consider other common infectious diseases like Malaria than hemorrhagic fever. By the time the bleeding diathesis appears, that would point to the diagnosis of Hemorrhagic fever. As to the type of virus causing the hemorrhagic fever, it is the confirmatory test that will determine the types. The virus can be isolated or the Ig M or Ig G can be isolated by ELISA. This is 88% sensitive and 90% specific. The virus can be cultured, too. in 7-10 days or can be detected by reverse transcription polymerase chain reaction (RT-PCR).

Treatment

Just as with Bolivian hemorrhagic fever, this virus is sensitive to Ribavirin, an antiviral drug. This is more effective if given early in the course of the disease. Of course, supportive treatment is of immense importance in the therapy to make sure that all the body organs are functioning well and to make sure no complication sets in.

Prevention

Since rodents are widely distributed throughout West Africa (they breed more frequently and produce numerous offspring), it is impractical to recommend a trapping method of prevention. Around and inside the house, rodent traps would be useful. However, there is no better measure to turn away the rodents than simply keeping your house always clean because a clean environment discourages the rodents. Store food in rodent-proof containers. For those people who consider the rodents a food source, they make direct contact in the preparation of food or the food may be improperly cooked. These are ways of transmission that can be prevented.

For the health worker, human to human contact or nosocomial route of transmissions are possible. These can be prevented by first isolating the patient confirmed to have the disease. In overseeing these patients, proper personal protective equipment such as gloves, masks, gowns, goggles, and other protective devices must be worn.

There is no available vaccine for Lassa Fever now. Research, however, is presently underway to develop a vaccine.

D. BOLIVIAN HEMORRHAGIC FEVER (MACHUPO VIRUS INFECTION)[9]

Bolivian Hemorrhagic Fever is also known as *Black Typhus* and *Machupo Virus Hemorrhagic Fever*. It is a zoonotic disease in Bolivia. There were sporadic reports of hemorrhagic illnesses in the rural areas of Beni Department, Bolivia, in 1959 but it was in 1962 when this condition was considered a new epidemic. In the following year, they were able to identify the virus in San Joaquin, Bolivia. They called the virus Machupo virus. Through ecological studies, they were also able to identify the reservoir as a rodent.

CAUSATIVE MICROORGANISM: *Machupo virus, under the family ArenaviridaeReservoir*: a rodent **Calomys callosus,** indigenous to the Northern region of Bolivia.

Pathogenesis and Symptoms

The Machupo virus stays in rodents without causing any infection. The virus is continuously shed off from the rodents through their secretions and excretions (saliva, urine and feces). These viruses are acquired by humans through inhalation of aerosolized secretions, by contact of the rodent's secretions with cuts or abrasions on the skin, or by eating meals contaminated by infected rodent excreta. In some regions, rodents are cooked as meals and eating uncooked rodents can be a source of the infection. Person to person transmission is possible but rare. There is a reported case in Magdalena, Bolivia where seven family members got sick from a lone source, and six of them died. This is proof that person to person transmission is possible.

The incubation period of Bolivian Hemorrhagic Fever is seven to 14 days. The initial presentations are the non-specific signs and symptoms like fever, chills, body malaise, muscle pain, headache, anorexia, and prostration. These are due to the viremic phase of the disease. Some patients may have a non-productive cough. Once the clotting mechanisms are affected, bleeding diathesis ensues. These signs of bleeding usually occur five to seven days after the onset of symptoms. These are manifested in the form of petechial hemorrhages on the skin and mucous membranes.

In severe forms, there is generalized bleeding involving different organs of the body. The gastrointestinal bleeding present as hematemesis, melena, or simply spontaneous bleeding from the mouth, nose, and anus. Respiratory involvement is manifested as hemoptysis or blood-streaked mucus, and urinary bleeding as Hematuria. Worse is in the brain. Bleeding in the brain is manifested as seizures or coma. The mortality rate is said to be 5-30%.

There were several postmortem examinations done on experimentally infected monkeys and researchers noted the following lesions: necrotizing enteritis, bronchopneumonia, encephalitis, hepatic necrosis, and subcutaneous hemorrhages in the lungs, intestine, liver, and lymph nodes.

Diagnosis and Laboratories

Initial diagnosis is presumptive considering that the symptoms are also present in some more common diseases like Malaria and Typhoid fever. At this time, the probability of having this Hemorrhagic Fever is high only if a vector has been identified or during an outbreak. With the appearance of the bleeding diathesis, the Hemorrhagic Fever is highly considered. Confirmatory diagnosis is done by isolating the virus, antigen capture by ELISA, or antibody Ig M or Ig G by ELISA.

Treatment

Treatment of Bolivian Hemorrhagic Fever is supportive by giving symptomatic treatment to comfort the patient; likewise, to make sure that fluid and electrolytes are in balance and to make sure that body organ functions are monitored to prevent organ failure. There were trials using the immune plasma taken from patients convalescing from Bolivian Hemorrhagic Fever (BHF) but the effectiveness of this therapy has not been quantified due to paucity of BHF survivors that can donate the plasma. Moreover, there are no established procedures in obtaining and storing immune plasma.

Intravenous Ribavirin, a broad-spectrum antiviral agent, has been tried in some reported trials and found out to be effective not only against Machupo virus but also against Lassa virus and *Junin virus*; however, there still needs to be more extensive research as to the efficacy and safety of this drug.

Prevention

There is no specific vaccine now against Machupo virus, although the present vaccine for *Junin virus* called Candid-1 showed some protection against the Machupo virus. Since the reservoir is known, control of rodent infestation is a significant preventive measure in endemic areas. This method has been proven in Bolivia. After the 1994 outbreak, there was a massive countrywide rodent control program. This came to reality by using numerous rodent trapping methods in endemic areas. Coupled with these were the use of barrier devices like gloves, masks, caps, and aprons among health workers. The results of this rodent control program was the termination of the outbreak.

On the part of the medical community, continuous education, and training of health workers as to the awareness is necessary, as is early recognition of the disease processes to isolate a carrier right away and prevent the spread of infection. Managing infected patients with potential contamination must be strict. Health care workers must implement the use of personal protective equipment.

Lastly, the virus has been found to be killed by Ultraviolet radiation, simple drying, 1% sodium hypochlorite or a 2% *Glutaraldehyde*.

E. DENGUE FEVER / DENGUE HEMORRHAGIC FEVER (DHF)[10]

Dengue fever is a common disease in tropical and subtropical regions of the planet. The Pacific-based El Nino Southern Oscillation (ENSO) was noted to have a positive relationship with the incidence of Dengue. This is because of water pooling and changes in household water storage practices. Between 1970 and 1995, the incidence of Dengue epidemic in the South Pacific was positively correlated with the La Nina condition (i.e., warmer and wetter).

Dengue fever and *Dengue Hemorrhagic Fever* (DHF) are thought to have the same serotypes of viruses, which are distinctly different in their clinical manifestations. Dengue fever can manifest as a Dengue fever alone but can lead to Dengue Hemorrhagic Fever in which complications set in. The word "hemorrhagic" is a misnomer because this condition is not due to hemorrhage per se but due to increase in capillary permeability leading to hypovolemia and shock.

This infection was first identified in 1779 when there were simultaneous epidemics in Asia, Africa, and North America. By 1950 and 1970 with several pandemics in Southeast Asia, Dengue Hemorrhagic Fever was considered the leading cause of death in children. In that year, there were only nine countries considered endemic to Dengue Hemorrhagic Fever, but this figure increased fourfold by 1995. At present, DHF is considered endemic in one hundred countries in Asia, America, Southeast Asia, Western Pacific, and Eastern Mediterranean.

In February of 2002, there was an outbreak of DHF in Rio de Janeiro with one million cases and sixteen deaths. In Venezuela, they had 890,000 reported cases in 2007.

The latest outbreak was in Rio de Janeiro, Brazil, which started in March 2008, when they had 23,500 cases with 30 deaths. By April 2008, the reported cases rose to 55,000.

One explanation of this continuous spread of this infection is that the vector mosquito thrives in urban areas, unlike other viruses that thrive only in rural areas. In urban communities with poor sanitation and poor water supply, coupled with substandard houses, the mosquitoes are expected to thrive and breed more, leading to more cases of Dengue Hemorrhagic Fever.

With increased population growth and increased urbanization, we would expect more breeding grounds for the mosquitoes. Even the United States has not been spared. From 1977 to 2004, a total of 3,800 cases of Dengue were reported. Most of these cases, however, originated from citizens coming from US territories endemic for DHF. Some came from American tourists or transient visitors in tropical countries.

According to WHO, some 2,500 million people-- which is about two-fifths of the world population-- are at risk of dying due to DHF. The current estimate of WHO is fifty million cases of DHF contacted every year.

CAUSATIVE MICROORGANISM: *Dengue virus of four different serotypes (DEN-1, DEN-2, DEN-3, DEN-4) of Genus Flavivirus Family Flaviviridae*.

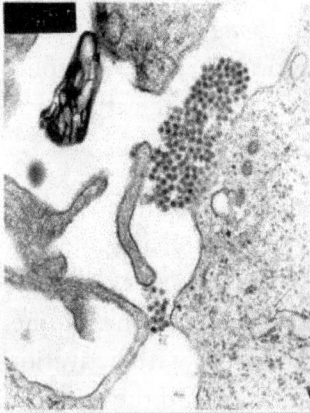

Fig. 8-6. *Dengue Virus*. **TEM.**[11]

VECTOR: *Aedes Aegypti, Aedes albopictus*. These mosquitoes feed during the day.

Pathogenesis

Mosquitoes as the vectors acquire the virus by biting an infected person. A mosquito sucks the blood of the infected person together with the virus circulating in it. After 8-10 days the mosquito transfers the virus by biting another human being. The female mosquitoes can also transfer the virus to its offspring by a transoral (via egg) pathway. In humans, the virus circulates in the blood within 2-7 days after introduction by the mosquitoes. This will correspond to the viremia causing the constitutional symptoms of the patients. Currently, the infected human is the source of the virus for the uninfected mosquitoes. Inside the human body, the virus enters the blood cells and further multiplies exponentially, causing more viremia. The presence of this virus causes activation of coagulation and fibrinolysis, which leads to bleeding diatheses such as petechiae and purpura. Further reaction of the virus with the platelets causes holes on the capillary wall, thereby affecting its permeability. Intravascular volume will decrease because there is escape of the plasma from intravascular spaces to interstitial spaces. If severe, this may cause hypovolemia leading to shock. The virus invades other organs like the liver, lymph nodes and thymus, causing some bleeding diatheses in these organs. In the liver, the hepatocytes are destroyed by the virus, causing more bleeding because as we know, some clotting proteins are produced by the liver.

Symptoms

Dengue Fever usually starts as a fever, headache, nausea, vomiting, loss of appetite, eye pain, muscle pain and joint pains. The joint pain is crippling, given other names like *Bone crusher disease* or *Breakbone fever*. Three to four days later, fever, pruritic rashes, purpura or petechiae appear first on the lower extremities and the trunk and later spread to cover all parts of the body.

In some cases, the symptoms are milder and there are no rashes; therefore, misdiagnosing Dengue fever as a mild viral infection or mild influenza is common. Travelers from endemic areas, who are misdiagnosed to have a mild viral infection, can bring the virus home, causing the spread of infection. In severe cases, those with activation of the coagulation cascade, fibrinolytic mechanisms, and the presence of thrombocytopenia (decreased platelets), the bleeding diathesis becomes more apparent. There are subcutaneous hematomas, bruises, spontaneous bleeding from body orifices. Coupled with increased capillary permeability, the person will develop hemoconcentration and later hypovelemic shock.

The World Health Organization has classified the Dengue Hemorrhagic Fever into four categories:

Grade I - thrombocytopenia + hemoconcentration without spontaneous bleeding

Grade II - thrombocytopenia + hemoconcentration with spontaneous bleeding

Grade III – thrombocytopenia + hemoconcentration + hemodynamic instability (Filiform pulse with pulse pressure less than 20 mm Hg)

Grade IV – thrombocytopenia + hemoconcentration + shock (pulse is zero, BP is zero)

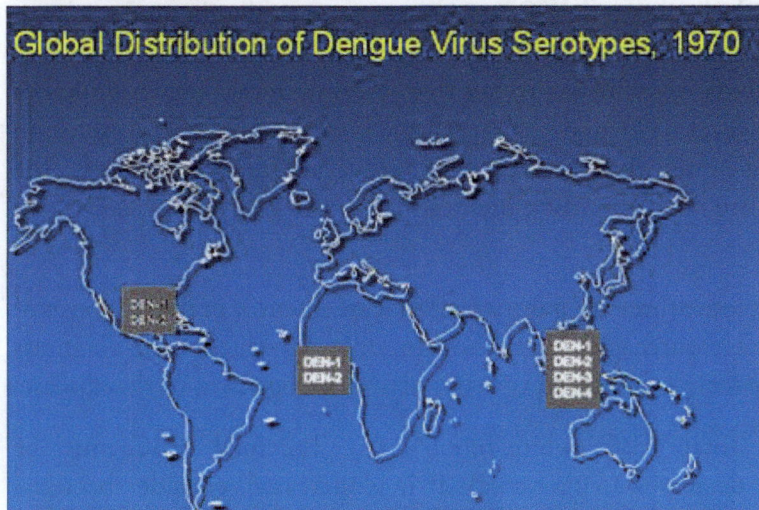

Fig. 8-7A. Global distribution of Dengue Virus Serotypes, 1970 (Courtesy of Duane Gubler)

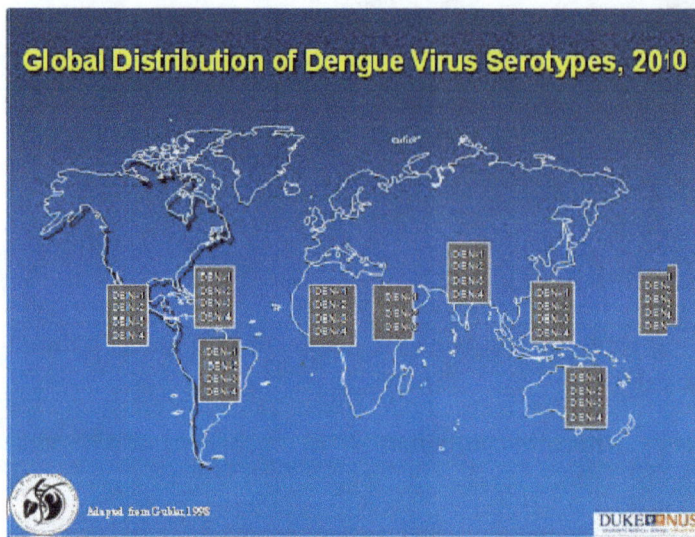

Fig. 8-7B. Global Distribution of Dengue Virus Serotypes, 2010.
(Courtesy of Duane Gubler)

Diagnosis and Laboratories

With the initial manifestations common to all other infections, it is not unusual to have an initial diagnosis of another common disease like Malaria or Influenza rather than Dengue fever. The diagnosis of Dengue fever or Dengue Hemorrhagic Fever is more on an exclusion basis, meaning after excluding all other common infections. The appearance of the bleeding diathesis is a clue that the case is Dengue fever.

There are three laboratory tests that are essential in the diagnosis of Dengue Hemorrhagic Fever. They not only give clues to the diagnosis but they serve as guides in the supportive management of this disease.

1. Total WBC (white blood cells). The total WBC must be decreased (leukopenia). If it is increased, consider bacterial infections, not viral infections.

2. Platelet count – The platelet count (thrombocytes) must be decreased. The level of decrease is also a guide as to the management of the disease, whether to give platelet transfusions or not.

3. **Hematocrit level** – Due to hemoconcentration, the Hematocrit Level must be high. A higher value by 20% is indicative of hemoconcentration or, if without a baseline Hematocrit level, a value greater than 45% is indicative of hemoconcentration.

If necessary to confirm the diagnosis, serologic tests can be done. These are the isolation of the viral antigens by Polymerase Chain Reaction (PCR) or by detecting the antibody Ig M using the MAC-ELISA (Ig M antibody capture by enzyme linked Immunosorbent Assay).

Treatment

There is no specific treatment for Dengue fever. The mainstay of treatment is supportive therapy. Presenting with constitutional symptoms, a person is given only supportive treatment that is to relieve him of his symptoms. Salicylates, however, like aspirin, are contraindicated because they cause bleeding and can cause Reye's Syndrome, especially in children. Fluid intake is encouraged. If not possible, intravenous fluid is administered. This is to prevent dehydration as well as to counteract the hemoconcentration.

Those patients with a low platelet count (normal is 250,000-400,000) and those with spontaneous bleeding must be hospitalized to prevent further complications. Constant monitoring of vital signs is especially important. Likewise, serial hematocrit and platelet counts should be taken as frequently as every two hours to monitor the progress of the disease. A platelet count lower than 25,000 necessitates platelet transfusion. A hematocrit of greater than 45% must be well hydrated with caution to avoid pulmonary edema, especially for older patients.

Patients in shock must be placed in ICU and given plasma expanders and platelet transfusion. Metabolic acidosis must be treated and oxygen must be given.

There are some in vitro (outside of living organisms) studies that show Mycophenolic acid and Ribavirin inhibit dengue virus replication but actual in-vivo (in living organisms) experiments have not been done as of this time.

Prevention

Since the vector is known, that is the mosquito, the main target of prevention is mosquito control. There are two methods: the larval control and the adult mosquito control.

Larval control simply means the elimination of breeding grounds for mosquitoes. In urban dwellings there are often old tires, old drums, flowerpots, plastic containers, and many other water receptacles. All these items catch water during rain and keep the water until it evaporates. The mosquitoes love these places and breed their larvae there. So, it would be helpful to eliminate all these water receptacles. Throw away those unused items; otherwise keep them indoors in garages so that they will not catch water. If stagnant water cannot be eliminated, use larvicidal spray. An example of a larvicide is *Pyriproxypine*. This is a safe and long lasting larvicide.

Adult mosquito control can be done by spraying insecticide such as DDT. (Although it may seem strange for an environmental book to recommend DDT, there are, unfortunately, some areas in the world where DDT is the only realistic way to control disease-carrying mosquitoes at present.) If it is necessary for somebody to be in an infested area, or if a subject is living in an infested area, personal protective items must be used. Around your dwellings, install mosquito screen or if not possible, use mosquito nets. When going out, wear clothes that cover all parts of the body as much as possible or use insect repellants containing NNDT or DEET.

There is no vaccine to date for Dengue fever. The Pediatric Dengue Vaccine Initiative was set up in 2003 with a goal to develop a vaccine against Dengue fever. As of now, research is still in the Phase I and Phase II in the testing of potential candidates for an effective vaccine.

2. WEST NILE VIRUS INFECTION[12]

West Nile virus can cause several diseases. These viruses are transmitted by mosquitoes to humans and a wide range of animals like horses, dogs, cats, rabbits, birds, and bats. This virus is commonly found in tropical countries like Africa, Asia, and the Middle East. Since 1999, West Nile Virus has been identified in the United States.

West Nile virus is believed to have been present as early as one thousand years ago. In fact, it was thought to be one of the probable causes of the early death of Alexander the Great. The first isolated virus, though, was obtained from a febrile woman in the West Nile District in Uganda where scientists were conducting research on Yellow Fever in 1937. Since then, the virus spread to neighboring African countries (Congo, Sudan, Egypt) and India. In 1950, in Egypt, they found out that 90% of individuals over forty were positive for West Nile Virus antibodies. In 1957, in Israel there was an outbreak of meningoencephalitis among older people.

In United States, the West Nile Virus was first isolated in 1999 when there was an increased incidence of Encephalitis among vertebrate animals like dogs, cats, and horses in the states of New York, New Jersey, and Connecticut. The virus later spread across the continental US to include Canada, Mexico, Central America, and the Caribbean peninsula. In 2001/2002, there was a media blast on the West Nile Virus infection due to rapid appearance of the virus in many areas.

CAUSATIVE MICROORGANISM: *West Nile virus, a positive sense single stranded RNA of the genus Flavivirus.*

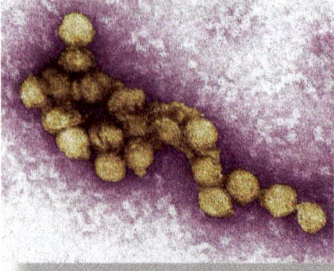

Fig. 8-8. West Nile Virus. Photographed by Cynthia Goldsmith, P.C Rollin[13]

VECTOR: Mosquitoes of the following species:

-*Culex pipiens* (common in Eastern US)

Fig 8-9 Culex pipiens[14]

-*Culex tarsalis* (common in Western and Midwestern US)

-*Culex quinquefasciatus* (Southeast US)

Transmission

These species of mosquitoes bite birds, and mammals, including humans. The most infected birds are the American crow, American robins, and jays. For unknown reasons, the virus thrives better in birds while it does not do well in other animals, specifically mammals. That is why the birds are called the "amplifying hosts. " These birds are the amplifying hosts while any infection in mammals is considered a "dead-end infection." When mosquitoes bite birds, they suck in these viruses to be transmitted again to other birds and mammals, including humans.

Once the mosquitoes bite humans, they inoculate their saliva with the virus on the skin. The virus then enters the blood stream, multiplying and spreading to various organs of the body to hibernate or to cause infection. Brain tissues and the meninges are the usual target organs of viruses to develop Encephalitis (inflammation of the brain tissue) and Meningitis (inflammation of the meninges).

Other proven routes of transmission are the following:

1. Blood transfusion. As a precautionary measure, in 2003, US blood banks started screening potential donors for the West Nile Virus. In the UK, blood donated by persons who came from the US and Canada within 28 days is screened for West Nile Virus.

2. Organ Transplantation: As of now, there is no research yet or survey of West Nile Virus coming from transplanted organs.

3. Intrauterine transfer: The West Nile Virus has been proven to be transferred from the uterus to the fetus through the placenta, but as to the effect of the virus on the fetus, there has not been any research.

4. Breast milk from lactating mothers. Although proven, this is an exceedingly rare method of transmission that should not be considered a factor in deterring mothers from breast feeding their babies.

Symptoms

Most people infected with West Nile Virus are asymptomatic, and they seldom have manifestations. Should they have some, they are usually mild. The symptoms usually appear three to 14 days after being bitten by infected mosquitoes. This mild infection is termed as " West Nile Fever" and the symptoms include fever, headache, vomiting, muscle pain. These may be accompanied by skin rashes, anorexia, and lymphadenopathies. These symptoms usually last for three to six days. In one percent of cases however, the infection may become severe enough to involve brain tissue (Encephalitis) and connective tissue covering the brain called meninges (Meningitis).

Presenting symptoms are severe headache, high grade fever, back aches, disorientation, tremors, and convulsions. These can lead to long term or permanent neurologic deficits like paralysis, aphasia, and convulsions.

Persons more prone to this condition are older people 50 years old and above. Pregnant mothers and lactating mothers must be observant, too, of the symptoms. Suspected West Nile fever necessitates medical consultation right away due to the possibility of placental transfer and spread from breast milk.

Diagnosis

Since the symptoms of West Nile Fever are quite common to infectious conditions, the diagnosis of West Nile Virus is by testing blood samples or cerebrospinal fluid samples. Any rise in the level of antibody against the West Nile virus in either body fluid is a confirmation of the disease. A *Ribonucleic acid test* (RNA) may also be requested. A positive RNA test is a confirmation of the disease. Blood sampling is done by venipuncture while the Cerebrospinal fluid is done by a lumbar puncture. A CT scan or Magnetic Resonance Imaging (MRI) to know the extent of inflammation of the brain and spinal cord is also necessary.

Treatment

Since West Nile Virus is only a mild infection there is no specific treatment. Most patients need no treatment. They may take some symptomatic medication like Tylenol for fever, headache, and body pain.

For severe cases, however (Encephalitis and Meningitis), patients are hospitalized and the treatment is to prevent further brain damage.

Some researchers have recommended Interferon Therapy. There is a better recovery rate for interferon takers rather than non-takers. Further studies are needed with this therapy, however.

Prevention

Like any mosquito-borne infection, the best way to prevent West Nile Virus is to eliminate the mosquitoes around one's home and neighbors' homes. For the endemic areas where total eradication is hard, use ways to avoid mosquito bites. Here are some guidelines:

-Remove all stagnant water in any containers outside home (plastics, old tires, drums, birdbath).

-Make sure your roof gutters are unclogged and not holding water.

-Remove water from swimming pools if unused.

-Report to local health authority any sick or dead birds.

-Follow any local government program in eradication of mosquitoes like DEET spray.

How to Reduce Mosquito Bites

-Avoid unnecessary exposure at night in endemic areas.

-If it is necessary to stay outside, use long sleeved shirts and long pants to cover your whole body as much as possible.

-Use mosquito repellant on your skin and on your clothes.

At this moment there is a vaccine for West Nile Virus only for horses. The human vaccine is still under study.

3. HEPATITIS A[15]

Hepatitis is an inflammation of the liver characterized by diffuse or patchy necrosis affecting all acini (small sac-like dilations in the gland). There are six types of Hepatitis (Hepatitis A, B, C, D, E, G) based on the etiologic viruses. I will limit only to Hepatitis A because this is the type that is transmitted by the fecal-oral route.

Just as with any other infections acquired from the fecal-oral route, climate changes leading to heavy rainfall cause increases in the amount of water and floods. In heavily populated areas, these floods lead to poor sanitation, poor waste and sewage disposal and scarcity of water supply. All factors combined with poor hygiene favor the transmission of Hepatitis A virus.

People who are traveling in heavily populated areas are also at risk of having Hepatitis A. Promiscuity and sexual practices like cunnilingus and fellatio are risks in acquiring the disease. Globally, about ten million people are infected with Hepatitis A every year. In the US, according to the CDC, there are between 125,000-200,000 cases of Hepatitis. Yearly there are between 84,000-135,000 people who experience these symptoms while the rest are asymptomatic.

Approximately one hundred people die of Hepatitis A. These deaths are not due solely from Hepatitis A but from Hepatitis A with concomitant more serious conditions like Hepatitis B or Hepatitis C or AIDS.

CAUSATIVE MICROORGANISM: *Group IV (+) single stranded RNA virus of Picornaviridae family*

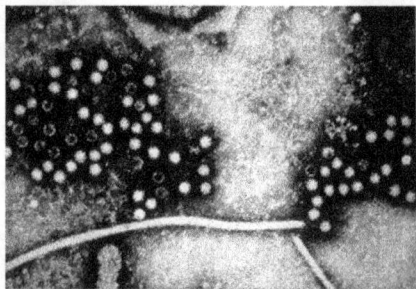

Fig. 8-10. Hepatitis E virus. TEM by CDC[16].

Transmission and Pathogenesis

Hepatitis virus is transmitted by the fecal-oral route, which is putting something into the mouth that has been contaminated with stools of patients with Hepatitis A. It can be water borne or food borne.

After ingestion, viruses find their way to the bloodstream through the epithelium of the oropharynx and intestine, then are carried with the blood flow to reach the liver. They stay and multiply in the Hepatocytes and *Kupffer Cells* of the liver, where they induce inflammatory reactions leading to diffuse or patchy necrosis. The viruses are later excreted through the bile and land in the intestine. They are subsequently passed out through the stools. The fecal shedding of the virus starts from the incubation period to a few days after the appearance of the symptoms. The patient is usually not infected at the time of diagnosis. These facts make Hepatitis A the least infective among the types of Hepatitis. Transmission is common in areas with poor sanitation, or overcrowded conditions. Most of these infections result from contact with infected household members or sexual partners. There was one report of an infected restaurant employee who transmitted the disease to his customers. Casual contacts as in the usual office, factory or school setting do not spread the disease.

The virus can survive in fresh water and salty water for several months. It is resistant to some solvents (ether, chloroform), acid (pH 1), drying, temperatures up to 60°F and detergent.

Symptoms

Symptoms of Hepatitis A vary from minor flu-like illness to fatal liver failure, depending on the causative virus and the immune status of the person. The symptoms occur abruptly. These are low to moderate fever, nausea, abdominal discomfort, and the universal sign of hepatitis which is jaundice (yellowish discoloration of the skin and sclerae of the eye), dark colored or tea colored urine. The symptoms usually last for one to two months and the average incubation period is 28 days.

Some patients are asymptomatic.

Diagnosis and Laboratories

All six types of hepatitis usually present the same manifestation (flu-like symptoms) at the early phase of the disease, so it is hard to determine what type of hepatitis the person has. At this time, determining the patient's history is particularly important to rule out other types like Alcoholic Induced Hepatitis, Drug Induced Hepatitis, or Hepatitis B. Liver enzymes (AST, ALT) are usually elevated at this time. These enzymes are indicative of liver injury. Other laboratory tests are the Prothrombin time (indicative of liver injury and liver dysfunction) and the Alkaline Phosphatase (indicative of cholestasis). The extent of the liver injury can be determined by ultrasound studies. The confirmatory test is the *Serologic Test* for the presence of Ig M anti HA. These antibodies appear in the early course of the disease.

Treatment

There is no specific treatment for Hepatitis A. In most cases, treatment is unnecessary. Most people with the infection will have short term illness that will resolve completely. Most patients do not require hospitalization; instead, they are advised to have bed rest for one to four weeks and to avoid intimate contact with others. Some doctors recommend multivitamins, high protein diet, and a low-fat diet in the course of the disease until recovery. Drugs and substances that can cause liver injuries should be avoided. These are alcohol, sedatives, and narcotics. Once recovered, the patient is immune for life. He can resume work after the resolution of the jaundice. Persons with severe malaise or severe jaundice or those with concomitant diseases are usually hospitalized, with intravenous fluid incorporated with vitamins.

Prevention

Personal hygiene is important in the prevention of Hepatitis A. The stool of a patient with Hepatitis A must be considered infectious. Always wash your hands with soap and water after coming from the toilet, before and after eating. Avoid intimate contact with others until after the disappearance of jaundice. For workers who are managing babies or children, wash your hands thoroughly after changing diapers.

As a prophylaxis, there are two vaccines that are presently available for Hepatitis A.

(1) The immunoglobulins (passive immunity) which provides short term protection. This is given to persons with active Hepatitis A. This must be given within two weeks after the exposure to the virus to attain maximum protection. (2) *Hepatitis A Vaccine, Havrix, Vaqta,* provides active immunity. It is recommended for people two years and older. This preventive vaccine is given to people who are most likely to acquire the disease or to those people who will be more likely to become seriously ill if infected with Hepatitis A.

4. HOOF AND MOUTH DISEASE[17] (HMD)

Hoof and Mouth Disease (or Foot and Mouth Disease) is an acute viral infection affecting cloven- hoofed animals. The causative virus has a predilection for cattle, pigs, goats, and deer. Other animals like elephants, hedgehogs and rats are susceptible too, but to a lesser degree. The phrase "Foot and Mouth Disease" is a misnomer because these animals have no "feet;" they have hooves instead. It is more appropriate to call this condition Hoof and Mouth Disease.

HMD is endemic in Asia, Africa, South America. Scandinavian and other European countries except for United Kingdom, which has been declared infection-free because of vaccinations. Some countries in this region have even stopped the vaccinations. Australia, New Zealand, Japan, and North America are infection-free too. The last outbreak of HMD in the US was in 1929.

HMD, having a short incubation period of three to six days, spreads rapidly, making it a highly communicable animal disease. It is hard to control the spread of infection. The 2001 outbreak in the UK resulted in severe economic implications. Suspected animals were quarantined, infected animals were slaughtered, all persons entering and leaving the farms underwent disinfection, and sporting events attended by farmers were cancelled. The British government lost a staggering $10 billion dollars during the outbreak. In US, with constant movement of livestock, infections can easily spread. That is why the US Department of Agriculture continuously exerts efforts to safeguard the livestock industries.

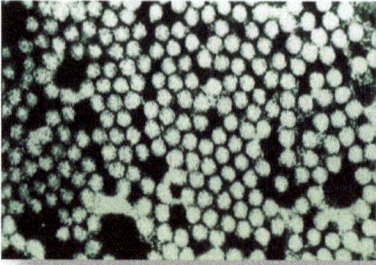

Fig 8-11 Aphthovirus[18]

CAUSATIVE MICROORGANISM: *A virus named from the disease " HMD virus" under genus Aphthovirus, Family Picornaviridae*

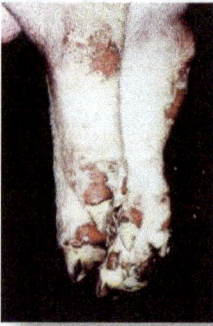

Fig. 8-12. Photo of ruptured blisters on pig feet because of Hoof and Mouth disease.[19]

Pathogenesis

Animal infection is most transmitted by inhalation. It may also be transmitted by direct contact with abraded skin or blisters on the mouths or hooves of the animals. The virus can be transmitted also by entering the eyes, nose, udder and even the uterus. Ingesting large quantity of viruses from food or feed is another way of acquiring the virus. Farmers and other animal handlers can be carriers of the virus. A farmer previously in contact with infected animals may have the virus on his clothes or skin, and it can be transferred to other animals once in contact. Contaminated pens, or vehicles used in animal transport, can be a source of the virus. Mating is another way of transmission, coming from infected bulls.

Once the viruses reach the host cells, they combine with the receptor sites causing the cell membrane of the animal to fold. Inside the cell, the protein coat of the virus dissolves, exposing the RNA of the virus. The viral RNA in the cells disturbs the translation of the proteins. Instead of synthesizing the normal protein, the cells now synthesize the viral RNA, thereby forming millions of young viruses.

Multiplication of the virus goes on until host cell lysis (*lysis* is when a cell breaks apart), releasing the young virus to infect again nearby host cells.

The viruses, especially in the viremic stage, can be found in all parts of the body. Those viruses transferred by aerosols multiply rapidly in the pharyngeal region, while other viruses enter the bloodstream through the pulmonary alveoli. Those viruses in the mouth, nares, muzzle, snout, interdigital spaces, and teats, develop blisters containing millions of young viruses. Once ruptured, the young viruses are disseminated to infect other nearby areas. The presence of blisters furthermore irritates the mucosa, leading to increased salivation as well as increased nasal discharges. Those viruses that reach the heart cause degeneration of cardiac muscles leading to myocarditis, then death.

One peculiar feature of HMD is that despite the severity of the disease and high morbidity (Ninety percent), its mortality rate is extremely low or insignificant. For whatever reason, the animals eventually recover and heal, except those that developed myocarditis, and infected young animals.

Symptoms

Hoof and mouth disease has a multi-organ involvement in that infected animals present several signs and symptoms. The most overt and the most common signs among the cloven-hoofed animals, are the skin lesions on the oral mucosa, snout, nose, and those lesions on the hoof. The lesions on these specific parts of the body are the bases for naming this condition as "Hoof and Mouth Disease." The skin lesions are initially erythematous areas on the skin which later become blisters. In a few days, the blisters rupture, spreading the young viruses to invade nearby areas.

Lesions on the mouth would increase salivation and drooling of the saliva, and increase nasal discharges, making the animal weak and abhor feeding. If secondarily infected, abscesses may be found, or muco-purulent discharges may occlude the mouth or the nostrils.

Blisters on the hoof make the animal weak and lame, refusing to rise or to move. Lactating animals may have blisters on the teats which may or may not be secondarily infected. These lesions reduce milk production of the animals, preventing them from nursing. Infected pregnant animals may have abortion or premature labor, giving birth to calves having the same lesions. Death usually occurs in young animals or those animals with complicated myocarditis.

HMD in Humans

HMD is more of an agricultural problem than a human health problem. It may be highly transferable to animals but not to humans. On rare occasions, however, humans may acquire the virus by direct contact with an infected animal or by laboratory accidents. Eating infected meat may cause minor oral lesions in humans. The viruses, once in the stomach, are destroyed by its acidity and thus end their cycle. The last reported case of HMD in humans was in 1966 in the UK. Symptoms in humans include erythematous ulcers on the oral mucosa and blisters on the skin. It is important to note that HMD is different from HFMD (Hand, Foot and Mouth Disease) which is common in humans. This does not only involve the feet but also the hands. The lesions are the same, but these are two different infections. The causative virus in HFMD is a coxsackie, a virus which is a member of enterovirus.

Diagnosis

Diagnosis of HMD is based on clinical signs, symptoms, and the course of the disease. Laboratory tests are not necessary. Although the blisters in HMD are indistinguishable from other vesicular disease of animals, the uniqueness of the course of the disease is enough to make a diagnosis. These unique features are the following:

-Sudden appearance of the vesicular lesions at the same time on specific parts of the animal. The lesions would appear on the mouth, nares, and hooves of the animals.

-The short course of the oral lesions. For whatever reason, the oral lesions heal right away unless secondarily infected.

-The long course of the hoof lesions, even without secondary infection.

-The concomitant mastitis of the animals during the disease.

-The sudden death of infected young animals and the increased incidence of abortion in pregnant animals.

Laboratory tests are done only if one wants to be certain of the diagnosis. Once diagnosed or even merely suspected, it is one's obligation to report to or notify the animal health officials.

Treatment

There is no specific treatment for HMD. The US Department of Agriculture would stick to the recommended guidelines in case of an incidence report, which is total eradication of the infection by burning or burying the animal carcass.

Cleaning and disinfecting animal holding areas, animal pens. and all equipment used in the livestock industry is necessary. Exposed veterinarians and animal handlers must use a disinfectant. MAFF (Ministry of Agriculture, Forestry and Fishery) has the list of disinfectants and their recommended concentration for public use. This can be obtained from the MAFF website.

Prevention

There are available vaccines used to control HMD. These vaccines, however, are not a perfect solution. Vaccinated animals can be reinfected again even with homologous strains of the vaccine. There is still a benefit of vaccinating a large population because the incidence of the disease goes down. With easy access due to modern rapid transportation, one can easily go back and forth to the livestock farms, bringing with them the virus. The US Department of Agriculture has very stringent guidelines in importation of animals, specifically those coming from the enzootic countries. The bottom line is that early detection is necessary to prevent wide spreading of this animal disease.

5. CHIKUNGUNYA FEVER[20]

Chikungunya Fever is another self-limited and rarely fatal infection. Its clinical picture resembles *Dengue fever* but can be differentiated from it by the absence of Dengue Hemorrhagic Syndrome and by the involvement of multiple joints. In Chikungunya Fever the person develops multiple *Arthritides* causing him/her to stoop on standing: a characteristic feature of this disease. In fact, the high-grade fever and the peculiar stooping posture are pathognomonic signs of Chikungunya Fever.

This disease was first described in 1952 by Marion Robinson and W.H.R. Lumsden in Makonde Plateau, between Tanganyika and Mozambique, Africa. In 1963 an outbreak of almost epidemic proportions was reported in India. Between 2005-2006, 237 deaths were reported due to this viral infection in Reunion Island (near Madagascar). Thereafter several outbreaks in India, Sri Lanka, Malaysia, Italy, Singapore, and Australia occurred. The latest reported incidence was in a town called Sullia in Southern India, with 1600 cases.

The word *Chikungunya* seems a strange word. There were two versions of its origin. Some authors claimed it is a Swahili (the lingua franca of the region) word that means *Chicken guinea*,or *Chikungunya*. The word is the *Mankonde* word meaning *"that which bends up"* in reference to the stooping of the person due to arthritis because of this viral infection.

CAUSATIVE MICROORGANISM: *Chikungunya virus, of genus Alphavirus, under Family Togoviridae*

VECTOR: Mosquito: *Aedes aegypti,*

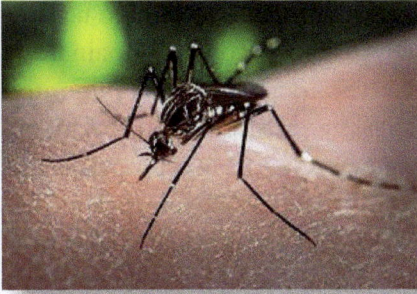

Fig 8-13 Aedes aegypti .Photo by James Gathany[21]

Aedes albopictus, a recently identified mosquito that transmits the mutated Chikungunya virus

Pathogenesis and Symptoms

The vector of this Chikungunya virus is the mosquito *Aedes agypti,* the same mosquito that transmits Dengue fever. This is the mostly likely the reason most of the symptoms are similar. The mosquito acquires the virus from infected persons and transmits the virus by biting another person (human – mosquito – human transmission). The mosquito injects some of its saliva with the virus while sucking the blood of the person. Inside the body, the virus multiplies exponentially to a considerable number (viremia), enough to cause the initial symptoms. These are fevers (greater than 40 °C, 104° F), headache, body malaise, myalgia, nausea, and vomiting. Later these are followed by petechial or maculopapular rashes. Hemorrhagic episodes are rare. Once the virus invades the joints, arthralgia or arthritis ensue. The pain due to arthritis causes the person to stoop on standing, a typical stance descriptive of this disease.

In severe cases, the virus invades other organs of the body like the brain and the spinal cord, causing headache, body weakness, and numbness. The viruses may invade the heart, causing chest pain. They can even penetrate the placental barrier, transferring the virus to the fetus. This mother to fetus transmission usually occurs at 12-18 weeks of gestation but the presence of the virus in the fetus is counteracted by the Ig G antibodies that pass freely through the placenta from mother to fetus.

Acute Chikungunya is not life threatening. The disease usually lasts for a couple of weeks. Those infected usually recover. The pain from the persistent arthritis, however, may be prolonged for several months or even a year, making it uncomfortable and disturbing to the patient. Few deaths are reported. Deaths have been due to inappropriate use of anti-inflammatory drugs and antibiotics. Some people would just take anti-inflammatory drugs as they wish, unaware that these drugs might cause gastrointestinal bleeding. In Chikungunya Fever, there is already thrombocytopenia (decreased platelets) which may cause bleeding, and this is aggravated by insidious use of anti-inflammatory drugs, be they steroidal or non-steroidal.

Recovery is the expected outcome. It is, however, based on age. Younger patients usually recover within five to 15 days, adults in one to two and one half months. Even pregnant women usually recover. No untoward results have been shown.

Diagnosis and Laboratories

Initially, the patient may be considered as a case of Dengue fever because of its commonality of symptoms. When the crippling arthritis appears, however, the consideration would then shift to Chikungunya Fever. In fact, the most common features of this infection are the high-grade fever and the crippling arthritis. It will be different in the presence of epidemics. When presented with initial fever and other constitutional symptoms, Chikungunya Fever must be the primary consideration right away.

Common laboratory tests for Chikungunya are the virus isolations done by getting a sample of whole blood of the person at the acute phase of the disease, exposing it to a specific cell and recognizing the virus' specific response. The result of this test is available for 1-2 weeks but its results provide the most definitive diagnosis. Measuring the Ig M antibody against Chikungunya by ELISA provides results in 2-3 days. However, this test may give false positive results from an O'nyong-nyong infection.

Treatment

There is no specific treatment for Chikungunya Fever. Mostly the patient is given drugs for symptomatic treatment like Ibuprofen, Naproxyn, Acetaminophen, Paracetamol to relieve pain and fever. It is also important to monitor the fluid and electrolyte balance to make sure that they are within the normal ranges. Supportive treatment with good nutrition and complete bed rest are supplementary.

Chloroquine phosphate at 250 mg/kg body weight has been tried for arthritis not responsive to aspirin and other anti-inflammatory drugs, and the results were promising. There were also reports that aside from the anti-inflammatory effect, Chloroquine has also anti-viral properties against the Chikungunya virus.

In India, they have discovered an **ayurvedic**[22] solution called *Triphala*. This is a concoction from three fruits, namely *Harada (haritaki)*, *Amla (amalaka)*, and *Baheda (bibhitaki)*. They have experienced how effective this concoction is in treating Chikungunya Fever. In another part of India (Cohen), powdered sunflower seeds taken together with honey is a good supplement in the treatment of this disease. This mixture together with the regular medicine has shown much better results in the relief of the arthritis which is the side effect of this illness.

Since this disease is self-limited, and recovery is extremely high, seldom has a patient needed hospitalization. The prognosis is good.

Prevention

There is no vaccine to date for Chikungunya Fever. In 2000, however, the US government started sponsoring research on a vaccine against this type of virus. They claim the results are very encouraging, but they are just on Phase II of the clinical trials.

Without the vaccine yet, the prevention as of this time is more geared towards eliminating the vector of the virus which are the mosquitoes. Here are some guidelines to follow: First, stay away from swampy areas or areas with stagnant water. If it is necessary to stay in these areas or close to these areas, follow ways to protect yourself from mosquito bites. Use clothing that covers all your body parts. Use mosquito coils or electric vapor mats to turn away the mosquitoes. Use mosquito nets or screens around your dwellings, and you may even apply repellant solution on the nets, like Permethrin and other insecticides.

Remove any old, unnecessary water receptacles in your own back yard such as old tires, flowerpots, old drums, water storage vessels, and plastic containers. These would catch water that will remain stagnant, allowing mosquitoes to breed.

Lastly, on the public health side, there must be national campaigns for the elimination, control and prevention of vector borne diseases, or to continue the program if there is an existing one. Of course, along with these programs, people must be educated and made aware of this disease. The people in the community must have a positive attitude about these programs and their cooperation must be expected.

6. O' NYONG – NYONG VIRUS INFECTION[23]

O'nyong-nyong virus infection is a self-limited viral infection. The virus has some similarities with the Chikungunya virus whose infection has the same signs and symptoms as that of O'nyong-nyong virus infection

The O'nyong-nyong virus was first isolated by the Uganda Viral Research Institute in Entebbe, Uganda, in 1959. Since then, this virus had caused several outbreaks in East African countries of Kenya, Uganda, Tanzania, Malawi, Mozambique. In Uganda alone from 1959 to 1962, it infected around two million people. The word "O'nyong-nyong" came from Nilotic language of Sudan and Uganda that means "weakening of the joints."

One thing about this infection that puzzled the researchers is the possibility of having a 30-50 year cycle of epidemic in Uganda. With the second reported epidemic in Uganda in 1996-1997 affecting about four hundred people (a hiatus of 35 years), and evidence of an outbreak in Uganda between 1904-1905, they have thought of an epidemic every 30-50 years. This is a hypothesis that is still to be proven.

CAUSATIVE ORGANISM: *O'nyong-nyong virus of genus Alphavirus and family Togoviridae*

VECTOR: *Anopheline Mosquitoes: Anopheles funestus, Anophleles gambiae*

Pathogenesis and Symptoms

For whatever reason, mosquitoes are not affected at all by the viruses inside them. Once the mosquitoes bite a person, they inject their saliva with the virus to the person. After one to two weeks incubation period, at which time the virus multiplies exponentially into large numbers (viremia), the initial signs and symptoms of high-grade fever, myalgia, headache, nausea, and vomiting appear. Further invasion of the virus to body organs like the skin results in *Maculopapular Rashes* (60-70%) of cases. Later the virus invades the joints, causing the so-called crippling arthralgia and arthritis without effusion. The arthritis usually affects the large joints, and in some instances, this symptom lasts for several months after recovery. This persisting arthritis is a peculiar sign of this viral infection. This is where the "weakening of joints" comes from.

Other organs that may be involved are the lymph nodes (lymphadenopathies), eye (painful reddish eye with or without discharge) and chest (pain). As this disease is a self-limited viral infection with a good prognosis, the patient usually recovers one to two weeks after the onset of the symptoms. This disease has no predilection for a particular age group or sex. All age groups and both sexes are equally vulnerable to this virus. There are no reported deaths due to O'nyong-nyong virus infection but there were two reports of abortions associated with this viral infection. Human to human transmission has not been reported.

Diagnosis

Knowing that this infection has a good prognosis, it is not of utmost importance to have a diagnosis of O'nyong-nyong right away. Presented with the non-specific generalized symptoms, it is more important for a medical practitioner to think about other possible common serious infections like Malaria or Typhoid fever rather than O'nyong-nyong virus infection. Clinical diagnosis of this infection is based on ruling out other more common serious illnesses. This means all the test results for the common illnesses are negative.

For confirmatory tests, the virus can be isolated from the humans by obtaining a blood sample at the acute phase of the infection and inoculating this sample in the brain of the mice.

Antibodies (Ig M) against O'nyong-nyong virus can be detected by titers of ELISA (Enzyme linked Immunosorbent Assay).

Treatment

There is no antiviral drug to date for O'nyong-nyong virus infection. The mainstay of therapy is supportive, that is symptomatic treatment by giving antipyretic for fever, analgesic for pain, antihistamine for rashes. It is also important to monitor basic body organs just to make sure that they are functioning well despite the viremia.

Prevention

Since the vector for this virus is identified and there is no human-to-human transmission, prevention is more geared to the eradication of the vector, **Anopheles** mosquitoes.

Avoid going or wading through lakes, swamps, or any stagnant body of water. Destroy any receptacles with stagnant water in the back yard. These are breeding grounds for the mosquitoes. Use insecticide spray on areas with stagnant water.

If it is necessary to be close to stagnant water, use protective shields or materials to protect your body from being bitten by the mosquitoes. Wear clothes that cover all body parts. Use mosquito nets in your dwellings. Apply insect repellants on your clothes and your exposed skin to turn away the mosquitoes.

The virus is sensitive to 70% ethanol, 1% sodium hypochlorite, 2% glutaraldehyde and some lipid solvents. They are also inactivated by moist, dry heat greater than 58 Degrees Centigrade. Survival outside the host is unknown. Yet, they live less than a day in the culture medium at 37 Degrees Centigrade.

There is no available vaccine to date against the O'nyong-nyong virus.

7. YELLOW FEVER[24]

Yellow fever is a viral infection that causes diverse signs and symptoms ranging from mild to severe, or even death. The word "yellow" refers to the yellowish discoloration called jaundice on the sclera, skin, and mucous membranes of the patient. This is also known as *Yellow Jack, BlackVomit,* and *American Plague.*

Yellow fever dates to the early 1700s when epidemics broke out in countries like Italy, Spain, France, England, and the United States. It has caused devastating effects on the population. In the United States, it caused several epidemics. In fact, it is believed that Yellow fever in some way shaped some chapters of American history. In 1708, there were several outbreaks of Yellow fever in Philadelphia that led to the decision of moving the US capital out of Philadelphia. In 1802, during the Haitian Revolution, it caused the deaths of half of the soldiers of the French army. In the 19th century, about 3,000 people died in Spain. In the construction of the Panama Canal, the French suffered significant loss of life, coining the name for Panama of "the white man's graveyard."

It was in the 1900s, when an American physician and major in the US Army, Walter Reed, following the research of Cuban doctor Carlos Finlay, showed that mosquitoes are the carriers of the disease. The etiologic virus, however, was not identified until 1928. During the Presidency of Theodore Roosevelt, Yellow fever cases were significantly diminished when he ordered US Army colonel William Crawford Gorgas to get rid of Yellow fever in the Panama Canal. Further eradication of Yellow fever came into being with the invention of vaccinations against the virus. In 1937, Max Theiler, a researcher at the Rockefeller Foundation, developed a vaccine affording 10 or more years of immunity from the virus. The availability of the vaccines plus effective control programs for *Aedes aegypti* almost eradicated Yellow fever.

At present, however, Yellow fever has emerged in Africa and South America. This is more common in deprived areas of the countries where wide scale vaccinations are not feasible. In 2001, WHO reported that there were 200,000 morbidities and 30,000 mortalities from Yellow fever annually.

There are three types of Yellow fever based on areas of occurrence:

1. *Sylvatic yellow fever* that is present in the tropical rain forest. The intermediate host here are the monkeys. Humans can be infected when entering these areas.

2. *Intermediate yellow fever* that is present in the humid savannahs of Africa. The intermediate hosts are monkeys and humans. Infections are sporadic and seldom result in death.

3. *Urban yellow fever* that is present in populated urban areas. Here the infection is usually introduced by migrants from endemic areas. With tick populations and poor sanitation, urban yellow fever has the potential to lead to an epidemic. The intermediate hosts are humans.

CAUSATIVE ORGANISM: *Yellow fever virus, a flavivirus of family Flaviviridae*

VECTORS: Mosquitoes of species: *Aedes aegypti, Aedes simpsoni, Aedes africanus* (in Africa)

Haemagogus genus (in South America)

Sabethes genus (France)

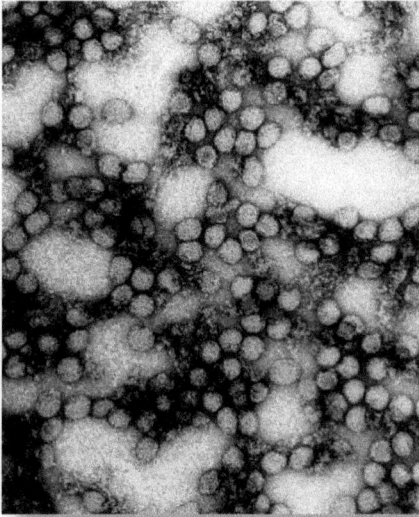

Fig. 8-14. Yellow Fever Virus. Photo by Erskine Palmer[25]

INTERMEDIATE HOSTS: Monkeys in sylvatic or jungle yellow fever. Monkeys and humans in intermediate yellow fever. Humans in urban yellow fever.

Pathogenesis

The causative virus is acquired by humans from the saliva of the biting mosquitoes. The mosquitoes leave the virus on the skin of humans while sucking the blood. The virus stays locally within the nearby lymph nodes. Viruses enter body cells by receptor mediated endocytosis. They multiply in the cytoplasm and are then released through the cell membrane. This stage represents the *acute phase* of the clinical symptoms. Once released in the blood, the viruses quickly disseminate by lymphatic routes to invade multiple organs of the body. In the kidney, the virus can cause fatty changes and eosinophilic degeneration that can lead to renal failure. In the liver, the virus can cause cell death and depletion of vitamin K dependent clotting factors leading to coagulopathy and bleeding. This stage represents the 'toxic phase" of the clinical symptoms. Other commonly affected organs are the heart, adrenals, and gastrointestinal tract.

An infected person may become the source of the virus if bitten again by the mosquitoes which transfer the virus to other humans, thereby disseminating the disease.

Symptoms

Incubation of Yellow fever is three to six days. The acute phase is manifested by the usual constitutional symptoms of fever, body ache, malaise, headache, shivering, loss of appetite, nausea, and vomiting. There is a fever – pulse dissociation (paradoxical slow pulse in the presence of fever). In a few days, the person is in remission, normalizing the symptoms. This stage represents the initial invasion of the virus in the local areas and neighboring lymph nodes. The person may completely recover. Fifteen percent of cases, however, lead to a severe form called "Toxic stage."

The toxic stage represents the viral invasion of multiple organs. This is manifested by high grade fever, abdominal pain, vomiting, oliguria, hematuria, and yellowish discoloration of the skin, mucous membranes and sclerae. Later, symptoms of consumptive coagulopathy appear which are ecchymoses, petechiae, and hematomas on the skin. The person may have melena, hematemesis, and worse spontaneous bleeding. *Hepatorenal Syndrome* may ensue, leading to a higher mortality rate which may reach as high as 50% among those in the toxic stage.

Diagnosis

At the acute phase, it might be difficult to have a definite diagnosis of Yellow fever considering that the manifestations are common in other infections like Hemorrhagic fevers, Influenza, Malaria, Typhoid fever. The only item pointing to Yellow fever is the residence time of the patient in the endemic area. This does not deter clinicians from starting the management of the patient.

Laboratory findings that can favor the diagnosis of Yellow fever are leukopenia, neutropenia, high liver enzymes (ALT, AST) levels, prolonged clotting, bleeding time, reduced fibrinogen and other clotting factor levels.

A patient in the toxic stage would show all the typical signs of Yellow fever that a clinician would hardly miss. The diagnosis of *Yellow Fever* is almost certain.

Treatment

Being a viral infection, Yellow fever has no definite treatment. At the initial acute phase, symptomatic and supportive treatments are started, giving antipyretics for fever and analgesics for pain. Antibiotics are of no use in viral infections.

In the toxic phase, with all the signs of multiple organ failure and coagulopathy, patients are placed in the Intensive Care Unit with IV fluid, plasma expanders or blood transfusions. With a high mortality rate in the toxic phase, all necessary management must be given. Intravenous steroids are usually given as well.

Prevention

Since 1930, with the invention of a vaccine by Max Theiler, Yellow fever was statistically abated. But at the present time, there is a resurgence of the disease in Africa and in South America. This is brought about by encroachment on the forests by humans, as in urbanization and deforestation, and the easiness of international travel that can easily spread the virus worldwide. Failure of large-scale vaccinations in poor countries adds to the burden.

Another aspect of prevention is the eradication of vectors. As with any other disease transmitted by mosquitoes, follow guidelines in eliminating the mosquitoes. Destroy mosquito breeding grounds like swampy areas, or stagnant water from pots, old tires, and other water receptacles. Wide scale spraying using DDT and other insecticides must be implemented by the government.

8. ROSS RIVER VIRUS INFECTION

Ross River Virus Infection is the commonest arthropod-borne viral disease in Australia. This infection occurs throughout most regions of Australia as sporadic epidemics; even urban areas such as Perth, Brisbane, Sydney, and Melbourne are not spared from this infection. It is estimated that there are around 4,800 cases of Ross River Virus infection every year in Australia. In 1979, outbreaks were noted in the neighboring areas of Australia which are the Fiji Islands, Cook Islands, Tonga, and Samoa.

This is a disease that is typically related to environmental changes. Climate change results in abnormal weather patterns like increased rainfall, increased seasons of high tide, and increased environmental temperatures. These abnormal patterns are conducive to the breeding of mosquitoes. Higher rainfalls and increases in high tide seasons correlate with increases in stagnant water, whether in a lake, swampy area, or open containers in our own back yards.

The first isolated virus came from a mosquito collected in the Ross River area in Townsville, Australia in 1959. This is where the virus' name came from. However, this disease was not recognized until 1971.

CAUSATIVE ORGANISM: *Ross river virus, an alphavirus.*

VECTOR: A mosquito named *Ochlerotatus vigilax.*

Pathogenesis

The life cycle of the virus includes mosquitoes as the vector and animals as the natural hosts. These animals are kangaroos, wallabies, marsupials, flying foxes, rodents, fruit bats and even horses. Humans are included in the cycle as an accidental host in epidemics. The Ross River virus is acquired by humans through a bite from an infected mosquito. There is no evidence that this virus is transferred from human to human.

The incubation period is usually three to 11 days, but some patients have the first symptoms after 21 days. Once inside the body, the virus can cause rashes of variable appearances lasting for seven to ten days. In the joints they cause *Arthralgias* and/or *Arthritis* for days or months. The commonly affected joints are the wrists, knees, ankles and inter phalangeal joints. Lymphadenopathies may be present and the commonly affected lymph nodes are those in the groin and axilla.

The prognosis of the disease is good. Everyone recovers from the illness, although at times the symptoms persist intermittently for months or even a year. The patient, however, will never be reinfected with the same virus because he/she will develop memory cells that will provide lifetime immunity.

Symptoms

Aside from the constitutional symptoms of fever, chills, headache, and muscle aches, in 60% of cases rashes appear, lasting for seven to 19 days. Joint involvement results in arthritis and/or arthralgia with difficult movements. In some cases, the infection renders the patient incapacitated and unable to work for a few months. Some patients may have paresthesia or the so called "pin and needle" sensations on the palm and the sole. Others may have enlarged tender lymph nodes on the groin and axilla.

Diagnosis

Diagnosis is based on serologic tests because the symptoms of the Ross River virus infection are like many other viral infections. There is no way for a medical practitioner to be certain that he/she is overseeing this type of infection. In case of epidemics, however, the diagnosis may be based on the clinical presentation of the patient.

Laboratory tests are just confirmatory for the disease. These tests are usually serologic that determine antibodies IgM or IgG against the Ross River virus.

Treatment

There is not any specific treatment for Ross River Virus infection. Antiviral agents are not necessary. Patients may be given medication for symptomatic relief only, such as antipyretics for fever, and analgesic or non-steroidal anti-inflammatory drugs (NSAIDS) for muscle pain, joint pains, and headache. Take note, however, that a person can get this disease once in a lifetime because of subsequent immunity.

Prevention

Preventive measures are mostly geared to protecting oneself from mosquito bites.

-Avoid staying outside the house after sunset and at dawn, especially during the seasons of autumn and summer when mosquitoes are greater in numbers. Mosquitoes usually go out one to three hours after sunset and at dawn.

-If you must be outside of the house, wear clothes that will cover the whole arm and leg and use repellant like DEET or find sprayed on your clothes.

-In the house, use fly screens on windows, doors, and any other openings.

-If screening the windows and doors is not possible, spray insecticide at night to keep the mosquitoes away or use mosquito nets around your bed.

-For those who are camping, be sure to have tents or at least mosquito-proof nets with you while at rest or sleep.

-In your own back yard, try to remove or destroy any stagnant water which provides good breeding grounds for the mosquitoes. Check for any empty receptacles that can later catch water that will remain stagnant.

9. AVIAN INFLUENZA[26]

Avian Influenza is influenza in birds. It is also called *Bird Flu*. This is caused by a species of virus called *Influenza A virus*. Initially the virus infects animals like pigs, but due to genetic mutations, new viruses are formed that later infect birds. To date the avian influenza virus is said to be species-specific, although at times it can still infect other animals. Some birds are noted to be carriers. For whatever reason, these birds do not manifest any signs of the disease despite carrying the virus in their intestines. Other birds are very vulnerable, especially the domesticated birds like chickens, ducks, and turkeys.

The virus can be transmitted also to humans. Though the incidence in humans is low, with the virus constantly mutating, the possibility of forming a type of virus virulent to humans or birds for that matter-- is highly probable. This may lead to an outbreak or pandemic if not immediately detected.

The first reported case of human avian flu was in Hong Kong in 1997. The causative virus was the subtype H5N1 and this was acquired from chickens. In 2004, there was a report in Thailand where a seriously ill child transferred the virus to the mother. In 2006, in Indonesia, one person acquired the infection from poultry. He later infected all the six members of his family. One of these six members later infected another family. As of now, the last reported case was on August 22, 2007 in Bali, Indonesia, where a chicken trader woman died of this condition. They found the causative virus as the subtype H5N1. In totality, as of now, there have been more than one hundred cases of human bird flu and 50% of these cases did not survive. This number might be insignificant compared to the amount of mortality in birds during epidemics. For whatever reason, this type of virus is not prone to cause infection in humans.

CAUSATIVE ORGANISM: *Influenza A virus* **with several subtypes. The most common of them is the subtype that infects birds and humans, and this is subtype H5N1.**

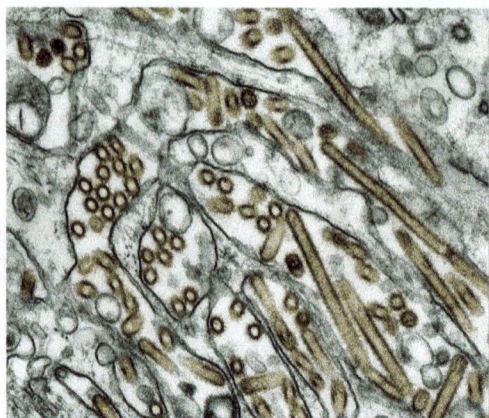

Fig. 8-15. *Avian Influenza.* **TEM Photo by Cynthia Goldsmith. Pathogenesis**

The main method of transmission is contact with infected birds. These birds release the influenza virus through their excreta, secretions, saliva, and nasal discharges. Even those birds recovering from the disease can still transmit the disease because they continuously shed the virus through their excreta and secretions. Other birds in direct contact with them can easily be infected. The virus can also be acquired indirectly by contacting contaminated materials, surfaces, objects, pens, bird enclosures, feeding receptacles, or even feed and water.

There are two forms of avian influenza based on their virulence. The low pathogenic form has mild symptoms only (ruffled feathers and decreased egg production). These symptoms may not even be detected. Some birds are asymptomatic even if they are carriers of the virus. The other form is the higher pathogenic form that can easily spread to flocks of birds. This form presents signs and symptoms referring to the several organs of the bird leading to a mortality rate of 90-100% in just 48 hours or a few days.

In humans, direct contact is also the main route of transmission. It can be birds to humans or humans to humans. The veterinarians and bird handlers, due to their constant contact with the birds, are prone to this type of disease. It can be acquired also by eating improperly cooked or raw meat of an infected bird. It has been found out that among the several subtypes of the virus, it is the subtype H5N1 that is highly transmissible to humans and the subtype that causes severe illness in humans. This virus is constantly mutating and that is why scientists are wary that, at any time, a new virus may be formed that can be too virulent to humans, leading to a pandemic.

Signs and Symptoms

In birds, signs of Low Pathogenic Avian Influenza (LPAI) vary depending on the virulence of the virus, the age and the species of the birds, or concurrent infection if any. The signs are referred to respiratory problems like sneezing, coughing, sinusitis, or lacrimation weakness; decreases in egg production and inferior quality of eggs are also seen.

In HPAI (high pathogenic avian influenza) which is a fatal form of influenza in birds, the onset is immediate and the course is noticeably short. The respiratory signs are moderate to severe inflammation of the airways, and exudate in the lungs. They also manifest vascular problems such as cyanosis, edema of the head, feet, and ulcers of the combs. Others develop diarrhea and paralysis leading to death. The mortality rate is high, approaching 100%.

In humans, infection with avian influenza virus is manifested by the classical flu-like symptoms of cough (usually dry and nonproductive), runny nose, sore throat, dyspnea, fever of more than one hundred degrees F, diarrhea, headache, and myalgia.

Patients that expire are usually those with complications like severe Pneumonia, Acute Respiratory distress, and sepsis with multiple organ failure.

Diagnosis

In LPAI, diagnosis is based on the history, signs, and presence of lesions on the birds. Testers may isolate the virus from the secretions and perform some laboratory examinations like AGID (*Agar gel immunodiffusion test*) to identify the antigen and to know the quantity of the antibody titer.

In HPAI, due to its short course and fatal outcome, the diagnosis is usually presumptive, based on the signs and body lesions of the birds. They can also inoculate the virus in chickens as a confirmatory test. This avian condition must of course be differentiated from other poultry diseases.

In humans, if you have signs and symptoms of the bird flu, see your health care provider but inform him/her of your visit so that the staff of the clinic can take precautionary measures to protect themselves from the infection. A test for detecting strains of bird flu recently approved by the US Food and Drug Administration called Influenza A/H5 Virus real time RT-PCR Primer can be done. Usually, the result is ready in a few hours. Other tests that can be performed are chest X-rays, nasopharyngeal culture, blood culture and other tests of organ functions.

Treatment

In birds, there is no effective treatment for LPAI. Antibiotics may be given as prophylaxis for secondary infections. Nutrition is particularly important, too, in the management.

In HPAI, especially in an outbreak, the spread of the virus is so rapid that in 48 hours one can have a mortality of 90-100%. It is a must to limit the spread in the area and the only method to do this is to kill all the birds in that area. This usually results in great economic loss but it is the only way to stop the spread.

In humans, treatment is by giving neuraminidase inhibitors which are *Oseltamivir* (*Tamiflu*) and *Zanamivir* (*Relenza*). These antiviral drugs can improve the status of patients, especially if given within 48 hours after the start of the symptoms. Even if the infection is caused by the most virulent subtype to date (subtype H5N1), the survival is greater if this medication is given right away. The other antiviral drugs that were given before, *Amantadine* and *Rimantadine* (M2 inhibitors) have been found to be ineffective due to emergence of resistance by the virus.

Antibiotics may have a role in the treatment not for the flu itself but for the prevention of secondary bacterial infections like *Pneumonia*.

Prevention

The world is not prepared for a pandemic with Avian influenza. With a constantly mutating virus as the causative organism, the emergence of a newly formed virulent virus is highly probable. The World Health Organization is working hard to prepare for this possibility. It has asked countries to make a contingent plan for an avian influenza pandemic, but only forty countries have done so. One measure that was proposed and has started now is to stockpile anti-viral medications good for several million subjects. This plan might raise eyebrows, but the idea here is that present medications may not totally eradicate influenza , but they may prevent or delay the spread of the disease in the event of a pandemic. This delay may help give us time in formulating a vaccine specific for that causative virus. Note that we cannot make a vaccine for the new virus without that new virus. We must first isolate the causative virus. It is of great importance for the health agencies to have good surveillance, early detection, and early isolation of the causative virus so that a vaccine can be formulated at the earliest time possible.

To date, we are developing vaccines for the most transferable subtype to humans and the deadliest among the subtypes. This is subtype H5N1. This will be effective if the causative virus of the pandemic is the same subtype. What if it is a new and different subtype? Here comes the problem. Our seasonal flu shots are of different subtypes, so they are of no use in a pandemic.

The USDA has no standard guidelines for the prevention of avian influenza. It is dependent upon the bird and animal industry to set their own measures. This is more on a voluntary basis on the part of the industries. The following guidelines have been set:

1. For areas with frequent Avian influenza, regular monitoring of antibody from blood or egg yolk is recommended. This will detect impending pandemics.

2. Veterinarians and animal handlers must be educated, and observant of any early signs of Avian influenza. If detected, they must report right away to the government health agency. On the other hand, they are also entitled to be informed of an impending pandemic so that they can prepare appropriate protective actions right away.

3. Once detected, it is the responsibility of owners to isolate their flocks, and sacrifice all the birds in the flocks. Additional measures must be done like limiting movement of animal handlers, observing sanitation, and fumigation of all equipment used.

A preventive vaccination is not practical considering that the birds are susceptible to many subtypes of the virus. As we know, the vaccine is species specific. The birds might be immune to a certain subtype because of the vaccination. If the causative virus of the next outbreak is a different subtype, they are again susceptible as before. In humans, aside from the above- mentioned guidelines, workers in the bird industries must wear appropriate personal protective equipment in the event of an outbreak or a pandemic. Avoid eating raw meat or undercooked meat of birds. Vaccination in humans likewise is not practical unless it is a vaccine specific for the causative subtype of the raging infection.

10. HUMAN PAPILLOMA VIRUS INFECTION[28]

Of the more than 130 different strains of human papilloma virus, about thirty of them can be transferred sexually. They cause conditions in the genitals ranging from a simple wart to a deadly cancer.

Human papilloma virus infection (HPVI) is one of the common sexually transmitted diseases in US. It was reported that in year 2000 there were 6.2 million new HPVI in the age range of 15-44 years old, and 74% of these were between 15-24 years of age. A study done between 2003-2004 claimed that at any given time, 26.8% of women ages 14-59 years old were infected by at least one type or strain of the human papilloma virus.

The National Cervical Cancer Coalition noted that 11% of women in the US do not have regular cervical cancer screening. It is estimated that 14,000 new cases of cervical cancer are diagnosed annually and about 3,900 of them die each year.

CAUSATIVE ORGANISM: *Human Papilloma Viruses* **are DNA-based viruses of family** *Papillomaviridae* **(HPV).**

There are about 130 identified strains, and approximately thirty strains can be transferred sexually. They usually infect the skin or mucous membranes of the genitalia of men and women. These sexually transmitted viruses have been categorized as "considerable risk" and "minimal risk" strains. Those in the high-risk group are the causes of cancer of the parts of the genitalia, primarily, the cervix. The low-risk group may cause genital warts or mild Pap smear abnormalities.

Pathogenesis

The HPV is transmitted by genital contact. Once the viruses are in contact with epithelial tissue through micro-abrasion, they attach to the integrin and lamina of the basal cell. They eventually enter the cell by endocytosis. Once inside the cell, they are transported to the nucleus by an unknown mechanism, and later alter the transcription stage of protein synthesis in the nucleus.

This alteration of transcription results in increased cell division with new cells more differentiated, like the upper layer of the epithelium resulting in increased layers of the epithelium. During the process of translation, a nucleic acid sequence of the virus is created, resulting in a new virus which later is sloughed off of the dead cells. These new viruses invade neighboring epithelial cells and the life cycle of the virus continues.

In the initial stage, viruses in the epithelial cells do not cause symptoms nor signs. This is called the latency period. This period is, however, very variable. It can be as short as three months or if several years, or even a decade. Even in the latency period, viruses can be transferred to others by sex, and there is no way to prevent the transmission because the person is asymptomatic and unaware of the presence of the virus in his/ her body.

For individuals with very good immunity, the viruses are eventually cleared and rooted out, while in others they continue to stay and develop in the epithelial cells and would show symptoms once the immune system weakens.

Human Papilloma Virus and other Diseases:

1. **Cancer:** The most common cancer associated with HPVI is cervical cancer. Less common cancers are anal cancer, vulvar cancer, penile cancer, and even oropharyngeal squamous cell cancer. The strains identified to cause malignancies are strains 16, 18, 33, 35, 45, 51. The cancer cells have the template of the sequences of the viral nucleic acid in the cell DNA.

In the case of cervical cancer, infection with HPV is not tantamount to the development of cervical cancer. A good immunity would eliminate the virus immediately. Moreover, with a long latency period, a person has enough time to prevent the growth and development of cancer, if she is observant and concerned enough about herself to submit herself to routine pap smears.

HPVI causes 25% of the cancers of the mouth and pharynx (tonsillar area) and most anal cancers. There is a 17-31 times higher incidence of anal cancer among gay people and bisexuals than among heterosexuals.

2. **Warts** are caused by the low-risk HPV. Here are the types of warts:

Genital warts: If a woman has genital warts, it does not necessarily mean that she has cervical cancer at the same time because the HPV types that cause warts (HPV 6, HPV11) are different from those causing cervical cancer. The presence of genital warts does indicate that the person may be vulnerable to the high-risk type of HPV.

The genital warts together with the anal warts (*condyloma acuminata* or venereal warts) are sexually transmitted HPVI. In early stages, these might not be noticed by the person, making him/her a carrier of the virus, but a routine examination will surely reveal the wart and a cause for concern.

Skin warts: Common skin warts are usually caused by HPV type 1 and type 2. They appear as cauliform-like outgrowths of the skin. They are not transmitted sexually and are non-cancerous. They are common on the hand, elbow, knee, and foot.

Plantar warts: These are found on the sole. They grow inward and that is why it is painful for someone to walk with plantar warts.

Ungual warts: These may be subungual (under the nail) or periungual (around the nail) and are hard to treat because of their proximity to the nail.

Flat warts: These are found on the forehead, face, and arms. They are common in children.

Respiratory papillomatosis: This is a recurrent growth of warts in the respiratory tract,on the larynx. These warts are caused by type 6 and type 11.

Juvenile Onset Recurrent Respiratory Papillomatosis: (JORRP) which is acquired by infants during birth from the genitals of the mother infected with HPV. This is a rare form.

Prevention

Cutaneous HPV can stay on any surface for a long time because it has a sturdy outer covering shell called a "capsid." Avoid contact with contaminated surfaces like inside the toilet. Once you have these types of warts, remove them as early as they appear, to prevent spread to other parts of the body and to other persons.

In the case of genital HPVI, a pap smear is a highly effective screening method to detect pre-cancerous lesions of the cervix. A female who undergoes this test regularly can prevent the development of cervical cancer. Since the HPV induces the cervical lesions at a slow pace, there is enough time to remove the abnormal lesions before they transform to a cancerous status. The pap smear is 70-80% effective in detecting abnormalities of the cervix. Other methods used are the *ThinPrep Pap* or *BD SurePath Pap*. They are more effective in detecting abnormal lesions at 85-95%.

CDC Recommendations on Pap Smear

The first pap smear must be done no later than three years after the first sexual intercourse or no later than 21 years old. Then have the pap smear done every year until age 30. After age 30, it is at the discretion of the doctor and the patient as to how often it will be done. For those with negative results on all previous pap smears and with minimal risk factors, pap smears can be done every two to three years until 65 years old. Studies show that there is an inverse relationship to the incidence of cervical cancer to the number of pap smears one received.

HPV Testing

HPV testing is a procedure that detects the DNA of the high-risk types of HPV. It is usually done if there are abnormal pap smear results. If both tests are done together, the sensitivity of the test can reach up to 100%. For men, however, there is no HPV test.

HPV Vaccine

Gardasil has been approved by the USFDA in June 2006 as a vaccine to prevent HPV infection. It is a vaccine for HPV types 6, 11, 16, 18. The types 6 and 11 cause 90% of the genital warts while types 16 and 18 cause 70% of the cervical and other genital cancers. The vaccine is recommended for females between 11-26 years old. The vaccine, however, cannot protect a female from any HPV that pre-infected her before the time of vaccination. Also, the vaccine will only protect her from infections of the types of HPV present in the vaccine. Since some other substantial risk HPV types can also cause cancer, the female has still to undergo pap smears even if she was vaccinated already.

Additional Factors

It has been found out that carcinogens from tobacco are concentrated in the cervix, resulting in increases in cervical dysplasia leading to a doubling of the risk of cervical cancer.

Condom use has lessened the incidence of HPVI. In a study of eighty-two female university students for a period of eight months, those females whose partners used condoms at a rate of 100% had an incidence of 37.8 % of HPVI while those females whose partners used condoms at a rate of 5% had an incidence of 89.3 %.

Microbicides like Carrageenan, used in sexual lubricant gel, have been reported to inhibit HPVI, but this is still under study.

Research is proceeding on the benefits of some vitamins and foods in lowering the risk of HPVI. Intake of more vegetables and fruits, especially papaya, has 54% decreased risk of HPV persistence. People with low levels of vitamin A have an increased risk of having carcinoma in situ. People who take more vitamin C have lower HPVI persistence. People who are taking more vitamin E have a shorter HPVI clearance. People who take more folic acid are reported to have less HPV positive results and their chances of converting to HPV negative results on repeat tests are greater.

There is no cure for human papilloma virus infection at present. The fact that prostitutes have a greater incidence of cervical cancer than nuns leads us to speculate that there is a relationship between sexually transmitted human papilloma virus and cervical cancer.

11. PORCINE REPRODUCTIVE AND RESPIRATORY SYNDROME (PRRS)

PRRS is also known as Blue Ear Pig Disease, a serious disease threatening the pig population of China at this very moment. It has already affected twenty-six of the 33 Chinese provinces.

This was initially believed to affect sows and piglets only, but now it includes boars and thus becomes more fatal. The seriousness of the disease was believed to be due to mutation of the virus. There are around 257,000 reported cases of PRRS but it is believed that there are millions, due to inaccurate reporting. This is a serious condition right now that causes the death of around 18,000 pigs. The Chinese government has started drastic measures to control the spread by sacrificing around 175,000 pigs and has started immunization.

This porcine infection was first reported in the United States in the mid-1980s. It was then called "Mystery Swine Disease." In 1991 in the Netherlands, the virus was identified. It was named *Lelystad virus* coined from the town where the research institution was located. In late 1992, there was an outbreak of PRRS in Texas that devastated the hog industry. There was a 25% mortality rate among the piglets. If they were not stillborn, they were mummified piglets. Some were so weak they survived only for a few days. Sows farrowed weeks early, delivering low birth weight piglets that were also infected by the virus and eventually killed. In early 1993, the only recourse to control the spread was to depopulate the herd. This porcine disease is now widespread worldwide.

It is a disease of pigs, but humans are economically affected due to the scarcity of pigs, with subsequent skyrocketing of prices of pork.

CAUSATIVE MICROORGANISM: *PRRS Virus,* **otherwise known as Blue ear disease virus (classified as *arterivirus*)**

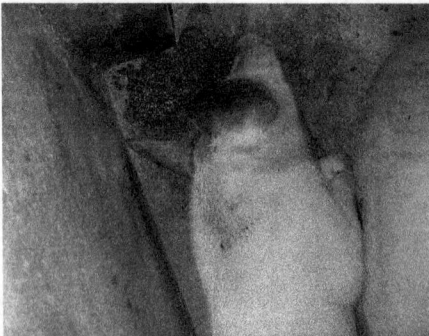

Fig 8-16 A pig with PRRS Virus (Blue-Ear Pig Disease)[29]

Transmission

Transmission is by pig-to-pig contact, from mucous secretions, urine, and manure of infected pigs. Even when there is only one infected live pig, the spread is very rapid, because these animals are usually in a large group or a herd. The chance of contact is exceedingly high. The most affected are the piglets and pigs that have not been weaned. It was found that piglets are still contagious four to five weeks after the PRRS while the older ones are still contagious two to three weeks after PRRS. The virus can be transferred also by a boar to gilts through artificial insemination.

Pathogenesis

The portal of entry of the virus is the respiratory tract. Once the viruses reach the alveoli, they invade the alveolar macrophages. These cells are sentinel cells in the lungs that kill invaders like bacteria, viruses, fungi, and other antigens. The virus, instead of being killed by the macrophage, stays unharmed and multiplies within the macrophages, later killing the macrophages. Since these are inflammatory cells, destruction of these cells lessens the immunity of the infected pigs, making them very vulnerable to the other viruses and bacteria. Thus, secondary infections ensue.

Symptoms

The manifestation of PRRS ranges from asymptomatic infections to a severe form. Diverse manifestations are due to different viruses that cause PRRS. There are about twenty strains of viruses. A previously infected pig is still vulnerable to the same infection caused by different viruses. Infected pigs may have milder symptoms because of some form of temporary immunity. The worst cases are in the virgin uninfected pigs. Without any specific immunity against the PRRS virus, they develop severe symptoms.

These manifestations are related to the respiratory system. They result in sneezing, nasal, and oral discharges, and cough. Other manifestations are fever, anorexia, reluctance to drink, lethargy, and the typical bluish discoloration of the ear. The sows produce no milk due to mastitis. They may farrow weeks earlier than normal and their piglets are of low birth weight, and some are aborted. If alive, the piglets are too weak to survive more than a few days.

For the boars, aside from the constitutional symptoms, they lose libido, have low sperm output, and produce small litters. In severe cases boars usually develop pneumonia and abscesses and eventually die.

Diagnosis

The diagnosis is based on the manifestations of the patients especially in an outbreak area. There are laboratory tests that are utilized to confirm the diagnosis. These are the Enzyme linked immunoassay (ELISA) and IFA (*Immunofluorescent Antibody test*). However, it takes 10 days for these tests to turn positive. Tests are done twice on an infected pig at a one-week interval.

Treatment

Once a PRRS is diagnosed in a herd, the first thing to do is to isolate the infected animals, and the infected ones. Clean and disinfect the pens, especially the nursery, and leave them empty for two to three weeks before the animals can reoccupy it. Those that are ill are given antibiotics.

Veterinarians usually sacrifice one infected animal and determine, by culture and sensitivity, the bacteria (usually *Pasteurella multocida*) responsible for the infection and the antibiotics best forcombating the bacteria.

Prevention

There are no pre-infection preventive measures except for the routine cleaning of the pens. What matters is to be aware and informed of the clinical manifestations of PRRS.

With this knowledge, one must be observant of the animals and report unusual behavior and presentations. Refer to veterinarians at once for any fever and respiratory symptoms.

If the outbreak is severe, it is worth sacrificing the animals rather than treating them, to prevent further spread of the infection.

Commercial vaccines are available and should be given right away. In a close-herd infection, it is better to determine the virus and later make a vaccine tailor-made against the specific causative virus.

12. OROPOUCHE FEVER[30]

Oropouche fever is a viral infection occurring in some tropical countries. It is caused by an arbovirus named *Oropouche orthobunyavirus* (OROV) and transmitted from a sloth to humans by biting midges and mosquitoes. The name *Oropouche* was coined from the Oropouche River in Trinidad Tobago where it was first described in 1955. With our desire to improve economic livelihood and the quality of life, we humans tend to invade the forest and other wilderness by constructing roads, building dams, extracting timber, or simply clearing it for agricultural crops and pastures. These activities unintentionally affect forest ecosystems and the cycles of inhabitants, resulting in more exposure to vectors and other insects. It was proven that deforestation in Latin America resulted in an increased incidence of infectious disease transmitted by vectors like mosquitoes, flies, and foxes. Oropouche fever is an example of a virus caused by these conditions.

In 1960, Oropouche was isolated from the blood of a sloth caught while the Belem-Brasilia Highway (Brazil) was being constructed. Since then, sporadic epidemics followed in the city of Belem, state of Para, in Brazil. In the Amazon, it was a common condition, second to Dengue fever. Between 1978 to 1980, there were 130,000 reported cases of Oropouche fever from this area alone.

CAUSATIVE MICROORGANISM: *Oropouche virus,* **a positive** *Sendai virus* **of family** *Bunyaviridae*

VECTOR: Midge (species of *Culicoides paraensis*) Mosquitoes (*Ochlerotatus serratus*)

IMMEDIATE HOST: Sloth (*Bradypus tridactylus*)

SYMPTOMS, DIAGNOSIS, AND TREATMENT:

The symptoms of Oropouche fever are like those from other viral infections. These are fever, body malaise, headache, *Myalgia*. In rare cases, symptoms of meningitis such as severe headache, vomiting and body stiffness may develop. The diagnosis is based on detecting the antibodies against the virus in the blood. The illness is self-limited and can be treated by taking analgesics and anti-inflammatory drugs. Aspirin, however, is not used for fear of prolonging the bleeding time. Patients are mostly treated on an outpatient basis. Seldom do complications set in.

13. HENIPAVIRUS DISEASE[31]

Henipavirus disease is caused by a virus under the genus of the family Paramyxoviridae order Mononegravirales. There are two types of viruses under this family that are known to cause this disease. These are the *Hendra virus* and the *Nipah virus*. These viruses were named after the places they first appeared.

In September 1994, in Hendra, Brisbane, Australia, a pregnant mare fell ill, in a group of 24 horses. The stable hand and a prominent horse trainer, Mr. Vic Rail, tried to nurse the sick mare. Unfortunately, the pregnant mare died. This was followed by the death of 12 other horses. After a week, the stable hand and Mr. Rail got sick with flu-like symptoms. The stable hand recovered while the horse trainer developed respiratory and renal complications and eventually died.

Retrospectively, news was learned of another horse death a month prior (August 1994). In this case, two dead horses were autopsied, with the owner in assistance. After several weeks, the owner was admitted to the hospital with a diagnosis of meningitis. He recovered from initial hospitalization but eventually died after one year. Researchers were able to isolate the *Hendra virus* from the brain of the owner. In the first case they found that the virus came from the nasal discharge of the pregnant mare.

In 1999, there was an outbreak of respiratory disease and neurologic disease on a pig farm in Nipah, Malaysia. This resulted in 105 human deaths and the sacrifice of over a million pigs. This was followed by eleven human deaths in Singapore, one of which was a worker exposed to pigs coming from Malaysia.

Retrospective studies were done in these countries and found out that the reservoir hosts of the viruses are a certain type of bat called the *Pteropod fruit bat* (flying fox). The viruses were isolated from the reproductive system and the urine of the bat, indicating that the transmission is from urine and birthing fluid of the bat.

Yearly thereafter, outbreaks of this disease were reported. Most of the outbreaks occurred in the Asian country of Bangladesh. A single report of an outbreak occurred in India in January 2001, with 66 cases and a mortality rate of 74% . The most recent incident reported was in February 2008 in the Rajbari province of Bangladesh, with eighteen cases and eight fatalities.

CAUSATIVE VIRUS: *Hendra* **virus and** *Nipah* **virus, genus of family** *Paramyxoviridae,* **order** *Mononegravirales.*

Pathogenesis

For unknown reasons, the virus stays in the bat without causing disease. There has been no study on how the virus enters the bat. The virus stays in the reproductive system and urine of the bat. With decreased bat habitat due to human encroachments like deforestation, real estate developments, and settlements, and with change in their food distributions, the bat would migrate to other areas closer to human population.

The *Hendra virus* is transmitted to horses. As a type of animal that constantly moves around the wilderness, there are chances for horses to encounter bat urine and birthing fluid. Horses infected with this virus develop respiratory signs like increased nasal discharges, lacrimations, and salivations. Autopsies of horses showed pulmonary edema and congestion.

Humans acquire the virus through direct contact with the infected horses. The viruses were found in the secretions and other body fluids of infected horses. The virus can also be acquired by eating improperly cooked or raw horse meat. The disease affects the respiratory system of humans. Autopsies showed pulmonary edema and congestion in the lungs, with pulmonary hemorrhage.

In the case of the **Nipah virus**, the index farm in Malaysia was found to be close to the bat territory, with some of the bat spillage going to pig farms. Infected pigs have a characteristic long dry cough described as "barking pig syndrome" or "one mile cough syndrome." Some pigs may develop the so called "porcine neurological syndrome." Just like the **Hendra virus**, humans acquire the virus by direct contact with secretions and other body fluids of the pig. Eating improperly cooked or raw pork is also a method of transmission. Infected humans developed more neurological symptoms than respiratory symptoms. In patients with severe disease, seizures may develop, leading to coma and eventually death. Those autopsied showed encephalitis and meningitis.

There is no report of bat transmission to humans nor humans transmitting to other humans. There were, however, unconfirmed reports of human-to-human transmission in Bangladesh.

Symptoms

Horses infected with the *Hendra virus* and *Nipah virus* have the same signs. The infected animals usually manifest respiratory signs in the form of increased nasal discharges, increased lacrimation, increased salivation, followed by weakness and failure to eat. The pig, though, has the typical long dry cough called "barking pig syndrome" or "one mile cough." Some animals manifest neurological symptoms like seizures and eventually die.

In humans, the body systems involved is more often the nervous system rather than the respiratory system. There are initially flu-like symptoms like headache, high grade fever, and muscle pain (myalgia). Other infected patients develop abdominal pain, cough, difficulty of swallowing, blurring of vision and vomiting. In severe forms of the disease people have seizures (60% of cases), eventually leading to coma that requires a respirator. Many later die. Others may just have respiratory involvement like the case of the horse trainer, Mr. Vic Rail. He initially had flu-like symptoms which later developed into a secondary respiratory infection, and later died.

Diagnosis

The clinical pattern of the disease process, the existing circumstances of its occurrence, and the signs and symptoms presented are sufficient as a basis for an educated diagnosis of this disease. For confirmatory tests, however, determination of antibody (IgG and Ig M) from the blood may be done by *Enzyme Linked Immunosorbent Assay* (ELISA) or by Real time Polymerase the chain reaction (RT-PCR).

In humans, since the initial symptoms are remarkably like flu-like symptoms, it is important to differentiate this from other common diseases presenting with the same symptoms. In serious cases, however, having these symptoms with existing circumstances suggestive of the disease (like the presence of an outbreak in horses or pigs), it is necessary to make an educated guess and to start treatment immediately. Laboratory tests in human infections are very necessary parameters to confirm the diagnosis, so blood tests for antibody (IgG, IgM) determinations must be performed. These tests are the ELISA and RT-PCR. The best diagnostic parameter, however, is the isolation of the causative virus.

Treatment

The treatment for this type of infection is more supportive or symptomatic, by giving analgesics for muscle pain and headache, and antipyretic medications for fever. No drug has yet been discovered to date that can cure this type of viral infection. Some health personnel, however, would recommend the antiviral *Ribavarin*. This drug has been shown to decrease the severity and duration of the illness, especially when given at the early stage of the disease course. Serious cases, like those with respiratory failure and comatose cases, need ventilation support.

Prevention

Since these viruses were recently discovered, studies have been lacking as to their habitat, geographical distribution, and reservoirs. These viruses might just be staying in the wilderness in the forest or any unsettled areas undisturbed by humans. They might be living in other creatures in the forest. So, there is much research left to do as to the epidemiology and the geographical habitat of these causative organisms. In the meantime, it is important that veterinarians, animal handlers and owners be observant of the animals, watching for any signs of infections. Any suspicious sign must be reported right away. The essence is early detection in preventing its spread. Once detected, it is necessary to sacrifice infected animals in the herd. Healthy animals must be relocated. Again, for human workers, it is necessary to always wear personal protective equipment in dealing with the infected animals. Observe proper sanitation.

Finally, fumigation of the animal pens, and any objects, equipment and feeding receptacles, must be done.

14. RIFT VALLEY FEVER

Rift Valley Fever is a disease of animals, especially livestock, which can cause enormous financial losses. It can be transferred to humans by mosquito bites. Again, when there are heavy rains due to climate changes, expect flooding and increased water levels. These will create breeding grounds for mosquitoes and can cause vector- borne infections including Rift Valley Fever. During the warm phase of the El Nino/Southern Oscillation (ENSO), heavy rainfalls are associated with outbreaks of Rift Valley Fever in Africa.

As early as 1915 in Kenya, there were reports already of certain diseases affecting livestock.

However, the virus was identified in 1931 after an epidemic among sheep in the farm village of Rift Valley in Kenya. (This is how the disease got its name.) Since then, several outbreaks were reported in the Sub-Sahara and North Africa. In 1977-78, there was an epidemic in Egypt where millions of people got infected and several thousands died. In 1998, an outbreak in Kenya killed four hundred people. In Saudi Arabia and Yemen, there was a reported outbreak in 2000. From 2006-2007, several sporadic outbreaks happened in Kenya, affecting the economy of the region. The popular food *Nyama Choma* (roast meat) in the region was not patronized because it was *believed* to cause the spread of the infection. At the end of January 2007, there were 148 casualties from the outbreaks. Massive vaccinations were resorted to by the Kenyan government. Cattle movements were banned. On March 12, 2007, the cases of Rift Valley Fever significantly decreased.

Rift Valley fever was also considered to be a potential biological weapon.

CAUSATIVE MICROORGANISM: A virus named after the disease, "Rift Valley Fever virus" of the genus *Phlebovirus* under the family of *Bunyaviridae*.

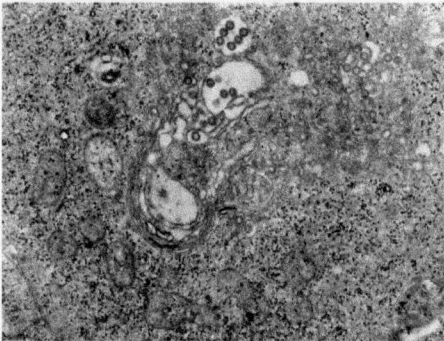

Fig. 8-17. Rift Valley Fever Virus. (CDC)[32]

VECTOR: A mosquito of the Aedes or Culex genera.

Transmission

Rift Valley Fever mainly affects domesticated animals like cattle, sheep, camels, and goats. For whatever reason, among these animals the virus has more affinity for sheep, causing more casualties among them. The virus is transmitted by the bites of mosquitoes, of Aedes species.

The mosquitoes bite infected animals and then bite other animals, thereby transferring the virus. There is also vertical transmission in the mosquito life cycle. The female mosquitoes can transmit the virus to their eggs. This is the reason the outbreaks can recur, because the infected mosquito eggs can survive for longer periods of time, even years.

Humans acquire the virus from the bite of infected mosquitoes. Other means of transmission are direct and indirect contact with infected animals like inoculation, contact with broken skin, or by inhalation. So, in outbreaks, expect more cases among slaughterhouse workers, butchers, veterinarians, and herders. Others may acquire the disease by eating poorly cooked meat or drinking unpasteurized milk coming from infected animals. There is no human-to-human transmission ever documented.

Symptoms

The incubation period ranges from two to six days. Infected humans usually manifest mild flu-like symptoms like fever, body malaise, joint pains, and headache. Some are even asymptomatic throughout the infection period. Others show manifestations of meningitis like stiffness of the neck and photophobia, the reason some would mistakenly consider this early part of the disease as meningitis. Usually, patients recover within a week after the onset of symptoms.

In severe cases, the infected humans manifest three distinct syndromes:

Ocular syndrome: (.5-2%): In this syndrome, the lesion is more confined in the retina and the patient has blurring of vision or decreased vision. If it affects the macula (the area of the retina with the sharpest vision), the person may end up blind (50% of those with retinal lesions).

Meningoencephalitis: (less than 1%): Here the virus has invaded the brain. The symptoms include intense headache, confusion, disorientation, hallucination, loss of memory and even convulsions or coma. Death rarely occurs, but residual neurologic deficits may last for several years or even permanently.

Hemorrhagic fever: Here the virus has invaded the liver and will be detected in the blood. The infected person manifests *Petechiae* (purple dots) and/or ecchymosis (hematoma) on the skin, gastrointestinal bleeding manifested as hematemesis (vomiting of blood) and melena (blood in the stool). Fifty percent of patients with this syndrome die; this usually happens 3-6 days after the first appearance of symptoms.

Diagnosis

For documentation of the disease, the diagnosis of Rift Valley fever can be confirmed by serologic tests like ELISA that show specific IgM antibody against the virus. Other tests are culturing the virus through inoculation on an animal, and antigen detection by the Polymerase chain reaction (PCR).

Treatment

No treatment is necessary for Rift Valley Fever since the symptoms are typical for a viral infection; besides, most cases either have mild symptoms or are asymptomatic. Just as with an ordinary viral infection, the patient can be given symptomatic treatment like analgesics and antipyretics. In severe cases, patients are hospitalized and treatment is directed to the syndrome they manifest.

Prevention

The focus of prevention is the animal vaccination. Both the live attenuated virus vaccine and the inactivated virus vaccine have been tried. The live attenuated virus given in one dose was proven to be effective but can cause abortion for pregnant ewes. The inactivated virus vaccine has fewer adverse effects but needs multiple doses to be effective. If vaccines are given, it should be at a time when there is no epidemic yet. Veterinarians, through their contact with the infected animals during outbreaks, may inadvertently transmit the virus to other animals, intensifying the outbreak.

Whether there is an outbreak or not, veterinarians-- or any animal workers, for that matter (slaughterhouse workers, butchers, animal husbandry workers)-- should always use protective clothing like aprons and gloves when processing animal meat. Likewise, health workers should follow the Standard Universal Precautions in handling patients. For ordinary consumers, refrain from eating poorly cooked meat and drinking unpasteurized milk. There are animals that remain asymptomatic yet are carrying the virus.

Banning movement of livestock can be implemented. It may not prevent the spread, but at least it can slow down the spread of the infection.

Lastly, use preventive measures directed at the vectors. Larvicide spraying on breeding ground is successful in limiting the vectors. Likewise, protective clothing must be worn if you are in an endemic area. Apply insect repellants on exposed skin and use mosquito nets or mosquito coils at night.

15. MURRAY VALLEY ENCEPHALITIS

Murray Valley encephalitis, previously known as Australian encephalitis, is a viral infection endemic in northern Australia. This is caused by two related viruses which are the *Murray Valley encephalitis virus* and *the Kunjin virus*. They are classified as different viruses because one causes a milder disease with a lower morbidity rate. This is an infection of birds but may affect humans through mosquito bites. As a mosquito-borne infection, this infection would be expected to increase during excessive rainfall due to climate change. More rainfall leads to more breeding grounds for mosquitoes.

In southeastern Australia, a certain disease was noted in 1917 with 114 cases. A similar disease course was noted in 1918 (67 cases) and 1925 (10 cases). They called it "Australian X disease." In 1951, again a similar disease course happened with forty-eight cases, of which nineteen people died. Currently researchers were able to identify the virus. In 1974, an outbreak of similar disease resulted in forty-two cases, mostly in the Murray Valley region of Australia, where the name of the disease came from. Since that year sporadic outbreaks have occurred in Australia, and most have been in the Aboriginal communities. The latest recorded outbreak was in the Kimberley valley in Australia in 1993 when extensive rainfall and flooding resulted to nine cases with four deaths.

CAUSATIVE MICROORGANISM: *Murray Valley encephalitis virus* and *Kunjin virus,* both genus *Flavivirus* under the family of *Flaviviridae.*

VECTOR: A mosquito called *Culex annulirostris*

NATURAL RESERVOIR: Water birds of the Order Ciconiiformes (herons, cormorants) [33]

The virus is maintained in a cycle between natural reservoirs (water birds) and the vectors (mosquitoes). The virus is transmitted by mosquito bites during the outbreak.

It has been said that the more mosquito bites a person has, the greater chance he will develop encephalitis. There is no document to prove that this virus is transmitted from human to human.

Symptoms

Murray Valley encephalitis is a rare but fatal disease. It is so rate that it occurs only at a rate of one to 500-800 infected by the virus. For symptomatic cases, the person shows the usual viral infection symptoms of fever, headache, and body malaise. Others may experience nausea, vomiting, dizziness, and diarrhea. For those with nervous system involvement, the usual manifestations are drowsiness or lethargy, irritability, confusion, and neck rigidity. Worse manifestations are convulsions, fits, stupor, and coma. Those patients who survive have either residual mental problems or functional limitations.

Diagnosis

The diagnosis of Murray valley encephalitis can be confirmed by several blood tests that show the presence of antibodies against the MVE virus or Kunjin virus. The blood extraction for the test should be done twice, at an interval of seven to ten days, within the acute or convalescent phase of the disease. There must be a four-fold rise of antibodies against the MVE virus or Kunjin virus to confirm the diagnosis. The cerebral spinal fluid can be used as a specimen for determining the antibody titer.

For severe cases, radiologic procedures like CT scans, MRI, and EEC can be used to show lesions in the brain. The cerebrospinal fluid can be examined for pleocytosis or the presence of Ig M specific for the MVE virus or the Kunjin virus.

Treatment

There is no specific treatment for Murray valley encephalitis. Since symptomatic cases present only mild manifestations, they are usually given symptomatic treatment just like what is given for a typical viral infection. Usually, the patient will recover in a few days.

Much more critical are those with severe neurologic symptoms. These patients should be hospitalized with supportive treatment, if necessary, they should be admitted in the intensive care unit with respiratory support.

Prevention

As the usual preventive method for mosquito borne-disease, eradication of the mosquito population is the primary goal. The use of insecticides or larvicides sprayed on the breeding sites of the mosquitoes must be instituted by the government.

Remove all water receptacles in the yards, especially during rainy days, because these serve as breeding sites. Of course, in situations when you are to stay in mosquito-infested areas, follow the guidelines to prevent insect bites. Wear clothes that cover all parts of your body as much as possible. On areas of exposed skin, you may apply insect repellant. Likewise, have mosquito screens on your house windows and doors. If this is not possible, use a mosquito net or mosquito coils during sleep.

16.   MARAYO FEVER

Marayo Fever is a recently documented viral infection in tropical South America. It presents as dengue-like illness with skin rashes, fever and arthralgia and is transmitted by mosquitoes. Again, a vector that thrives more during rainy sessions.

In February 2008, there was a study done among the residents of the Pau D'Arco settlement in Santa Barbara, Brazil. In this study, they isolated the causative virus from infected persons and from the mosquitoes. They inoculated the serum of this infected person and the pooled mosquitoes to newly born mice. After a few days' of observation, they were able to isolate the same virus from the mice by complement fixation and immunofluorescent assays.

CAUSATIVE MICROORGANISM: *Marayo virus of genus Alphavirus and family Togaviridae*

VECTOR: Mosquitoes (*Haemagogus janthinomys* – main vector
Other genera are *Wyeomyia, Aedes, Sabethes, Limatus* – which are minor vectors).

In the study, thirty-three out of the thirty-six infected persons manifested the following symptoms: fever (100%), *Arthralgia* (89%), *Myalgia* (75%), headache (64%), articular edema (58%), rashes (49%), retroocular pain (44%). Less common symptoms were itchiness, dizziness, anorexia, lymphadenopathies, and vomiting. There was no mention as to the treatment in the study. The disease, however, would last only for a week with good prognosis.[34]

17. BLUETONGUE DISEASE

Bluetongue disease is a viral infection of wild and domesticated ruminants. It affects primarily sheep and occasionally goats, deer, or (rarely) cattle. This infectious disease cannot be transferred by direct contact. It can only be transferred by bites of an insect-vector.

Bluetongue disease peaks at midsummer and ends in late fall. Since 1998, there is a trend observed by researchers on the spread of Bluetongue disease. The spread is northward on the globe. This has been attributed to the global warming that results in warmer temperature in winter times, favorable for the existence of the virus. In 2006, there were cases of Bluetongue disease in Germany, Luxemburg, Belgium, and the Netherlands. By 2007, Bluetongue disease was detected along the Germany-Czech Republic border, and in Great Britain, Switzerland, and Denmark. By 2008, the Swedish provinces of Smaland, Halland, and Skane have been affected by Bluetongue disease. By 2009, in Norway, cows from two farms showed immune responses to Bluetongue virus.

CAUSATIVE MICROORGANISM: Bluetongue virus, of the genus *Orbivirus,* under family Reoviridae

VECTOR: Midges (*Culicoides imicola, Culicoides variipennis sonorensis, Culicoides brevitarsis*)

Transmission and Pathogenesis

Bluetongue disease usually peaks in midsummer and ends in late fall. A perplexing observation, however, is that the blue tongue virus (BTV) can survive the winter season.

It would appear again in late spring and peaks in summer. Researchers believed that the virus could last in midges (small flies) as they overwinter. Likewise, the virus can stay a longer time in affected animals during wintertime.

Upon the bite of infected midges, the virus is introduced to the animal. As the T-cells of the animal invade the bite area, they engulf the virus. The virus multiplies inside the T-cells. Since these T-cells are immune cells and constantly circulate throughout the body, the viruses are spread. The animal then develops viremia (viruses in the blood). Symptoms of animals with BTD are in the oral cavity and the hoof of the animals. This does not mean that these are the only body parts affected. The virus may cause symptoms in other body parts, but they are usually minor symptoms. Some viruses, however, become inactive or reposed once inside the animal and they would be reactivated once taken up by the midges. When the midges bite the infected animals, the midges will assume the virus and transmit it to other animals.

For transmission to happen, the weather must be suitable for the development of vectors and the animals must be in the viremic state.

Aside from the bites of the midges, BTV can be transmitted through the placenta to the fetus. This has been proven in Northern Europe. Researchers at the Institute for Animal Health (UK) have reported three cows recovered. In the late Fall of 2007, the cows were exported to Northern Ireland in January 2008. After a month, they gave birth to calves with the virus. BTV also can be transferred by using instruments for the butchering the animals at the time of viremic stage.

Symptoms

The incubation period of Bluetongue disease is seven to 20 days. The main symptoms are not those seen on the tongue (despite the name) but those seen on the mouth. In fact, the manifestations on the tongue occur only in severe cases. The animals manifest fever, excessive salivation, nasal discharges, and an inflamed and swollen face, especially the lips. The swelling may extend to the lower jaws. In extreme cases, there are widespread hemorrhages on the oral mucosa and nasal mucosa and the tongue. The tongue becomes cyanotic. That is why it is called *Bluetongue disease*. Because of the effects on the mouth, the animals become emaciated.

Other manifestations are foot lesions. The cuticle above the hoof may be inflamed, causing lameness. The sheep move by knee-walking while the cattle roll from one side to the other, causing another name, "Dancing disease." Torticollis (stiffness of the neck) is occasionally observed in some animals.

Note that these manifestations are commonly seen in sheep, which are the most affected among the ruminants. If the infection is severe, the sheep may die in a few weeks. For those sheep that recover, convalescence is terribly slow, lasting for several months. In other animals like cattle and goats, there are no manifestations in spite of viremia.

Mortality rate is low. The cause of death is emaciation due to inability to eat, weakness and lack of mobility.

Treatment and Prevention

There is no recommended treatment for Bluetongue disease. Anti-inflammatory and antipyretic medications may be given.

There are several ways to control the spread of the infection. First is movement control and quarantine of imported animals, especially those coming from areas endemic for bluetongue disease, to make sure that they are free from disease. Second, vector control through housing the animals during the peak time for vector activity. This can be achieved by covering the animal shelters with nets impregnated with insecticides. Third, disruption of the breeding cycles of the vectors by eliminating breeding grounds. The midges would breed in moist areas and those containing animal dung. Remove all dung regularly. Destroy any stagnant water. For animals that need constant watering, change the water weekly. Of course, spraying the possible breeding areas with insecticides and larvicides is recommended.

Vaccination has been tried in Europe and the results were encouraging. Right now, there is an available vaccine in United Kingdom produced by Intervet Fort Dodge Animal Health.

[1] https://commons.wikimedia.org/wiki/File:SARS-CoV-2_without_background.png#/media/File:SARS-CoV-2_(CDC-23313).png Viewed 24 April 2021.

[2] www.wikipedia.org/wiki/Ebola

[3] https://commons.wikimedia.org/wiki/File:Ebola_virus_virion.jpgViewed 5 Jan 2022.

[4] www.en.wikipedia.org/wiki/Marburg_virus Viewed 5 Jan 2022.

[6] https://commons.wikimedia.org/wiki/File:Marburg_virus.jpg Viewed 6 Dec 2022

[7] www.cdc.gov/ncido/dvd/spb/mnpages/dispages/lassaf.htm

[8] https://commons.wikimedia.org/wiki/File:Lassa_virus.JPG Viewed 7 Jan 2022.

[9] www.cdc.gov/ncided/ed/vollno#/kilgore htmn

[10] www.en.wikipedia/wiki/Dengue_fever

[11] http://upload.wikimedia.org/wikipedia/commons/b/b0/Dengue.jpg

[12] www.en.wikipedia.org/wiki/West_Nile_virus

[13] https://commons.wikimedia.org/wiki/File:West_Nile_Virus_Image.jpg Viewed 9 Jan 2022.

[14] https://commons.wikimedia.org/wiki/File:Culex_pipiens_2007-1.jpg Viewed 10 March 2022.

[15] www.en.wikipedia.org/wiki/Hepatitis_A

[16] https://commons.wikimedia.org/wiki/File:Hepatitis_E_virus.jpg Viewed 9 Jan 2022.

[17] www.livestock.uiuc.edu/diarynet/

[18] https://en.wikipedia.org/wiki/Aphthovirus Viewed 12 March 2022.

[19] https://en.wikipedia.org/wiki/Foot-and-mouth_disease Viewed 9 Jan 2022.

[20] www.en.wikipedia.org/wiki/Chikungaya

[21] https://commons.wikimedia.org/wiki/File:Aedes_aegypti_CDC-Gathany.jpg Viewed 10 March 2022.

[22] Ayurveda is a Sanskrit term literally meaning "Science of Life." It is a natural system of medicine thatis practiced in India. http://www.holisticonline.com/ayurveda/ayv-introduction.htm Viewed Aug 22, 2009.

[23] www.en.wikipedia.org/wiki/O'nyong'nyong_virus

[24] www.en.wikipedia.org/wiki/yellow_fever; www.emedicine.medscape.com/article/232244-overview

[25] https://commons.wikimedia.org/wiki/File:YellowFeverVirus.jpg 9 March 2022.

[26] www.health.nytimes.com/health/guide/disease/; www.cdc.gov/flu/avian/gen.info/facts.htm;www.en.wikipedia.org/wiki/Avian_flu

[27]https://commons.wikimedia.org/wiki/File:Colorized_transmission_electron_micrograph_of_Avian_influenza_A_H5_N1_viruses.jpg Viewed 5 Jan 2022.

[28] www.en.wikipedia/wiki/Human_papillomavirus

[29]

[30] www.en.wikipedia.org/wiki/Oropouch_fever

[31] www.en.wikipedia.org/wiki/Henipavirus

[32] https://commons.wikimedia.org/wiki/File:Rift_Valley_fever_tissue.jpg Viewed 5 Jan 2022.

[33] www. ajtmh.org/cgi/reprint/67/3/319.pdf https://en.wikipedia.org/wiki/Betaarterivirus_suid_1#/media/File:Pig_blue_ears.JPG Viewed 10 March 2022

[34] www.cdc.gov/eid/content/15/11/1830.htm - viewed Jan 13, 2009.

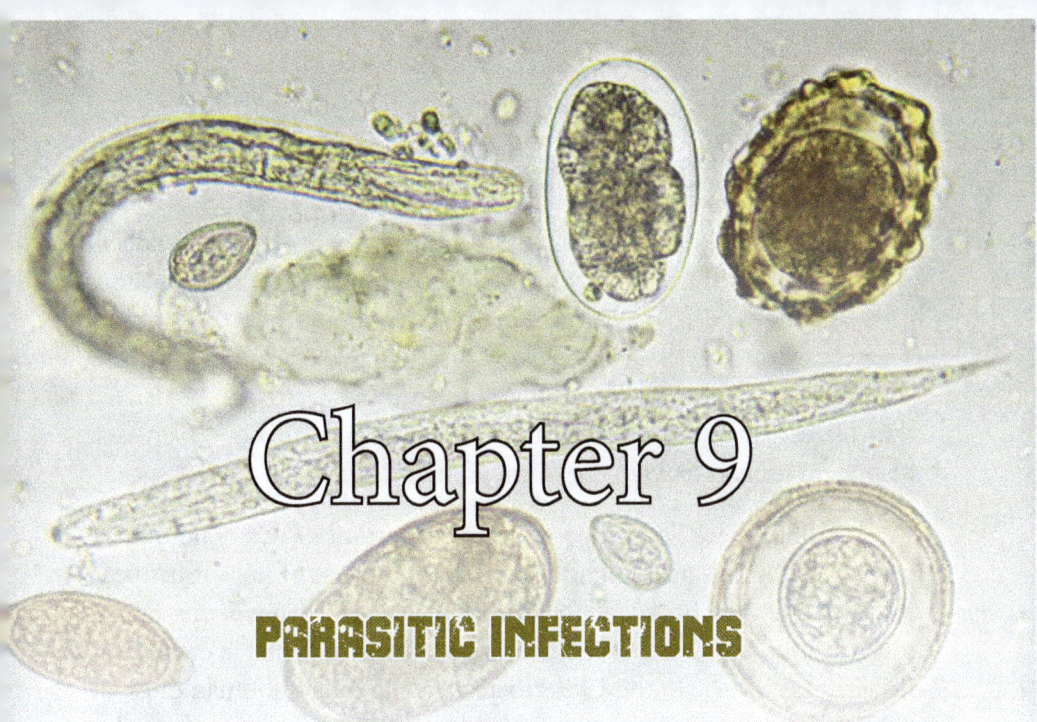

Chapter 9

PARASITIC INFECTIONS

EXAMPLES OF PARASITES THAT CAUSE INFECTIONS

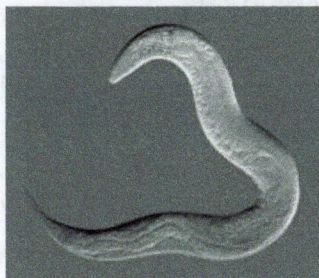

Fig. 9-1 A Nematode worm ***Caenorhabditis elegans*** Photo by Bob Goldstein[1]

Nematodes are a large group of unsegmented roundworms. The majority are good but we only hear about the bad ones which cause disease. There are thousands of known nematodes and some have been implicated in animal and plant diseases. They range in size from microscopic to over a meter in length.

1. DRACUNCULIASIS

Dracunculiasis is a preventable parasitic disease. It is also known as Dracunculosis, Dracontiasis, or Guinea worm disease. This disease is believed to have been in existence since ancient times. In fact the symbol of medicine, the Greek Askleepios or the Roman Aesculapius, which is a one headed snake wrapped around a totem, was inferred to be derived from Dracunculiasis because the best way to treat this disease, even as at the present time, is to wrap the worm around a stick.

In 1980, the CDC proposed total eradication of this condition by the year 2009. This goal might be attained because as of now, it has been estimated that there has been a 99% decrease in the incidence of this disease. Jimmy Carter, a former US President, was very instrumental in eradicating this terrible disease. (In the 1980s there were 3.5 million human cases; by 2023 there were only 14 cases worldwide). (Go on Youtube and watch a video about Jimmy Carter and Guinea worm eradication)

If successful, it will be a landmark medical accomplishment to date. Currently, this disease is limited to sub-Saharan Africa and in Sudan, where because of the civil war, it becomes impossible to totally eradicate this disease.

CAUSATIVE ORGANISM: *Dracunculus medinensis*, **a nematode (roundworm)**

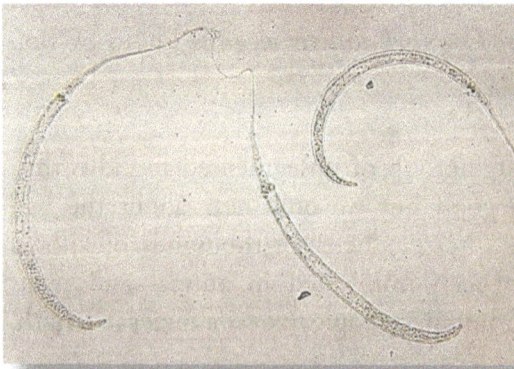

Fig. 9-2. Dracunculus medinensis larvae sp. CDC/ Provided by Mae Melvin[2]

INTERMEDIATE HOST: fresh water crustacean or copepods (water fleas)

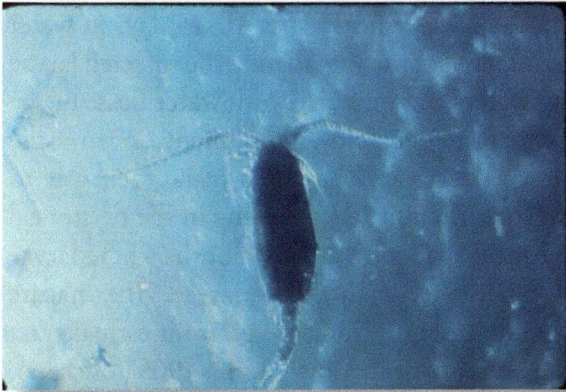

Fig. 9-3. An example of a copepod. A type of organism that acts as a vector of guinea worm infection. Photo by David Arieti and Moshe Gophen.

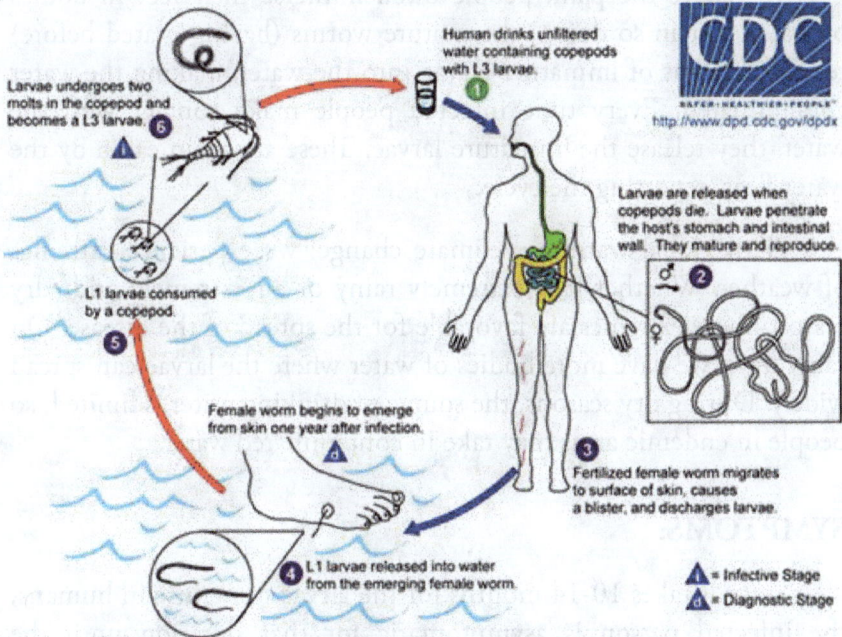

Fig 9-4. Life cycle of *Dracunculus medinensis*[3]
Public Health Image Library #1342

PATHOGENESIS:

The larvae of **Dracunculus medinensis** stay in the fresh water. They are eaten by Copepods or water fleas. After 10-14 days in water fleas, these larvae become infective. Humans acquire the infected larvae by inadvertently drinking water contaminated with water fleas. In the human stomach and duodenum, the acid environment destroys the water fleas but not the infective larvae. They then make their way to the small intestine, penetrating their mucosa, staying in the peritoneal cavity or in the retroperitoneal spaces. In 10-14 months, these larvae mature and mate, resulting in the death of the larvae. The mature worms, which are all females, migrate from the gastrointestinal tract to the subcutaneous tissue, mainly of the lower limbs (90% of cases). They grow as long as two to three feet and exit from the skin. At first the person will have painful blisters in the area of exit. After 24-72 hours, the blister ruptures, forming a small ulcer where one end of the worm emerges.

To relieve the pain, people often immerse their feet in bodies of water, but in so doing, the mature worms (having mated before) release millions of immature larvae into the water, making the water contaminated. Every time infected people make contact with the water, they release the immature larvae. These are again eaten by the water fleas, repeating the cycle.

With global warming, climate change, we experience extremes of weather. Whether it is extremely rainy or an extremely arid/ dry season, these extremes are favorable for the spread of the disease. On rainy days, we have more bodies of water where the larvae can spread widely. During dry seasons, the source of drinking water is limited, so people in endemic areas may take in contaminated water.

SYMPTOMS:

Since it takes 10-14 months for the larvae to mature in humans, the infected person is asymptomatic for that duration until the appearance of the symptoms. A few days before the worm emerges from the skin, the infected person develops fever and allergic types of symptoms. The affected skin develops a blister which is painful, swollen and pruritic.

Eventually this blister ruptures to become an ulcer where one end of the worm stays. For those people who have no access to medical care, this ulcer is secondarily infected by bacteria, eventually leading to abscess or cellulitis formation. In some cases, lymphadenopathies may develop on the affected limb.

DIAGNOSIS:

Clinical manifestation of Dracunculiasis is very typical. In areas endemic with this disease, laboratory confirmation is not necessary. The emergence of a worm-- described as a white filament, in the center of a painful ulcer, which can be wrapped around a stick and slowly pulled out-- is a definitive diagnosis. Some would want to examine the yellowish discharge for the presence of *rhabditiform* larvae under the microscope. The infected person may develop some hematologic findings but these findings are non specific and common in other parasitic infections. In the CBC (complete blood counts), the person may have leukocytosis which is mainly due to eosinophilia. Immunoglobulin levels may rise depending on the stage of the disease, but again these are nonspecific findings. There are no serological tests for Dracunculiasis. Some radiologists can recognize the density of the worms on X-ray films, especially if the worms are already calcified.

TREATMENT:

The core in the treatment of Dracunculiasis is the so called "Stick Therapy," which is wrapping the worm around the stick and pulling it out a few centimeters per day. The pulling should be done slowly and carefully not to break the body of the worm; otherwise a portion of it will be left in the person, causing more allergic reactions and severe pain. Antibiotics (Metronidazole) may be given as an adjunct therapy to prevent secondary infections. Patients may take analgesics like Aspirin or Ibuprofen to relieve the pain and swelling. If a surgical set up is available, the worms can be removed by incision of the skin.

PREVENTION:

Since transmission to humans is through the oral route by drinking contaminated water, it is a must for the people in endemic areas to follow control measures such as:

-Do not drink water from ponds or rivers. Instead, drink water from underground sources such as boreholes or hand-dug wells.

-Use a filtering cloth or a nylon mesh filter to remove the larvae and the copepods as well.

-Treat the drinking water with an approved larvicide (ABATE) to kill water fleas.

-For patients suffering from the disease, never enter ponds and other bodies of water used for drinking purposes.

ERADICATION:

The Bill and Melinda Gates Foundation and the Carter Center have sponsored the so- called Dracunculiasis Eradication Program (DEP) whose primary goal is to eradicate Dracunculiasis worldwide. As of 2005, Asia has been completely free of the disease.

No transmission has been reported in 9 out of 20 endemic countries. In 1984, there were 3.5 million cases reported, in 2003 over 30,000, but only about 16,000 in 2004. With these very encouraging results, researchers are confident that this parasitic infection can be globally eradicated in 2009.

2. ONCHOCERCIASIS

Onchocerciasis is a parasitic infection (nematode) transmitted by a blackfly. The WHO considers Onchocerciasis as a global infection with 18-40 million afflicted worldwide.

99% of these are found in Africa while the rest are found in Yemen and 6% in the Latin American countries of Mexico, Guatemala, Ecuador, Columbia, Venezuela and Brazil. Onchocerciasis is also called "River Blindness" because this infection is mostly common in areas where there are rapidly flowing streams in remote agricultural areas in Africa.

One of the main organs afflicted by this infection is the eye. About 5,000 infected people have visual impairments and another 270,000 are blind.

One peculiar feature of this parasitic infection is that one bite of the blackfly would not be enough to cause the infection. Multiple bites by the blackfly are required to develop the infection. The more blackfly bites a person has, the greater chance he will get the infection. For this reason, Onchocerciasis is rare in casual travelers but common in those people that reside longer in endemic areas, such as adventurous travelers, missionaries, and Peace Corps volunteers.

CAUSATIVE ORGANISM: A nematode worm *Onchocerca volvulus,* of the Family *Flaridae;* female worm size is 60 cm whilemale is 2-3 cm.

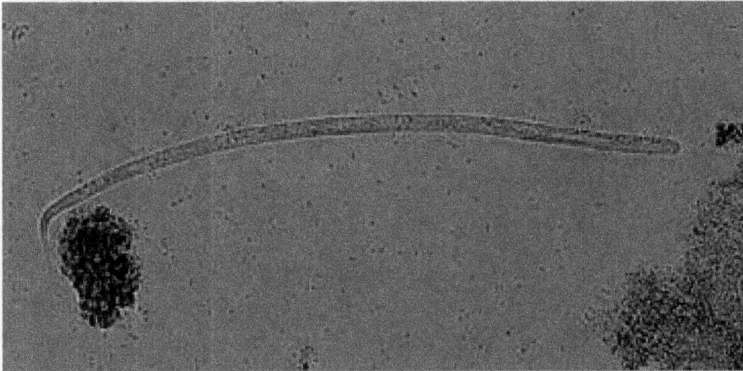

Fig. 9-5. *Onchocerca volvulus-* [4]

VECTOR: Blackfly, *Simulium sp.*

PATHOGENESIS:

Humans are the only definitive host of the parasite, *Onchocercia volvulus*. When the blackfly bites an infected person, it sucks the microfilariae (small worms) in the subcutaneous tissue with the blood meal. From the blackfly gut, the microfilariae migrate to the thoracic flight muscles as the microfilariae mature to the first larval stage (J1), then they mature into the second larval stage (J2) at which time they migrate to the proboscis of the blackfly and stay in its saliva. In the saliva, they attain their third larval stage (J3). It takes about two to three weeks for the microfilariae to pass through these three larval stages.

When the same blackfly bites another person, the third stage larvae are injected by the blackfly into the blood of the person being bitten. These will migrate to the subcutaneous tissue, where they mature into adult worms within 6-12 months, causing subcutaneous nodules. The smaller adult male worms will mate with the larger female worms, producing 1000-3000 eggs per day. The eggs develop inside the female adult worm and later become microfilariae. These are released one by one by the adult female worms to the subcutaneous tissues. The microfilariae are the stage of the worm when the blackflies suck them upon biting the host. The adult worms last for 15 years in the subcutaneous layer of the host.

These subcutaneous nodules called **Onchocercomas** are firm, painless lesions that usually appear on the trunk, face and head. They can also be found on the pelvic areas and lower extremities. Adult worms usually remain in the nodules which are encapsulated by a fibrous coat, limiting access of the immune system. Dead adult worms remain in the nodules. The microfilariae, however, migrate to different parts of the body through direct extension through the blood or through the nerves. Some of them die as immature worms. These dead microfilariae cause intense inflammatory reactions with the formation of cell- mediated immunity and antibody- mediated immunity. Circulating immune complexes are deposited in the perivascular part of different body organs causing granulomatous inflammations.

Skin involvement can present as Acute Dermatitis, Chronic dermatitis, Dermatitis or Lichenified Dermatitis. Involved lymph nodes show granulomatous inflammation, fibrosis, or even atrophy. They migrate to the eye and stay in the aqueous humor or vitreous humor. Inside the eye, they can invade the sclerae, cornea and even conjunctivae causing Punctuate keratitis, which later becomes Sclerosing keratitis which can turn a transparent cornea to an opaque one. Opacity of the cornea leads to blindness.

SYMPTOMS:

The incubation period for Onchocerciasis is 9-24 months. The initial symptoms are mild pruritus followed by papular rashes erosions then lichenification due to frequent irritations and scratching. Later on the skin is described as "leopard skin" due to the hardness and the leathery texture. On the skin, usually there are areas of hyperpigmentation around the hair follicles and these are surrounded by hypopigmentation.

Eye lesions are generalized in the whole eye, although it is the cornea that is more involved. Persons may have Punctuate keratitis (35%) or Sclerosing keratitis (10-15%).

These keratitis forms, added to dead microfilaria, cause more inflammatory reactions on the cornea, leading to opacity and neovascularization. This will block the passage of light rays going inside the eye, leading to blindness. Inside the eye, the microfilariae affect the iris, causing a pear-shaped iris or iris atrophy. When the ciliary body is involved, this results in glaucoma, and the lens may have a cataract. In the posterior chamber, the microfilariae may affect the choroid and retinal layer of the eye, causing chorioretinitis.

Lymph nodes are also affected, especially the ones in the groin and the sex organs. The lymph nodes develop granulomatous lesions resulting in the so called "hanging groin" and elephantiasis of the genitals.

Lastly, in severe cases, general debilitation ensues, impairing the person's mobility and affecting his/her social acceptance.

DIAGNOSIS:

Clinically, the presence of the triads of subcutaneous nodules, chronic dermatitis, and eye impairment in a debilitated person coupled with elephantiasis are indicative of Onchocerciasis. In an endemic area, it is not hard for a medical practitioner to diagnose this parasitic infection.

There are several laboratory tests that can be done to confirm the diagnosis. The standard lab test is the "skin snip" or "scleral punch" which quantifies the microfilarial load /milligram of tissue. This test, however, has some pitfalls. It is an invasive procedure that can transmit blood borne diseases, including HIV. It is painful as well. Moreover, this test is less sensitive in low transmission areas or in those areas where treatment has been started already.

A simple nodulectomy can be done and the specimen is sent for microfilarial identification. The adult worms are coiled, threadlike and white. The living microfilaria are coiled while the dead microfilaria are straight and opaque.

A non-invasive and more reliable test is the "Oncho-dipstick Assay." The sensitivity of this test using urine is 100% and using tears is 92%. Specificity of the test on both specimens is 100%.

Other good laboratory tests are the ELISA which is 97% sensitive and Immunochromatographic test (ICT) which is 86% sensitive.

Another test is the Mazzotti test which is a combination of 1.6% diethylcarbamazipine (DEC) with a Nivea lotion applied on an area of skin, 10 x 10 cm square. A papular reaction on the skin is a positive result. This test, however, has some side effects like pruritis, and discomfort.

TREATMENT:

The mainstay of treatment is drug therapy using Ivermectin (Meclizan or Stromectol) given a single dose of 12mg or 150 micrograms / kg body weight. The benefit with this drug is that a single dose is effective against the microfilaria for two years. Some quarters, though, recommend repeat doses as early as 18 months. In three days, the microfilaria population is reduced by 83%, and 99.5% after three months.

Suramin (Antrypol) is the only drug effective against the adult worm. This is given as an intravenous medication weekly for six weeks. It has several side effects, though, which is why this drug is given only to patients that are not improving with Ivermectin.

Amocarzine (CGP-6140) and diethylcarbamazepine (DEC) are drugs previously used, but due to their side effects, they have been phased out.

PREVENTION:

1. Onchocerciasis Control Program in West Africa (OCP). The United Nations Development Program (UNDP), the Food and Agricultural Organization (FAO), WHO, and the World Bank created the OCP in 1974. The mission of this program was to eradicate the vector blackfly in seven West African countries. This was done by weekly aerial spraying of insecticides in those suspected breeding areas of the blackfly over a period of 14 years. This program, which ended in 2002, attained remarkable success.

2. In 1995, following the success of OCP, the African Program for Onchocerciasis Control (APOC) was created. Here the objective was to have a sustained and continuous distribution of Ivermectin to the endemic communities. This program's target was 59 million people in 17 other non- OCP countries. This program was done with continuous aerial spraying of the insecticide in isolated areas.

3. In 1992, Onchecerciasis Elimination Program for the Americas (OEPA) was created with the theme "The Elimination through sustainability of Mectizan treatment." The program treated 270,622 people in America in 1998.

A vaccine has not been developed yet because the antigen of *Onchocercia volvulus,* aside from being very complex, shows cross-reactivity with other filarial diseases. Lastly of course, people in endemic areas, especially the adventurous travelers, volunteers and missionaries, must follow the guidelines for preventing bites from mosquitoes such as using repellants of any form. The use of mosquito nets in the place of residence is very important.

Currently, Onchocerciasis is not yet eradicated in spite of all the programs. Still, there are very encouraging results from these programs, and they need to be continued.

FILARIASIS

3. SUBCUTANEOUS FILARIASIS

4. SEROUS FILARIASIS

5. LYMPHATIC FILARIASIS

Filariasis is an infectious disease in tropical countries caused by parasitic nematodes. This group of diseases is classified into three categories according to areas of the body where the parasites stay. They are the following: the Subcutaneous filariasis, where the parasites stay in the subcutaneous layer of the skin; the Serous cavity filariasis, where the parasites stay in the serous cavities of the abdomen; and Lymphatic filariasis or the so called Elephantiasis, where the parasites stay in the lymphatic system. Among the three types it is the Lymphatic filariasis that is of most medical interest.

There are more than 120 million people afflicted with Filariasis worldwide and around 1 billion are at risk of infection. The lymphatic type, which is most common among the three, accounts for 90 million infections, forcing the World Health Organization to initiate a global program for eradication of lymphatic filariasis.

There is no endemic area for Filariasis in United Sates. Once in Charleston, South Carolina there was prevalence of the nematode worm, **Wuchereria bancrofti,** due to the presence of mosquito vectors in that area. Common sources of infection in the US are returning travelers coming from endemic areas. Some were the returning missionaries and returning Peace Corps Volunteers.

HISTORY:

History mentions Elephantiasis alone, being the most common type of Filariasis.

As early as 4000 years ago, Elephantiasis was believed to be in occurrence, as was leprosy. It was, however, in the 16th century when the symptoms were first demonstrated by Jan Huyghen von Linschoten during his exploration. Similar reports from other explorers to Asia and Africa followed. The clear understanding of the disease came a century later.

In 1866, Timothy Lewis was able to establish the association between microfilaria and elephantiasis. In 1876, Joseph Bancroft identified the causative adult worm that was later named after him. In 1877, Patrick Mason theorized that the life cycle of the worm involved an arthropod vector. In 1900, George Low discovered that the worms can be found in the proboscis of the mosquito and transmitted to humans by injecting its saliva upon sucking blood.

CAUSATIVE ORGANISM;

Lymphatic filariasis – caused by *Wuchereria bancrofti, Brugia malayi, Brugia timori*

Subcutaneous filariasis – caused by *Loa loa* (African eye worm), *Mansonella streptocerca, Onchocerca volvulus, Dracunculus medinensis* (guinea worm)

Serous cavity filariasis – caused by *Mansonella perstans, Mansonella ozzardi*

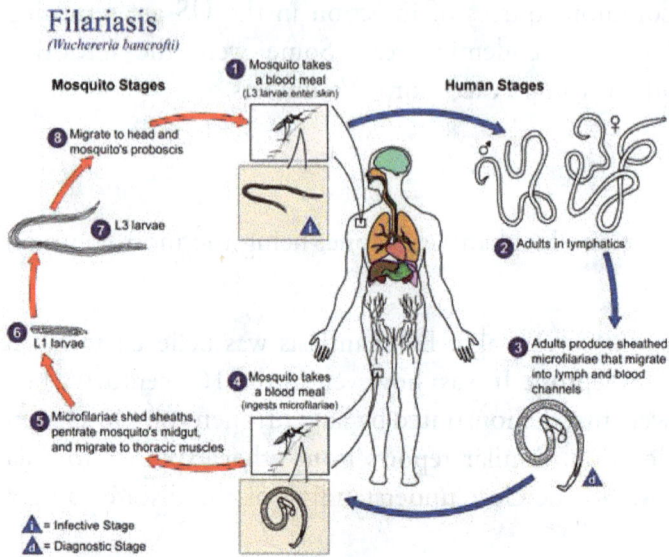

Fig. 9-6. Life cycle of Wuchereria -CDC. [5]

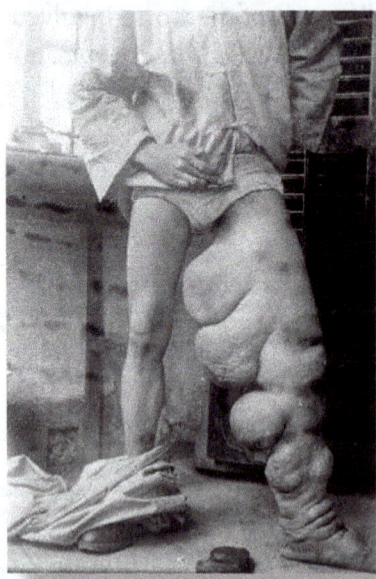

Fig. 9-7. A case of elephantiasis caused by *Wucheria sp.*[6]

VECTOR:

Lymphatic filariasis - transmitted by mosquito of genera **Aedes, Anopheles, Culex, Mansonia**

Subcutaneous filariasis – is transmitted by blackfly of species **Simulium**

Serous cavity filariasis – is transmitted by midges (small flies)

TRANSMISSION AND PATHOGENESIS:

The adult male and female filarial worms stay in the lymphatics, blood and subcutaneous tissue of humans. They mate and then the female worms release thousands of microfilarae.

The vectors suck the microfilariae during a blood meal. In the mosquito, the microfilarae shed their sheaths and migrate to the midgut and to the thoracic muscles of the vectors where they develop into three stages of infective larvae. These infective larvae find their way to the proboscis of the vector where they are released to human skin upon another blood meal of the vectors. After a year, the larvae molt in two stages and mature to become adult worms.

In the case of lymphatic filariasis, these worms stay in the lymph nodes (mainly the femoral and epitrochlear nodes) and lymphatic vessels. They cause migration of plasma cells, eosinophilia, and macrophages. Due to their prolonged presence, there are repeated inflammatory reactions causing damage to the lymph nodes, resulting in the blockage of lymphatic flow and leakage of protein-rich lymphatic fluids that cause veruccous (wartlike) and thick skin. There are secondary infections by streptococcus and fungus that cause enlarged feet mimicking those of elephants, thereby the name Elephantiasis. **Wuchereria bancrofti** and **Brugia malayi** may lodge in the lung tissue, causing hyper responsiveness called Tropical Pulmonary Eosinophilia. This is also called Occult filariasis.

In subcutaneous filariasis, the worms stay in the subcutaneous tissues and the eye. They cause dermatitis and skin nodules which are more prominent in bony prominences. In the eye, they cause pannus formation, optic atrophy and corneal fibrosis leading to blindness.

Serous cavity filariasis is the mildest among the three types. The worms stay in the abdomen and seldom cause symptoms.

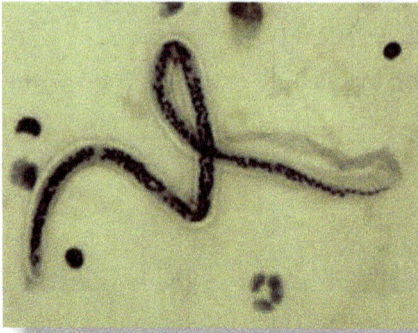

Fig 9-8 Wuchereria bancrofti[7]

SYMPTOMS:

Lymphatic filariasis is otherwise called Elephantiasis. Here the patient presents with swollen genitals and a swollen scrotum. Patients have difficulty in mobility and prefer to sit on their enlarged scrotum. In addition, the backflow of lymphatic fluid causes the legs to enlarge and the leakage of the lymphatic fluid causes the verrucous and thick appearance of the skin mimicking elephant legs. Other symptoms are testicular and inguinal pain, slight fever, exfoliation of the affected skin and inguinal lymphadenopathy. In subcutaneous filariasis, the symptoms are limited to the skin and the eye. The skin manifests with erythematous, pruritic, edematous, with scablike eruptions. The skin thickens, described as hanging groin and leopard skin. The eye manifestations are corneal fibrosis, keratitis, choroiditis, glaucoma, iridocyclitis and optic atrophy. All of these eye findings can cause blindness. Patients with serous cavity filariasis are usually asymptomatic. If ever they have symptoms, they are mild fever, pruritus, and mild abdominal pain.

DIAGNOSIS:

Presentation of a patient with enlarged legs and scrotum or genitals in an endemic area would be enough to consider primarily a Filariasis condition. Of course, confirmation is by identification of the worm in thick blood smears by using the Giemsa stain. The blood smear is taken at night because at lower temperatures the microfilarae circulate (nocturnal periodicity) and besides, the mosquitoes bite more at night. Microfilaria can be indentified from the chylous fluid (a **milky-appearing peritoneal fluid that is rich in triglycerides**) from the lymph node aspirate. A polymerase chain reaction can demonstrate the filarial antigens. Other diagnostic procedures are CT scan, MRI , and even a simple X-ray may at times detect the presence of calcified worms in the lymph nodes.

TREATMENT:

Pharmacological treatments for Filariasis are the combination of Albendazole and Evermictin or Abendazole and Diethylcarbamazepine (DEC). This pharmacological regimen, however, is not enough in a patient with elephantiasis. More important in these patients are the treatments of the concomitant secondary infections (bacterial infections and fungal infections). So continuous hygiene to the affected limbs and genitals must be observed. Measures to promote lymphatic flow must be done.

In a wider perspective, the primary goal of the eradication of filariasis is to break the cycle of the organism by eliminating the vector and eliminating the microfilariae in humans. Studies show that single doses of DEC have a good one-year effect in eliminating the microfilariae. More so if combined with Albendazole. A single dose of the two drugs is effective in eliminating the microfilariae in 99% of patients.

The WHO Global Program for Elimination of Lymphatic Filariasis recommended single doses of the two drugs (Albendazole and DEC) yearly for four years. Another recommendation is the use of ordinary table salt fortified with DEC to be taken for one year.

6. SCHISTOSOMIASIS[8]

Schistosomiasis, also called "Bilharziasis," is a parasitic disease to consider if you are traveling in endemic areas. This disease is widespread in South Africa, particularly in sub-saharan areas, like Malawi, and the Nile river in Egypt. In South America, it is common in Brazil, Suriname and Venezuela. In the Caribbean, it is common in the Dominican Republic, Guadalupe, and Montserrat. In the Middle East, it is common in Iraq, and Saudi Arabia. Lastly in Asia, the incidences of this disease are still in great numbers, particularly in the center of Cambodia, Laos, Philippines, Indonesia and the Mekong Delta.

With global warming, resulting in floods, we expect more water bodies or increases in the water level of these bodies. This factor will provide more areas for snails to thrive and multiply. Since the snail is the intermediate host of the schistosoma worm, we would expect rises in the incidence of this disease with global warming.

In December 1988, 16 US tourists visited Okavango in Botswana. Within five weeks in the place, 11 of them got the disease. They went swimming, wading, bathing and washing in the forest waters. Although their symptoms were generalized and non-specific, the laboratory tests confirmed the diagnosis. The 11 tourists had Schistosoma ova in their fecal specimens. None were found in the urine. Fifteen tourists however, turned out positive for antibodies to *Schistosoma* in their serum.

In May 1989, eight US tourists traveled to Cote d'Ivoire and seven of them got the disease after swimming, bathing and wading in freshwater. Right after contact with freshwater, they developed pruritus. Two to three weeks later they manifested the symptoms of schistosomiasis. Laboratory confirmation was done. Four of them turned out positive in that they had ova in their stools while six of them were positive for antibodies for *Schistosoma.*

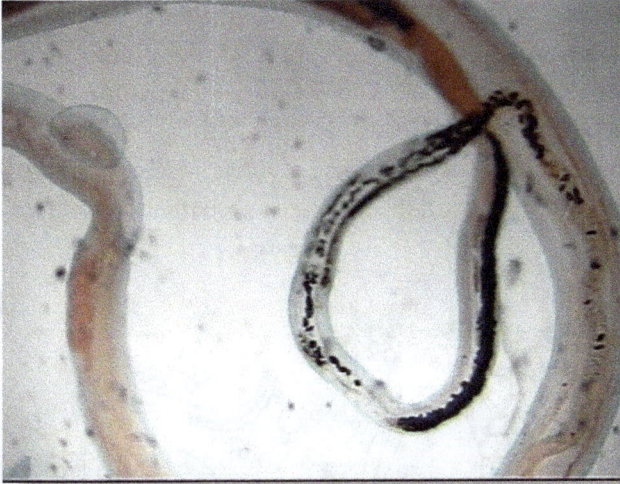

Fig. 9-9. Picture of a female *Schistosoma* worm living in the groove of a male. Photo by David Arieti

CAUSATIVE MICROORGANISM: Parasitic trematodes of genus Schistosoma *(Schistosoma mansoni, Schistosoma haematobium, Schistosoma japonicum, Schistosomia mekongi)*

INTERMEDIATE HOST: Fresh water snail

PATHOGENESIS:

Eggs are brought out from infected people through feces and urine. Once in freshwater, the eggs hatch and release the miracidia (a free-living larval stage). These miracidia can last for two to three weeks in freshwater, within which they infect susceptible snails. In snails, they undergo two generations of sporocysts until they mature into cercaria which later exit from the snail to seek human hosts. The cercaria can last for 48 days in freshwater.

Once attached to the human skin, they migrate to the body tissues, particularly the veins. ***Schistosoma japonnicum*** are found in the superior mesenteric artery which supplies the large intestine. *Schistosoma mansoni* are found in the superior mesenteric artery draining from the small intestine. The *Schsitosoma haematobium* are found in the venous plexus of the urinary bladder. In those venules, the female cercaria deposit eggs that progressively move to the lumen of the intestine and the urinary bladder. They are subsequently brought out with feces or urine.

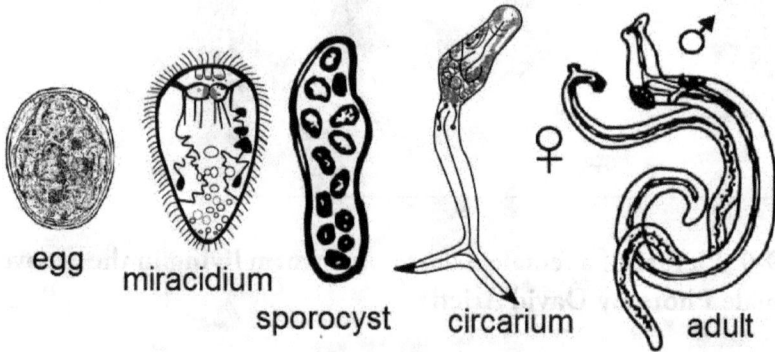

egg   miracidium   sporocyst   circarium   adult

Fig 9-10 Lifecycle stages of *Schistosoma japonicum*[9]

PATHOLOGY:

A few hours after attachment of the cercaria to the skin, mild maculopapular rashes may develop; however, the deposition of the eggs may cause marked peripheral eosinophilia and an increased circulating immune complex. This is manifested as acute serum sickness also called **"Katayam Disease" or Acute Schistosomiasis**. There is also **Chronic Schistosomiasis** which is more common. In the chronic type, granuloma formations and fibrotic changes happen in the tissues they invade. Unshed eggs may go back to the portal circulation and settle at their small tributaries. Here they produce granulomas and fibrotic changes. Granulomas in the intestines cause bloody, cramping diarrhea that eventually changes into Inflammatory Colonic Polyposis. Some cercaria may reach the liver by portal blood vessels and cause peri- portal fibrosis leading to portal hypertension. In the case of **S. haematobium**, eggs are deposited in the venous plexuses of the urinary bladder thus causing granuloma formations. This may cause hematuria and bladder ulcers and can even lead to small cell carcinoma of the urinary bladder.

Other organs affected are the subject's spinal cord (causing transverse myelitis) and the brain.

SYMPTOMS:

Usually a person infected is asymptomatic during the exposure to cercaria but some people may have maculo papular skin lesions within hours after exposure. After several weeks, at the time of egg deposition, the person may have fever, sweat, chills, headache, fatigue, and gastrointestinal discomforts. It is interesting to note that the symptoms of Schistosomiasis are due more to the body's reactions to egg deposition rather than the body's reactions to the invasion of the worms.

LABORATORY TESTS:

Complete blood count (CBC) shows eosinophilia. The presence of Schistosoma eggs in stool and urine specimens is diagnostic of the disease. In blood, the presence of antibodies to Schistosoma is supportive of the diagnosis but it cannot determine if it is an active or inactive disease.

Other tests are the colonoscopy with a biopsy of intestinal granulomas. Cystoscopy with biopsy for urinary bladder granulomas, chest X-rays for lung involvement, liver ultrasound, liver biopsy for liver involvement, and MRI and CT scan for nervous involvement.

TREATMENT:

Once a definitive diagnosis is made, treatment is initiated right away. Drugs cure the infection in 60-98% of cases. It is harder to treat those patients with granulomas and fibrosis and those with other organ complications. For young patients, however, symptoms were reported to improve significantly with treatment. Here are some drugs:

1. Praziquantel (Biltricide), for **S. haematobium** and **S. mansoni**; 40 mg/kg/day divided BID (twice a day) for 1 day; for **S. japonicum** and **S. mekongi**, 60 mg/kg/day divided TID (Three times a day) for one day

2. Oxamniquine (Vasil), is not available in the US, and serves as an alternate treatment for **S. mansoni** only; it is not given to those people with **S. japonicum and S. haematobium**.

Of course if the patient has complications (bleeding varices due to portal hypertension, hemoptysis with dyspnea due to pulmonary hypertension, neurologic manifestation due to transverse myelitis), treatment is directed to the complication first. Once stabilized, the definitive therapy is initiated.

PREVENTION:

People living in endemic areas are often more difficult to educate than travelers to these areas. Some preventive measures are the following:

For residents in the areas:

1. It is important to observe proper hygiene, sanitation and sewage disposal to prevent fresh water contamination.

2. Drugs like molluscicides, to decrease the number of susceptible snails, must be given.

3. Mass treatment of the infected populations must be done promptly to minimize if not prevent contamination.

For travelers in the areas:

4. Avoid swimming, bathing, wading in freshwater in endemic areas. Preventative measures include swimming in the ocean, rather than in freshwater.

5. Make sure the water you drink is safe. If in doubt, better boil the water for at least one minute.

6. Bath water must be stored for more than 48 hours because the cercaria cannot live more than 48 hours in fresh water. To be sure, heat the water for at least five minutes at 150 °F.

7. Vigorously dry the skin with towels, or apply rubbing alcohol after contact with contaminated water to minimize if not totally eliminate miracidia attached to the skin.

8. Bring plenty of bottled water or bottled sodas for the trip.

9. Extensive health care preventive counseling is needed prior to traveling to endemic areas.

7. CRYPTOSPORIDIOSIS

Cryptosporidiosis or "Crypto" is another disease that can be attributed to heavy rainfall and flooding. Global warming, causing an increased frequency of rainfall and extensive flooding, correlates with the increased incidence of Cryptosporidiosis. These floods will wash out the waste water such as raw sewage and water coming from livestock farms, spreading the microorganisms to other areas.

Cryptosporidiosis was rarely heard of until 1976 when it was recognized as a cause of a type of diarrheal disease. In 1982, there were reports of an increased incidence of this disease among AIDS patients. Because of this, Cryptosporidiosis has attained much attention in the medical community. In 1993, there was an outbreak in Milwaukee with 400,000 cases. Since then, concerns have been focused on the safety of the water supply, especially the drinking water.

In developing countries, 5-10% of acute diarrheal disease is due to Cryptosporidiosis. In the US, there was a .27-1.6 incidence of Cryptosporidiosis per 100,000 population in the year 2005. It is considered the most common parasitic cause of food borne diarrheal disease in the United States. As of July 24, 2007, the CDC has reported 18 outbreaks of Cryptosporidiosis in US.

Who is vulnerable to Crypto? It is a fact that a person with competent immunity may be asymptomatic or may have symptoms for a few days only. Even then, he/she is still considered a carrier passing out the parasites in stools. Such a person can easily spread the infections to others, particularly through poor hygiene. People with compromised immunity are very vulnerable to this infection. These are the patients with AIDS, patients with cancers, patients with transplanted organs, and patients under steroid maintenance. As we know, steroids suppress the inflammatory functions of the blood. Also prone to this infection are travelers, tourists, backpackers, and mountaineers who are fond of hiking and trekking to the countryside. Likewise included in this group are people who are fond of eating raw vegetables and fruits, uncooked or improperly cooked foods.

People who don't care much about hygiene are also at high risk, such as children in day care who are fond of putting anything they can hold onto in their mouths, and old people in nursing homes who are careless about sanitation.

CAUSATIVE MICROORGANISM: A protozoan of genus *Cryptosporidium*

There are 3 species that infect humans and animals:

Cryptosporidium hominis – infects only humans
Cryptosporidium parvum – infects humans and animals
Cryptosporidium canus – infects dogs and humans

PATHOGENESIS:

Cryptosporidium is found anywhere, be it in the water, soil, plants grown in the soil or any objects that made contact with human or animal wastes. The life stage of the parasites in the environment are the oocytes. One thing peculiar with this parasite is the hard protective covering shell allowing it to stay outside the body for longer periods of time. This makes the parasites resistant to chlorine- based disinfectants. So chlorinated water that we drink is not totally free from such organisms. It may contain oocytes of the Cryptosporidium.

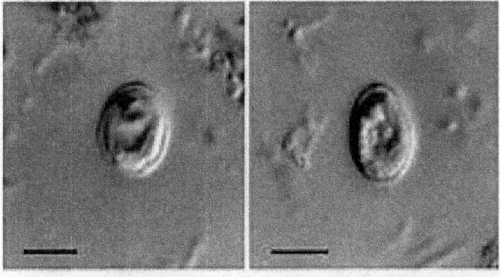

Fig. 9-11. *Cryptosporidium*. CDC.[10]

The disease is transmitted by the fecal-oral route. Humans may accidentally ingest the oocysts by drinking contaminated water. These may be recreational waters (swimming pools, Jacuzzis, hot tubs, or even water fountains) or contaminated lakes, rivers, springs, ponds or swampy areas. In fact, the most common cause of diarrheal outbreak in recreational water is Cryptosporidiosis. This can also be acquired by accidental swallowing of the oocytes picked up from surfaces like bathroom fixtures, toys, tables and other objects. It can also be acquired from eating uncooked or improperly cooked food, raw fruits and vegetables contaminated by Cryptosporidiosis.

Lastly, it can be transmitted by person to person contact, common in day care centers where we have groups of children. They are fond of putting anything in their mouths, even those things picked from the floor. Infected diapers and toys are the usual sources of the parasites. In nursing homes where the residents are not that attentive in their personal hygiene this infection is common. It is also common in promiscuous individuals who can have several partners, especially those performing anal sex, fellatio and cunnilingus.

Once the oocytes enter the body by ingestion or inhalation, excystation occurs in the stomach where the protective shells are removed, releasing the sporozoites. These sporozoites migrate to epithelial linings of the gastrointestinal tract and the respiratory tract. In the epithelial cells, the sporozoites multiply asexually (schizogony) and sexually (gametogony) producing microgamonts (male) and the macrogamonts (female). These gametes later fertilize to produce sporulated oocytes. This is also the stage when they are excreted together with the stool from the human and animal hosts. A single bowel movement can release millions of oocytes and this continuous shedding of the oocytes starts the appearance of the symptoms that may last 50 days after the disappearance of diarrhea. The oocytes last for longer periods of time in the environment because of their protective shells. With the fecal-oral means of transmission, the cycle is continuous.

Histological studies show that the linings of the gastrointestinal tract and the respiratory tract of infected individuals have villous atrophy of the endothelial cells. There is also intestinal crypt hyperplasia with lymphocytic infiltrations.

SYMPTOMS:

People with competent immunity may remain asymptomatic throughout the course of the infection. This is worse from the point of view of an epidemiologist, because the person is unaware of being a carrier of the parasites.

The incubation period of Cryptosporidiosis is two to ten days (average seven days). The main symptom is diarrhea. This may be watery, mucoid or voluminous, with or without accompanying abdominal pain, usually devoid of red blood cells and white blood cells. Diarrhea is usually continuous. It may be intermittent or persistent. It may last as long as ten days, or longer in immunocompromised patients. Voluminous diarrhea can be as much as 15 liters in a day. Most of the time, dehydration ensues. Persistent diarrhea may lead to intestinal mucosal injuries causing malabsorption syndromes and later weight loss.

Other accompanying symptoms are abdominal pain, nausea and vomiting, fever and body malaise. Less common are jaundice, joint pains, dizzy spells, or fatigue.

DIAGNOSIS AND LABORATORIES:

It is hard to have a definitive diagnosis based on the history alone. Presented with the usual gastrointestinal symptoms, there are so many microorganisms that can cause the same manifestations. However, if the patients belong to high risk groups like AIDS patients, cancer patients, patients with transplants, patients on steroid maintenance, children in day care centers or residents of nursing homes, Cryptosporidiosis is considered right away.

Microscopic identification of the oocysts is a must in the diagnosis. This is not seen in routine laboratory examinations. The stool needs to be treated by Sucrose Floatation method or Formalin ethyl method before using the Kinyoun-acid fast stain. Here the oocytes can be identified with their color red. The most reliable, most specific and widely available commercial method is the detection of the antigen by Enzyme Immunoassay.

The oocytes in the stool or in the intestinal biopsy specimen can be seen by use with the electron microscope.

TREATMENT:

For immune-competent individuals, the infection lasts only for a few days with complete recovery. They may be given some anti-diarrheal drugs like Nitazoxanide and oral rehydration therapy. Some cases don't even need medication. Complete bed rest and soft diets are enough management. Problems arise in immunocompromised individuals where this condition may be life threatening. In AIDS patients, diarrhea may extend up to 30 days. The length of diarrhea is usually inversely related with the CD4 lymphocytes of the patients.

Aside from immunocompromised patients, those individuals in the extremes of life are vulnerable to Cryptosporidiosis. Infants whose immunity is not well developed and the elderly whose immunity has declined tend to be more susceptible. In these types of individuals, diarrhea is usually severe, resulting in dehydration. In patients admitted to the hospital, the treatment is initially focused on fluid and electrolyte balance. Any imbalance can lead to detrimental results like organ failure. In chronic situations, the lining cells of the microvilli of the intestine are injured by these microorganisms, decreasing the lactase production of the cells, leading to lactose intolerance. So it is not rare to have lactose intolerance as a complication of chronic Cryptosporidiosis. The diarrhea, however, can be minimized by having a lactose-free diet.

PREVENTION:

Prevention is mostly geared to water purification. Since chlorination does not destroy the oocysts, other methods can be added like Flocculation, Filtration, Ozonation, and Ultraviolet radiation treatment of water. These methods might be expensive but are more beneficial in the long run. The authorities must be strict and aggressive in implementing the rules on the use of recreational water. The authorities must exert power to close any recreational water establishment that is reported to have fecal contamination or even suspected fecal contamination.

On the part of individuals, prevention is more geared to personal hygiene and sanitation. Hand washing is the simplest and the most important form of sanitation. Wash the hands thoroughly with soap and water before and after eating, before and after preparing or cooking food, and after coming from the toilet. Wash hands thoroughly after touching anything that might be contaminated by human or animal stools like diapers, and after cleaning pets and other animals.

Do not drink water if you are not sure of the source. Never use ice if you are not sure where it comes from. If there is no other water to drink, one must boil the water thoroughly before drinking. Do not eat raw vegetables and fruit if you do not know the source. Do not eat improperly cooked or uncooked food. If you are swimming never take a chance by drinking pool water accidentally.

Avoid practicing unprotected anal sex, fellatio and cunnilingus, especially if the partner is having diarrhea.

For those suffering from Cryptosporidiosis, limit movement. As much as possible stay at home. Flush the toilet thoroughly after using it. Clean your hands properly with soap and water and never attempt to go swimming in the pool.

For travelers, it is safer to bring bottled mineral water or carbonated drinks enough for the whole trip. If you want to eat fruit, better peel it yourself to be sure it is not contaminated. As the saying goes, " Peel it, eat it, or forget it."

[1] https://commons.wikimedia.org/wiki/File:CelegansGoldsteinLabUNC.jpg
Viewed 24 April 2021.

[2] https://commons.wikimedia.org/wiki/File:Dracunculus_medinensis_larvae.jpg
Viewed 5 Jan 2022.

[3] https://en.wikipedia.org/wiki/Dracunculiasis Viewed 9 Jan 2022.

[4] https://commons.wikimedia.org/wiki/File:Onchocerca_volvulus_mf1_DPDx.JPG
Viewed 7 Jan 2022.

[5] https://commons.wikimedia.org/wiki/File:Filariasis_01.png Viewed 9 Jan 2022.

[6] https://commons.wikimedia.org/wiki/File:Wuchereria_induced_elephantiasis.jpg
Viewed 9 Jan 2022.

[7] https://commons.wikimedia.org/wiki/File:Wuchereria_bancrofti_1_DPDX.JPG
Viewed 21 Jan 2022

[8] www.cdc.gov/ncidod/dpd/parasites/schistosomiasis/
factsht_schistosomiasis.htm; www.cdc.gov/epo/mmwr/preview/mmwrht-
ml/00001570

[9] https://en.wikipedia.org/wiki/Trematode_life_cycle_stages Viewed 1 June 2022

[10] https://commons.wikimedia.org/wiki/File:Cryptosporidium_muris.jpg
Viewed 7 Jan 2022

Chapter 10

FUNGAL AND PROTISTA INFECTIONS

Fungi are *Eukaryotic organisms* that were once considered plants in the 1950s. Through new biochemical techniques it was decided to place Fungi in a new kingdom by itself. This group includes Mold, Yeasts, and Mushrooms.

They include *Chitin* (a derivative of glucose) in their cell walls, a substance also found in the Exoskeletons of Arthropods such as Insects, Crabs, and Centipedes. Many fungi cause disease in humans and in other organisms such as plants as well.

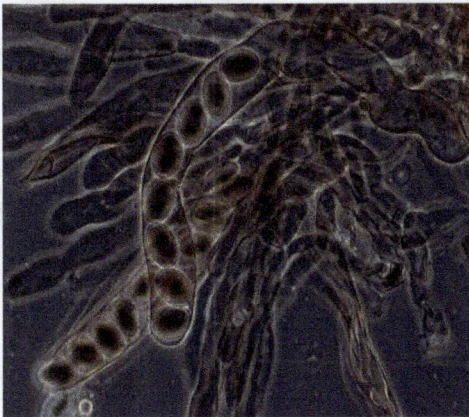

Fig. 10-1 Spores of Morchella eleta. **Photo by peter G. Werner.**[1]

PROTISTA (Protozoa)

Protists are Eukaryotic organisms that is not a plant, animal, or fungus. Rather, they share characteristics of the other groups so for convenience they are grouped as such.

FUNGAL INFECTIONS

1. SPOROTRICHOSIS

Sporotrichosis is a sub-acute chronic infection of the skin and subcutaneous tissue. People touching thorny plants like Roses, Sphagnum Moss, as well as baled hay, are prone to Sporotrichosis. The fungus usually enters through small cuts, abrasions, or puncture wounds by plant spines, sharp edges, or baling wire on the skin. Some would call this *Rose thorn disease* or *Rose-garden disease* because the Rose is one of the thorny plants where the fungus stays. Outbreaks have been reported among nursery workers, handlers of sphagnum moss, children playing on baled hay and rose gardens.

Although the distribution of this disease is worldwide, there is no reported global incidence. In Highland, Peru, the incidence was one case / 1000 people. In the US, it is estimated that there are around 200-250 cases of Sporotrichosis each year.

The *Causative fungus* involves the skin. But with the emergence of AIDS, multiple organs were also involved, including the Lungs (*Pulmonary sporotrichosis*), Joints (**Osteoarticular sporotrichosis**), the Brain (**Sporotrichosis meningitis**), and Paranasal Sinuses.

Causative Microorganism

A fungus named *Sporothrix schenckii* is the Causative fungus. This type grows in the barks of trees, shrubs, and plant debris. It can also be found in plant products like Sphagnum Moss, Timber, and Mulch Hay.

Pathogenesis

Sporotrichosis is quite common in people managing plants with their hands, like farmers, gardeners, horticulturists, and even veterinarians when managing infected animals. The main method of transmission is contact with the fungus through abrasions, cuts, punctures from thorns, barbs, and wires. This infection commonly affects the fingers, hands, and arms.

At the point of contact, papules appear in one to ten weeks. These become reddish and later develop pus and ulcerate. These lesions spread to nearby tissues, following the Lymphatic Channels, resulting in multiple ulcerations. In Microscopy the lesions are made up of Histiocytes, Neutrophils, and giant cells at the center with surrounding Lymphocytes and Plasma Cells. If left untreated, or in patients with compromised immunity such as AIDS, the infection extends by lymphatic routes or hematogenous routes to infect other body parts. Common body parts affected are the joints together with their surrounding structures such as Tendons, Ligaments, Bursae, and Synovial Cavities (*Osteoarticular sporotrichosis*). On rare occasions, fungus may be inhaled in the form of *Conidia* resulting in the Cavitary Lesions in the lungs (Pulmonary sporotrichosis). These lesions as seen by X-rays are like those seen in *Tuberculosis* and *Histoplasmosis*. This pulmonary involvement is usually seen in patients with Chronic Obstructive Pulmonary Disease (COPD). In people with compromised immunity (such as AIDS patients), Sporotrichosis will cause multiple organ involvements including the Eyes, Paranasal Sinuses, Oral Mucosa, Larynx, and Brain.

Symptoms

Cutaneous sporotrichosis is manifested initially as papules which later enlarge and redden, resulting in pustules with ulcers and draining pus. They spread to nearby tissues, following the Lymphatic Channels. Most of the time, patients are *Afebrile* (no fever) and may experience minimal pain. Frequently, patients try several courses of Antibiotic Treatment for Skin Lesions. However, it is often to no avail.

Manifestation of Pulmonary sporotrichosis symptoms are similar to those of COPD, including persistent coughing and difficulty in breathing. Therefore, it is hard for a Medical Practitioner to diagnose a Pulmonary sporotrichosis concomitant with COPD.

Osteoarticular sporotrichosis usually involves multiple joints. They are painful with movement limitations. Severe forms can result in permanent functional impairment of the joints.

Laboratories

Serologic tests involve measurement of antibodies for Sporotrichosis. They are available, but, because of variability of sensitivity and specificity, they are seldom used. The Polymerase Chain Reaction (PCR) is not available for routine use.

The best laboratory method is the isolation of the *Sporothrix schenckii* from pus or infected body fluids like Sputum, Synovial fluid, Cerebrospinal fluid, and Subcutaneous tissues using *Sabouraud dextrose agar* or *Inhibitory Mold agar* as culture medium.

Imaging techniques like X-rays or Computerized tomography for suspected Pulmonary sporotrichosis and Osteoarticular sporotrichosis are only supportive. They are not specific enough for a conclusive diagnosis.

Diagnosis

With the slow progression of the disease and subtle symptoms, it is hard for a clinician to have a definitive diagnosis of sporotrichosis as other infectious diseases present the same Skin Lesions. Differential diagnosis includes: Leishmaniasis, Leprosy, Sarcoidosis, Syphilis, or Tuberculosis of the Skin.

A definitive diagnosis rests on culturing the causative fungus from the exudates or infected body fluids. Serologic and imaging techniques are supportive, but not specific for the diagnosis of Sporotrichosis.

Treatment

There are several treatment options for sporotrichosis but it is drug therapy that is the mainstay of the treatment. The following drugs are effective:

- *Itraconazole* (*Sporonox*): This is the drug of choice for the treatment of sporotrichosis. This drug inhibits the synthesis of Ergosterol, which is a main part of the fungus cell membrane.

-*Fluconazole:* Another Antifungal Drug. This is used as an alternative drug if Itraconazole is not tolerated.

-*Amphotericin B:* This antifungal is given intravenously. This drug plus 5-*Fluorocytosine* are given for Sporotrichosis meningitis. Some patients, however, cannot tolerate this drug because of side effects like fever, nausea, and vomiting.

-*Saturated potassium iodide solution treatment* for 3-6 months has been reported to cure *Cutaneous sporotrichosis*.

-Surgery is the treatment of choice for extensive bony involvement or cavitary lesions in the lungs.

Prevention

The most common method of acquiring the Causative organism for Sporotrichosis is through cuts or abrasions on the skin while touching plants, plant products, and plant debris.

So, use gloves and wear long-sleeved shirts while working with plants, Hay Bales, Rose Bushes, Pine Seedlings and Sphagnum Moss. Additionally, use protective equipment while holding animals or human Skin Lesions to minimize transmission.

PROTOZOAN INFECTIONS

1. MALARIA

Malaria is still an infection of great concern today. It remains one of the worlds most deadly diseases. In spite of advances in medicine and the invention of new anti-malarial drugs, it is still a major problem in tropical and subtropical countries. This includes Africa, Asia, the Middle East, South America, and Central America.

According to the World Health Organization (WHO), 40% of the world's population is at risk of Malaria. Those afflicted are usually people living in the world's poorest countries. It is estimated that there are 515 million Malarial cases every year with 1-3 million deaths.

Malaria is classified as one of the so-called *Tourist Bugs*. Try traveling in those endemic areas. Travel Agents and the local authorities will not allow you to step on this land without taking Chemo Prophylactic Drugs. Furthermore, the travel agents will advise you to bring Insect Repellants, Mosquito Coils, Mosquito Nets, Antimalarial Spray, and Protective Clothing.

Malaria was previously called *Ague* or *March fever* because it commonly occurs in swampy areas. The word *Malaria* however was coined from the medieval Italian term *mala aria*, meaning *Bad air*.

It was Charles Louis Alphonse Laveran, a French army doctor, who first observed the parasites inside Red Blood Cells in 1880. A year later, a Cuban doctor by the name of Carlos Finlay working with Yellow fever claimed that mosquitoes were responsible for transmitting Malaria to and from humans. It was Sir Ronald Ross who studied Malaria extensively. He was able to isolate the Malarial parasites from mosquitoes and was able to prove the transmission of Malaria by the mosquitoes to birds.

Curiously, however, treatment for Malaria had been used even before the discovery of the disease. A concoction from the bark of the Cinchona Tree was used as early as the 16th century for high grade fever. It was in 1820 when the active ingredient *Quinine* was extracted from the Cinchona Bark by French Chemists Pierre Pelletier and Joseph Caventou. Today, several drugs are used in the treatment of Malaria, and *Quinine* is reserved for those drug-resistant cases due to its side effects.

How is Malaria affected by Climate Change? Environmental temperature and the surface of water are factors for the breeding of the Mosquitoes. The higher the temperature and humidity, the better it is for mosquitoes to breed. Another factor is rainfall. The greater the downpour (especially in the El Nino phenomenon), the more substantial the breeding of Mosquitoes. However, extremely hot, and very dry conditions are not conducive for the survival of Mosquitoes.

Causative Microorganism

Plasmodium falciparum, most common causative agent responsible for eighty percent of all Malarial cases, found throughout Africa, Asia, and Latin America.

Plasmodium vivax, found in Tropical and Semi Tropical Countries.

Plasmodium ovale, found in West Africa.

Plasmodium malariae, found Worldwide, but in patchy distribution.

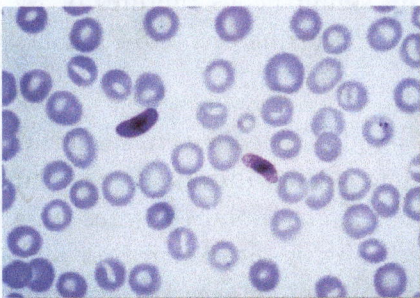

Fig. 10-2. *Plasmodium falciparum*[2]

VECTOR: *Female anopheles mosquitoes* (Male mosquitoes do not bite.)

Pathogenesis

When mosquitoes bite a person who has the *malarial parasite*, the mosquitoes ingest the blood together with the *Plasmodium* gametocytes. Once ingested, the gametocytes stay in the mosquitoes' salivary glands where they differentiate into male and female Gametes. They later migrate to the mosquito's stomach, where they unite to produce the *Ookinete* that penetrates the mosquito gut linings and later transforms into *Oocysts*. In fact, it takes one week for them to mature and upon rupturing, release thousands of sporozoites, infective agents, which later find their way to the mosquito's salivary glands. At this time, they are ready to be injected into another person.

Once the mosquito bites another person, it injects its saliva together with the sporozoites before it sucks the blood. Inside the humans, the sporozoites travel in the bloodstream to end up in the liver. About 30 minutes after injection, in the hepatocytes (liver cells), the sporozoites reproduce asexually for 6-15 days (this is the *Exoerythrocytic stage*) producing thousands of *merozoites* that upon rupture of the hepatocytes go back to the circulation and infect red blood cells. Inside the red blood cells (this is the erythrocytic stage), the merozoites undergo rapid multiplication, again asexually, until the red blood cells rupture, causing *Parasitemia*. This is clinically manifested by high grade fever, chills, and drenching sweats. Again, merozoites in the blood will infect other red blood cells, thereby repeating the clinical symptoms. Usually there are simultaneous ruptures of the red blood cells causing a wave of fever occurring every 48-72 hours, depending on the type of parasites.

In the case of *Plasmodium vivax* and *Plasmodium ovale*, some of the parasites are not transformed to merozoites. Instead, they remain dormant in the liver as hypnozoites for several months to three years, after which they may be reactivated. This is the explanation for a long incubation period and late relapse in *Plasmodium vivax* and *Plasmodium ovale*.

The parasites are not easily destroyed by the immune system because they are intracellularly located in the hepatocytes and in the red blood cells. Only those in the blood circulation are destroyed as they pass through the spleen. This is the reason Malaria is chronic.

In the case of *Plasmodium falciparum*, the phagocytic function of the spleen seldom happens because the red blood cells with the parasites in them are sequestered and do not pass through the spleen. The red blood cells infected with *P. falciparum* display an adhesive protein on each surface, causing the red blood cells with the parasites to stick to the capillary walls. These sticky masses can block the flow of blood, causing a hemorrhage in malaria. Likewise, the sticky masses can stay longer in other body organs, as in the brain causing cerebral malaria, and can stay longer in the placenta causing *placental malaria*.

Eventually most of the merozoites develop into male and female gametocytes and are again ready to be sucked by the mosquitoes, and the cycle is repeated.

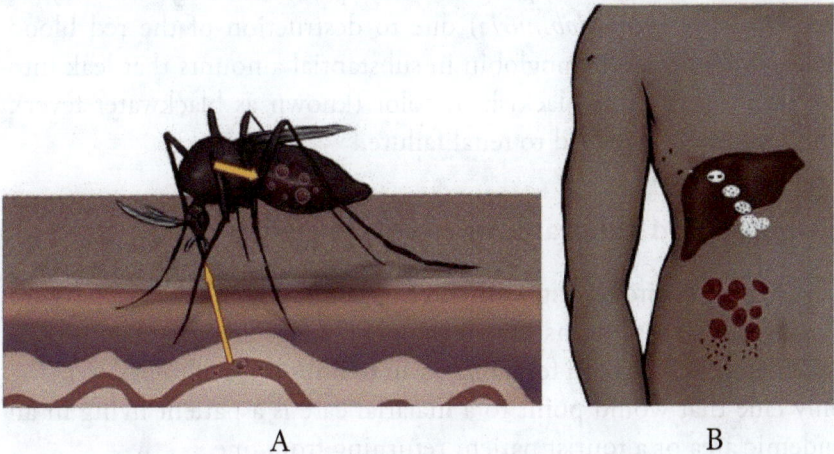

Fig. 10-3 A. A Malaria-carrying mosquito sucking blood of a human. B. Hypno sporites in the Liver. (Illustration by Kevin Bley)

Symptoms

The three most important symptoms of malaria are high grade fevers, chills or shivering, and drenching sweats. These symptoms correspond to episodes of *Parasitemia*. Other symptoms are headache, bodily malaise, nausea, vomiting, arthralgia, and anemia. The latter is caused by *RBC hemolysis*. The typical presenting symptoms of malaria are sudden coldness followed by chills or shivering, fever and then profuse sweating that may last for four to six hours. This course is repeated depending on the species of the parasites. In *Plasmodium vivax* and *Plasmodium ovale*, the course is repeated every two days while in *P. malariae* it is repeated every three days and in *P. falciparum* the course is not periodic. It may repeat every 36-48 hours or may have continuous fever. Those parasites that spread to the brain (cerebral malaria) manifest as severe headache, seizures, or coma. The liver is usually enlarged (hepatomegaly) due to prolonged stay of hypnozoites in the hepatocytes. The Spleen is usually enlarged (*splenomegaly*) due to continuous phagocytosis of the infected red blood cells by the macrophages in the spleen. Hemoglobin may be seen in urine (*hemoglobinuria*) due to destruction of the red blood cells, which release hemoglobin in substantial amounts that leak into the urine, turning it blackish in color (known as blackwater fever). This condition may lead to renal failure.

Diagnosis and Laboratories

A person presenting with high grade fever and headache with constitutional symptoms has all the possible differential diagnoses to consider. It will be hard for a physician to limit himself to malaria. The only clue that would point to a malarial case is a patient living in an endemic area or a tourist patient returning from one.

Once the chills and the drenching sweats appear, malaria is the most likely diagnosis. It does not mean that you must wait for chills and sweat before ordering the *malarial smear*. It is up to a medical practitioner to request pertinent laboratory tests at the very start. Specific laboratory tests to request are based on the doctor's educated guess in the sense that there is no confirmatory evidence yet.

Specifically, this is where the competency and the experience of the doctor are major factors. The more experience he or she has, the easier it is to diagnose the disease.

The most economic and reliable confirmatory test for malaria is the blood smear. Two of these smears taken at 6–12-hour intervals are enough. A thick blood smear film is better than a thin one in identifying the malarial parasites. However, in the thick smear, the parasites can sometimes be distorted, making it hard to identify the species of the parasites. A thin blood smear film will be better for species identification.

In areas where the staff are not well trained in microscopy, a simple *Dipstick Test* that requires only a drop of blood has been developed. This is called the *Immunochromatographic Test* (also called *Malaria Rapid Diagnostic Test, Antigen Capture Assay*, or simply *Malarial Dipstick*). This test detects Antigens in a drop of blood in 15-20 minutes. Another test that has been developed is *Optimal-IT Assay* that does not only detect the Antigen, but it can also differentiate a *P. falciparum* from non-*P. falciparum*.

Treatment

It is important for the clinician to identify the causative species of Malaria because their treatment and prognosis differ. A person with *P. falciparum* malaria needs immediate treatment, while those with *P. ovale* and *P. vivax, and P. malariae* are treated as outpatient cases. In areas where the diagnostic tests are not available, it is wise to consider all malarial cases as caused by *P. falciparum* until proven otherwise. Infections due to *P. vivax, ovale, malariae* are milder. Patients are usually sent home with antimalarial drugs. They are also given symptomatic treatment like antipyretics and NSAIDS and are advised to increase fluid intake. Be cautious, however in giving NSAIDS for those patients with suspected bleeding. NSAIDS can aggravate the bleeding and at chronically high doses, can initiate renal insufficiency. The real problem is the infection due to *P. falciparum*. In most cases the condition is an emergency that requires immediate hospitalization.

What medication to give depends on several factors such as type of species, the status of drug-resistance in the area where the infection was acquired, and other accompanying conditions like pregnant and pediatric patients. For *P. ovale, P. vivax* Primaquine is the drug of choice given at 30mg base PO QID for 14 days.

P. malariae is sensitive to all anti-malarial drugs but the drug of choice is Chloroquine. For non-resistant *P. falciparum* infections, Chloroquine is given at 600mg "STAT" followed by three doses of 300mg at 6,24, 48 hours afterwards. For resistant *P. falciparum* infections, or an unknown type, Quinine sulfate is the drug of choice given at 542 mg base PO TID for three to seven days. Quinine, however, is contraindicated for a pregnant patient.

There have been reports that a beta blocker, *Propranolol*, has shown to be beneficial in combination therapy. It has been found that this drug reduces the ability of the *P. falciparum* to enter the red blood cells and reduce duplication of the parasites. A study at Northwestern University in December 2006 suggests that Propranolol, given in combination, reduces the dosage of the other antimalarial drugs by five to ten fold.

A pitfall in the antimalarial drugs is the development of drug resistance. Researchers have made several drug combinations that were effective, but later found to be ineffective due to development of resistance. Right now, the best drug combination for malaria is the ACT (*Artemisinin-based Combination Therapy*). This is used as the first line of treatment in many countries. This drug combination destroys the parasites and reduces the number of gametocytes faster, shortening the treatment from weeks to days.

Another combination drug is the ASAQ (*Artesunate-Amodiaquine Winthrop*). This recently used drug simplifies treatment by taking one pill for infants, children, and adolescents. For adults, it is two pills for only three days.

The drawback with these combination drugs is their cost. They are more expensive than conventional drugs. Severe cases and those with complications like cerebral malaria or blackwater fever are better managed in the ICU. Those with life threatening Hemolytic anemias are transfused with packed red blood cells.

Prevention

Malaria is still considered one of the most lethal diseases in the world. Its prevention is a gargantuan task for the WHO. Some economists would say that the prevention of malaria is cost effective because treatment would cost a staggering three billion US dollars every year.

The mainstay of prevention is vector control. This is the elimination or at least reduction of the mosquito population. This can be done by spraying insecticides. There are several recommended by WHO; the most widely used is still DDT. For mosquitoes resistant to this, pyrethroids and permethrins are used. These insecticides are sprayed on the breeding grounds such as swampy areas and stagnant waters. Also, they can be sprayed indoors with IRS (Indoor Residual Spray).

Educating people as to their share of responsibility is primary. It is imperative to ensure that no stagnant water is in their immediate surroundings. Throw away unused tires, plastic containers, old flowerpots, drums, or water receptacles that can catch rainwater.

In areas where mosquitoes are in great quantity, use mosquito netting and screens around your dwellings. When going out with exposure to mosquitoes, use clothes that cover the whole body. Furthermore, spray yourself with insect repellants. For travelers or short-term residents, *prophylactic drugs* can be given. Right now, the recommended ones are *Mefloquine (Lariam)*, *Doxycycline*, and the combination of *Atovaquone* and *Proguanil HCl (Malarone)*. These drugs must be taken two weeks before landing in the endemic area and four weeks after leaving it. There is no vaccine for Malaria to date.

2. AFRICAN TRYPANOSOMIASIS

African trypanosomiasis, otherwise known as *sleeping sickness*, is endemic in Sub- Saharan countries in Africa. It is called this because infected people have disturbances in their sleep cycle; they sleep in the daytime and have insomnia in the night-time. It can infect both humans and animals. It is caused by a Protozoan called *Trypanosoma brucei* and transmitted by a specific fly (Tsetse fly).

Due to inadequate surveillance and remoteness of the endemic areas, there were no definite data on Human African Trypanosomiasis. As a result, in 1986 WHO sponsored a convention about African Trypanosomiasis. From that convention they estimated that around seventy million people were exposed to sleeping sickness. There were three epidemics recorded in history. The first occurred from 1896-1906, mostly in Uganda and the Congo Basin. The second happened in 1920. This did not last long, however, because of systematic screening of people at risk by that time. Continuous surveillance and screening were done and proven to be successful. In 1960 and 1965, the incidence of the disease fell to insignificant levels. During this time, local authorities believed that they had successfully treated the disease so they relaxed the screening. Unfortunately, they were proven wrong when the epidemic came back in 1970. This was the third time. Since then, there have been sporadic epidemics. In 1998, there were estimated to be 37,991 cases, but as of 2004, this fell tremendously to a low level of 17,616 cases. On the other hand, cases of sleeping sickness are exceedingly rare in the United States. In fact, it is estimated at one case every year and mostly in travelers from East African countries.

This disease is believed to have been present as early as the 14th century in Africa. The causative organism and the vector were found around 1902-1903 by Sir David Bruce.

The first medication used was *Atoxyl*, an arsenic-based drug. This was developed by Paul Ehrlich and Kiyoshi Shiga in 1910. This was discontinued later due to side effects, including blindness.

Causative Microorganism: *flagellate protozoans*

Trypanosoma brucei gambiense (common in Western and Central Africa)

Trypanosoma brucei rhodesiense (common in Eastern and Southern Africa)

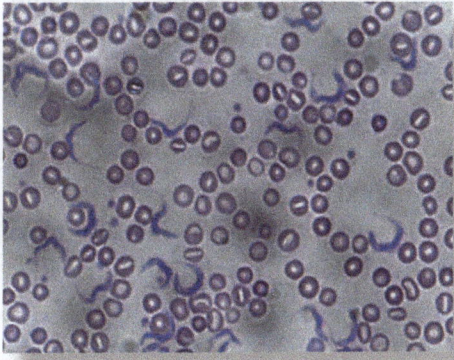

Fig. 10-4. Trypanosomes found in human blood. They are the small, purple squiggly things roaming around the red blood cells. Photo by David Arieti.

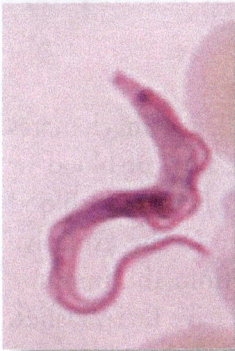

Fig 10-5 A single Trypanosome.[3] ID number #613

VECTOR: Tsetse fly (***Glossina palpalis, transmit Trypanosoma brucei gambiense, Glossina morsitans, transmit Trypanosoma brucei rhodesiense***)

Transmission

When the *tsetse flies* bite an infected reservoir for a blood meal, they suck the trypomastigotes (undulating leaf-like flagellated form of the parasites). In Tsetse Flies, the trypomastigotes stay in the Midgut for two to three weeks where they multiply by binary fission (splitting in two like most bacterial cells). They leave the midgut and transform into epimastigotes and proceed to the salivary glands of the flies while continuing their multiplication by binary fission until they are transformed into a trypomastigote form again. Once the flies bite another person, they inject the parasites into the person in exchange for blood meals. Inside the human body, the trypomastigotes are carried by the flow of blood or lymph to other parts of the body to infect distant organs. These parasites are again taken up by the flies completing their cycle.

A human pregnant mother can transfer the parasites to her baby because the parasites can pass through the placenta. Accidental pricks from contaminated needles can happen in the laboratories. These occurrences are rare, but possible.

Pathophysiology

At the bite sites, inflammatory reactions called skin chancres develop. The trypomastigotes then travel with the flow of blood or lymph to spread to other parts of the body. Once they are in blood and lymphatics the parasites cause parasitemia due to the extensive glycoproteins on their surface that protect them from the immune response. In this stage of parasitemia, patients may have fever, rashes, and bodily malaise. Those parasites that reach the nervous system can cause perivascular infiltration into the brain and spinal cord interstitium leading to neurologic deficits, behavioral changes or meningoencephalitis with edema and bleeding, even coma or death. Those that reach the liver cause fatty degeneration, portal infiltration and Kupffer Cell Hyperplasia. Those that reach the heart cause extensive cellular infiltration and fibrosis of this organ. There is tissue leading to pancarditis (inflammation of the heart), arrhythmias, then cardiac failure. Those that remain in circulation infiltrate blood vessels attracting phagocytic cells leading to endarteritis (inflammation of artery linings). Some would infiltrate the spleen and lymph nodes leading to fibrosis of the said organs.

Symptoms

The clinical manifestation of Human African Trypanosomiasis has two phases: the first phase is the hemolymphatic phase which starts with skin lesions called *skin chancres.* They appear at the site of the bite five to 15 days after. When the parasites enter circulation, the person manifests fever, headache, joint pains, and body malaise.

Infiltration of the lymph nodes causes them to enlarge and swell. On the back of the neck are the so-called *Winterbottom's sign.* This is also known as Human African Trypanosomiasis. Being carried in the circulation, the parasites reach other body organs. Those that infect the heart would manifest as chest pain, difficulty in breathing, and easy fatigability. Those that infect the gastrointestinal tract manifest as loss of appetite, loss of weight, and wasting syndrome.

The second phase is the neurological phase when the parasites reach the nervous system and pass through the blood brain barrier to invade the nervous tissue. This presents as behavioral changes, confusion, mood swings and neurologic deficits. There are alterations in sleep cycles. That is why it is also called sleeping sickness.

Diagnosis

Microscopic identification of parasites from body fluid is the diagnostic method of choice for human trypanosomiasis. The fluid specimens can be an aspirate from a skin chancre, enlarged lymph nodes, bone marrow, or even from blood. At later stages, it can be from cerebrospinal fluid. This can be done by a wet preparation or a smear with Geimsa stain. For those infected with **Trypanosoma brucei rhodesiense**, isolation is possible by inoculating the parasite in rats or mice.

Three *serologic tests* can be used to detect the parasites. These are the (1) micro-CATT using the dried blood, (2) wb-CATT using the whole blood, (3) wb-LATEX using whole blood. Among the three, wb-CATT is the most efficient for diagnosis, while wb-LATEX is the most sensitive test.

Treatment

The different drugs used for treatment of Human African Trypanosomiasis are placed into two groups. The first group is for Phase I of the disease. These are more effective and less toxic. They consist of 1) Pentamidine IV, which has been used since 1939. It is proven to be effective. even today. This is the best treatment for *Trypanosoma brucei gambiense.* 2) The drug is Suramin IV, which has been used since 1920. At present, this is given to patients with *Trypanosoma brucei rhodesiense* infection.

The second group of drugs is used for Stage II, which is brain involvement. The drugs of choice here, however, must pass through the blood brain barrier. Melarsoprol IV has been used since 1940. It has the side effect, though, of encephalitis (3-10%) or even death (10-70%). Because of its effectiveness, however, it is still used today. The next drug is Eflornithine IV, which has been in use since 1990. It is effective on *Trypanosoma brucei gambiense*.

Research has been going on with an Oral Medication, known as DB289. This drug is still in phase III of clinical trials. Additionally, there is ongoing research for a trypanosomiasis vaccine.

Prevention

There are two approaches in the prevention and control of Human African Trypanosomiasis. (1) Entomological approaches that target the tsetse fly population. They primarily disrupt the life cycle of this insect; (2) The medical or veterinary approach targets human and animals. This is primarily monitoring, screening, surveillance or giving prophylaxis and treatment for the disease. This is accomplished by mobile clinics and screening centers that travel daily in endemic areas.

3. AMERICAN TRYPANOSOMIASIS, ALSO CALLED CHAGAS DISEASE[4]

Chagas disease is a tropical infection caused by a protozoan named *Trypanosoma cruzi.* It is transmitted by bugs and some forest mammals. This infection is exclusively found in poor, rural areas of Latin America. These consist of Mexico, Central America, and South America. There are approximately 8-11 million people infected with this disease in these Latin American Regions. In the United States, there have been sporadic reports of Chagas disease, but never in epidemic proportions.

CAUSATIVE MICROORGANISM: flagellate protozoans:

> *Trypanosoma cruzi* **(mainly)**
>
> *Trypanosoma brucei gambiense*
>
> *Trypanosoma brucei rhodesiense*
>
> *Trypanosoma rangeli* **(transmitted by humans)**

Vectors

A bug of subfamily *Triatomine* under family *Reduviidae (Triatoma infestans, Rhodnius prolixus, Triatoma dimidiate)*

This bug is called by many names: Assassin bug, Kissing bug, Chipo, Chupanga, Chincoro, Vinchuca (Argentina, Bolivia, Paraguay), Barbeiro (The Barber in Brazil) Pito (Columbia)

History

In 1909, a Brazilian doctor, Carlos Chagas, was able to describe a unique infection complete with its pathogen, vector, host, signs, and symptoms. He believed that the pathogens were transmitted by the bites of the bugs. In 1915, Emile Brumpt, et al, had disproved this belief. She claimed that the infection was acquired from the feces of the bugs. The bugs would bite first and later defecate in the areas of the bites. In 1926, a doctor by the name of Salvador Mazza did research on the same disease in Argentina. With his research, the disease gained importance in the medical field. However, the real impact of the disease was never appreciated until the year 1960. In Argentina, Chagas disease is called *Mall de Chagas Mazza,* in honor of Dr. Mazza.

Transmission

The bugs usually stay in the walls and crevices of roofs. They go out at night when people are asleep to feed on the blood of humans. If the person they bite is already infected, the bugs get the parasites. These microorganisms taken by the bugs, are in the form of trypomastigotes. Once inside the bugs, the trypomastigotes are transformed into amastigotes in the midgut of the bugs. This is a non-infective form. They multiply by binary fission in the midgut and then migrate to the hindgut to differentiate again into the trypomastigotes, the infective form. At night, these infected bugs go out to feed on blood from humans. For some unknown reason, the bugs prefer to bite the face of the individual (that is why they are called kissing bugs), ingest the blood and defecate on the site of the bite, transferring the parasites. As a result, the person may experience itchiness from the bite. Then he or she often scratches the area, causing the trypomastigotes to enter the human body through the bites. They enter the tissue cells at the area of the bite and are later transformed as the amastigote form.

Next, the parasites continue their multiplication inside the cells and later are transformed again to the infective form of trypomastigotes. Afterward, they are carried by the flow of blood and lymphatics to other parts of the body to invade new host cells.

While in the blood, they can be acquired by the biting bugs to complete the cycle.

Trypanosoma cruzi has been divided into two types based on their Vectors. Type I, the *Domiciliary type,* is the one involved in the transmission to humans and domesticated Animals. Those involved in the *Sylvatic cycle* (Wild Mammals and other animals) fall in the Type II category. These include opossums, armadillos, raccoons, monkeys, wood rats, and coyotes. They acquire the parasites by eating (*Carnivorous*), the *Triatomine bugs.*

However, birds, amphibians, and reptiles are resistant to the parasites.

Other means of transmission in humans is by blood transfusion. In endemic areas, this infection once posed problems but subsequently they were solved using a newly developed serological assay for *Trypanosoma cruzi*. These parasites also can be transmitted by organ transplantation. There were reports of this infection acquired from organ transplantation in Latin America and the United States.

Pathogenesis

At the site of the bites, there are inflammatory reactions leading to a lesion called *Chagoma*. The parasites, aside from the blood, can be found also in muscles, subcutaneous tissues, and lymph nodes, causing interstitial edema, lymphatic infiltrates, and reactive hyperplasia. The trypomastigote form may invade distant organs like the heart and brain.

In the heart, the parasites invade the *myocardial cells* causing patchy areas of *necrosis,* which may lead to *Acute myocarditis*. The patient can usually overcome the initial myocarditis. If the parasitemia continues, it can result in thinning of the heart's ventricular wall. Thinning of the ventricular walls will lead to dilation of the heart that will end up in *heart failure (Dilated Cardiomyopathy)*. Both heart ventricles will be dilated, with the right ventricle usually bigger than the left ventricle.

In the brain, the parasites can be found in the neurons and in the cerebrospinal fluid causing *meningoencephalitis*. The gastrointestinal tract is not spared. The organs here, commonly involved are the esophagus and the colon. The parasites invade the muscular layer of the gut, decreasing the myenteric nerve plexuses within these layers, leading to preganglionic and intraganglionic fibrosis. These will eventually cause the paralysis of the muscular layer and later dilation of the gastrointestinal tract. This is called megaesophagus and megacolon.

Symptoms

The incubation period of *Chagas disease* would be difficult to determine because the bug bites happen at night when the people are asleep. The next day, they have no idea if they were really bitten or not. There is a rough estimate for the incubation period of seven to 14 days.

There are two stages of Chagas disease. One is the Acute Stage and the other is the Chronic Stage.

Acute Stage

This lasts only for three to eight weeks. The infection is usually mild and asymptomatic. Some people, however, develop constitutional symptoms like headache, body malaise, fever, rashes, and loss of appetite. The two most recognized signs of *Chagas disease* are the *Chagoma* which is a local inflammatory reaction on the site of the bite and the *Romana sign*. This is a swelling of the eye on the side of the face, close to the bite. In most cases, the Acute Stage resolves itself, making the patient symptom-free. This stage has a mortality rate of five percent. In immunocompromised patients, however, the infection may continue to the chronic Stage.

Chronic Stage

30% of those suffering from the Acute Stage develop into the Chronic Stage. Here the most common organs affected are the heart, gastrointestinal tract, and the nervous system.

In the heart, the disease can cause myocardial damage leading to cardiomyopathy, arrhythmias, and later death. In the gastrointestinal tract, the patient may develop *megaesophagus* and *megacolon* causing difficulty of swallowing, and persistent diarrhea ending up in severe weight loss. In the nervous system, patients may develop dementia, and some neurologic deficits.

Diagnosis

The parasites can be detected using a microscope from fresh blood treated with an anticoagulant. The blood is prepared in a thin and thick smear with *Giemsa stain*. It can be cultured using media like NNT, and LIT. *Immunoassays* can be used to differentiate several strains. Other tests to use are ELISA, *Compliment Fixation Test, Fluorescent Assay, Radioimmunoassay*, and the PCR (*Polymerase Chain Reaction*).

Treatment

The treatment of Chagas disease is directed to two goals: to eliminate the parasites and to resolve or minimize the symptoms. The antiparasitic drugs are *Benznidazole* and *Nifurtimox*. There were reports of resistance to these two drugs. Some studies claimed that these drugs do not really eliminate the parasites. Yet, no other drugs have been accepted as a treatment.

Symptomatic Treatment is directed to the signs produced by the involvement of the organ systems. Those with infected hearts may be given drugs for the arrhythmias.

Pacemakers may be implanted if the sinoatrial node is affected. For chronic heart disease with heart failure, a heart transplant may be an option because all the heart muscles are affected. For megaesophagus and megacolon, surgery is the treatment of choice. There have been studies showing that slow doses of the immunosuppressive drug, *Cyclosporine*, can improve the survival rate of Chagas disease.

Prevention

Prevention is directed more at eliminating the vectors. These can be accomplished by spraying dwelling areas and sleeping quarters with insecticides. There are some paints combined with insecticides that can be sprayed onto the walls and the roof to eliminate the bugs. Foremost of all is the cleanliness of the dwelling, especially in rural areas and crowded, urban areas.

For travelers, in endemic areas, it is advised to dwell in well-constructed facilities like hotels. If not possible, an air-conditioned room or screened room will suffice. Bugs usually stay in the poor, warm dwellings. If going outdoors, make sure to wear protective clothing. Try to always wear clothing that would cover the whole body and apply some repellant lotions on exposed skin. Unfortunately, there is currently no prophylactic drug or vaccine available for Chagas disease at this time.

4. THE ANIMAL TRYPANOSOMIASIS

A. NAGANA or NAGANA PEST

Animal African Trypanosomiasis (AAT)

Souma or Soumaya in Sudan

When Tsetse Flies bite a vertebrate animal other than humans, they may further spread the disease. Specifically, in vertebrate animals, the parasites infect the blood causing fever, weakness, lethargy, weight loss, and anemia. Animals may even die if left untreated. cattle and horses are commonly affected. This animal disease is more common in Southern and Central Africa.

Not all vertebrate animals are susceptible to trypanosomiasis. Certain types of cattle, like N'Dama[5], are said to be resistant to this type of infection. Likewise, some wild animals are resistant to this disease.

Fig. 10-6 N'Dama herd[6]

This disease in animals is caused by several species of *Trypanosoma*. These are also caused by bites of the same tsetse fly that bites human. Again, the usual symptoms and signs of Nagana are fever, weakness, lethargy, weight loss, and anemia. There are enlargements of lymphoid tissues, the liver, and the spleen. The parasites may invade the spinal cord, rendering the animal paralyzed.

B. **SURRA or MAL DE CADERAS (Brazil)**
    DERRENGADERA (Venezuela)
    MURRINA (other Latin American countries)

The word *Surra* means *Breathing Sounds* through the Nostril. This was the first type of Trypanosomiasis described by Griffith Evans in India in 1880. It is a chronic disease mainly affecting horses and camels. Other affected vertebrates are donkeys, mules, deer, llamas, dogs, cats, cattle, and buffalo. This animal disease is common in North Africa, Middle East, China, Indian subcontinent, Southeast Asia, and South America. There is no report so far on the Pacific Regions. In endemic areas, this disease is of economic importance because draught animals are vulnerable to infection, and dairy animals do not reach their production potentials.

This disease is caused by ***Trypanosoma evansi*** and transmitted by biting flies (***Tabanids, Hippoboscidae***), and vampire bats (***Desmodus rotundus***) in South America.

Incubation periods for Surra are five to 60 days. Initial Parasitemia in animals would manifest as fever, dullness, lacrimation and loss of appetite. At later stages, loss of weight is apparent, shrinking of the humps (Camels) and weakness, and abortion in pregnant animals. Some may have bleeding from *Orifices*, like eyelids, nostrils, and the anus. death may result within two weeks to four months.

In dogs and cats, the disease is usually fatal with nervous system signs that can mimic Rabies. Surra is diagnosed by identifying the parasites from blood smears, especially if the blood is concentrated and obtained from deep vessels. Diagnosis can also use *Serologic Tests* like *Complement Fixation, Indirect Hemagglutination,* Enzyme linked *Immunosorbent Assay* (ELISA) and the *Cold Agglutination Test* (CATT). Treatment is by administration of *Diminazen Aceturate* (3.5 mg/kg). Regrettably, no Vaccine is available.

C. Dourine

Dourine is a type of Animal Trypanosomiasis confined to the *Equine Family (Horses).* The most susceptible to these organisms are Horses, followed by donkeys and then mules. *Dourine* is widespread throughout the world, as far as Canada and Russia in the North, and Chile and South Africa in the South.

Causative Microorganism *Trypanosoma equiperdum* and transmitted by sexual means without a Vector.

Symptoms

The symptoms of Dourine are divided into three stages:

Stage I: is manifested by swelling of the genitals with patches of depigmentation on the penis and vulva. The mares may have bloody vaginal discharges.

Stage II: is manifested as urticarial lesions on the chest, flank, rump, and neck.

Stage III: results in paralysis of the extremities, causing incoordination of movements.

Diagnosis

Diagnosis is by Mucous Discharge Smears from the Vagina and from swollen genitalia, or *Urticarial Lesions.* Serologic tests can be done, too.

The following six diseases are caused by Trypanosomes and have different names and disease characteristics.

D. **BALERI OF SUDAN**

E. **GAMBIAN HORSE SICKNESS (of central Africa)**

F. **KAODZERA (Rhodesian trypanosomiasis)**

G. **TAHAGA (camel disease in Algeria)**

H. **PESTE-BOBA (Venezuela)**

I. **GALZIEKTE (gall sickness of South Africa)**

5. LEISHMANIASIS

Leishmaniasis is a parasitic disease common in tropical and temperate regions of the World. It is transmitted by sandfly larvae, which thrive well in warm, moist organic matter such as house walls, waste, and old trees. As the environment turns warmer due to climate change, we expect more infections of this kind.

There are two types of Leishmaniasis: (1) Visceral leishmaniasis also known as *Kala- azar*, *Black fever*, or *Dumdum fever*. Here the Causative Parasites are in the blood cells and invade the Liver and Spleen. (2) *Cutaneous leishmaniasis* known by many names such as: *Alleppo* Boil, *Biskra*, and Button, Chiclero ulcer, and Pian bois, Uta. This type is usually a *Sequela* of the *Visceral leishmaniasis*.

In 1824, a certain disease caught the attention of doctors in Jessore, India (now Bangladesh). It was known then as *Kala-azar*, which means *Black fever* in Hindi. It was called this because there were blackish skin discolorations on the extremities and abdomen of infected patients. The causative organism was isolated, however, much later by the scientific works of two doctors: Dr. William Leishman (from Scotland) and Dr. Charles Donovan (from Ireland). They published their works simultaneously. So, the organism was named *Leishmania donovani* in honor of the two doctors.

At present *Leishmaniasis* is still endemic in West Bengal, in Africa especially in The Sudan, Kenya, Somalia, Middle East and Southern Europe. Because of the continuous unrest and Civil War in some parts of Africa, the disease remains and even spreads to other areas brought about by fleeing refugees. In one village named *Dauri,* in the center of the epidemic in Southern Sudan, only four out of a thousand people survived.

At present, Leishmaniasis is found in about eighty-eight countries. Ninety percent of cutaneous *leishmaniasis* is found in parts of Afghanistan, Algeria, Iran, Iraq, Saudi Arabia and Syria, Brazil, Peru, while 90% of Visceral Leishmaniasis is found in parts of India, Bangladesh, Sudan, Nepal, and Brazil. Cases of Leishmaniasis in the US are rare and most of them are the cutaneous type. They are usually noted from travelers coming from Latin America. Of importance today is the emerging concern of *leishmaniasis* as a co-infection in HIV positive patients.

CAUSATIVE MICROORGANISMS: Visceral leishmaniasis is caused by:

> ***Leishmania donovani, L. chagasi*** (domesticated dogs serve as reservoir), ***L. infantum*; Cutaneous leishmaniasis is caused by** *L. tropica L. major, L .aethiopica, L. mexicana, and L.braziliensis.*

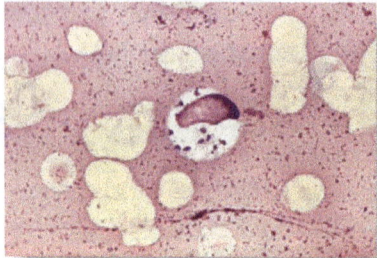

Fig. 10-7. *Leishmania donovani* in bone marrow cell. Photo by L.L. Moore Jr. CDC[7]

VECTOR: Sandflies of the genera *Phlebotomus* and *Lutzomyia*

Transmission and Pathogenesis

Leishmaniasis is transmitted by a female Sandfly, who usually feeds at night while the prey is asleep. When the Sandfly bites an infected animal or human, it sucks blood carrying with it the causative parasite, which is in the amastigote form. In the stomach of the sandfly, the amastigote form is transformed into the *promastigote form*. The latter reproduces asexually in the sandfly gut. Later they migrate to the *proximal gut* of the sandfly and are ready to be injected into the next prey. Once the sandfly bites another person, it would regurgitate its saliva together with the promastigote form of the parasites. In the blood, the parasites invade the *macrophages*. For whatever reason, these parasites are resistant to the *lysosomal enzymes* of the macrophage.

So, instead of killing the parasite, the macrophages remain helpless with the multiplying parasites until it is *lysed* (broken apart) by sheer pressure of the increasing parasites. These new daughter parasites spread out in the blood stream to invade new macrophages and the cycle continues.

As the parasites circulate within the blood they invade the organs, the spleen and liver, causing the enlargement of these organs in the *clinical disease phase*. As more parasites attack the *immune cells*, the patient is vulnerable to other microorganisms such as co- infections with pneumonia, tuberculosis, and dysentery. Leishmaniasis seldom causes flaring up of the disease and death of the patient. It is usually the opportunistic co- infections that make it worse and lead to death.

Patients who recover may be symptom free for several months or a few years. However, they may have the secondary form of the disease which is *Cutaneous Leishmaniasis*. The skin lesions may start as a small measle-like papule at the site of bite. It spreads gradually increasing its size and later coalesces to cause disfiguration on the skin. It may appear as a *swollen lesion*, just like *leprosy*.

Symptoms

Cutaneous Leishmaniasis is manifested as a single or multiple *ulcerating or non- ulcerating wide lesion*. Some would call it a volcano with rising edges and an ulcer (resembling a crater) at the center. Some are *Leprosy-like lesions* that are very disfiguring, especially on the face. Usually, these skin lesions are painless unless secondarily infected. Other symptoms are fever and *lymphadenopathies*. The course of the disease is variable from a few weeks to 20 weeks. The outcome, of course, depends on the immunity status of the patient. The skin lesions may heal spontaneously or may become worse and lead to *visceral leishmaniasis*.

Visceral leishmaniasis usually manifests as prolonged intermittent fever, anemia, leucopenia, Splenomegaly, and hypergammaglobulinemia. These are the cardinal signs of visceral leishmaniasis. The spleen is hard, and non-tender. There may be accompanying hepatomegaly. Other symptoms are generalized lymphadenopathies, blackish discoloration of the skin, extremities, abdomen, and forehead. In severe cases, there are signs of bleeding like petechiae, epistaxis, and gum bleeding. In addition, jaundice and *ascites* may appear.

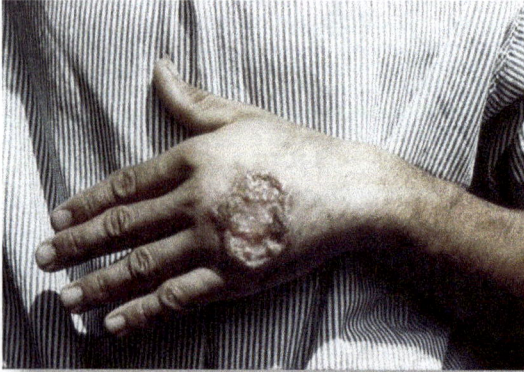

Fig. 10-8. Cutaneous lesion in the hand of a person with Leishmaniasis. Photo by Dr. D.S. Martin[8]

Diagnosis

In cutaneous leishmaniasis, the presence of the skin lesions, plus a history of staying in an endemic area are a good basis for an initial diagnosis. To confirm, there are several tests to resort to. Most commonly used is microscopic identification of the parasites from skin lesions. Other methods are an *in vitro culture* of the parasites, *animal inoculation* of the parasite and the polymerase chain reaction (PCR), which has higher specificity and sensitivity.

In visceral leishmaniasis, the presence of the cardinal signs plus a history of staying in the endemic areas is enough basis for an initial diagnosis. The parasites may be isolated from the blood, nasopharyngeal secretions, bone marrow aspirate and splenic aspirates. Serologic tests can also be done like the *K39 Dipstick Test*, which is easily done by health workers in the endemic areas. Another test currently in trial in Asia and Africa is the *Latex agglutination test* (KAtex). *Direct agglutination, ELISA,* may also be resorted to.

Treatment

Treatment of leishmaniasis has been discouraging because of many factors. Some of these factors are not being financially feasible, including *multiple syndromes, drug toxicities, and the emergence of drug resistance.* Research for new drugs is done in universities sponsored by charitable organizations. In visceral leishmaniasis, the treatment is *pentavalent antimonials* like *Sodium Stibogluconate* and *Meglumine antimoniate.* Amphotericin B has been used in India. *Oral Methionine* has also shown promising results.

For cutaneous leishmaniasis, the treatment of choice is *Sodium stibogluconate* or *Aminosidine* with *Methylbenzethonium.* Other drugs used are *Ketoconazole* and *Pentamidine.* In Saudi Arabia, they use *antifungal fluconazole.*

Prevention

The main goal of prevention is to eradicate the vector. Destroy breeding sites of the Sandfly, and spray areas with insecticides or larvicides. If you are in an endemic area, refrain from outdoor activity, especially at night when the sandflies are more active. If it is necessary to be outdoors, cover all your skin as much as possible by using long sleeved shirts tucked in under the pants, with boots. Exposed Skin must be protected with insect repellants. Indoors, have well-screened windows and doors, or air-conditioned rooms.

6. HUMAN BABESIOSIS

Human babesiosis is a parasitic infection similar to malaria. The clinical course of the disease ranges from asymptomatic infection to a fatal and deadly one. Those individuals who succumb to babesiosis are the elderly, *asplenic,* and *immunocompromised* (those with steroid maintenance therapy, cancer patients, AIDS patients). The disease can be inflicted upon domesticated animals, wild animals, and humans.

Babesiosis has a biblical reference. In the book of Exodus, "Murrain" (later found out to be hemoglobinuria) was mentioned among cattle and other domestic animals. It was in 1888 when a person by the name of Babes first described this *Febrile hemoglobinuria* among the cattle in Romania. Babesiosis was named after her. In 1957, the first human infection due to this condition was reported by Skrabalo in former Yugoslavia.

There are sporadic cases of babesiosis in Europe and most of them are *bovine* (related to cows) babesiosis. There were several types of human babesiosis; but all of these were with patients who had *splenectomies*. In the United States, it was in 1969 when a case of babesiosis was confirmed in Nantucket Island in Massachusetts. Later, the disease became more frequent and affected several states. At the present time, the following areas in the United States are endemic for human babesiosis: Nantucket Island, Martha's Vineyard, and Cape Cod all in Massachusetts; Block Island in Rhode Island; Shelter Island, Long Island, and Fire Island in New York.

Another thing of importance in babesiosis is that the causative microorganism is spread by the same vector that spreads Lyme Disease. Therefore, a person bitten by an infected. single tick can have two different infections concomitantly. Lyme Disease, however, is the most frequently reported *arthropod borne infection* and the incidence of Lyme Disease is far greater than human babesiosis. In 1994, there were 13,000 cases of Lyme Disease reported from forty-three states and there were only a few hundred cases of human babesiosis.

CAUSATIVE MICROORGANISM:

> *Babesia microti* (mostly common in US)
>
> *Babesia divergens* (mostly in Europe)
>
> *Babesia bovis* (mostly in Europe)

VECTOR: Ixodes Ticks (hardheaded Ticks):

> *Ixodes scapularis or Ixodes dammini*
>
> *Ixodes ricinus*

RESERVOIRS: White Footed Mouse (mainly), Meadow Vole, Chipmunks, Norway Rats, Cottontail Rabbits, Short-tailed Shrew

Pathogenesis

Learning the pathogenesis of babesiosis would be confusing if you do not first learn the life cycle of the ticks involved. Tick life cycle is an integral part in the pathogenesis of human babesiosis. The tick life cycle takes almost two years. Starting from the adult stage, ticks lay eggs in late fall and spring. In summertime, eggs hatch into larval forms, which later develop into *nymphs* by next spring (one year lapsed). They remain in this form in summer and then molt into adults by fall to lay eggs again by late fall and spring. This completes the life cycle.

The *Babesia* species live in the white footed mouse (their reservoir), without causing any symptoms. They develop from larvae to nymphs, and then to adults. These require blood meals from animals. For an uncertain reason, the larvae and nymphs would prefer to feed on the White Footed Mouse, while the Adult Tick would prefer to feed on White Tail Deer. Deer, however, are not considered a reservoir of Babesia Species, but considered as an essential host of the tick. So, the ticks acquire the **Babesia** species once they bite the infected mouse for a blood meal (for whatever reason, the **Babesia** species live in the white footed mouse without causing any symptoms). This microorganism remains in the tick until it bites another animal, or a human, again. These ticks, if not on the reservoir, would remain on shrubs, bushes, grass, leaves of plants and even on the ground, ready to stick onto any passersby, be they animals or humans.

Pathophysiology

To better understand the pathophysiology, it must be known that the life cycle of the **Babesia** species has three stages: gametocytes, sporozoites, and merozoites. Ticks acquire the **Babesia** species in the *gametocyte form*. Gametes later fuse to form zygotes. The latter then enters *epithelial cells* of the gut of the ticks and then transfer to the salivary glands of the ticks. Here the zygotes form sporozoites. This is the stage that is injected to humans by tick bites during their meals.

The sporozoites reach the bloodstream and invade the erythrocytes. They mature inside the erythrocytes to form the trophozoites, later becoming merozoites. The merozoites multiply inside red blood cells until there are too many for them to hold and they rupture (*Hemolysis*). This causes the merozoites to invade other red blood cells, resulting in *hemoglobinuria*. Other red blood cells that have not ruptured yet are destroyed by the spleen with the **Babesia** species becoming phagocytosed. This is how important the spleen is in defending against human babesiosis. An asplenic (without a spleen) person with babesiosis usually ends up in an acute stage with an extremely poor prognosis.

Lastly, humans with the microorganisms in their blood are said to have decreased complement in their immune system and a diminishment of the production of *suppressor T-cells* and *cytotoxic cells*, adding to the vulnerability of the person to the disease.

Symptoms

The incubation period for babesiosis is one to three weeks and the symptoms range from silent asymptomatic infection (in healthy individuals) to a severe malaria-like condition with hemolysis (infection in elderly, asplenic patients, those with steroid maintenance and AIDS Patients).

Those patients with overt infections would usually manifest with high grade fever (up to 104° F) with chills and excessive sweating that would lead to body weakness, fatigue, headache and body malaise. Since there is hemolysis, patients would manifest jaundice and dark urine. As the microorganisms disseminate and invade other organs, the liver is manifested as hepatomegaly and jaundice; the spleen is manifested as splenomegaly; the kidney manifested as hematuria, hemoglobinuria, and renal failure; the lungs are manifested by *cyanosis, dyspnea* and gasping for breath; and the blood as *disseminated intravascular coagulation* manifested as *petechiae* and *ecchymoses*. *Erythema migrans (rash)*, which is the telltale sign of *Lyme disease*, is not present in *babesiosis*.

Diagnosis

Signs and symptoms of human babesiosis may not be a good basis for the clinical impression. There are two more common conditions that have similar courses and presentations of disease. These are Lyme Disease and malaria. In fact, Lyme Disease can occur together with human babesiosis. Diagnosis is more based on the demonstration of the parasites on peripheral blood smears. This is done by Geimsa-stain blood smear or the Wright Stain Blood Smear. The *intraerythrocytic parasites* may be mistaken as *Plasmodium falciparum* Malaria, but the presence of the *Tetrad of Merozoites* or the *Maltese Cross* in the smear is indicative of human babesiosis.

For those with negative blood smears, but with strong intuition of human babesiosis, Serologic Tests can be done. These are the Immunofluorescent antibody assay (a titer of 1:256 or greater is diagnostic of human babesiosis); polymerase chain reaction is required if there is an extremely low level *parasitemia*.

Treatment

Most cases of human babesiosis are asymptomatic or with mild symptoms. Only few cases are severe requiring treatment. The mainstay of treatment is drug therapy. The agent of choice is a combination of *Clindamycin* and *Quinine*. Clindamycin is given at 600 mg TID (three times a day) while Quinine is given 650 mg. TID, both for 7-10 days. Alternate combination would be Atovaquone (750 mg BID) with Azithromycin (500- 1000 mg per day). Other drugs have been used including Tetracycline, Primaquine, Sulfadiazine and Pyrimethamine. There is some improvement with these drugs with a slight decrease in Parasitemia. Therefore, they are not extensively used.

Compromised patients, those with asplenia and people on steroids, usually have severe parasitemia. They are extremely ill and have marked hemolysis. They are usually treated by exchange transfusion *with Antibabesial Chemotherapy.*

Prevention

Preventive measures would start from avoiding endemic areas or tick infested areas, especially during the months of May through September. These are considered peak transmission months. If you really need to be out in infested areas, you may use insect repellants like DEET, 10-35%. This concentration is enough to provide adequate protection. Since it takes 24 hours for Ticks to stick to the skin to transfer the organism, it would be of significant help if you look meticulously at your outfits and your skin every day and remove ticks right away.

On the government side, public health departments must initiate programs in eliminating tick populations. This can be done by application of *Acaricides* on host nests. They can also apply acaricides on cattle to reduce transmission to humans.

Below is an actual description of a case of Human babesiosis by the late Dr. Randolph Swiller, an author of the first edition of this book.

Babesiosis, a Fulminant Case

This was a 70-year-old white male, who was taken to the E.R. by his family. He was having chills, fever (103), rigors, and Nuchal (back of neck) rigidity. This patient had suffered from L.T.P (low platelet count). He had been treated with Prednisone, which was ineffective. He then had a splenectomy, and his LTP slowly subsided.

This gentleman was living in Shelter Island, N.Y. and on the day of his admission, he had been sleeping on a boat, when he was awakened by above symptoms (lethargy, rigors, fever, and a stiff neck). He was also prostrate, and confused, in near delirium. On evaluation his blood was analyzed: He was azotemic, and his red blood cells (RBC) had what appeared to be *Ring Forms*. There was a suspicion of *Falciparum Malaria*, but this was ruled out by absence of other signs of malaria. On further exam, Tick bites were noted all over his chest and abdomen. A suspicion of Systemic Babesiosis was strongly entertained, and supported by hemolytic anemia, bilirubinemia, and abnormal LFTs (Liver function tests).

Therapy consisted of exchange transfusions, which were multiple. He was seen by an infectious disease consultant, who started the patient on *Pentamidine*. After 24 hours his fever decreased, and his R.B.C.s were free of Ring Forms, or other debris. His status was much improved.

Discussion: This case was observed in 1979. Babesia is commonly seen in Minnesota, as well as in Northeastern U.S. (New England, coastal N.Y., N.J., and Maryland). The tick is usually ***Ixodes dammini,*** which serves as a vector. In most cases (Immunological Stability), a low-grade fever, weakness, and flu-like symptoms occur. The above patient was asplenic (no Spleen), and therefore a target for massive disease. The parasite caused early encephalitis, which was stopped by supportive treatment, as well as Pentamidine. Nowadays, the agents of choice are Quinine and Clindamycin. Further diagnostic tests include P.C.R. (polymerase chain reaction) as well as I.F.A.(Indirect immunofluorescence reaction).

7. GIARDIASIS

Giardiasis is also known as *Beaver Fever* or *Backpacker Diarrhea*. This infection occurs worldwide with a prevalence of 4-40% with high morbidity rate in Eastern Europe. In United Kingdom, it is considered the most common intestinal parasite during the past two decades. Giardiasis is also one of the most common waterborne diseases in US. The incidence is also high in economically underdeveloped countries, especially in those areas with poor sanitation.

This is not only a disease of humans. The *causative microorganism* can be found in the intestines of many animals, especially dogs. Research in veterinary medicine suggests that the protozoa has been present in five to ten % of all dogs in North America.

It was Leeuwenhoek in 1600 who first described the *causative protozoa*, but Vilem Dusan Lambl and Alfred M. Giard in 1859 recognized the protozoa in the stool.

The name of the protozoa was coined from their last names. In recent years, Giardiasis has been reported to be transmitted sexually, including among *some* gay men with multiple partners.

CAUSATIVE MICROORGNISM: Giardia lamblia (sometimes called *Giardia Intestinalis* or *Giardia duodenalis*)

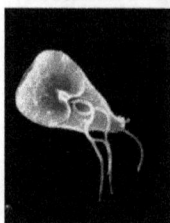

Fig. 10-9. *Giardia lamblia* –Photo by Janice Haney Carr.[9]

Pathogenesis

The life cycle of *Giardia lamblia* is made up of two stages: a Motile Stage in the Intestine (Trophozoites) and a Non-motile Stage released with the feces (cysts).

Giardiasis is transmitted by the fecal-oral route. The Protozoa is acquired by ingesting contaminated food and drink. It can also be acquired from sexual practices. In the stomach (due to its acidic secretions) and in the upper intestine (due to its acidic pancreatic digestive enzymes), the Protozoa undergo excystation releasing the trophozoites. They multiply rapidly by binary fission, doubling in numbers in nine to 12 hours. These trophozoites stick on epithelial cells of the intestine, altering its cell's motility and the cell's secretions of mucus. If in large numbers, they can occlude the intestinal mucosa and compete with the host cells for nutrition, and worse, can cause epithelial injuries.

These injuries result in the symptoms of diarrhea, abdominal pain and at times blood- streaked stools.

As the trophozoites are carried away by fecal flow, they encounter the neutral pH of the large intestine, causing them to undergo encystation, that is, walling off the Trophozoites by a Protein. These are now called *cysts* which are excreted together with the feces.

Millions of these cysts are released in every bowel movement. If the infected person has poor sanitary practice and poor toilet habits, he/she can spread the protozoan cysts on any surfaces of the toilet, or in soil, water bodies, and even more so in foods.

The Cysts are acquired by another person by putting something in his/her mouth that has been in contact with feces of persons or animals infected with Giardiasis. It can also be acquired by accidental swallowing of recreational waters such as in swimming pools, Jacuzzis, rivers, and ponds that have been contaminated by feces of infected persons or animals.

Eating raw vegetables-- especially poorly prepared salads, raw fruits and raw or improperly cooked food-- is another means of acquiring the cysts.

This infection is easily spread in areas with greater concentration of people and poor sanitary conditions like Prison Cells, Mental Institutions, Nursing Homes, and Day Care Centers.

It is also common in areas with high concentrations of the gay community; specifically, this refers primarily to people with multiple sexual partners. Travelers to endemic areas, hiking and backpacking in the wilderness, are also prone to this infection.

Symptoms

About 15% of persons infected with **Giardia lamblia** are asymptomatic. They however continuously pass out the cysts and some trophozoites in their stools. About 50% of those infected develop symptoms. These symptoms usually appear 1-3 weeks after ingestion . They correspond to the invasion of the protozoa on the linings of the intestinal mucosa causing epithelial damage, altered mucus secretions and changes in the motility of the microvilli. They also compete with the other cells involved with the absorption of food.

Diarrhea is the most common symptom of this disease. It is present in 90% of those infected with the **Giardia lamblia**, The stools are described as greasy, tending to float. Usually, blood and mucus are absent. 50% of those individuals develop abdominal pain, bloating, and flatulence. Some infected people may develop transient lactose intolerance due to temporary deficiency of lactase in the intestine, steatorrhea, and malabsorption of Vitamin B12 and Vitamin A leading to loss of weight in 60% of cases. People who are asymptomatic, or those with mild symptoms, may not seek consultation, making them carriers of protozoan cysts for several months.

Diagnosis and Laboratories

History and physical examinations are not enough of a basis for diagnosis. It is so hard to determine the real disease without laboratory results. However, a case of chronic diarrhea with weight loss occurring in an Endemic Area, or in someone living in a family with members having the disease, or in a child from a day care center, is highly indicative of Giardiasis.

Confirmatory diagnosis is reached by microscopic identification of the cysts or Trophozoites (in some cases) in the stools of the infected persons. Since the excretions of the protozoan cysts and trophozoites are intermittent, chances are they cannot be seen in one specimen. It is recommended to give the specimen in three separate days. In high incidence setting, like in an epidemic or in a day care center, screening is done by the ELISA or immunofluorescent assay of the protozoan antigen. This test has an 88-99% sensitivity and 87-100% specificity.

If the microscopic identification of the protozoa and the ELISA are negative and if Giardiasis is highly suspected, the patient may undergo upper endoscopy with duodenal aspirate examination or biopsy of the duodenal mucosa.

Treatment

Those with mild symptoms do not warrant treatment because their body immunity can easily defeat the infection. Those with moderate to severe symptoms are treated with antibiotics. The recommended treatment of Giardiasis is a single dose of 2000 mg of *Tinidazole*. This single dose drug is better complied with by the patient as it shortens the course of treatment and leads to less distress. The multiple dose drug *Metronidazole* is given at 250 mg three times a day by mouth for five to seven days. This is the drug of choice among multiple dose drugs. Other medications are: *Albendazole*, given at 400 mg four times a day by mouth for three days; *Furazolidone* is given at 100 mg four times a day for seven days.

Treatment is usually on an outpatient basis. Seldom is a patient hospitalized. There is no specific diet preference unless the patient develops transient lactose intolerance.

Prevention

Again, being a disease transmitted by the fecal-oral route, personal hygiene is especially important. Handwashing with soap and water before and after eating is necessary. Thoroughly wash your hands after coming from the toilet. If you are changing diapers, even if you are wearing gloves, make sure you wash your hands with soap afterwards. If you are cooking or preparing food, make sure your hands are thoroughly clean. Refrain from swimming if you are having diarrhea.

Do not drink water or do not use ice if you are not sure of their sources. Avoid drinking recreational water while swimming. If there is no other source of drinking water, you can boil the water before drinking. You may try chemical treatment by chlorination or iodination (adding Iodine). For those travelling in endemic areas, it is wisest to bring bottled water or canned sodas.

[1] https://en.wikipedia.org/wiki/File:Morelasci.jpg Viewed 19 Jan 2022.

[2] https://commons.wikimedia.org/wiki/File:Plasmodium_falciparum_01.pn Viewed 13 Jan 2022.

[3] https://en.wikipedia.org/wiki/File:Trypanosoma_sp._PHIL_613_lores.jpg Viewed 25 June 2022.

[3] www.cdc.gov/chagas/factsheets/detailed.htm

[5] https://en.wikipedia.org/wiki/N%27Dama

[6] https://commons.wikimedia.org/wiki/File:N%27Dama_herd_in_West_Africa.jpg Viewed 8 June 2022.

[7] https://commons.wikimedia.org/wiki/File:Leishmania_donovani_01.png Viewed 13 Jan 2022.

[8] https://en.wikipedia.org/wiki/File:Skin_ulcer_due_to_leishmaniasis,_hand_of_Central_American_adult_3MG0037_lores.jpg Viewed 13 Jan 2022.

[9] https://commons.wikimedia.org/wiki/File:Giardia_lamblia_SEM_8698_lores.jpg Viewed 13 Jan 2022.

Chapter 11

OTHER ARTHROPOD-BORNE AND NON-MICROORGANISMAL CONDITIONS

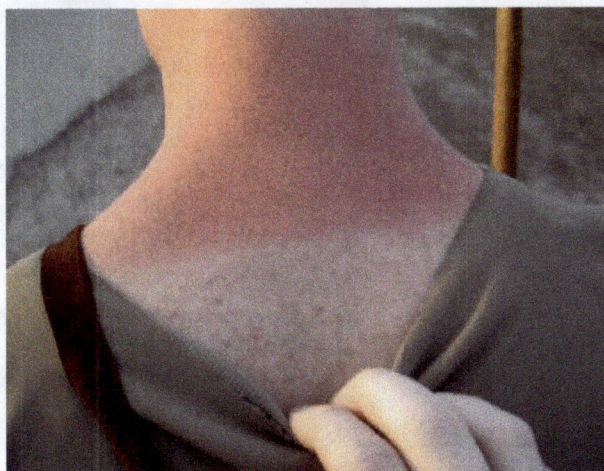

Fig 11-1

Sunburn Photo by Phil Kates[1]

This chapter deals with arthropod transmitted diseases as well as those diseases caused by non organisms such as asthma and sunburn which can be quite dangerous.

ARTHROPOD-BORNE

473 | David Arieti

1. TUNGIASIS

Tungiasis is a skin disease mainly affecting the foot, especially toenails. It is caused by a burrowing *ectoparasite*, the sand flea. This is common in poor, tropical countries like Africa, South America, the Caribbean, and the continent of Asia. The flea is also known by other names like *Chique, Flea, Jigger,* and *Nigua*. Additionally, it is referenced as *Bicho de pie* in Latin America.

This infection was first reported by Gonzalo Fernandez de Oviedo y Valdez from the crew of Christopher Columbus' ship named The Santa Maria when it was shipwrecked in Haiti. Due to many expeditions during those times, fleas were spread to other countries in the African Continent, Caribbean, Latin America, and West Indies. Reported cases in United States are from travelers from endemic countries.

At present, this *causative organism* has been spread in eighty-eight countries worldwide.

The infection is usually associated with poor communities where sanitation and body care are poorly observed and practiced. In Nigeria, Trinidad, Tobago, and Brazil, the incidence can reach as high as 50% of the population.

Specifically, the sand flea thrives in areas with warm conditions, moisture, and high humidity. With increased warming from Climate Change, these microorganisms are expected to multiply substantially, thereby increasing the incidence of the disease.

CAUSATIVE MICROORGANISMS: An Ectoparasite: Sand flea *Tunga penetrans,* also known as *Sarcopsylla penetrans or Pulex penetrans.*

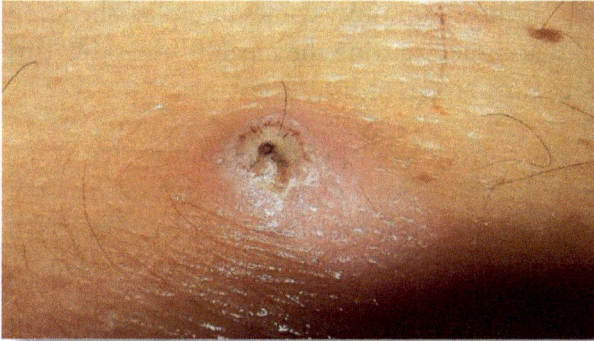

*Fig 11-2 **Tunga penetrans** in a human.*[2] Photo by Gilberto M Palmer

Transmission and Pathophysiology

Tunga penetrans thrive in hot areas like sand, stables, farms, dry soil, beaches, and wooded areas. adult fleas, male and female, stay in those warm areas. Once in contact with humans, male and female Fleas anchor on the skin and feed on the *Warm-Blooded Hosts*. They can anchor on various parts of the body. With their poor jumping ability (a few cm), they mostly affect the foot, especially the interdigital areas and the toenails.

Male and female fleas, however, show no interest in copulation while on the skin. Female fleas then burrow deeper in the dermis using their posterior legs to push the head deeper, while their anus, copulatory organs and their stigmata remain protruding on the epidermis. This also serves as a passage for breathing. After two to three days, the penetration is completed. When the female fleas exude their excreta, this attracts male fleas for mating.

Copulation occurs; that causes the death of the male fleas. The impregnated fleas increase in size, stretching the overlying skin, making it appear as a white halo with a black dot ulceration in the center (protruding parts of the flea). This skin lesion is usually secondarily infected, forming papules and pustules. Then the eggs are hatched, initially sticking to the skin, and then released onto the ground. After hatching her eggs, the female flea starts to lose signs of life and eventually dies. The host immunity comes into play, shedding the parts of the dead flea and then starting the repair of the skin. This causes the skin lesions to turn brown or black.

The eggs hatch about a week later, producing pupae that make a cocoon and thrive in the warm soil or ground. The warmer the environment, the more favorable it is for the fleas to thrive. After 9-15 days the pupae become adults and are ready to invade human skin.

In severe cases, patients develop chronic ulcerations with nail deformities, auto- amputation of digits and lymphadenoma of the surrounding areas which can lead to gangrene of the affected areas. Tetanus may ensue if the person has no immunization.

Symptoms

When female fleas burrow in the skin, the initial symptoms would be erythema at the site and itchiness. Pain at the boring site is seldom felt because the flea produces keratolytic enzymes that make penetration easier. When the impregnated female fleas expand, they produce a white halo on the overlying skin surrounding the black dot representing the protruding hind parts of the flea. The size of skin lesions can range from four to ten mm in diameter. At this stage, the patients may feel excruciating pain, especially at night, and excessive pruritus. The tendency of the person is to continuously scratch the lesion, which results in some abrasions and ulceration. These are usually invaded by bacteria, resulting in secondary infections. If not properly treated, complications set in like chronic ulcers, auto-amputation of digits, then gangrene. In the later stages when the female flea dies, the skin lesions turn Blackish or Brownish.

Diagnosis

The diagnosis of Tungiasis is mainly based on clinical manifestations. Presented with a very typical skin lesion of the foot, a person from an endemic area indicates this kind of parasitic infection. Even an untrained doctor who has seen a case once, can easily recognize the manifestations of Tungiasis. Biopsies may be done. However, they are not required for diagnosis.

Treatment

The main treatment for Tungiasis is the removal of fleas. If they are close enough on the surface, they may be removed with forceps. One must be careful not to break the flea and make sure all flea body parts are removed. If it is too edematous or bloody, it might not be right to use forceps. In this case, minor surgical procedures can be done to remove the flea. Once extracted, a topical antibiotic is applied, followed by *Prophylaxis* to prevent Tetanus. Other methods of treatment include: a topical antiparasitic drug like *Ivermectin (Stromectol)* is applied at the skin lesion; Oral antiparasitic drugs like *Niridazole* are used for embedded fleas; application of Petrolatum solutions like *liquid paraffin* are used to suffocate fleas; application of *Cryotherapy* is also used to kill the fleas.

Prevention

Preventive measures start with personal hygiene. Frequent washing of feet with soap and water will help. When exposed outdoors, wearing shoes and other protective clothing will certainly prevent fleas from attaching to the skin. Likewise, hygiene in the household must be observed.

It is better to have a cement floor rather than sand or soil. Frequent washing of the floors would also help eliminate the fleas.

An effective *nontoxic anti-repellant* for fleas is *Zanzarin*, a derivative of Jojoba Oil, Aloe Vera, and Coconut Oil. It has been proven to be effective in a cohort study when the incidence of tungiasis dropped to 90%. Another method is the spraying of the pesticide DDT. This is an effective pesticide for malaria parasites but also for other arthropods like the flea. In fact, in one of the programs in Mexico called *Campana Nacional para Erradicacion de Paludismo* in 1950, not only *Anopheles mosquitoes* were significantly eliminated, but other arthropods like fleas were also significantly reduced. So, the in-house spraying of DDT was recommended.

2. TICK PARALYSIS

Tick Paralysis is a *nonpathogenic condition* transmitted by ticks. It is caused by the toxin produced by ticks as they attach to the skin. This substance attacks the nervous system causing *neurologic deficits*. The tick paralysis incidence in North America occurs more in the late spring and summer (months of April to June). This is the period when the ticks come out of hibernation.

Tick Paralysis is more of a problem for veterinarians than physicians. There have been thousands of animals (mainly cows and sheep) that perished due to tick Paralysis in some parts of the world. This disease, however, is exceedingly rare in humans. In fact, there are not many scientific papers about tick paralysis in the United States. In Australia, there were reports of twenty fatal cases due to this condition in 1989. The causative tick in that area is species ***Ixodes holocyclus.***

CAUSATIVE ORGANISM: *Dermacentor andersoni* (Rocky Mountain Wood Tick)

Dermacentor variabilis (American Dog Tick)

Pathogenesis

There are forty-three species of ticks implicated in tick paralysis. It is however these two species (***Dermacentor andersoni* and *Dermacentor variabilis***) that are mostly associated with tick paralysis.

Once ticks stick on the skin (usually the Scalp) of the host (whether it is a human or another mammal), they suck blood. Female ticks must attach longer to suck enough blood for breeding eggs. It is during this time that female ticks form the *Neurotoxin* in their Salivary Glands that later is introduced into the host together with the saliva while sucking blood. It is on the fifth- seventh day of attachment that the toxins are maximally produced. The more toxin produced, the more severe the symptoms are. Note, however, that the toxins are only available when the tick is attached to the skin. Once ticks are removed from the skin, the symptoms subside.

Symptoms

Symptoms start to appear two to seven days after attachment of the ticks. The person would experience restlessness, weakness, and irritability. These symptoms will be followed by numbness and then paralysis, starting from the lower extremities going up the upper extremities. In severe cases it can reach the face, paralyzing facial muscles and the tongue. It can lead to respiratory failure due to paralysis of respiratory muscles, and eventually death.

Some patients may present mild symptoms but their knee reflexes, deep tendon reflexes are hypoactive. Electromyographic studies may show low amplitude of the muscle action potentials.

Diagnosis

Diagnosis of Tick Paralysis is mainly based on medical history and the course of the disease. The medical history of exposure to tick-infested areas (camping grounds, mountains) plus numbness and paralysis are indicative of tick paralysis. Also consider this condition if pets of the patients have spent some time outdoors. Finding ticks on the skin further substantiates the diagnosis of the condition, and the disappearance of the symptoms after removing the ticks is a confirmation of the diagnosis. There is no laboratory test specifically for tick paralysis.

Worth mentioning at this point is the differential diagnosis of *Guillain Barre Syndrome*. Clinical presentations of tick paralysis are quite similar to Guillain Barre Syndrome, so this should be considered once a tick is no longer found.

Treatment

The mainstay in the treatment is the removal of the tick. Examine the whole body thoroughly because the ticks are small and may hide on hairs. The commonly affected areas of the body are the head and neck. For those patients with respiratory failure, a mechanical respirator supplied with oxygen is necessary.

Here are some guidelines in removing ticks:

-Use tweezers to remove the tick. Make sure you are able to remove all body parts of the ticks. Hold the tick with the tweezer as close to the skin as possible, and pull it back firmly and slowly.

-Never use any nail remover liquid, or a lighted match, in removing the ticks. These can harm the skin more.

-Wipe the area with antiseptic solution after removing the ticks.

-Since the ticks carry other harmful organisms, you may preserve the tick by placing it in a clean small container and put it in the freezer. Should you develop other symptoms, have the tick brought to a laboratory for identification. This may be important in the diagnosis of your disease.

Prevention

The best preventive measure is to avoid areas known to be infested with ticks. If camping or mountaineering, avoid areas with thick and tall grasses and shrubs. If it is necessary to go out into these areas, wear clothes that cover all your body as much as possible. Wear high cut shoes, long pants, a long-sleeved shirt, and gloves. It is of help if you wear light colored clothes to easily identify ticks crawling on the clothes. As additional precautions, tick repellants (DEET) can be applied on the clothes or skin. If possible, apply an insecticide (Permethrin) on the clothing. This is more effective in eliminating the ticks. It is, however, never used on skin.

Non-Microorganismal Illness

1. ASTHMA

One of the quite common conditions associated with environment is asthma. With the decreasing quality of air from the high ozone levels, traffic pollution, extreme temperatures, increased pollen counts, what more can asthma sufferers expect but exacerbations of their condition. Probably the only people who will benefit from this situation are the pulmonologists and the pharmaceutical companies producing the anti-asthmatic drugs.

Recently, there has been concerns in developed countries about asthma because of increased incidences, especially among children. It is believed that one out of four children in the developed countries have asthma. Worldwide, it is estimated that three hundred million people are suffering from asthma. In the US alone, there are about 4,000 deaths attributed to asthma yearly.

Several researchers have shown that there is a positive relationship between asthma and the low quality of air. The increased exposure to pollutants correlates with the increased incidence and exacerbation of asthma in children. Extreme temperatures are another cause. In winter, chilly air can trigger bronchospasms, narrowing the airways and leading to asthmatic attacks. Hot, arid, and humid climates can trigger asthmatic symptoms. They can also enhance the growth of molds and dust mites that can trigger asthmatic attacks. In a study by the nonprofit Asthma and Allergy Foundation, they ranked the best cities for asthmatic persons. Most of their findings were related to environmental factors. Three of the best cities were on the coastal areas. The air blowing in these coastal areas eliminates pollen and other allergens in the air, while the valleys with less air movement have continuous allergens and pollen in the air. These areas would be difficult for an asthmatic to live in.

Classification

Asthma can be classified in many ways but the most commonly used classification is the one based on the frequency of symptoms and two clinical parameters: namely forced expiratory volume at one second (FEV1) and the peak expiratory flow rate. This classification has the best clinical and therapeutic implications. It categorizes Asthma as: Intermittent (attacks that occur less than once a week); Mild persistent attacks (greater than once a week but less than once a day); Moderate persistent attacks (daily and greater than once a week with nighttime symptoms); and Severe persistent attacks (daily and frequent nighttime symptoms).

Another commonly used classification is the one based on etiology. This classification categorizes asthma as follows:

1. **Extrinsic asthma** (also known as *Allergic* or *Atopic asthma*). This type is triggered by allergic reactions from inhaled allergens like pollen, mold, and dust mites, leading to excessive production of immunoglobulin E. Patients with this type have the usual eczema during childhood. This is the most common form of asthma affecting 50% of the twenty million asthmatic patients.

2. **Intrinsic asthma** or *non-allergic asthma*. This is triggered by factors other than allergic reactions. This type may be caused by infections (from virus, bacteria), cold air, dry air, exercise, smoke, and hyperventilation. It can also be caused by psychological factors like emotional stress and anxiety. Based on the aforementioned etiologies, intrinsic asthma can be subclassified into *Infective*, *Psychological* (due to anxiety, emotional stresses), and *Occupational,* if triggered by chemicals in the industries like Metallic Dusts (Platinum Salts), Flour, Dusts from Grains, and Biological Detergents. [3]

Causes

Several studies have been done to determine the causes of asthma. Most have been confirmed, while others were not. There are two main groups of causes of asthma.

1. **Environmental**. There is a prevailing theory worldwide. This is the *Hygiene Hypothesis* that the increased incidence of asthma and allergies is an unintended outcome of the modern hygienic practices in preventing disease in childhood. In studies in West Germany and East Germany, children living in less hygienic environments, day care centers, and from big families have low incidences of asthma. Research also has shown that upper respiratory tract infections seem to protect against asthma risk while lower respiratory tract infection tend to increase the risk of asthma.

   Another claim was that babies born by caesarean section were more prone (20%) to asthma in their childhood than babies born by vaginal delivery. This is due to the modified bacterial exposure in caesarean sections that modify the immune system. Smoking, especially in pregnancy, is associated with high asthma risk and other respiratory infections. Psychological stresses have been implicated to trigger asthma by modulating the immune system to increase the airway inflammatory response to allergens and irritants. The use of antibiotics early in life was shown to increase the incidence of asthma because they modified the *gut flora*, and the immune system. There were studies linking the use of paracetamol to the prevalence of asthma.

Exposure to indoor allergens like dust mites, which are prevalent in Western styles of housing, increases the incidence of asthma in infancy and childhood. In the United States, United Kingdom and Taiwan, obesity has been noted to have positive correlations with asthma. Low quality of air due to traffic pollution and high ozone levels is positively correlated with asthma. In low-income populations, especially in industrial countries, asthma is more prevalent. Asthma has been associated also in areas with plenty of cockroaches.

Asthma incidences in different countries have great disparities. These may be due to differences in socioeconomics, genetic makeup, environmental risk factors, and advancements in technology. Not surprisingly, however, the incidence of asthma in the United States is higher compared to other countries. This may be due to the hygienic hypothesis.

Pathophysiology

Asthma is basically airway thickening resulting in decreased airway diameters. This is caused by three pathologic findings which are: 1) *Inflammation of the Bronchial Mucosa*, 2) *Hypertrophied Bronchial Mucus Glands*, and 3) *Bronchoconstriction* due to tightening of the Bronchial Muscles.

When allergens enter the bronchial tree, they stick to the linings and end up being eaten by the APC (Antigen Presenting Cells). A portion of the antigen combined with the MHC (Major Histocompatibility

Complex) on the cell membrane of the APCs. When the APCs reach the lymph nodes; they stimulate the immune cells. In most people, these cells are not stimulated. They ignore these allergens and do not produce antibodies. For other people, however, these fragments of the allergens stimulate the humoral immunity producing antibodies against the allergens. Aside from the antibodies, memory cells are also produced that remain permanently in the body. When the same allergens reenter the body, memory cells recognize them and the immune reaction starts right away. Frequent exposure to the same allergens activates more memory cells and more antibodies are produced.

Antibodies attack the allergens in the bronchial walls, causing inflammatory reactions. Inflamed mucosa become *Edematous* leading to thickened bronchial walls and decreased airway diameters. Inflammation likewise results in overproduction of the mucosal glands secreting more thick mucus that further decreases airway diameters.

At the same time, the allergens stimulate the *Parasympathetic Nerve Endings* in the Bronchial Walls. Impulses are transmitted to the respiratory center in the medulla via afferent fibers of the Vagus nerve and then back to the bronchial airway by the efferent fibers of the Vagus nerve.

The efferent impulse stimulates the bronchial muscles to contract resulting in further bronchoconstriction.

Signs and Symptoms

The signs and symptoms of asthma can be categorized into the *Steady state* and the *Acute state*. The symptoms in steady state are seen in chronic asthmatics. These are the chronic *throat clearing type* of cough, nocturnal cough, shortness of breath on exertion, and chest tightness.

There is no dyspnea, however, at rest. The acute state symptoms are those shown by patients with acute asthmatic attacks. These are dyspnea, even at rest, wheezes in the lungs and chest tightness. Coughs may or may not be present in acute attacks. If present, they are usually productive of clear sputum. Other manifestations are the flaring of the *Alae nasi*, *Intercostal retraction*, elevation of the shoulder due to contraction of the *Sternocleidomastoid, Scalene, and Pectoralis Minor Muscles* that lead to overexpansion of the chest. On examination, the patient has T*achycardia, rhonchus lung sound.* In S*tatus asthmaticus* (continuous severe attacks not responsive to medication), patients turn Bluish due to lack of oxygen in the blood. This *Hypoxemic state* can lead to chest pain, numbness of the entire body, cold extremities and even loss of consciousness. If prolonged, this becomes life threatening and may lead to *Respiratory arrest.*

Diagnosis

In most cases, asthma can be diagnosed by simple inspection. If a patient comes to the clinic gasping for breath with flaring *Alae nasi*, *Intercostal retraction*, and elevation of the *Shoulder on Inspiration*, chances are this patient is having asthma. You can further confirm it by asking a simple question as: "Are you an asthmatic?"

For more objective bases of diagnosis, clinicians would do *Pulmonary function tests (Spirometry or the Peak Flow Meter)*. A spirometer is a device that measures the volume of air and speed of air as inhaled and exhaled. A patient with asthma would have less volume of air inhaled and exhaled within a certain time period. The National Asthma Education and Prevention Program (NAEPP) recommends spirometry on the first diagnosis, after treatment, when symptoms stabilize, when symptoms deteriorate, and every one to two years on a regular basis from then on.

A peak flow meter is a hand-held device that measures the amount of air breathed out. Definitely an asthmatic patient would usually have less amount of air in a particular time. However, peak flow meters are more useful in daily self-monitoring of the effects of medications. Patients are advised to write the results of this meter on a daily basis and show them to the doctor on consultation.

Treatment

The best way to treat asthma is to target the culprits which are the triggers. Once determined, avoidance is the main course of treatment. This must be a two-way management plan which means a good collaboration between the doctor and patient. Patients must be aware and should understand their condition. They must have the interest to accomplish the goal, which is to reduce exposure to the triggers, or to be compliant to medication and program exercises (*Buteykomethod*).

Medical treatment of asthma includes: the long-term control medications called *Preventers*, the quick-relief medications or *Relievers*, drugs for *Allergy Induced Asthma* and emergency treatments.

The *Preventers* are usually taken every day. These include: (1) Inhaled corticosteroids such as *Fluticasone, Budesonide, Triamcinolone, Beclomethasone*. Since they are inhalants and their effects are localized in the lungs, they have less complications compared to the Oral corticosteroid whose effects are systemic. (2) Long term beta -2-agonists (LABAs) such as *Salmeterol* and *Formoterol*. They are usually given in combination with the inhaled steroids. (3)*Leukotrienes* modifiers such as *Montelukast, Zafirlukast* and *Zileuton* (4) Cromolyn and nedocromil. They decrease allergic reactions. (5) Theophylline. It relaxes the muscles around the airways leading to *Bronchodilations*.

The *Relievers* (also called quick-relief medications or rescue medications) are only used in Asthmatic attacks. They act rapidly and last for several hours. These are: (1) Short acting beta -2- agonist such as *Albuterol*. This drug relaxes smooth muscles of the Bronchus. (2) *Ipratropium* also relaxes the bronchial smooth muscles (3) corticosteroids, oral and intravenous (*Prednisone, Methylprednisolone*). They cause less inflammation of the bronchial walls.

Drugs for *Allergy-induced* Asthma: These are the (1) desensitization shots that are given for a long period of time in a gradually sliding dose. They decrease the immune reactions of the body to particular allergens.(2) *Anti Ig E monoclonal antibodies* given by injection, too, and also reduce the reactions to a particular allergens.

The *Emergency treatments* for Asthma include the (1) Oxygen inhalation to treat the Hypoxia (2) *Intravenous steroids* (3) *Nebulized short acting beta -2-agonist with ipratropium* (4) *Nonselective beta agonists* like *Epinephrine, Isoproterenol* and *Metaproterenol* (5) *Intubation and Mechanical Respirator* for those with *Respiratory arrest*.

There were several studies using *Non-medical treatments* for Asthma like Acupuncture, Air Ionizers, Dust Mite Control Measures, Air Filtration, Osteopathic, Chiropractic, and Physio Therapies. Yet, all these methods did not show evidence of efficacy. The only non-medical method recommended (by the British Guidelines) is the *Buteyko Technique*. This is based on the principle that Asthmatic Patients *chronically over breathe*. So, the main goal of this technique to breathe less. This is a series of breathing exercises like nasal breathing, hold-breathing, and relaxation.

Prevention

The best method to prevent asthma is by *Allergy Desensitization*, also known as *Allergy Immunotherapy*. This is made up of a series of injections of the allergens in increasing doses to desensitize the patient. It takes several weeks to months. If done correctly, it can reduce the need for Asthmatic Medications by 50%. If done early in the course of the disease, it can even result in remission (aka *Asthmatic cure*).

Smoking has been known to induce asthmatic symptoms. It irritates the bronchial lining mucosa thereby producing more mucus. It retards the movement of the cilia of the epithelial cells lessening their protective functions. Smoke causes increased severity of symptoms, decreased lung function and a decreased response to medications.

Emission from automobiles, wood smoke, gas stove smoke can trigger asthmatic attacks. Air cleaners and room air filters have been shown to prevent asthmatic attacks. Breathing cold air or dry air can exacerbate asthma. Sports that expose the participants to these types of air, can aggravate asthmatic symptoms. These are skiing, cycling, long distance running, and mountain biking. sports in a normal environmental temperature or in warm water can minimize asthmatic symptoms. These include weightlifting, diving, and swimming. There was an interesting report that in the 1996 Olympic Games in Atlanta Georgia, revealing that 15% of the athletes were Asthmatic and 10% of them were taking asthmatic medication.

Dietary supplements like Vitamin C have been shown to increase pulmonary functions. *Magnesium sulfate* also helps, although this element is still under study. Of course, we have the Preventers Group of Medications (as discussed in the section of Treatment) that can be taken to prevent asthmatic symptoms.

2. SUNBURN[4]

Sunburn is a burn on the skin and underlying tissues from brief (acute) overexposure to *Ultraviolet Rays*. Its formation depends on a person's length of exposure to the sun, as well as ability to produce Melanin. This substance is produced by the melanocytes in the Epidermis and its effect is to protect the underlying tissue from damage by Ultraviolet Rays (UVR).

The source of UVR is primarily the sun. This is directly connected to Global Warning. As our societies progress, we consume more oil and petrochemicals. The byproducts of these materials decrease the ozone layer that protects the Earth from the sun's harmful UV Rays. The less of an ozone layer we have, the more exposed we are to the UV Rays.

The rays that cause sunburn do *not* only come from the sun. They can also transfer from welding arcs and tanning lamps. Sunburn can also be caused by drugs that render the users more sensitive to UV Rays. Some antibiotics, tranquilizers and contraceptives can have these effects. The UV Rays cause damage to the DNA of the cells, initiating several immune reactions by the cells to repair the damaged DNA, as well as to increase Melanin Production to prevent further damage. This makes the exposed body areas darkened. A lighter form of sunburn is called a Suntan. This is simply a mild redness and tenderness of the affected areas, while severe forms can cause painful erythema, swelling, and blisters. The severe forms can be very painful and debilitating enough to require hospitalization.

Symptoms

Sunburn is usually acquired between the hours of 10 a.m. and 3 p.m. It is advisable to refrain from exposing yourself to the sun within these times of the day. The symptoms may start one hour to one day post exposure. In cases of exposure to Non-shielded Welding Arches, it can be as early as 15 minutes post exposure. Initially the person develops redness for 30 minutes to several hours. This is followed by pain and tenderness that can extend several hours to three days. after which peeling of the skin starts. The peeling usually ends by one week. However, in some situations the peeling can extend up to several weeks.

Treatment

Application of cold compresses can soothe raw and hot areas. Skin moisturizers can be applied to hydrate the skin. Products containing Vitamin E or Aloe vera are good moisturizers. For painful conditions, *Topical Lidocaine* or B*enzocaine* can be applied. This *Anesthetic care* in the form of ointment and sprays is approved by the FDA. You can also use an *Oral analgesic* like NSAIDs (Non-steroidal anti-inflammatory drugs) for pain. For those with blisters, burn creams or ointments with or without antibiotics can be used. Steroid creams like *Hydrocortisone* can be given for severe sunburns. The healing process can take days to weeks. For those with fever and headache, Acetaminophen will be sufficient.

Prevention

The most obvious prevention is to stay away from direct sunlight, particularly between 10 a.m. to 3 p.m. on sunny days. If exposure is necessary, wear long sleeved shirts, with a wide hat or umbrella. You may apply commercially prepared sunscreen or sunblock. The higher SPF(Sunburn Protection Factor), the lesser the UV Rays are striking the skin. Use a *Broad spectrum* (effective on both UV Rays A and B) 10SPF or 20 SPF Sunscreen. Apply this 15-30 minutes before exposure to the sun to be repeated 15-30 minutes after the start of the exposure.

Studies showed that eating foods containing more *Beta-carotene* and *Lycopene* (ketchup and other tomato products) makes the skin more resistant to UV Rays. If the cause is occupational, protective clothing, welding helmets/shields, or eye goggles must be worn. The UV has been said to cause *Pterygium* and Cataracts.

3. HEAT WAVE[5]

One of the lethal effects of Global Warming is the increased incidence of *Heat Waves*. Severe ones can cause *Hyperthermia* or *Heat Stroke*, which could lead to death. Increased demand for electric power due to continuous use of cooling systems can cause wildfires in areas with drought. This can cause disastrous effects on livestock and crops, resulting in economic maladies.

There is no common definition of Heat Waves because temperature varies in different parts of the globe. The temperature of one in temperate countries may be normal temperatures in the tropical countries. In the Netherlands, Belgium, and Luxemburg, Heat Wave is defined as a period of temperature exceeding 25 C (77° F) for five consecutive days with three of these days having a temperature of more than 30 °C (86° F). In Denmark, a Heat Wave is a period of temperature exceeding 90 F(32.2° C) for at least three consecutive days. The World Meteorological Organization gave a definition of one as a temperature above the average maximum by 5 C (9° F) for five consecutive days.

There have been several significant incidences of Heat Waves in history. For example, in 1955 in Chicago, there was a Heat Wave that claimed 739 lives in only one week. In 2003, Heat Waves in Europe claimed 70,000 lives inclusive of the 15,000 deaths in France alone. This phenomenon lasted for two weeks. The temperature then reached a fatal level of 104 °F. In July 2006, 140 people died from a Heat Wave in North America. In 2007, Heat Waves were reported in different regions of Australia. Here in the US, between 1979 to 1998, the CDC reported about 7,400 deaths due to exposure to excessive heat.

HEAT WAVES AND HUMAN HEALTH

Heat Waves are caused by a natural phenomenon aggravated by Global Warming. They are formed initially from areas of high air pressure in the atmosphere. Since air pressure at sea level is comparatively low, the movement of the high-pressure air in the atmosphere goes downward, preventing the formation of clouds. As the high-pressured air descends, it is compressed, making it warmer.

In urban areas, Heat Waves are aggravated by emissions of *Hydrocarbon Residues* from automobiles and plant generators. This makes cities more prone to Heat Waves than rural areas.

When a person's body temperature reaches the maximum normal (99.67 °F), it starts to dissipate heat by increasing blood circulation to the skin. This increasing sweating. Now there are temperature receptors in our body that stimulate the *Hypothalamus* (temperature center of the body), which will send impulses to the *Medulla* to increase the heart rate. These Homeostatic Mechanisms of the body increase the circulation in blood. Coupled with this is the fact that heat itself dilates the blood vessels of the skin, diverting more blood to the skin. At this time, heat stimulates the *Sudoriferous Glands* to increase production of perspiration.

Sweat does not actually decrease body temperature. In fact, it is the convection of sweat into vapor that decreases body temperature and keeps the body cool. The heat used in converting the sweat into vapor is deducted from the body, thereby lowering body temperature. Ninety percent of cooling functions is attained in the skin. When the heat is more than the body can manage, illness ensues. Below are common heat related disorders:

-sunburn or erythema is simply redness of the skin, which may be accompanied by headache. Usually, application of ointments like Vitamin A is enough blisters; it is better to leave them as they are and not pop them out because the fluid inside is sterile. When there are Blisters, it is better to leave them as they are and not pop them out because the fluid inside is Sterile.

- **Heat edema** is manifested as swollen hands and puffy feet. This is due to an increase in *Aldosterone secretions*, which retain sodium and water, increasing the volume of blood. This is a *homeostatic mechanism* of the body when it is faced with a decrease in fluid volume from excessive sweating. Retention of sodium and water, plus the dilation of the blood vessels, increases body fluids that migrate to the extravascular space. This is mainly on the dependent parts of the body like the hands and feet. This edema is usually resolved once the person is adapted to warm weather. No treatment is needed in this condition. Elevating the Edematous part or wearing stockings on the Edematous dependent parts of the body may relieve this condition.

-**Heat rashes**: also called *Prickly Heat*. As heat increases, some ducts of the Sweat Glands are blocked. Continuous secretion of the sweat glands result in tortuous, dilated ducts until they rupture. Once they do, *Pruritic Vesicles* are produced. They have an Erythematous Base. These are common on body sweat glands. If this condition persists, bacteria may set in. causing *Secondary Bacterial Infections*. Antihistamines are given for Pruritus: Chlorhexidine lotions are applied to remove Desquamated Skin; Antibiotics are given for infection. Do not forget the advice to wear loose clothing in hot weather.

-**Heat cramps** are due to *Involuntary Muscle Spasms* producing severe pain. This condition usually happens after strenuous exercises or activities. It usually involves muscles of extremities and the abdomen. This condition is due to *Hyponatremia*. A person with continuous profuse sweating must drink water with *Electrolytes* to replace those lost through sweating. Consuming plain water only would lead to *hyponatremia*, resulting in heat cramps. So, the overheated person should drink Electrolyte Solutions like *Gatorade*, a popular drink for athletes.

-**Heat syncope**: is *Orthostatic hypotension* during heat exposure. This is due to *Abrupt* hypovolemia, because of excessive sweating and vasodilation of the blood vessels. These conditions also decrease the blood flow to the brain. This person has *Blacked Out*, collapsed with *Pallor*, and has a weak, thready pulse. Confronted with this condition, the first thing to do is to put the individual in the Trendelenburg Position (Shock Position) where the brain is lower than the body. This sends more blood flowing to the brain by gravity. If the patient is conscious, he or she can take electrolytes in water; otherwise, an intravenous administration of an *Isotonic Solution* is mandatory.

- **Heat exhaustion** (a forerunner of Heat Stroke). This is severe dehydration and severe electrolyte imbalance manifested by headache, dizziness, myalgia, weak thready pulse, pallor, and clammy skin. The person, however, has no neurologic deficits: a feature that differentiates this from heat stroke. usually, the person is hospitalized and given fluid replacements by intravenous infusions. After discharge, the person is advised to have compete bed rest for several days.[6]

-**Heat stroke** *(Hyperthermia)* This condition is due to continuous severe high temperature and *Humidity*. This is an emergency condition that needs immediate treatment. Otherwise, fatal results may ensue. The person is described to have dry and hot skin, no sweating, with a rapid strong pulse. Patients may be unconscious with Neurologic Deficits. This condition is more common at the extremes of life. Also, it occurs in infants and the elderly. This is due to poor *Vasomotor Reflexes* in these age groups. Also prone are those people taking *Diuretics, Antihypertensive drugs,* and *Antipsychotic drugs.* Confronted with this condition, the first thing to do is call for emergency help by dialing 911. Place the person in a cold area, remove clothes and give a cold-water sponge bath. Do not give fluids. If an air conditioning unit is available, place the person near the cold air. If not available, an electric fan will do.

What To Do in Case of Heat Wave

1. Refrain from any strenuous activities. If an activity is necessary, reschedule it for the coolest time of the day.

2. Wear light colored clothing. This will reflect rather than absorb the light rays.

3. Avoid a low protein diet because protein increases metabolism, producing more heat and utilizing more water. In other words, it would further aggravate the situation.

4. Take plenty of fluids to keep the body cool. For those patients with Cardiopulmonary, Gastrointestinal, or Renal problems, have a doctor's clearance before giving fluids. In these conditions, water intake is strictly monitored.

4. HEAT CRAMPS, 5. HEAT EXHAUSTION, 6. HEAT STROKE

These three conditions exist along a continuum and share common attributes. They are categorized as such based on the degree of heat and different signs and symptoms. Their Etiologies, Pathogeneses, and treatments are basically the same. I opted to define each condition separately and later will discuss their common features as a group.

Heat Cramps are involuntary, continuous painful contractions of some body muscles associated with sudden activities or exercise. They mainly involve *Hamstring Muscles*, (*Biceps femoris, Semimembranosus* and *Semitendinosus*) and Calf Muscles (Gastrocnemius, Soleus). Other muscles that might be involved are the Abdominal and back.

Heat Exhaustion creates the feeling of being exhausted. This is one stage higher than Heat Cramps. In Heat Exhaustion, the person may or may not have prior Heat Cramps. The person sweats heavily with cold, wet skin and possible electrolyte imbalances resulting in systemic symptoms such as fatigue, weakness, dizziness, and *Tachycardia*. If not treated, Heat Exhaustion can lead to Heat Stroke.

Heat Stroke is the most severe form of heat-related illness. The patient may or may not have previous Heat Cramps or Heat Exhaustion. This is a *Medical Emergency*. Here, the body temperature is more than 104° F. The patient is not sweating with hot, Reddish skin. Patients may already have neurologic symptoms like delirium, loss of consciousness or seizures.

Heat related illnesses are common in countries with higher room temperature coupled with some event that requires exposure for an extended period, just like the Annual Muslim Pilgrimage in Mecca, Saudi Arabia. In 2003, Europe was not spared heat related illness. There were 14,800 reported deaths in France. In the United States, 4,780 heat related deaths were reported from 1979-2002 with Arizona as the state with the highest death rate. In 1990, in Chicago, there were eighty deaths from heat related illness where 47% of those deaths occurred in people older than 65 years old.

Etiology

The common causes of these three conditions are Heat and Dehydration. Heat Gain by the body is far greater than the Heat Loss. Heat Cramps are believed to be caused by poor conditions, heat, and dehydration rather than electrolyte imbalances. A person who suddenly performs moderate to heavy activities or exercise, without warming up, will have a greater chance of having Heat Cramps.

Heat Exhaustion is due more to prolonged exposure to a hot environment, resulting in sweating and dehydration without replenishing the lost electrolytes and water in the sweat. It does not necessarily mean that the person is doing activity or exercise. He or she may or may not have prior Heat Cramps. Moreover, there are comorbid conditions in Heat Exhaustion. These are *Endocrinopathies* (*Hyperthyroidism, Diabetes, Pheochromocytoma*), heart disease and some systemic skin conditions. Some drugs can also enhance Heat Exhaustion like *Diuretics, Amphetamines, Beta Blockers,* and *Anticholinergics.*

Heat Stroke is caused by extreme high temperatures and high humidity such that a normal person cannot dissipate the heat from the body to the environment, thereby causing rise of body temperature. Normal heat produced by the body from its metabolism cannot be dissipated and instead causes a rise in body temperature. Severe dehydration, which usually occurs before or at this stage, will further aggravate Heat Stroke because the person does not sweat anymore due to lack of water and electrolytes, thereby no dissipation of Body Heat. Heat Stroke, being a stage higher than Heat Exhaustion, has the same comorbid conditions as Heat Exhaustion.

Pathogenesis

These three conditions have common pathogenesis. They differ only in degrees of temperature, signs, and symptoms. It will be more appropriate to discuss them as a continuum.

Body temperature is an interplay of heat produced by the body, plus the heat acquired by the body from the environment. Another factor is the heat lost by the body to the environment. The human body is only adaptive within a range of temperature. Any extreme temperature is detrimental to the body and must be corrected right away. Prolonged exposure leads to failure of the *Temperature control mechanism* resulting in a *High core temperature*.

Increased Core Temperature stimulates *Heat Receptors* of the *Hypothalamus* and *Peripheral Organs*. The body reacts by shunting the blood to the skin, increasing ventilation, and maximizes sweating so that more heat is dissipated from the body. During heavy exercises or activities, the heat production of the person can reach fifteen times the *Resting Rate*. This necessitates rapid Homeostatic reactions as mentioned above. For people who are not preconditioned and suddenly perform excessive activities or exercises, those patients with a *Compromised cardiopulmonary status* (*Congestive heart failure, Myocardial infarction, Pulmonary congestion*) are very vulnerable to heat related illnesses. Likewise, patients taking Diuretic drugs and other drugs that affect blood volume are prone to heart-related diseases.

If core body temperature continues to be high, sweating continues until the person becomes dehydrated. This is followed by diminished blood volume. Once this stage is reached, the body has to shift the blood flow to the Splanchnic Organs to maintain enough blood going back to the heart. The body has to prioritize the return flow to the heart to keep the heart pumping; otherwise, the heart fails to pump.

This reaction diminishes the blood flow to the Periphery, thereby lessening the sweating and lessening the dissipation of heat resulting in increases in core temperature or hyperthermia.

Human cells are not easily destroyed by heat because they produce *Heat Shock Proteins* that protect the cells from Heat Denaturation. However, after 45 minutes to six hours exposure to core temperatures of 107.6 °F (42 °C), denaturation of cells begins, resulting in cellular damage. People who do not acclimatize or those who have not been preconditioned are said to have less production of these heat shock proteins so they are prone to heat related illnesses. This is the reason heat cramps are quite common in non-acclimatized and non- preconditioned people who suddenly perform heavy exercise or activities. They usually lack these heat related proteins to protect the cells from damage.

Further heat leads to Heat Exhaustion. At this time, there is still enough blood volume to sustain continuous sweating. Although the temperature is already high, the Thermoregulatory Mechanisms of the body still exist. For persons with compromised heart, lung, and blood volume status (Congestive heart failure, Myocardial infarction pulmonary congestion), Heat Exhaustion usually follows. Those patients taking diuretics and other drugs that affect blood volume are also vulnerable to Heat Exhaustion. At this stage, the body's Thermoregulatory Mechanisms fail to function anymore. So, no heat is dissipated; instead, the heat remains in the body to accumulate. This is aggravated with high humidity of air. The higher the humidity, the lesser is the evaporation. It is believed that at 100% humidity, evaporation is zero.

Symptoms

Heat Cramps are described as involuntary, forceful, painful muscle spasms usually involving the calf muscles (Gastrocnemius, Soleus) and the hamstring muscles (Bicep femoris, Semitendinosus, Semimembranosus). In some instances, abdominal and back muscles are involved. These muscles spasms are brief. At times, they are intermittent but self-limited. The patient may or may not have profuse sweating.

Heat Exhaustion may or may not have previous Heat Cramps. This is manifested by fatigue, weakness, dizziness. With profuse sweating. the patient may develop Orthostatic hypotension and later develop *Syncope*. Other symptoms are Tachycardia, Nausea, Vomiting, Headache, and Irritability. The patient's skin is cold, wet with *Pilo erection* and temperature as high as 106° F (41° C).

Heat Stroke is a progression of Heat Exhaustion. The main feature of this condition is a body temperature greater than 106 °F. The patient's skin is Hot, Reddish and with Anhidrosis (no sweating). This is not because of the failure of the sweat glands to produce Sweat In fact, it is because there is no more Sweat to produce. The patient has signs and symptoms of *End-organ damage*. With less blood going to the Brain, patients may *have Obtunded mental status, Impaired judgment, Hallucinations, and Cerebellar dysfunctions*. More severe symptoms include *Muscular rigidity* or *Seizures*. With respiratory alkalosis, the patient may develop *Acute Respiratory Distress Syndrome (ARDS)*. With diminished blood flow, patients may develop A*nuria, Hematuria*, or Acute Renal Failure. With compromised blood flows and low preload to the heart, the patient may develop Heart failure and Arrhythmias. Lastly, Coagulation Mechanisms are impaired, resulting in Disseminated Intravascular Coagulation with multiple hemorrhages in different body organs.

Laboratories

Since the possibilities of End-organ damage are present, especially in Heat Stroke, it is imperative for clinicians to check the status of the main body organs. To check for Liver functions, certain enzymes are determined. These are enzymes produced by the Liver. They are the *Aspartate aminotransferase* (*AST*) and Alanine aminotransferase (ALT). Elevated counts of these enzymes are signs of liver involvement. The *Complete Blood Count (CBC)*, Prothrombin time (PT), partial thromboplastin time (PTT), platelets and fibrinogen levels are requested to check for coagulation status. The urinary system can be checked by ordinary urinalysis, blood urea nitrogen level and blood creatinine levels. *Arterial blood gasses* are important in the evaluation of respiratory functions. It is also a parameter for tissue oxygenation and acid-base balance. Another lab test of prime importance in heat related illnesses is *Blood Electrolyte Levels(Na, K, Cl)*. These substances must remain within the allowable window, otherwise it may lead to detrimental results.

A *Computerized scan (CT)* and Chest X-rays may be requested; but these are done to rule out other conditions that would result from heat related illnesses.

Treatment

The primary goal in the treatment of heat related illnesses is transferring the patients to a cool environment, correcting fluid, and electrolyte imbalance if any, and checking for body organ functions. If necessary, correct the malfunctioning of the body organs.

With Heat Cramps, put the patient in a cold environment like a shady area, or air-conditioned room. Remove clothing as much as possible and place the patient in front of a fan to increase evaporation. Sprinkle Tepid Water or apply ice packs on the neck, armpit or groin and cover with wet sheets of clothing. Let the patient take cold drinks. Water would be enough to correct dehydration up to a point, but further fluid intake would necessitate Electrolyte solutions, sport drinks like Gatorade. Have the patient rest and monitor the *Vital Signs*. If the probability of Heat Exhaustion and Heat Stroke is great, the patient must be transferred to the hospital right away.

In Heat Exhaustion, the same procedures are followed as with Heat Cramps if the patient is in a hot environment. Since a patient already has some cardiovascular symptoms, it is necessary for the patient to have an IV line right away for *Rehydration*. He/she must be given isotonic solutions initially and then given hypertonic or hypotonic solutions depending on blood electrolyte levels. The time element is critical in this case. The patient's water deficit must be corrected within three to six hours, otherwise causes of elevated temperature must be considered.

In Heat Stroke, cooling management must be given aggressively. Aside from the above- mentioned measures (for Heat Cramps and Heat Exhaustion), the ABCs of emergency must be attended. Airways and breathing must be maintained. Patients are given Oxygen inhalation. IV lines are connected right away to maintain adequate circulating fluid volume.

There are two ways of lowering body temperature: ice water or slush immersion, and the evaporative cooling.

Ice Water or Slush Immersion has the advantage of lowering the core temperature faster. This is immersing the patient in ice water. However, there are pitfalls in this method. Aside from being uncomfortable to the patient, it may cause peripheral vasoconstriction shunting the blood from the skin to the central body, lessening the evaporation of the heat. Moreover, it is difficult to monitor vital signs when the patient is in the bathtub and there is a danger of hypothermic overshoot.

The lowering of the core temperature must be abrupt but not so fast as to cause peripheral vasoconstriction. The ideal rate of decrease is 0.2°C/min and the minimum must be no lower than 38° C to prevent hypothermic overshoot.

Evaporative Cooling is well tolerated and effective. It may be slower in lowering the body temperature, but it is safer than ice water immersion. Here the patient is undressed right away, sprayed with tepid water, and placed in front of a large fan or in an air-conditioned room. In this method, the patient can be monitored anytime and hypothermic overshooting can be averted.

Alcohol sponge baths are not recommended. They may pose more danger to the patient. Likewise, antipyretic drugs are not used here. Take note that the cause of the elevated temperature is external, not internal. So, the hypothalamic temperature center is not set at a higher point as in fever; that is why these medications will not work. These may increase the danger to the patient because of their side effects in coagulation.

Hospital care includes correcting fluid and electrolyte imbalances and monitoring organ functions. The kidney is especially important in this condition. Both kidneys must receive enough blood supply to keep them working, just like the heart. Inadequate blood supply to the kidneys will result in renal failure. Patients may be given osmotic diuretics (Mannitol) to prevent renal failure. Inadequate blood flow to the brain may result in seizures and this can be treated by benzodiazepine like Diazepam or Lorazepam. Other useful drugs are Chlorpromazine. This is given to shivering patients to minimize the production of heat internally.

Prevention

Heat related illness is a preventable condition. The mainstay of prevention is keeping oneself in a cold environment. For ordinary people, stay as much as possible in a cold area if the weather is too hot. Minimize outdoor activities. If it is necessary to be outdoors, drink plenty of fluids before the activity. Do not wait for your thirst before taking these fluids. Take about 400-500 cc of fluids beforehand and take around 200-300 cc at frequent intervals during activity. Use loose, light weight and light-colored clothing outdoors. Avoid alcohol and coffee as these can lead to further dehydration. Encourage yourself to perform exercise in cold weather and take cool baths to acclimatize yourself in the crisper environment.

For athletes, *Acclimatization* is particularly important. That is, doing exercises for 90-120 minutes every day in hot environments (not extremely hot) for at least a week. Then increase the duration and intensity of the exercises as the weeks pass by. Athletes are also encouraged to wear light, loose clothing during their exercises.

Heat related illness is preventable, and as such, significant morbidity and mortality is more a reflection of unpreparedness of Federal Agencies rather than ignorance of people. In areas vulnerable to heat related illness, the federal office must have a *Health Response Plan* that is always ready for activation. There must be outreach programs to At Risk Groups like the Elderly in Nursing Homes.

7. TROPICAL SPRUE[7]

Possibly the first recorded case of *Tropical Sprue* was in 1759 when William Harvey described Chronic Diarrhea among visitors to Barbados. Subsequently this disease was observed in the tropical countries around the world as it is called "Tropical Sprue." This is a *Digestive system disease*. Patients with Tropical Sprue have *Abnormal mucosal linings* of the *Small Intestine*, thereby failing to absorb nutrients, minerals, vitamins, and folic acid.

This disease is said to be common in tropical countries about 30 degrees above and below the equator. Common countries affected are Southeast Asia, India, Haiti, Cuba, Puerto Rico, and the Dominican Republic. There are no reports of Tropical Sprue in the United States except in Americans who travelled or stayed in tropical areas.

Etiology and Pathology

The fact is that, in Tropical Sprue, the *Microvilli,* minute fingerlike projections along the linings of the small intestine, are gone or atrophied. These minute structures are important because they increase the absorptive area of the small intestine. So, patients with this disease have poor absorption or have malabsorption of nutrients. The elevated blood levels of *Enteroglucagon (* a newly discovered molecule in the human intestine) is a sign of *Intestinalmucosal injury*. Initially, the upper parts of the Small Intestine are affected. However, because this is a progressive condition, it spreads later to the lower part of this organ. Yet, it seldom affects the Colon or Stomach.

How this absence of Microvilli comes about is still not fully understood. One theory is previous microbial infection of the small intestine. It is believed that these histological changes are started and propagated by Enterotoxins from *Coliform bacteria*. Some have said it is due to previous viral infections, while others would claim that it is both.

With these alterations in *the Small Intestinal Linings*, absorption of nutrients is impaired. This is even classified under the malabsorption syndromes. The commonly involved nutrients are fats, minerals (Calcium, Phosphate), Vitamins (A,D,E,K, B12), Folic acid and Albumin.

Others consider non-infectious conditions to be the cause of Tropical Sprue such as ileal disease, pancreatic disorders, and scleroderma.

Symptoms

Patients with Tropical Sprue experience chronic diarrhea with foul-smelling stools due to the presence of fats in the stool. The fats are not absorbed and eliminated with the stool. This will be followed by weight loss. Because of lack of absorption of Folic acid and Vitamin B12, substances which are necessary for red blood cell production, patients develop *Pernicious anemia* (lack of Vitamin B12) and *Megaloblastic anemia* (lack of Folic acid). Due to impaired absorption of albumin, patients may have edema and swollen Legs. Other symptoms are fever, body weakness and easy fatigability.

Laboratories

Complete blood counts would show Anemia, either Megaloblastic or Pernicious types. Blood levels of Minerals (Calcium, Phosphate), Electrolytes (Potassium, Magnesium), Albumin, Cholesterol, and Iron are decreased.

Stool collection tests are supportive to the diagnosis. This is measuring the fat content of the stool in three days after a diet of 80-100 grams of fats. A result of six grams/24 hours of fat in the stool is positive for this disease.

D-xylose test is another supportive test. This is done by administering twenty-five grams of D- xylose orally, and after 1 hour, a blood level of the sugar less than 20 mg/dL and after 5 hours a urine level of less than 4 grams are positive for the disease.

X-rays of the Small Intestine (UGI Series or Small Bowel Series) may show thickening of mucosal folds and segmentation of barium in the Small Intestine.

The real and definitive test is of course the small intestinal biopsy showing villous atrophy with an increase in intestinal crypts with mononuclear infiltrates.

Diagnosis

Given a case of a patient living or visiting a tropical country, with Chronic Diarrhea, Weigh Loss, Pallor, and Body Weakness, a primary impression of Tropical Sprue is highly probable. Of course, all the laboratory tests, specially stool collection tests of the D-xylose, are incredibly supportive of the diagnosis.

The best confirmatory procedure is the small intestinal mucosal biopsy showing the Atrophied villi with increased intestinal crypts and increased mononuclear infiltrations.

Treatment

The course of treatment is usually 3-6 months. For patients with persisting symptoms, the course may extend to one year. Usually, the Antibiotic Tetracycline is given at 100-mg /day in four divided doses for three to six months. This is given in combination with folic acid at 5 mg/day orally to reverse the deficiency of Folic acid. Vitamin B12 is added if there is Vitamin B12 deficiency. Minerals like Calcium and Phosphate must be administered with constant blood level tests to correct the deficiency.

Seldom are patients with Tropical Sprue admitted to the hospital unless they are in severe dehydration due to hronic diarrhea or Cachexia (a disorder with extreme weight loss).

Prevention

Prevention is mainly directed to travelers going to tropical countries. They must be educated and be aware that there is such a thing as *Tropical Sprue* and they may acquire it.

They must follow guidelines for travelers in preventing *Coliform infections* such as avoiding raw foods, or cooked food, but from unknown sources. Never drink water from or with ice unless you know where the ice was made. It is better to drink Bottled Mineral Water or Canned Sodas. *Antibiotic prophylaxis* before going to tropical areas has no place in the prevention of Tropical Sprue.

8. MALIGNANT MELANOMA

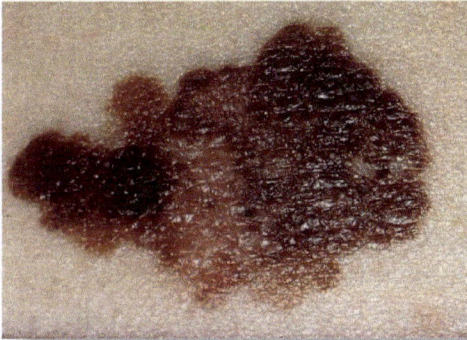

11-3 Malignant melanoma[8]

Malignant melanoma is a tumor arising from the pigmented areas of the body, predominantly the skin. Melanocytes, the pigment producing cells of the body, can be found not only in the skin but in the Choroid and Retina of the Eye, Ears, Gastrointestinal Tract, Leptomeninges, Oral mucosa and Genital mucosa. It comprises about four percent of total skin cancers but it accounts for 75% of deaths due to skin cancers.

About 40-50% of Malignant Melanomas develop from *Pigmented moles,* while the rest arise from the melanocytes of the normal skin. A person has an average of thirty moles in the body and most if not all of these moles are insignificant. However, a person with more than fifty moles in the body has greater chance of developing Malignant Melanoma, more so if these moles are atypical.

The incidence of Malignant Melanoma has increased Worldwide with Australia and New Zealand as the highest. Each year there are around 160,000 newly diagnosed Malignant Melanoma cases worldwide with 48,000 deaths. In the United States, the incidence of Malignant Melanoma has tripled during the last 20 years. The current lifetime risk is one case for every 60 Americans.

Causes and Risk Factors

There are two risk factors in the development of Malignant Melanoma.

1. **Intrinsic factor**, generally referred to the *Genotype of the Person*. It has been found that mutations in the CDKN2A, CDK4 and other genes have been present in the Chromosomes of Melanoma-prone Families. A person's risk is increased if a member of his family has this Skin Cancer. A person who has had a Malignant Melanoma before has greater chances of getting another one later.

2. **Extrinsic factor**, Sun exposure. It has been found that *Ultraviolet Ray A (UVA)* and *Ultraviolet Ray B (UVB)* can cause Malignant Melanoma. These rays can damage the cell's genes thus creating a *Mutation*. When these cells undergo *Mitosis*, a new generation of cell is produced. Usually, the mutated cells undergo rapid *Uncontrolled mitosis*, leading to the formation of the tumor.

Signs and Symptoms

Symptoms of Malignant Melanoma can be summarized by the mnemonics "ABCDE." Others omit the last letter "E" considering the "ABCD" as enough guidelines. A=Asymmetrical lesion; B= Border is irregular; C= Color is multiple; D= Diameter of the lesion greater than 5mm is most likely to be malignant. E= evolution of the lesion of elevation from the skin lesion. Some lesions ulcerate, itch, and bleed easily. Slow- healing lesions are signs of late-stage melanoma.

It is important to take note that some lesions are colorless (Amelanotic melanoma) while some have only the superficial layer of skin discolored (Lentigo melanoma). This is common in the elderly.

Skin cancer is more common in white male Caucasians living in sunny places. They can occur in any area of the body with melanocytes but more commonly in those areas exposed to sunlight. In fair-skinned individuals, they are more common on the back. In women, however, they are more common on the legs between the knee and the ankle.

Skin cancer does not spare dark-colored individuals. Though rare in these individuals, if ever they have one, the lesion is usually found on the palm or sole.

Diagnosis

Skin lesions with variegated colors (e.g., brown, black with shades of red, blue, and white), irregular elevations that are visible or palpable with angular indentation and notches in the exposed area of the body are highly suspicious of Malignant Melanoma. Some lesions are accompanied by ulcerations and bleeding, Visual examination can be followed by Dermatoscopic exams (an instrument that illuminates more to show the underlying pigments and vascular network). Of course, the real basis of the diagnosis is the skin biopsy. The biopsy does not only confirm the diagnosis, but it also determines the depth of the invasion of the cancer cells.

If the lesion is small, an excision biopsy with a frozen section is done to make sure the line of incision is free from malignant cells.

For big lesions, a punch biopsy is done; that is, only a portion of the lesion is obtained for confirmatory diagnosis and to determine the extent of the lesion.

For those suspected with metastasis, several diagnostic tools are used. One is the Serum LDH (Lactic dehydrogenase) level determination. Although a metastatic melanoma may have normal LDH, an exceedingly high value is suggestive of metastasis. Other diagnostic tools are the CT scan, MRI, PET scan to determine the extent of the Metastasis.

Treatment

Surgical excision is the mode of treatment for Malignant Melanoma. The extent of the excision is something that is still debatable as of now. Some would recommend 2 cm from the border of the lesion. Others would recommend 1cm to 5 cm. A 2004 United Kingdom prospective study suggests that the effectiveness of 1 cm and 3 cm extensions from the border are the same for lesions 2 cm deep.

Now the recommended excisions are the following:

5 cm. extension from the border for carcinoma-in-situ.

1 cm extension from the border for lesions 1 cm deep or less.

2 cm extension from the border for lesions 1-4 cm deep.

For those suspected with *Metastasis*, *Lymph node dissection* is done. Prospective randomized trials, however, have shown no benefit for lymph node dissection. For High-risk melanoma, high doses of interferon are given as adjuvant therapy. This treatment produces a good prognosis but has its side effects. So, the benefit of this adjuvant therapy must be weighed against toxicity and compliance. Note that the duration of therapy is a yearlong and it causes flu-like symptoms.

Radiation therapy may be used for lesions that are non-respectable, especially distant metastasis. This method may reduce recurrence but has no effect on the survival rate.

Vaccines have been developed for Malignant Melanoma but these are of benefit only for advanced cases (stage III and stage IV) but not for prevention.

For prognosis, there are several factors to consider: the depth of involvement, the presence of histologic ulceration and the lymph node involvement.

Prevention

There are few preventive measures against Malignant Melanoma, except to minimize exposure to the sun. These measures are more appropriate for those with family history of skin cancer and for those people with plenty of atypical moles. If it is necessary to be exposed to the sun, use protective clothing such as long pants and long-sleeved shirts, and wear a hat. People have often been advised to use sunscreens, but their effectiveness in preventing melanoma is still debatable.

Be observant of any skin lesions, especially if they are atypical and fast growing. Remember the "ABCDE" in recognition of abnormal skin lesions. If in doubt consult a dermatologist or surgeon.

NON-MELANOMA SKIN CANCERS:

9. BASAL CELL CARCINOMA
10. SQUAMOUS CELL CARCINOMA
11. MERKEL CELL CARCINOMA
12. KAPOSI SARCOMA

There are skin cancers that do not develop from melanocytes. These are grouped and called non-melanoma skin cancers. Usually, these types of skin cancers arise from the keratinocytes, the most abundant cell type in the epidermis. These skin cancers are also called Keratinocytes cancers or Keratinocytes carcinoma. There are two more common

Keratinocytes cancers:

1. **Basal cell carcinoma** – As the name implies, they arise from the basal cells of the stratum basalis of the epidermis. They usually occur on body parts exposed to the sun, especially the neck and the head. Eighty percent of skin cancers fall into this category. Basal cell carcinoma are indolent skin cancers. They rarely metastasize, thereby can be easily excised in toto with no malignant cells left. However, they can grow concomitantly in other exposed parts of the body or can recur on the same site from previous excision.

2. **Squamous cell carcinoma** – This skin cancer is not as common as the Basal cell carcinoma. Only about 10-20% of skin cancers fall in this category. This type of skin cancer arises also from areas of the body frequently exposed to the Sun. Although this type is rare, it is invasive. It can metastasize to the surrounding fatty areas and distant lymph nodes.

3. **Merkel cell carcinoma**. This is a rare skin cancer. It arises from the Merkel cell of the epidermis.

4. **Kaposi sarcoma**. This is a skin cancer arising from the dermis. This type of skin cancer used to be rare and occurred only in

elderly people of Mediterranean descent; however, because of the emergence of AIDS, this skin cancer has become quite common.

RISK FACTORS:

There are conditions that increase a person's chances of developing non-melanoma skin cancer.

1. **Frequent exposure to Sunlight.** The longer the lifetime exposure to the sun the greater the chance of having this skin cancer. Ultraviolet (B) rays can cause sunburn and are more associated with Basal cell carcinoma while ultraviolet (A) rays go deeper into the skin causing premature aging and wrinkling.

2. **Fair skin** – Skin with less pigments has less protection from ultraviolet rays and burns easily.

3. **Gender/Age** – Non melanoma cancers are more common in males 50 years old or older.

4. **Individual history** – People with xeroderma pigmentosa and albinos are prone to this type of skin cancer. Likewise, people with compromised immunity (steroid takers) or takers of drugs that make their skin sensitive to light are also prone to this type of skin cancer.

5. **Precancerous conditions**- Actinic keratosis and Bowens disease are two conditions considered forerunners of non-melanoma skin cancers. Anyone with this type of skin lesion must be observant and must have regular referrals to a physician.

6. **Previous skin cancer**- It has been found out that 35%-50% of people diagnosed to have Basal cell carcinoma will develop the same skin lesions within a five-year period. They must be observant of their skin and have regular checkups.

SYMPTOMS:

The following skin changes has been considered pathognomonic for non-melanoma skin cancer:

For Squamous cell carcinoma:

-a wart like growth

-an elevated growth with central depression and rough edges

-a red scaly patch with irregular border that bleeds easily

-a non-healing sore for several weeks

For Basal cell carcinoma – two or more of the following are considered pathognomonic for Basal cell carcinoma:

-a scar like white, yellow waxy area with ill-defined borders

-a Reddish growth with central depression and elevated borders

-a shiny pink pearly white translucent elevated growth on the skin

-a Reddish, elevated, irritating area with crust

-an ulceration with crusts that oozes or bleeds

Diagnosis

The diagnosis of non-melanoma skin cancers is based mainly on skin biopsies. Since these types of skin cancers are indolent, rarely metastasizing, the excision biopsy with frozen section is done. This is not only diagnostic, but also therapeutic as well. The surgeon must make sure that the line of excision is free from cancer cells.

For those lesions suspected to be metastatic, a staging system is used. This staging system basically determines the extent of the metastasis. This requires additional diagnostic paraphernalia like chest X-rays, liver scans, bone scans, brain scans or simply blood tests. This is immensely helpful in the selecting the best mode of treatment for the patient, likewise in the prognosis of the disease.

Treatment

Simple, small lesions can be treated by a dermatologist or surgeon. Metastatic lesions, however, require multispecialty. Aside from the dermatologist and surgeon, they need a medical oncologist and radiation therapist.

Here are the treatment modalities:

1. **Surgery** - Most skin cancers can be excised simply without further treatment. It can be done by the dermatologist or surgeon. They must make sure however that the line of excision is 100% free from cancer cells. This is determined by frozen section method during the surgery.

2. **Topical chemotherapy** – For lesions limited on the top layer of the skin and on pre-cancerous lesions, antineoplastic topicals are used. Some use an immune response modifier called Imiquimod (Aldara). These topicals are applied daily for several weeks. They may cause skin irritations but these are transient only, fading after the treatment.

3. **Radiation therapy** – For lesions difficult to excise, radiation therapy with high energy rays are used. They may cause dryness and discoloration of the skin but these fade away after treatment.

4. **Combination of surgery, chemotherapy, radiation therapy**. Usually recommended for advanced skin cancer.

Prevention

Prevention is basically avoiding exposure to the sun. The ill effects of ultraviolet rays are cumulative. The greater the lifetime exposure, the greater the chances of having skin cancer. Sun exposure, if moderate, is beneficial to the body because it activates the vitamin D in the skin, but excessive Sun exposure, especially between 10 a.m.-3 p.m., is harmful. It may cause sunburn. Those whose work requires extensive exposure to the sun, such as farmers and construction workers should use clothing labeled with UPF (ultraviolet protection factor), use UV protective eyeglasses, and wear wide hats to shade the head and neck.

If you use sunscreen, make sure it has fifteen or higher SPF (Sun protection factor). Apply it often, especially after perspiration. Lastly, there is no better measure than to inspect your skin regularly for any unusual lesions.

13. Photodermatoses

Photodermatoses is an *Immunological Reaction* triggered by sunlight. These are sometimes called *Solar urticaria* or *Sun allergy*. Certain diseases, however like *Systemic Lupus Erythematosus* and *Porphyria* can predispose the skin *to Photodermatoses* once it is exposed to Sunlight.

Diagnosis, Treatment, and Prevention

There is no specific diagnostic test for Photodermatoses. Doctors mainly suspect a person as having the condition if rashes appear in areas exposed to sunlight. Persons with severe forms of Polymorphous light eruption may benefit from oral steroids or *Hydroxychloroquine*. As a prevention, a person sensitive to sunlight should wear protective clothing, avoid exposure to sunlight as much as possible and use Sunscreen if he/she really needs to be outside.

15. OTHER EFFECTS OF UVR ON HUMAN HEALTH

-CONDITIONS AFFECTING THE EYES

- *Acute photokeratitis* – also called *Radiation keratitis* or *Snow blindness* (burning of the Cornea, which is the clear front surface of the eye by *UltravioletLight*-UV B)

-*Photoconjunctivitis* – inflammation of the *Conjunctival surface tissue* on the white of the eye.

-*Climate droplet keratopathy* – degeneration of the cornea causing Blindness.

- *Pterygium* – this is a wedge-shaped Fibrovascular growth of Conjunctivae that extend onto the Cornea. *Pterygia* are Benign Lesions.

-*Uveal melanoma* – the uvea is the part of the eye consisting collectively of the Iris, Choroid, and Ciliary body. These parts contains melanocytes that secrete the pigment, Melanin. Tumors arising from these parts are called *Uveal melanoma*.

-*Acute solar retinopathy* – disease of the Retina acquired from exposure to the short wavelengths of light. It is common after viewing a Solar Eclipse.

- *Macular degeneration* – this is caused by central deterioration of the central portion of the Retina (the inside back of the eye that records the images we see and send them to the brain via the *Optic Nerve*). The *Macula* is responsible for focusing central vision in the eye, and it controls our ability to read, drive a car, recognize faces or colors and see objects in fine details.

-Cancer of the cornea and conjunctivae

- Cataract (Lens opacity)

- *Effects on Immunity and Infections*

- *Suppression of the Cell mediated immunity* – This form of immunity utilizes the T cells in the human body.

-*Impairment of the Prophylactic immunization* – Prophylactic immunity is the artificial immunity with vaccines.

-*Activation of a Latent virus infection* – Latent viruses are those that lie in wait to cause a problem. They are dormant until activated by something.

-**Increased susceptibility to infection.**

[1] https://commons.wikimedia.org/wiki/File:Sunburn_Treatment_Practices.jpg Viewed 18Jan 2022.

[2] https://commons.wikimedia.org/wiki/File:Bicho-de-p%C3%A9_1.jpg

[3] https://www.aafa.org/display.cfm.Allergic Asthma, p.1

[4] www.en.wikipedia.org/wiki/Sunburn

[5] www.en.wikipedia.org/wiki/Heat_wave

[6] www.emedicine.com/EMERG/topic236.htm

[7] www.emedicine.com/med/TOPIC2162.HTM

[8] https://commons.wikimedia.org/wiki/File:Melanoma.jpg

Chapter 12

Harmful Algal Blooms (HABs) and their Toxins[1]

Algae are one of the most important organisms on the planet Earth. Normally they are the best friends of ponds, lakes, oceans, ponds, rivers, and places on land as is the case with lichens (a structure where algae and fungus live together such as on rocks and trees) all because they supply nutrients and oxygen to zooplankton (little animals such as copepods that eat algae known as phytoplankton: see Fig 12-1), amphibians, fish, and other organisms. In other words, they are at the base of aquatic food chains. Put another way, algae make life on earth possible because of base nutrition and the production of oxygen.[2]

What are algae? Algae are a large and diverse group of eukaryotic and prokaryotic aquatic organisms where many photosynthesize but not all. The Eukaryotic algae include the flagellated forms such as dinoflagellates, certain green algae, the Diatoms, Euglenoid types, Seaweeds such the large Kelps, and other types. The prokaryotic algae formerly called Bluegreen algae are bacteria, so their new collective name is Cyanobacteria (Cyan is greenish blue). The cyanobacteria can appear cyan or other colors such as red, yellow, black, or even red. They also photosynthesize like plants do.

Fig 12-1 Photo of various types of zooplankton. Copepods are the most abundant animal on earth (See Number 1 in photo.) If you look you will see the various types of zooplankton that eat algae. Photo by the following: Adriana Zingone, Domenico D'Alelio, Maria Grazia Mazzocchi, Marina Montresor, Diana Sarno, LTER-MC team

In this chapter I discuss algal blooms, give a brief description of the major groups of algae responsible for toxic algae, their reproduction and a detailed discussion of diseases caused by their toxins in humans. Many algae are also involved in animal deaths. This chapter deals with human toxicities.

Fig 12-2 Photo of an algal bloom.[4]

What are harmful Algal blooms (HABs) and why are they discussed in this book? Algal blooms are defined as rapid increases in a population of algae. There may be over one million cells per liter of water (1,000,000 or 10^6 expressed in scientific notation). Blooms can appear in lakes, rivers, and oceans. These blooms are often referred as "Red Tides" because of the colors of the algae but many blooms do not appear red in color.

The reason I cover this topic in this book is because blooms may be caused by climate change, and they can cause lots of damage to aquatic and other organisms including humans. HABs cause millions of dollars in damage every year. The damage includes many socioeconomic, ecosystem and human health issues.

Socioeconomic issues include loss of scallops, Alaskan shellfish, fish, commercial wild shellfish industry, losses to watermen such as fishers, seafood dealers and seafood restaurants and more aquaculture losses occur due to certain algae such as the diatom *Chaetoceros* which caused mortality in many salmon due to the algae lodging in their gills, causing mucus production, suffocation, and death. [5]

Ecosystem damage is caused by certain species that outcompete and overgrow which reduce light penetration to benthic (bottom) communities and may result in low oxygen (hypoxia).[6]

Many species of algae produce toxins. Toxins are secondary metabolites. Primary metabolites are those which are needed for reproduction and growth, whereas secondary metabolites are those not involved in reproduction and growth such as antibiotics, pigments, and toxins.

Toxins associated with algal blooms are Hepatotoxins (toxic to the liver), Neurotoxins (Nerve toxins), skin toxins and intestinal toxins. Varied species of algae can produce the same toxins.

Toxic algae diseases discussed in this chapter are the following with the page numbers:

7. Microcystin poisoning (MCYST)   592

8. Palm island mystery disease (CYN)   598

9. Pfiesteriosis   601

10. Ciguatera food poisoning (CFP)   610

11. BMAA (beta-methylamino-L-alanine)   616

Blooms occur when the following conditions are met: increases in phosphorous, nitrogen, plant nutrients, runoff from agricultural land and industrial activities into the waters, which run off from agriculture and industrial activities into the waters, temperature increases, pH changes and increases in solar energy (sunlight).[7]

A little ecology about the three major groups that produce toxins while they bloom are included below. While this is not a biology book per se, we must give the reader some information regarding these organisms.

This chapter will cover the major groups of algae that result in HABs. These groups of algae are the following: The *Cyanobacteria*, bacteria like *Prokaryotes*, the *Diatoms*, and the *Dinoflagellates*. The latter two are both *Eukaryotic*, which means they possess a clearly defined nucleus.

The Major Bloom Forming Algae Groups

Cyanobacteria

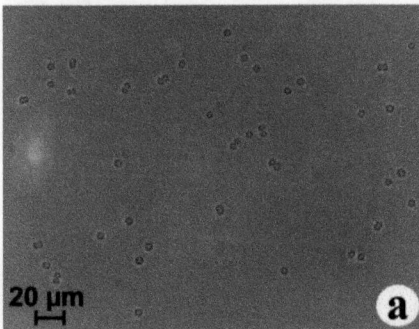

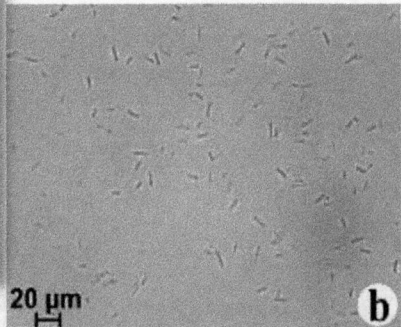

20 μm   **a**

20 μm   **b**

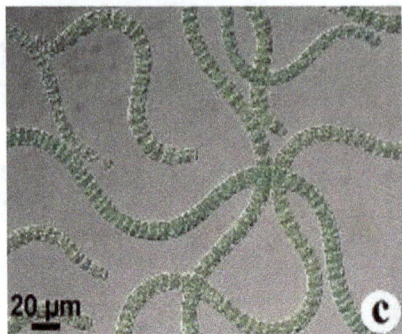

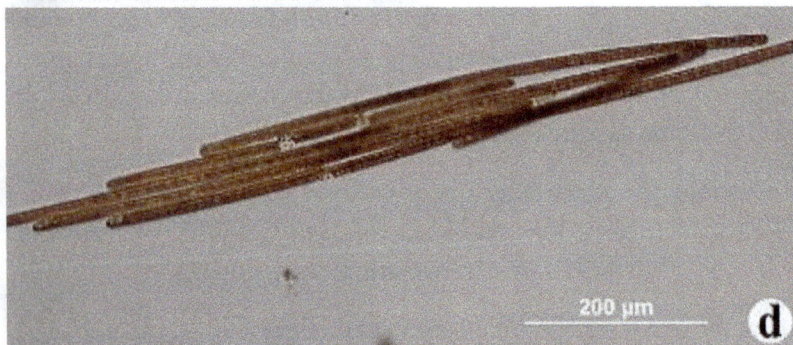

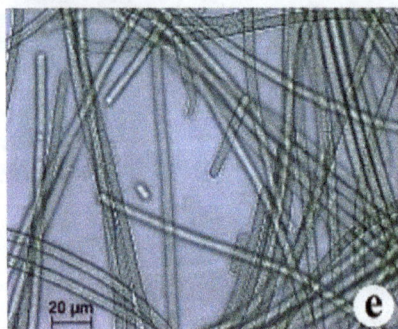

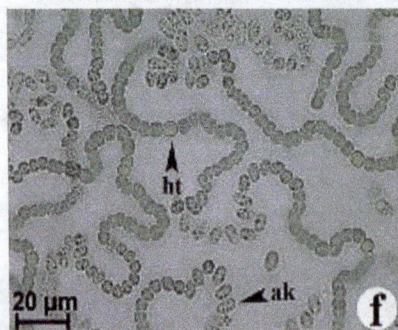

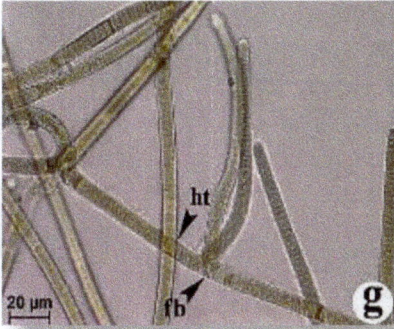

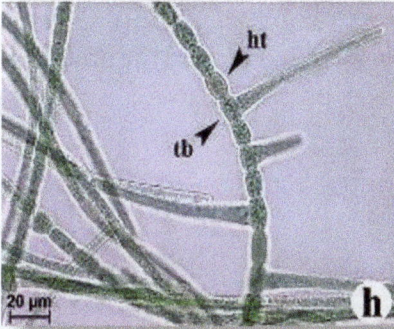

Fig 12-3 Photos of Cyanobacteria[8]

Photos by Alberto A. Esteves-Ferreira, João Henrique Frota Cavalcanti, Marcelo Gomes Marçal Vieira Vaz, Luna V. Alvarenga, Adriano Nunes Nesi and Wagner L. Araújo.

Unicellular: (a) *Synechocystis* and

(b) *Synechococcus elongatus*
• Non-heterocytous:

(c) *Arthrospira maxima*,

(d) *Trichodesmium* and

(e) *Phormidium*- False- or non-branching heterocytous

(f) *Nostoc-* and (g) *Brasilonema octagenarum*- True-branching heterocytous

(h) *Stigonema*

(ak) akinetes (fb) false branching (tb) true branching

These are a group of Prokaryotic photosynthetic organisms which fix nitrogen and live in a variety of ecosystems from soils, water and as symbionts with other organisms such as with lichens. One of their most prominent claims to fame is the fact that the chloroplasts of plants are cyanobacteria which contributed to the origin of plants.[9]

Cyanobacteria could either be filamentous, single cell, or colonial living in a mucilaginous matrix. Many are involved with toxic algal blooms and cause algal diseases which are discussed in detail in this chapter. See Fig 12.3.

It should also be pointed out that warmer temperatures make good conditions for the cyanobacteria to multiply like crazy causing toxic algal blooms. Below is a photo of *Microcystis aeruginosa*, the producer of the toxin, Microcystin, which is a very potent hepatotoxin (effects the liver) and might cause cancer.

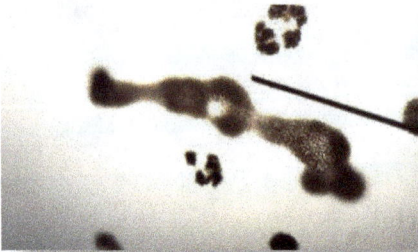

Fig 12.4 *Microcystis aeruginosa.* **Photo by David Arieti**

The general mode of cyanobacterial reproduction is as follows:

The cyanobacteria grow when temperatures increase. Vegetative cells form long chains and float near the surface. When the bloom expands nutrients are depleted. When nutrients are depleted, specialized cells called *Heterocysts* are formed. They can fix the nitrogen from the air into a type of nutrient. There are about eighty species out of the thousands that cause toxic blooms. Algal toxins can be classified as being *Hepatotoxins* (Liver toxins), *Neurotoxins* (Nerve toxins), *Gastrointestinal* and *Skin toxins*.

When temperatures fall in the autumn there is less energy for them, thus they form a type of cell called an akinete. The akinetes sink to the bottom and remain in the sediments for long time periods.

When environmental conditions are right the cells will grow. The cells produce gas vacuoles that allow the organisms to float to the surface. Here they will photosynthesize, and the colony grows.[10]

Diatoms

Diatoms are unique forms of algae that have capsules made from silicon called Frustules (Glass Houses: see fig 12-5). They come in all shapes including boat-shaped (Pennate) and circular ones called (Centric). They are all Eukaryotic. You might say that they resemble petri dishes, which are used to grow bacteria. They can live virtually everywhere whether it be soils, streams, lakes, rivers, and oceans. They account for 40% of Photosynthesis on the planet. Therefore, they are especially important for those of us who like to breathe oxygen.

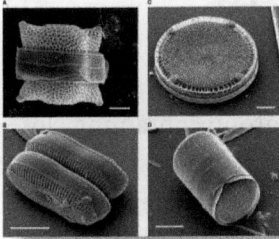

Fig 12-5 Examples of Diatoms with their frustules[11].These photos were made with the Scanning Electron Microscope (SEM). Photos by May Ann Tiffany, San Diego State University

Reproduction of Diatoms is by Asexual means. Some Diatoms such as *Pseudo-nitzschia* form long chains. Two cells are produced each genetically identical; but one is slightly smaller.

Eventually this results in a reduction in the population size. When the diatom reaches about half its original size after many divisions, it reproduces sexually when two Parent Cells align and produce *Gametes* (Sex cells). The latter meets and fuses. The fused gametes form an *Auxospore*. This protects the developing Pseudo-nitzschia cell until it is fully grown.[12]

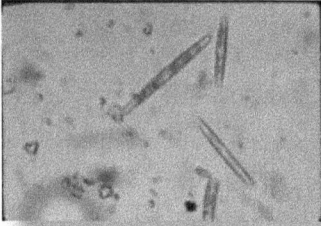

Fig 12-6 Typical Diatoms. Photo by David Arieti

Fig 12-7 The Diatom , Cymbella hantzschiana.[13]

Dinoflagellates

The *Dinoflagellates* are a major Eukaryotic algal group containing close to two thousand known species. These algae have characteristics unique to this group. They have two flagella, one is trailing like a sperm cell and the other is transverse, coming from the side. Dinoflagellates have a unique type of nucleus called a ***Dinokaryon*** in which chromosomes are fibrillar in appearance and are continuously condensed. They lack histones (protein structures which are in chromosomes) and which are present in most other known Eukaryotic organisms including *Protists*.

Some species are known for their bluish glow called Bioluminescence. It is thought that there may be a few hundred toxins produced by Dinoflagellates, which also produce substances that have therapeutic effects such as being antiviral, antibacterial and having antioxidant activity. Approximately 80 species of dinoflagellates produce toxins. They are unique in that many species are both *Photosynthetic* and *Heterotrophic* (they eat). The main types of illnesses by Dinoflagellates are Respiratory Paralysis and Gastrointestinal Problems such as *Diarrhetic Poisoning*.

Reproduction of the Dinoflagellates is described as follows: Cysts of Dinoflagellates lay dormant buried in sediment until environmental conditions are exactly right for them to Germinate (grow). Swimming cells will then emerge from the Germinating cysts. When abundant nutrients are present the cells will reproduce exponentially. One cell can divide many times. Within a week there will be several.

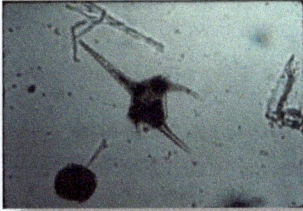

Fig 12-8 Two species of Dinoflagellates: *Peridinium gatunense* (left) and *Ceratium hirundinella*-Photo by David Arieti

Fig 12-9 Toxic algae bloom in Lake Erie Photo by Jesse Allen and Robert Simon[14]

ALGAL DISEASES

AMNESIAC SHELLFISH POISONING

Domoic Acid (DA)

Amnesic shellfish poisoning is also known as *Domoic acid poisoning* because it is caused by Domoic acid. It can be a life-threatening syndrome manifested by neurologic and gastrointestinal symptoms. After these initial symptoms, some patients may develop *Dementia*. It was initially reported in Cardigan in Prince Edward Island in Canada, but now it is an ongoing problem in the state of Washington and Oregon, as well as other parts of the world.

Cases/Outbreaks

Europe[15]

Belgium - In 2000 and 2001, there were 151 and 154 samples respectively assessed for ASP, but no toxic events were detected in Shellfish.

Denmark - There were several investigations done in Denmark that showed a diatom species (*Pseudo-nitzchia seriata*) was present in colder areas of Northern Hemisphere that produced domoic acid (DA). Other species of *Pseudo-nitzschia* were evaluated for DA. In 1993, there was algae bloom caused by these types of Diatoms. DA was detected but in low concentrations.

France - In May 2000, a bivalve (*Donax trunculus*) was assessed with DA levels above the regulatory limit. The causative Diatoms were the Pseudo-nitzschia types. There was one episode reported in April 2002, but the toxin level was low.

Ireland - In December 1999, high concentration of DA up to 3,000ug/g was found in the scallops hepatopancreas on every Irish coast. In 2000 and 2001, the ASP toxin were again detected above regulatory limits in scallops. In 2002, for the first time, ASA toxin was detected in mussels.

Italy - In 2002, DA was detected at levels above regulatory limit in the scallop (Pecten maximus).

The Netherlands - In the Dutch Wadden Sea, Diatoms of the species Pseudo-nitzschia were detected between November 1993 and July 1994 although there was no shellfish poisoning recorded.

Norway - In 2000 and 2001, mussels and scallop were found to have DA but not above regulatory limits.

Portugal – In 1996, DA was detected in every species of bivalves around the Portuguese coast. In 1999, DA was detected in 960 samples. DA in shellfish are detected several times a year, in the spring and autumn months. Levels recorded can be twice as high as the accepted limits and are not unusual. This DA was also detected in other shellfish such as oysters, furrow shells, razor clams, and sardines. However, in sardines, the DA was restricted in the gut and not on the muscle tissue.

Spain - The first detected DA was in mussels in October 1994. These mussels came from Galicia in northwestern Spain. From September to December 1995 and 1996, DA was detected in scallops. In 2002, there were toxin episodes in Galicia and Andalucía. Due to the presence of Diatoms Pseudo-nitzchia, the production area was closed.

The United Kingdom of Great Britain and Northern Ireland - In July 1999, on the west coast of Scotland, a scallop harvesting area of 8,000 square miles was closed due to the presence of ASP toxin. In 2000 and 2001, ASP toxin were again detected above the accepted limit among the scallops, restricting fishing in the affected areas. In Northern Ireland, scallops were found to contain ASP toxin above regulatory limits. The United Kingdom Food Standards Agency banned scallop fishing in the affected areas.

Figure 12-10: Occurrence of ASP toxins in coastal waters of European ICES countries From 1990 to 2000 (represented by yellow dots).

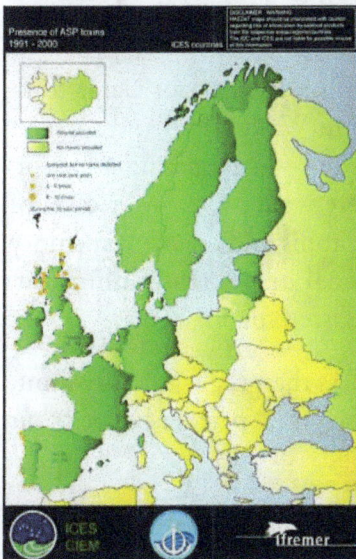

Source: Food and Agriculture Organization of the United Nation, 2004, Marine Biotoxin, Amnesic Shellfish Poisoning, Reproduced with permission

North America

Canada - Canada is where the first ASP was reported in 1987. During November and December 1987, there was an outbreak in the Cardigan Region of Eastern Prince Edward Island.

There were three deaths and 108 cases of acute intoxication after eating blue mussels. The toxin was identified as DA and the Diatoms as the *Pseudo-nitzschia* species. In the years after 1987, algae blooms happened but were less extensive. Since 1988, phytoplankton samples have been collected in four stations in the western Bay of Fundy. Usually, algae blooms occurred in late summer. The highest concentrations were recorded in 1988 and 1995 during late August and early September, leading to the closure of harvesting areas.

United States of America

Alaska - There was no problem with ASP even though toxic *Pseudo-nitzschia* species had been identified. In 1992, approximately 3,000 samples of commercially valuable Shellfish and Finfish were evaluated; the highest value was 11.1 mg/g 17 samples above the regulatory level.

West Coast - In October and November 1991, razor clams from the surfing zone on Pacific coast beaches in Washington and Oregon contained DA with levels as high as 154 mg / g, triggering a ban on clam harvesting In Washington state, there were twenty-four cases of reported illness with gastrointestinal symptoms and one with memory loss. In the autumn of 1994 in Hood Canal in western Washington, there was bloom of *Pseudo-nitzschia* that lasted for six weeks.

Shellfish assessed for DA contained less than the regulatory limit. In Penn Cove, Washington, a bloom of the same species occurred, but the level of DA was less than the limit.

In the autumn of 1991, several species of *Pseudo-nitzschia* were detected in Monterey Bay, California but there was no ASP reported.

In May and June 1998, over four hundred California sea lions (*Zalophus californianus*) died along the central California coast. The death was believed to be caused by DA produced by *P. australis*. In April and May 2002, many marine mammals and birds were dying along the California coast. About seventy dolphins, more than two hundred sea lions, and two hundred seabirds died. No human illness was reported.

East Coast - In January and February 1991, DA producing Diatoms were isolated along the Nantucket coast in Massachusetts, but the levels were half of the regulatory limits. A multi- year study in Louisiana showed the presence of *Pseudo-nitzschia* species in the water, but there were no reported cases of ASP.

Gulf of Mexico - Except for isolated cases, there is no direct evidence of ASP toxin in the shellfish in the Gulf of Mexico. Therefore, ASP is not considered a public health hazard in this area.

Figure 12-11: Occurrence of ASP toxins in coastal waters of North American ICES countries from 1991 to 2000 (represented by yellow dots)

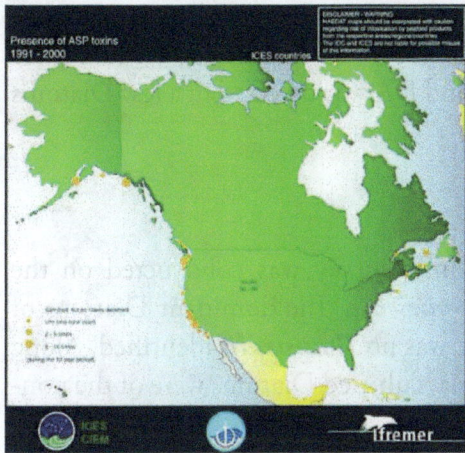

Source: Food and Agriculture Organization of the United Nation, 2004, Marine Biotoxin, Amnesic Shellfish Poisoning,Reproduced with permission

http://www.ifremer.fr/envlit/documentation/dossiers/ciem/aindex.htm

Central and South America

Argentina - In the winter of 2000, two massive mortality episodes of seabirds were reported in Mar de Plata. The dominant species was P. australis and the toxin were identified in Mussels.

Chile - Since 1967, Shellfish Samples have detected DA levels exceeding the regulatory limit. Up until 2001, there had been no ASP intoxications recorded.

Mexico - 150 dead brown pelicans were found in Cabo San Lucas on the tip of Baja California Peninsula in January 1996. These birds were fed with Mackerel, *(Scomber japonicus)* with DA from Diatoms.

During January and February 1997, many marine organisms died in the Gulf of California. They were 766 common loons, *Gavia immer*, and 182 sea mammals. In their examination, they found DA and its isomers in some the stomach of some dolphins.

Asia

Japan - ASP toxin screening has been implemented in Japan since 1991. DA had not been detected except in August 1994 when there was a red tide in Hiroshima bay and DA was detected in a few Diatoms of *Pseudo-nitzschia* species.

Oceana

Australia – A wide taxonomic survey was cond ucted on the potentially toxic diatom *Pseudo-nitzschia.* The dominant Diatoms of this species were not toxic. There was no *P. australis* identified. Along the coast of Tasmania and Victoria, cultured Diatoms were of the non-toxic type.

New Zealand - During the summer of 1992 and 1993, DA was detected at low levels from the diatom samples in the Bay of Islands, the Hauraki Gulf, and Bay of Plenty. The highest level of DA was found in early 2003. These were found in green mussels from Marlborough Sound, scallops from Tauranga Harbor and Whangaroa Bay.

CAUSATIVE ORGANISMS

The causative toxin *Domoic acid* (DA) is produced by the diatom of *Pseudo-nitzschia* species and by a seaweed, *Chondria armata*. These species are widely distributed around the word and are: *Pseudo-nitzchia multiseries, P. australis, P. delicatissima, P. pungens, P. seriata, P. multiseriata, P. turgidula, and P. fraudulenta.* These Diatoms are consumed by different shellfish, most notably by Mussels and Scallops which are eventually harvested for human consumption.

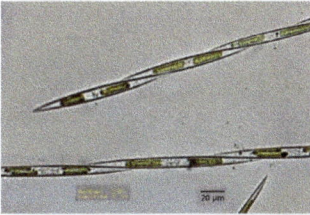

Fig 12-12- Pseudo-nitzchia fraudulenta[16]

Fig 12-13-Chondria sp.[17]

MECHANISM OF ACTION

Domoic acid is an amino acid like *Glutamate*, the body's most prominent Excitatory Neurotransmitter in the nervous system. It has been noted that the affinity of Domoic acid to Glutamate Receptors (a-amino-5-methyl-3-hydroxyisoxazolone-4-proprionate - AMPA) in the brain is one hundred times more than glutamate. The presence of Domoic acid increases activation of these receptors causing increased brain activity that can precipitate seizures and damage the amygdala and hippocampus of the limbic system. Aside from these, activation of AMPA increases calcium into the brain cells. This can cause toxicity to the brain cell by signaling apoptosis. The hippocampus which plays a role in memory and is particularly at risk due to its abundance of glutamate receptors. This is the reason patients can develop *Amnesia* after having the illness.[18]

Domoic acid is heat stable and will not be destroyed by cooking or freezing. As little as 1.9mg/g concentration of domoic acid can cause CNS toxicity in humans. This level is of concern since some muscles of the shellfish contain as much as 128 mg of domoic acid per 100-gram tissue.

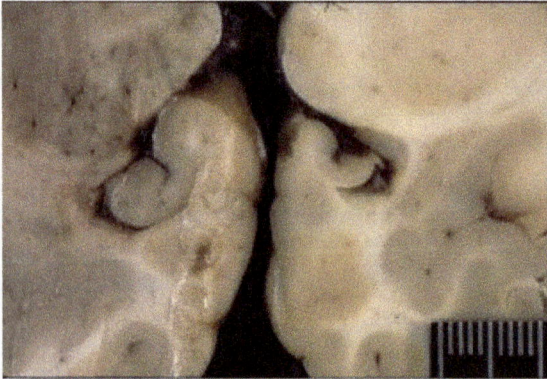

Fig 12-14 Hippocampus of Sea Lions. Left:Normal California Sea Lion brain-Right: California sea lion brain affected by domoic acid exposure.[19] Used with permission from the Marine Mammal Center.

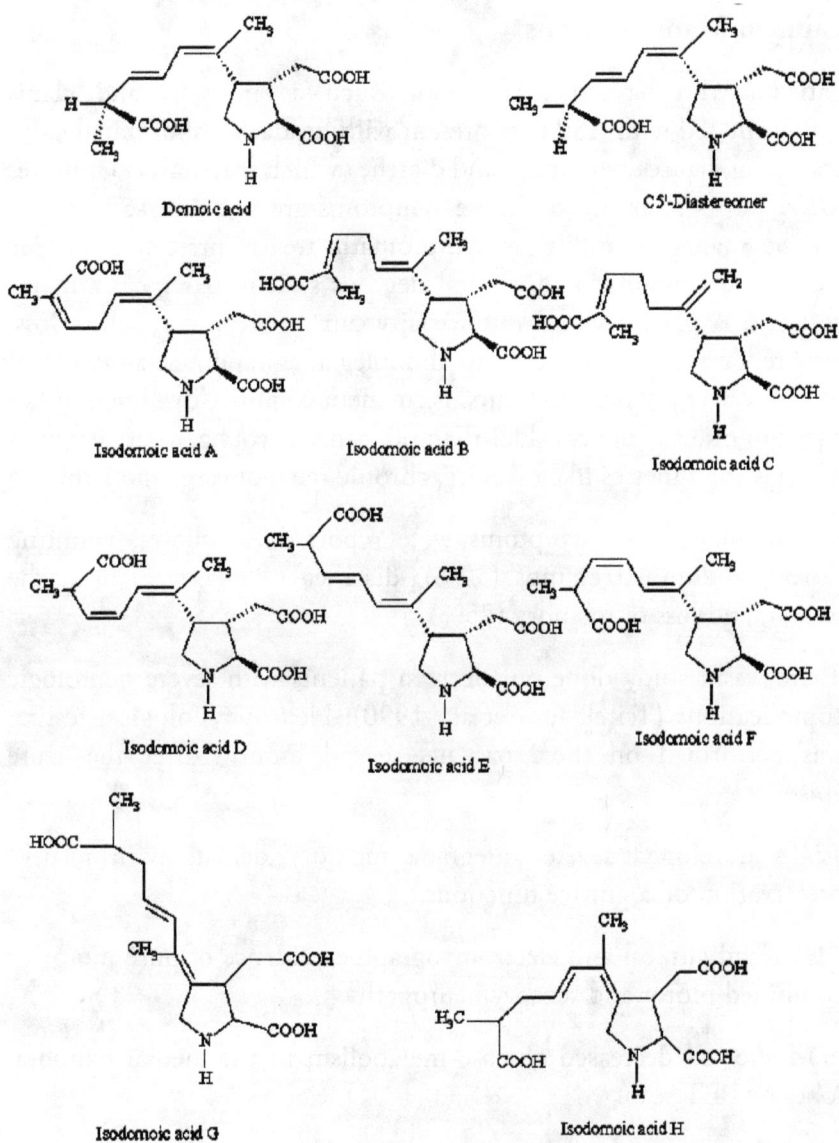

Figure 12-15: Chemical structures of domoic acid and its isomers[20]

Source: Food and Agriculture Organization of the United Nation, 2004, Marine Biotoxin, Amnesic shellfish poisoning, Reproduced with permission[21]

Clinical Manifestations

Patients who have taken seafood such as mussels, and clams contaminated with ASP may present with headache, abdominal pain, cramping, nausea, vomiting, and diarrhea which may start 30 minutes to 24 hours after intake. These symptoms are not that serious and can be managed readily by symptomatic treatment. Since domoic acid is like glutamic acid, central nervous system toxicity is a major concern. A patient with more toxicity can progress to memory loss, hyporeflexia, hemiparesis, ophthalmoplegia, agitation, seizure, coma, or shock over the next 48 hours. Permanent cognitive dysfunction can occur in older patients (older than 60) and in younger patients with pre-existing illnesses like diabetes, chronic renal disease, and stroke.

The frequencies of symptoms were reported as follows: vomiting (76%), abdominal cramps (50%), diarrhea (4%), severe headache (43%), and loss of memory (25%).[22]

There was a study done on fourteen patients with severe neurologic complications (Teitelbaum et al., 1990): Neurophysiological testing was performed on these patients several months after the acute episodes.

12/14 developed severe antegrade memory deficits with relative preservation of cognitive functions.

11/14 had clinical and electromyographic evidence of pure motor or combined motor and sensory neuropathy.

4/14 showed decreased glucose metabolism in the medial temporal lobes by PET scan.

All fourteen patients developed confusion and disorientation within 1.5 to 48 hours after consumption. Acute coma was associated with the slowest recovery while seizures became more frequent up to two months but ceased by four months.

For those four fatalities, their brain revealed necrosis and loss predominantly in the hippocampus and amygdala.

Diagnosis

Since hospitals and clinics do not have laboratory assay for ASP toxin, diagnosis is based on clinical manifestation and history. Presented with these arrays of common clinical manifestations, a practicing physician will find it hard to consider ASP. He must think of many diseases and conditions with the same manifestation. The only probable clue pointing to poisoning is a history of intake of shellfish within minutes or to a few hours. More so if there is an outbreak of ASP in the area.

Since the concern of this type of poisoning are the neurologic deficits, laboratory and ancillary procedures are geared toward the brain. A CT scan, EEG, or MRI can be requested to determine abnormalities in the brain, specifically in the hippocampus and *Amygdala*.

For any suspected ASP, a sample of the food must be obtained, frozen, and sent to the Department of Health. Currently, there are no private laboratories offering Domoic acid testing.

The amnesia in ASD can be differentiated from Alzheimer's disease due to its relative preservation of intellect and higher cortical function. The lack of confabulation with preserved frontal function will distinguish ASP from Korsakoff syndrome.

Treatment

There is no clear consensus for the treatment of ASP. The management is more symptomatic or supportive. Antipyretic for fever, antiemetic for nausea and vomiting, oral hydration fluid or intravenous fluid for diarrhea. Patients with seizures are given *Phenobarbital* and *Diazepam*. If a patient is resistant to *Benzodiazepines*, *Propofol* or *Barbiturates*, he or she can be given Adjunct Therapy. For Amnesia, some treatments include Cognitive Therapy, Neurofeedback, and Nutritional Supplements. There is no antidote available for Domoic acid toxicity. However, there are animal studies that *Pyridoxine* (vit B6) decreases the level of Domoic acid.

Like any other poisoning, a case must not be considered isolated as this maybe be followed by an outbreak. Report the case to the health authorities and proper follow up must be done to prevent spread. Every effort must be made to obtain the contaminated source for laboratory testing.

Regulation and Monitoring

Europe

Member states in the European Union set 20 mg/g as the regulatory limit. Samples with more than this value are destroyed.

Denmark - Shellfish monitoring was initiated in 1993.

Ireland - A Biotoxin Monitoring Program started in 1984. Initially, it was done on DSP poisoning but now it includes domoic acid and *Azaspiracid poisoning*. They have weekly Shellfish Testing using Mouse Bioassay and reports are sent to regulatory agencies, health officials, shellfish producers and processors by web-based communications.

North America

Canada - Regulation started in 1988 with 20 mg/ kg of mussel as the regulatory limit. Harvesting areas are closed once the toxic level exceeds the regulatory limit.

Phytoplankton monitoring is also done at 4 stations in the Western Bay of Fundy.

United States - There is a non-official guideline of 20 mg/ kg of bivalves. For cooked crabs, the guideline is set at 30 mg/kg. Closure is done once the level of 20 mg/kg is reached. Exported shellfish are accompanied by health certificates.

Central and South America

Argentina - Argentina has a national monitoring program for mussel toxicity. regional laboratories are in each coastal province and there is one fixed station in Mar de Plata.

Brazil - Had a pilot monitoring program for one year but a national monitoring program has not been initiated.

Chile - Chile has a national monitoring program for shellfish and phytoplankton.

Uruguay - Regularly monitors mussel toxicity and phytoplankton.

Oceana

Australia - Monitoring and regulation on mussels and algae were initiated in 1993.

New Zealand - Monitoring for shellfish started in 1993 and their guideline limit is also 20 mg of DA / kg of Shellfish Meat. The New Zealand Biotoxin Program monitors shellfish and Phytoplankton using Mouse Assay.

PARALYTIC SHELLFISH POISONING (PSP)[23]

Paralytic shellfish poisoning (PSP) is a food borne disease acquired by eating shellfish contaminated with Diatoms, Dinoflagellate, and Toxic algae. These shellfish are clams, mussels, oysters, geoduck, and scallops. In crabs, the toxin is found in the guts of crabs, but not in the crab meat. If you do not want to suffer from PSP, remove the guts before cooking the crab and do not drink the broth. There are many organisms that produce *Saxitoxin*. The majority are either Dinoflagellates or Cyanobacteria

Table-12-1 List of the organisms that produce Saxitoxin [24]

NAME OF TOXIN	GENUS	GROUP	TOXIN TYPE
SAXITOXIN (STX)	Alexandrium ....[25]..	Dinophyceae	Neurotoxin
	Gymnodinium	Dinophyceae	
	Anabaena	Cyanophyceae	

Cylindrospermopsis [26]	Cyanophyceae	
Lyngbya	Cyanophyceae	
Planktothrix	Cyanophyceae	
Pyrodinium bahamense	Dinophyceae	
Aphanizomenon	Cyanophceae	
Cyanophceae		

This is a photo of one of the many organisms that produce Saxitoxin

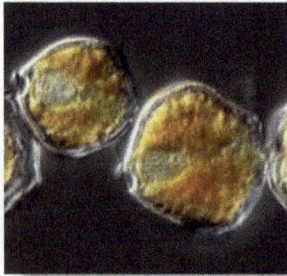

Fig 12-16
Alexandrium ostenfeldi. **Photo by Nancy Lewis.**[27]

Viewed 16 May 2020.

On June 5, 1990, on the Island of Nantucket, Massachusetts, six people fishing developed an illness after eating Blue Mussels (*Mytilus edulis*) harvested in deep water (1). They manifested the following symptoms one to two hours after eating the shellfish: numbness of the mouth, tongue, throat, face with *Perioral edema* and *Paresthesia* of the extremities. These symptoms lasted for 14 hours. Low back pain was also noticed, which started 24 hours after the onset of neurologic symptoms and lasted for 3.3 days. Of these six, four people recovered right away; but two were hospitalized. Then one lost consciousness; but was discharged after three days.

Outbreaks of PSP[28]

There is an increasing distribution around the globe of paralytic shellfish poisoning (Hallegraeff,1993). In 1970, PSP was well known from temperate waters of Europe, North America, and Japan. Arriving in 1990, PSP was documented in South Africa, Australia, India, Thailand, Brunei Darussalam, Malaysia, Philippines, and Papua New Guinea.

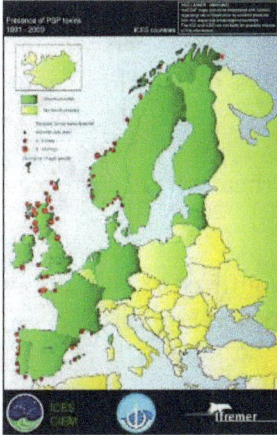

Figure 12-17: Occurrence of PSP toxins in coastal waters of European ICES countries from 1991 to 2002 (represented by red dots)[29]

Source: Food and Agriculture Organization of the United Nation, 2004, Paralytic Shellfish Poisoning Reproduced with permission

Countries with PSP outbreaks:

Denmark - At the east coast of Jutland, PSP episodes in 1987, 1990, 1996, 1997.

France - Atlantic coast in 1992, PSP toxin was found in mussels. In 1998, toxin was found in clams, oysters, mussels, and their production facilities were closed for two months. In 2000, two production facilities in Britain were closed.

Germany --- In 1972, mussels were monitored for toxins. In 1987, three cases of PSP poisoning were reported in Lower Saxony. In March 1992, PSP producing algae had been isolated.

Ireland - In July 1992, PSP was noted in Cork Harbor and persisted for one week. In 1999, *A. tamarensis* was detected and in 2002 PSP toxins were above the limit in mussels, oysters.

Italy - HAB had been recurring in the Adriatic coast but there have been no reported cases of PSP.

Monitoring station in Emilia Romagna in1994 to 1996 had greater than eighty ug / 100 gram of mussels' meat but no public health cases was reported.

The Netherlands -- In the North Sea in 1989, Alexandrium sp, an ASP producing algae was identified. Yet, no PSP cases were found.

Norway --- Among the Scandinavian countries, Norway has the earliest recorded cases of PSP. They have recorded PSP six times (1901, 1939, 1959, 1979, 1991, 1992) with a total of Thirty-two victims with two deaths. Surveillance conducted in 1994 led to closures of several fishing Communities.

Portugal --- Along the north coast of Roca Cape, there were occurrences of PSP from 1996 to 1990.

In 1992, there were cases of PSP along the south coast of Lisbon and coast of Algarve with a concentration of 100-500 ug / 100 gram of shellfish meat. In October 1994, PSP was reported among nine patients (six women and three men) after eating mollusks (Mytilus edulis) from the west coast of Portugal.

Spain --- Among the European countries, Spain is the country most concerned with PSP because mussel aquaculture is an important industry in the Northern Atlantic coast of Galician Rias.

In 1976, people in several countries in Europe including Germany, France, Switzerland, and Italy developed PSP after consuming mussels exported from Spain. There were a total of 120 people affected in Western Europe; however, there was no mortality. In 1993, this unfortunate episode lasted for an unusually longer period. They found out that the culprit was the Dinoflagellate (Alexandrium catenella that contaminated the mussels. In 1994, there were outbreaks along the Atlantic shore of Spain. Beginning in 2000, because of the toxic events, the harvesting of bivalves has been prohibited around Galicia. In 2002, production of scallops were closed due to the presence of G. catenella.

Sweden --- In Sweden, contamination of mussels usually occurs at the end of spring and beginning of summer. In 1985 and 1988, PSP toxin had been detected in mollusk meat.

United Kingdom of Great Britain and Northern Ireland

The initial cases of PSP poisoning were reported in 1968, along the Northern coast of England. There were seventy-eight people hospitalized but no mortality. From 1978 to 1980, sporadic toxins were reported along the Northern coast, spreading toward Scotland. In August 2001, the United Kingdom Standard Agency prohibited scallop fishing along the sea adjacent to Northern Ireland.

Morocco - From October to November 1994, HAB was associated with PSP outbreaks. In November 1994 ,there was 6,000 µg eq/ 100 gram of PSP toxin in mollusk meat.

South Africa -- In 1969, six cases of PSP poisoning were reported and seventeen cases in 1979. There were two fatalities in 1984.

Tunisia --- In 1998, more than seven hundred tons of cultured sea bass and sea bream and several species of wild fish were found dead along the lagoons of Burger and Gar El Melh.

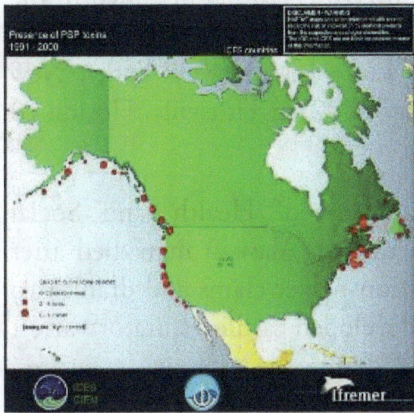

Fig. 12-18: Occurrence of PSP in North America (represented by red dots)

Source: Food and Agriculture Organization of the United Nation, 2004, Paralytic Shellfish Poisoning

Reproduced with permission

Canada --- In 1993, four crew members from an expedition became sick and one died after consuming shellfish in the waters of British Columbia. PSP has been documented as early as 1880 in the St. Lawrence Estuary (Quebec) and the Bay of Fundy (between New

Brunswick and Nova Scotia). Overall, from 1880 to 1995, there were 106 incidents reported with 538 cases and thirty-two deaths. In May 1999, two workers at a Salmon farm in Herbert Inlet, British Columbia developed symptoms after consuming wild scallops.

United States -- East Coast

On the east coast, the first reported case of PSP was in the coast of Maine near the Canadian border in 1958. In 1972, there was HAB along the coast of .... Maine, New Hampshire, and Massachusetts causing PSP in 33 people. No one among them died. In 1979, mussel beds were closed at Narragansett Bay, Rhode Island. In the following years, there were reports of PSP in a small embayment in Connecticut and Long Island.

From 1973 - 1987, the Massachusetts state health department reported nineteen outbreaks. These were caused by the consumption of mussels, clams, oysters, scallops, and cockles.

In June 1990, the same Health Department documented PSP among six fishers in a fishing boat in the George Bank area off the Nantucket coast. They ate cooked blue mussels (***Mytilus edulis***) obtained in the deep water in by the island of Nantucket. Other areas on the east coast with occasional PSP outbreaks are: Long Island (New York), New Jersey, and Connecticut. Since January 2002, in Titusville, Florida 10 cases of PSP were reported.[30]

On June 26, 1990, Alaska Department of Health, and Social Services (ADHSS) has documented that an Alaskan man died after eating shellfish (2). He ate 25.-30 steam butter clams and drank two teaspoons of butter clam broth. He developed cardiorespiratory arrest two hours later and eventually died despite resuscitative measures. His stomach contained 370 ug/100 g of toxin (maximum safe level is eighty ug/ 100 g). The butter clam broth he drank contained 2650 ug /100 g. Two other companions had shared the butter clam broth. One manifested numbness of the face and hand followed by tingling sensations one hour later while the other one was Asymptomatic.

West Coast

Alaska- PSP has occurred for centuries with the first reported case in 1799. From 1973 to 1994, there were sixty-six recorded outbreaks involving 146 individuals, eight of them developed Paralysis of the limbs, eight required mechanical ventilation and two died. Most of these outbreaks occurred in the island of Kodiak and southeastern Alaska between the month of May and June. Most of these cases were related to eating mussels.

In the Strait of Juan de Fuca (international boundary between Canada and USA), PSP caused several deaths in 1942. In the Puget Sound region of Washington recurrent outbreaks of *Alexandrium* were reported in the late 1970. The Northern California and Oregon area had these perennial problems of PSP outbreaks.

This PSP was followed by three more episodes in the Alaska Peninsula and Kodiak island in that month of June 1990. This outbreak had thirteen cases among twenty-one suspected individuals (62%).

Epidemiologic study done showed the following:

- consumption of butter clam -- 7 / 13 – (54%)
mussels ------ 6 / 13 -- (46%)

- number of shellfish consumed --- 3-30 (median four shellfish)

- onset of symptoms -- 0 to 2 hours (median 1 hour)

- duration of symptoms --- 1 to 24 hours (median 7.5 hours)

- sought medical care --- 7 persons / out of 13 (54%)

- mortality -- 1 /13 (8%)

- PSP toxin test done on four sites along the Alaskan Peninsula where the shellfish were harvested:

- butter clams from Volcano Bay and King Cove -- 7750 ug / 100 g

- mussels from Sand Point and Kodiak --- 1925-12,960 µg / 100g

From 1976 through 1989, there were 42 PSP outbreaks (accounting for ninety-four cases) which were documented in Alaska. Thirty-one of the forty-two outbreaks occurred between May to July.

In 1942, due to several deaths from PSP, the Strait of Juan de Fuca was closed. In 1970, the San Juan and Bellingham coasts were closed. PSP was also an annual problem along the coast of Northern California and Oregon.

Argentina - The first HABs in this country was in 1980 in the Valdes Peninsula. Two members of a crew of a ship died after eating mussels. In 1991/1992. HAB was documented in the northeastern shore of Beagle Channel. This was followed in the summer of 1993, in the Buenos Aires shelf known as Rincon.

Brazil - Beginning in 1992, there have been reports of HAB with shellfish toxicity in the Tiera del Fuego every year. In 1998, *A. catenella* algae were observed in off the coast of Santa Catarina State.

Chile - As early as 1886, there were reports of PSP from eating mussels among the natives of Ushuaia.

From October 1972 to January 1997, there were 329 PSP cases in the southern regions of Chile. Twenty-six people died. PSP and DSP had severe public health and economic impact in Chile until 2001. In March 2002, there were eight cases of poison and one death after consumption of shellfish in the southern Chile prohibiting the harvesting of shellfish in this area including the Ancud Community.

Guatemala- There was an outbreak of PSP in 1987 with 187 cases and twenty-six deaths after the consumption of clam soup. Fifty percent of the fatalities were children.

Mexico - In 1979, there were reports of PSP poisoning after eating local mussels. There were twenty cases with three fatalities. In November 1989, along the coast of Chiapas and Oaxaca, there were ninety-nine cases of PSP poisoning with three fatalities. Between March 1993 and April 1994, shellfish harvesting was closed along the Gulf of California following increased density of dinoflagellate *(A. catenella)*.

Trinidad - The first PSP recorded in this area was in 1994. The PSP toxins were found in the meat extract of the mussel Perna viridis.

Uruguay - In 1980, 60 cases of PSP poisoning were reported. They were not able to determine the causative algae species.

Venezuela - 171 cases of PSP were reported in 1979 and 9 cases were reported in 1981.

China - The first PSP cases reported were forty separate episodes involving 423 cases and twenty-three fatalities in the Zhejiang Province between 1967-1979. The next reported case was in Donghan (south of Fujian Province) with 136 cases and one fatality. From 1990 to 1992, there was a survey done in Guangdong Province with the result of the presence of PSP toxins in thirty- three edible marine organisms.

Timor- Leste - There was a report of a man who died after ingestion of the crab *Zosimus aeneus.*

Hongkong - There were three outbreaks of PSP in 1992. No details were available as to the number of cases and fatalities.

India -- PSP was reported in 1981, with ninety-eight cases and one fatality.

Japan -- First reported PSP case was in 1992, in Hiroshima Bay. In March, 1997, in the island of Fukue, twenty people were poisoned after eating oysters.

Malaysia -- In 1977, there were 201 cases of PSP reported after eating local clams. Most of these cases were confined in the west coast of Sabah in Borneo. In the Malayan Peninsula, the first reported PSP poisoning was in early 1991 when three people became ill after eating farmed mussels from Sebatu in the Strait of Malacca. In September 2001, there were six people who became ill after eating clams harvested from the coastal lagoon on Kelantan on the east coast of Malaysian Peninsula. One of them died.

Philippines - PSP is a public concern in the Philippines. It caused economic loss in the fishery sector of the country, especially in shellfish industry. Between 1983 and 2005, there were reports of 2,161 cases and 123 fatalities in twenty-seven coastal waters. The 1983 outbreak in the Central Philippines resulted in a 2.2 million loss with a dramatic decline in demands for fishery products. The 1988 outbreak in Manila Bay, caused economic damage since the price of all seafood products dropped to 40% of the original price.[31]

There are reports dated Dec 27, 2016 from BFAR (Bureau of Fisheries and Aquatic Resources) of two deaths and forty cases in the town of Mari pipi, Tacloban City, Philippines. [32]

Taiwan Province of China - In January 1986, there were reports of two mortalities and thirty morbidities after eating Soletellina dipos. In February 1991, eight cases of morbidity were reported after eating Soletellina dipos. The PSP in this area were found among purple clams, Xanthidae crabs and gastropods.

Thailand - In 1983, 62 cases of PSP were reported with one fatality in Thailand after the consumption of local mussels. A freshwater puffer poisoning case was reported in 1990.

Australia - There were reports of elevated levels of PSP toxins in the Port of Philip Bay and Western Port Bay in Victoria from 1987 to 1997.

New Zealand - In January 1983, there were more than 180 reports of respiratory irritation from air-borne toxins in sea spray. From January 1993 to July 1996, samples of shellfish taken around the coastal area was found to have toxic levels above regulatory limits.

CAUSATIVE AGENTS

Diatoms, Toxic algae, Dinoflagellates (Alexandrium catenella, A.tamarense, A. minutum, A. fraterculus, A.fundyense, A.cohorticula, Pyrodinium bahamense, Gymnodinium catenella) that produce a toxin called "Saxitoxin". This is a water soluble but heat stable substance unaffected by standard cooking or steaming.

Fig 12.19: Chemical structures of PSP toxins

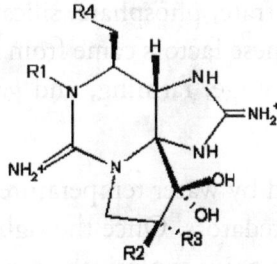

R_1	R_2	R_3	carbamate toxins	N-sulfo-carbamoyl toxins	decarbamoyl toxins	deoxy-decarbamoyl toxins
H	H	H	STX	GNTX5(B1)	dcSTX	doSTX
H	H	OSO_3^-	GNTX2	C1	dcGNTX2	doGNTX2
H	OSO_3^-	H	GNTX3	C2	dcGNTX3	doGNTX3
OH	H	H	neoSTX	GNTX6(B2)	dcneoSTX	doneoSTX
OH	H	OSO_3^-	GNTX1	C3	dcGNTX1	doGNTX1
OH	OSO_3^-	H	GNTX4	C4	dcGNTX4	

Source: Food and Agriculture Organization of the United Nations, 2004, Monse et al, 1988, Quilliam et all 2001, Paralytic Shellfish Poisoning Reproduced with permission Molecular Action of Saxitoxin

Saxitoxin blocks the Na+ (sodium) receptors on the external portion of the Na channels in the cell membrane of the nerves and muscles. As we know, the nerve impulse itself is the Na influx across the Na channels. Blocking the Na influx stops the flow of nerve impulse thereby paralyzing the muscles. What if it affects your respiratory muscles (diaphragm and intercostal muscles)? It may result in death. A Natural Reservoir consists of shellfish, mussels, clams, and other Marine Fish.

Transmission

These Dinoflagellates and other types of Phytoplankton are considered Microscopic Plants in watery environments both salty and fresh.

They depend on ocean currents for transport. The growth of the phytoplankton depends on the presence of carbon dioxide, sunlight, and nutrients such as nitrate, phosphate, silicate, calcium. and a low concentration of iron. These factors came from human inputs into the sea such as untreated sewage, farming, and gardening products like fertilizers.

Their growth are affected by water temperature, salinity, water depth, wind, and the type of predators. Once the right set of environmental factors are present, they grow explosively, a phenomenon known as "bloom" or "harmful algae bloom" (HAB). Rainfall, sunlight, and elevated temperatures increase the ocean current, and are favorable for the phytoplankton to spread out. These microscopic plants in enormous numbers produce a biotoxin, specifically Saxitoxin (which is one of the most toxic compounds known to man) which has deleterious effects on the marine environment. During the HAB, fish, and Shellfish consume the Phytoplankton without apparent harm. Any marine mammals, humans, fish, or birds that consume these shellfish and small fish acquire the toxin and may develop PSP.

Clinical Presentation

As early as five to 30 minutes after ingestion of contaminated shellfish, a person may develop a tingling sensation around the mouth that eventually spreads to the face and neck. Numbness may follow. As the toxin spreads from the mouth to the throat, the patient may develop dysphagia, sense of constriction on the throat, or worse, aphasia. Within three to 12 hours in severe cases, these symptoms lead to paresis and eventually paralysis. If it affects the respiratory muscles (diaphragm, intercostal muscles), the patient cannot breathe, leading to death. If the patient is fortunate enough to last for more than 12 hours without respiratory paralysis she starts to recover and improve. Usually, she becomes symptom free after a few days.

Aside from the paralysis, the patient may experience other symptoms, such as nausea and vomiting, headache, and dizziness. There is no loss of consciousness, however. The highest mortality rate was reported in Guatemala in 1987 with a 14% fatality rate. Children among the fatalities comprised 50%. This is due to the higher vulnerability of children to the toxin and the poor emergency access and inadequate medical services in the area. The overall global mortality rate is 8.5-9.5% (Meyer in 1953.Ayers and Cullum in1978).

Diagnosis

Presented with a patient with complaints of a tingling sensation on the mouth with numbness would not point to PSP as these symptoms are common in other forms of oral poisoning.

However, the history of intake of shellfish and the quickness of the appearance of symptoms right after intake will lead you to think of PSP. More so if these shellfish were taken from areas suspected with *Red Tide*.

Laboratory confirmation is the best basis of the diagnosis. The recommended diagnostic method is the *Mouse Bioassay*. However, this diagnostic test cannot differentiate the PSP toxins from other *toxins*. A "mouse unit" is the minimum amount of toxin that can cause the death of an 18-22 g of a white mouse in 15minutes. The lethal amount for humans is 5,000 - 20,000 mouse units (equivalent to 1-4 mg) depending upon the age and physical status of the patient.

Enzyme linked immunosorbent assay (ELISA) and Radioimmunoassay has been developed for saxitoxin and their results have good correlations with the result in mouse bioassays.

Treatment

In severe cases, just like with other poisonings, the first thing to do is to remove the obnoxious material in the gastrointestinal tract. This is done by gastric lavage that washes the stomach of the ingested toxin. Insert a gastric tube through the nose or the mouth up to the stomach and pour in a saline solution. Let it stay for few minutes and aspirate the fluid. This is done several times. This remedy must be given as early as possible as the time element is important in poison treatment. Knowing that ingested materials usually stay in the stomach for three to four hours, a patient with a history of intake of shellfish within four hours has a greater chance of recovery as the toxin is still in the stomach not in the whole body.

Other medicines that can be used are activated charcoal or diluted bicarbonate. These substances attract the toxin thereby preventing them from being absorbed in the stomach and intestine. Any respiratory difficulty must be treated with ventilator and blood gas levels that must be corrected. It has been reported that 75% of severely affected people die within 12 hours. Drugs used ordinarily for poisoning may have some beneficial effects like DL amphetamine (Benzedrine). Others like anticholinesterase agents and anti-curare agents are non-beneficial. There are some studies that show these drugs may be harmful to PSP. Metabolic acidosis must be corrected accordingly. Knowing that this poison will circulate in the body and eventually end up in the kidney and may cause kidney failure, hydration will dilute the toxin in the body and furthermore enhance the elimination of the toxin. Kidney function must be checked.

Areas affected by this Red Tide have traditionally developed some local treatment. In the Philippines, a concoction of coconut milk and sugar is given. Studies show that this combination has some detoxification property. Less severe cases without respiratory failure may manifest headache, nausea, vomiting, and/or dizziness. These are treated symptomatically, analgesic for headaches and intravenous fluid for vomiting.

Most important in the treatment of PSP is the elimination of human contact with shellfish. Any incident must be reported to the authorities who then should conduct laboratory testing on sample shellfish. If proven true, the local health units must initiate measures to contain the area and if necessary close the area for commercial fishing. Surveillance must follow. In the USA, eight hundred ug or more of PSP /kg by mouse assay is tantamount to closure of the area.

Prevention

Depuration is the process of removing the contaminated shellfish from the area and transfer them to water free of algae and allow them to self-depurate (purify). However, this process is very tedious and costly as some shellfish (*Crassostrea, Plactopectin, Spisula,* and others) require several months to detoxify. In the clam Saxidomus gigantea, the elimination of toxin may take more than a year. They studied that the rate of elimination of toxin is dependent on what part of the shellfish, the toxin is stored. Shellfish that stored the toxin in their gastrointestinal tract are easily detoxified while those shellfish which stored toxin in their tissue will take time to detoxify.

Detoxification by instantaneous electric shock accelerated the toxin excretion in scallops. Acidification of the water to detoxify butter clams was not successful. Chlorination of water resulted to in the alteration of the flavor of the shellfish, thereby the market value. Ozonation has been tried with some success for shellfish recently contaminated by toxin and it prevents further accumulation of the toxin. However, this method is not feasible in those areas where the shellfish has been contaminated for a longer period.

Cooking the shellfish reduces the toxicity but does not totally remove the toxin. This may be effective for those shellfish with low levels of toxicity. Boiling the oysters for ten minutes at 98 degree centigrade can reduce the toxicity to 68-81%, but not enough for extremely toxic shellfish. Pan frying is better than simple boiling, as the toxin stays in the soup which people can drink.

At present, large-scale detoxification is not commercially feasible. The fishing community depends more on extensive monitoring programs with immediate containment procedures in the event of incidence. There must be regular inspections of seawater bodies, especially during the bloom seasons. The presence of this harmful algae like the dinoflagellates is a warning sign to evaluate the level of toxin in the shellfish. In case of contamination, implement measures to prevent consumption of the shellfish. Any incidence should be reported to the proper authorities right away.

In Chile, since 1997, aside from constant monitoring, teaching strategies has had been applied, such as training workshops to fishers and extensive dissemination of information about harmful algae blooms to the community including the teachers and the students.

Worldwide Regulations (1)

European Union set the limit for PSP using mouse assay at 80 ug of STX/ 100 grams of the meat of shellfish.

Morocco, Africa -- at 80µg/ 100 grams of meat of shellfish

Canada -- 80µg/ 100 g of meat of mollusks

    -- 160µg/100 g of meat of soft-shell clams and mussels

    ---500 µg /100 g of meat of butter clams after removing the entire siphon

    -- 300 µg/ 100 g of meat of butter clams after removing the distal portion of siphon

USA --- 80 µg / 100 g of tissue of bivalves

Argentina --- 400 MU / 100 g of mollusks

    --160 µg/ g for snails. Argentina has a station in Mar del Plata that monitors mussel toxicity in each coastal province.

Brazil has started a monitoring initiative but does not have national monitoring program

Chile -- 80 µg / 100 g of meat of mollusks

Guatemala -- 80 µg/ 100 g of meat of mollusks

Mexico --- 30 µg/ 100 g as regulator limit

Panama --- 400 MU /g of bivalves

Uruguay ---400 MU / g of mollusks

Venezuela -- 80 µg/ 100 g of mollusks

China, Hongkong Special Administrative Region -- 400 MU / 100 g of shellfish

Japan -- 400 MU / 100g of bivalves

Malaysia -- In 1990, the Malaysian Department of Fisheries has started the shellfish toxicity monitoring program that has decreased the incidence of PSP. An additional monitoring facility was established in the east coast of Malaysian Peninsula after PSP cases were reported.

Philippines -- has set a lower limit of forty ug/100 g

Singapore   80 mg/ 100 g for bivalves

Republic of Korea   400 MU / 100 g for bivalves

Australia   80 mg / 100 g of shellfish meat

New Zealand   80 µg / 100 g of shellfish meat

DIARRHOEIC SHELLFISH POISONING (DSP)
(Diarrhoeic is the British Spelling; Diarrhetic is the American spelling)

Diarrhoeic shellfish poisoning (DSP) is a food borne disease like *Paralytic shellfish poisoning* (PSP) but it causes gastrointestinal symptoms, not neurologic symptoms. It is caused by eating contaminated bivalves such as mussels, scallops, oysters, or clams. The toxins in DSP accumulate in the fatty tissue of the bivalves. Patients develop gastrointestinal symptoms such as nausea, vomiting, abdominal pain, and diarrhea which start as early as 30 minutes to a few hours after ingestion and last for three days.

DSP was first reported in Netherlands in 1960. This was followed by outbreaks from Japan in the late 1970s. Similar incidents were reported in Europe, South America, and Far East. [33]

The organisms responsible for DSP are ***Dinophysis and Prorocentrum***.

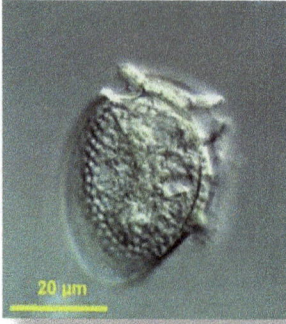

Fig 12-20 *Dinophysis acuminata*. Photo by Fjouenne[34]

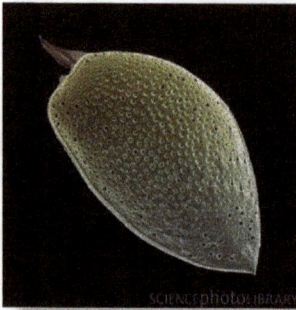

Fig 12-21 Prorocentrum sp. (Scanning electron micrograph)[35]

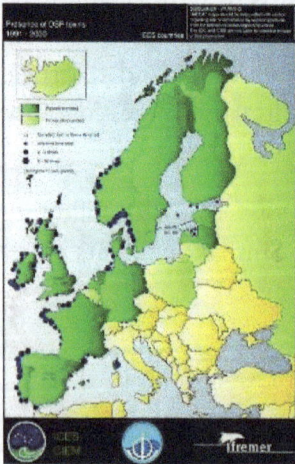

Fig 12-22

Occurrence of DSP toxins in coastal water of European ICES countries from 1991 to 2000 (represented by blue dots)

Source: Food and Agriculture Organization of the United Nations, 2004, Diarrhoeic Shellfish Poisoning, , [36]

Reproduced with permission

OUTBREAKS OF DSP

EUROPE

Belgium - In February 2002, 430 cases of DSP were reported in Antwerp. The blue mussels that they ate came from Denmark. The remaining imported mussels were not sold.

Croatia - In 1994, toxin analysis of mussels *(Mytilus galloprovincilis)* from Central Adriatic Sea yielded DSP toxins. In the summer of 1995, DSP toxins were also found in mussels in Kastela Bay.

Denmark - There were reports of DSP poisoning involving 415 persons in the North Danish coasts.

In 2001, these toxins were found in high concentrations in commercially fished mussels. In 2002, there were reports of people who became sick after eating Danish mussels.

France- From 1978 onwards, several areas in France (Normandy, South Brittany, West Brittany, Mediterranean coast) reported DSP poisoning. In 1984 and 1985, 10,000 cases and 2,000 cases, respectively, developed symptoms of DSP after eating mussels raised in France. In 2000, several areas on the Atlantic coast and Mediterranean coast were closed due to DSP toxins.

Germany - In 1987, DSP toxins were detected in mussels from the Wadden Sea. In 2000, two elderly women were reported to have DSP intoxication.

Ireland - The first DSP were recorded in 1980. An elevated level of DSP toxin was found in the Southwest coast of Roaring water, Dunmanus, Bantry, Kenmare and Dingle Bay in 1988.

Aggressive detection of DSP toxins resulted in the closure of thirty shellfish harvesting areas in August 2000.

Italy - On the northern and central Adriatic coast, there were reports of DSP from 1989 onwards. Monitoring of the Italian shellfish banks began in 1989. Due to cases of DSP, the Emilia Romagna region was closed from August to December 2000 and closure of the Veneto region from October to December 2000.

Netherlands- The first reported case of DSP was reported in 1960. The Warden Sea coasts have had several cases of DSP since 1961. In the summer of 2001, a bloom of D. acuminata occurred in the Dutch Wadden Sea contaminating the mussels.

Norway - There was an outbreak of DSP in the Oslo Fjord area in 1979. From October 1984 to October 1985, there were 300 to 400 cases of DSP in the southeast Norway.

Portugal – Since 1987, DPS toxins have been detected in Portugal but no human poisoning has been reported. In the summer of 2001, there was a DSP outbreak after eating razor clams harvested in Aveiro Lagoon. In 2002, due to algae bloom (***D.acuminata***), the entire northwest coast producing mussels and other bivalves were closed.

Spain - The first confirmed DSP was in 1978. From April to December 1995, coastal areas in Galicia were closed due to an elevated level of DSP and PSP toxins detected. In 2002, toxic events occurred in Galicia and Andalusia causing long closure periods and in Catalonia causing a short closure period.

Sweden - In 1971, mussel farming was started in Sweden. There were no toxic events but in 1983, when they observed DSP among the people who ate mussels. During the winter 1989 and 1990, harvesting mussels was stopped due to the high concentration of DSP toxin. It is common in Sweden to close the production areas of mussels during the period of September to March.

United Kingdom of Great Britain and Northern Ireland

First reported incidence of DSP in the United Kingdom was in 1997 when 49 patients developed symptoms of DSP after eating mussels in two restaurants in London. In 1992, DSP cases were reported in the south coasts. The causative organism was ***Dinophysis acuminata.***

NORTH AMERICA

Occurrence of DSP toxins in coastal waters of North American ICES countries from 1991 to 2000 (represented by blue dots)

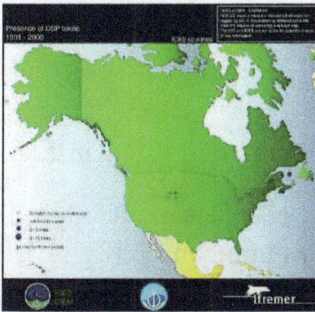

Fig. 12-23

Source: Food and Agriculture Organization of the United Nations, 2004, Diarrhoeic Shellfish Poisoning, Reproduced with Permission.

Canada - In 1989, DTX1 DSP toxin was isolated from mussels in the Prince Edward Island. In August 1990, 13 out of 17 persons in Eastern Nova Scotia developed gastroenteritis one to eight hours after eating boiled or steamed mussels. In October 1993, in Bonavista Bay, Newfoundland, several persons developed DSP after eating. mussels.

United States of America - Before 1980, there were sporadic cases of DSP in the New York and New Jersey areas. There were four reported cases of DSP like illnesses between 1983 and 1985 in Philadelphia and Long Island, New York after eating clams and mussels. In 1989, there was an elevated level of D. acuta that discolored the water in Long Island. No cases of DSP were reported.

In 2010, a pilot study was created by the Washington Department of Health and the FDA Gulf Coast Seafood Laboratory to monitor the presence of Dinophysis species and its associated toxins. They found out that the Dinophysis counts were above threshold, but the toxins (OA, DTX) were below the FDA guidance level.[37]

The US Food and Drug Administration set the limit for PSP toxins at 80 micrograms per one hundred grams of shellfish tissue.

The World Health Organization (WHO) says that 100,000 cells/mL is a moderate human health risk, but there are currently no standards for cell or toxin concentrations in the United States.

Central and South America

Argentina - In 1999, 40 cases of DSP were reported in Patagonia.

Brazil - In the region of Florianopolis, several persons developed gastrointestinal distress and diarrhea after consuming mussels.

Chile - In 1970 and1971, there were sporadic blooms of *Dinophysis* causing gastrointestinal disorders. One hundred twenty people became ill after consuming mussels in January 1991. DSP in Chile caused severe public health and economic impacts until 2001, when fishing beds were closed from 44o SL southwards and nationwide monitoring was established.

Mexico -In 1993 and 1994, there were increased densities of *dinoflagellate D. caudata* in the Gulf of California (*Punta arena, Playa Escondida, Amorales,* and *San Ignacio*). Shellfish extracts were positive for DSP in samples from Bahia Conception in the Gulf of California. No cases of DSP were reported though.

Uruguay - In 1990, several persons developed diarrhea and gastrointestinal distress after consuming mussels. In January 1992, DSP was detected in shellfish along the coast of Uruguay. In the area of La Paloma, a partial ban was implemented on shellfish harvesting.

ASIA

China – DSP caused by diverse types of shellfish were widely distributed along the coast of China. In 1996 and 1997, elevated levels of DSP toxins were found in twenty-six out of eighty-nine samples. The highest level was found in ***Perna viridis*** from the region of Shenzhen (5). No human poisoning was reported though.

Japan - DSP was first documented in Japan in June 1976 and 1977 when 164 persons suffered severe diarrhea and vomiting. Between 1986 and 1982, about three hundred cases of DSP were reported.

India - There was a two-year study in India that showed DSP toxins were present in several shellfish samples (5).

Philippines - While most of the shellfish poisoning in the Philippines is of the Paralytic type, there were some reports of DSP. In 1995, five species of Dinophysis had been detected to have DSP toxins but no human cases of poisoning happened.

Oceana

Australia and New Zealand

A pipi ***(Donax deltoides)*** shellfish poisoning with fifty-six hospitalized cases was reported in New South Wales, Australia in December 1997. Another reported poisoning due to pipi shellfish was reported when an elderly woman became ill after consuming pipis from a local beach.

Between the period of September 1994 and July 1996, there were outbreaks of human DSP involving thirteen cases in the coastline of New Zealand (5). Samples collected on a weekly basis showed toxic levels above the regulatory limits.

Africa South Africa - During autumn in 1991, DSP cases were reported in the west coast of South Africa and in the autumn of the following year, DSP poisoning were reported in the south coast. The causative organism was ***Dinophysis acuminata.***

Causative Organism

The causative organisms are marine dinoflagellates that belong to the genera Dinophysis and Prorocentrum. There are seven Dinophysis species which are: D.fortii ((in Japan), D.acuminata (in Europe), D. acuta, D. norvegica (in Scandinavia), D. mitra, D. rotunda and D. tripos. The Prorocentrum species are the following: Prorocentrum lima, P. concavum, and P. redfieldii. These algae, under favorable conditions, can grow in large numbers causing an algae bloom. These dinoflagellates produce the DPS toxins. The toxins can be divided into three groups depending on the chemical structures. The first group are the acid toxins. This group includes Okadaic acid (OA) and its derivatives named dinophysistoxin (DTXs). The second group are the neutral toxins which are polyether-lactones of the pectenotoxin group (PTXs). The third group is the sulphated polyether and its derivatives yessotoxins (YTXs)

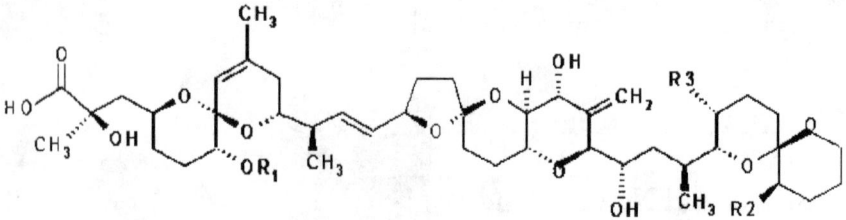

Fig 12-24

R1	R2 H	R3 H	CH$_3$
okadaic acid (OA)			
dinophysistoxin-1 (DTX1)	H	CH$_3$	CH$_3$
dinophysistoxin-2 (DTX2)	H	CH$_3$	H
dinophysistoxin-3 (DTX3)	acyl	CH$_3$	CH$_3$

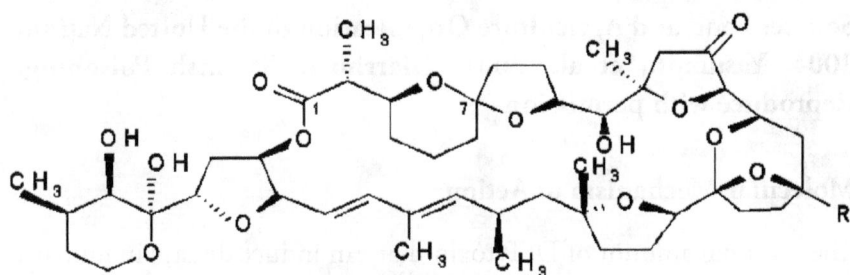

Fig 12-25

R		C-7
pectenotoxin-1 (PTX1)	CH 2OH	R
pectenotoxin-2 (PTX2)	CH 3	R
pectenotoxin-3 (PTX3)	CHO	R
pectenotoxin-4 (PTX4)	CH 2OH	S
pectenotoxin-6 (PTX6)	COOH	R
pectenotoxin-7 (PTX7)	COOH	S

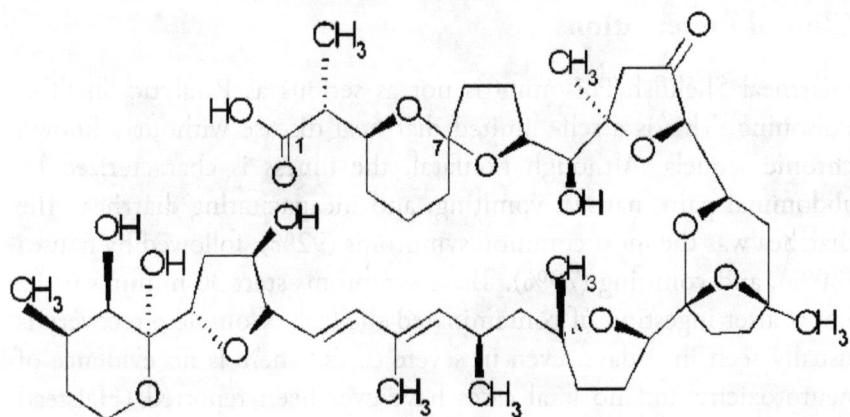

Fig 12-26

C-7

pectenotoxin-2 seco acid (PTX2SA) R

7-epi-PTX2SA S

Source: Food and Agriculture Organization of the United Nation, 2004, Yasumoto et al., 2001, Diarrhoeic Shellfish Poisoning, Reproduce with permission

Molecular Mechanism of Action

The minimal amount of DSP toxin that can induce disease in humans is 12 MU. One mouse unit (MU) is the amount of toxin that can kill one mouse with a body weight of approximately twenty grams in 15 minutes.

One example of the DSP toxin is Okadaic Acid (OA). OA being an acid toxin is lipophilic. It promotes phosphorylation that controls the sodium secretions by the intestinal cells. The presence of OA in the intestinal lumen promotes more sodium secretions of the intestinal mucosa. Since sodium is an osmotic electrolyte, it tags with it water. Therefore, sodium and water are copiously secreted by the intestinal mucosa in the presence of OA resulting in copious diarrhea. The increased amount of fluid in the intestinal lumen will cause an increase in peristalsis causing the abdominal pain and nausea and vomiting.

Clinical Presentations

Diarrheal Shellfish Poisoning is not as serious as Paralytic Shellfish Poisoning. This is a self- limited diarrheal disease without a known chronic sequela. Although nonfatal, the illness is characterized by abdominal pain, nausea, vomiting, and incapacitating diarrhea. The diarrhea was the most common symptoms (92%), followed by nausea (80%), and vomiting (79%). These symptoms start 30 minutes to 12 hours after ingestion of contaminated shellfish. Complete recovery is usually seen in 3 days, even in severe cases. There is no evidence of neurotoxicity and no fatal cases have ever been reported (Halstead 1988, Viviani 1992).

Diagnosis

Medical history is particularly important when considering DSP as the diagnosis. Since patients manifest the usual gastrointestinal symptoms of abdominal pain, nausea, vomiting, and copious diarrhea, a physician must consider many gastrointestinal infectious conditions. The only clue to DSP is a history of ingesting shellfish. More so if the shellfish came from areas known to have outbreaks of DSP. Remember, there are no neurologic manifestations. These manifestations will differentiate this type of shellfish poisoning from the Paralytic type (PSP). Confirmation of the diagnosis might be difficult. One way of confirming the diagnosis of DSP is the mouse bioassay (commonly used in Japan). This is done by injecting the toxic extracts into the abdominal cavity of the mouse and observe the mouse for 15 minutes. A DPS toxin level more than 50 MU are considered toxic to human consumption. Other countries consider a 15 MU level toxic to humans.

Management and Treatment

Treatment is symptomatic and supportive with regards to diarrhea and accompanying fluid and electrolyte losses. In general, hospitalization is not necessary; fluid and electrolytes can usually be replaced orally. Patient must be on a "nothing by mouth" except for the fluid and electrolyte replacements for several hours until their diarrhea subsides. Abdominal pain, nausea, and vomiting are usually not treated unless they are severe enough to cause other symptoms. Other diarrhetic illnesses associated with shellfish consumption, such as bacterial or viral contamination, should be ruled out (Aune & Yndstad 1993).Just like any other shellfish poisoning, the presence of one case must not be considered lightly. This might be an initial case of an impending outbreak. Any suspected case of DSP must be reported right away to the appropriate public health authorities. These cases must be monitored properly. Necessary precautions must be implemented to prevent the spread.

Prevention

Depuration is a purification process wherein harvested shellfish are placed in a land-based plants containing clean estuarine water to purge or expel the gastrointestinal content enhancing the separation of the contaminant from the shellfish. This is an effective way of removing the toxin from the shellfish but not effective in removing other contaminants like norovirus and hepatitis A. In Japan, depuration decreased the DSP toxin from 4.4 to 2.5 MU in one week and then to 0.5 MU by the next week.

The main strategy to prevent DSP is effective monitoring of mussels with respect to the DSP toxin so that the contaminated shellfish will not be available for consumption. Frequent monitoring of the seawater around aquaculture facilities or shellfish farms for the presence of phytoplankton strains known to produce toxins. This must be done routinely regardless of an outbreak or not. Then data on the occurrences, types, and concentration of algae species must be collected. These will serve as the basis for determining what type of toxin expected for a particular period. Some countries monitor only the two most common algae species, while other countries have an extensive list of species to monitor. Once the algae exceed certain concentrations, the harvesting area is closed. In Italy, closures are done if the toxin is detected only in shellfish.

Regulations and Monitoring

Europe- In 1996, EU-National Conference Laboratories Meeting on Analytic Method and Toxicity Criteria agreed that mouse assay (MU) is the preferred method for the detection of DSP toxins. Tolerable levels are 80-160 mg of OA per kg of shellfish meat. This is equivalent to 20-40 Mu/kg of whole shellfish meat.

In March 2002, the European Commission laid out the following rules:

1. Maximum levels of OA, DTXs and PTXs together in edible tissue of mollusks, echinoderms, tunicates, and marine gastropods shall be 60 mg equivalent/ kg meat.

2. The maximum level of YTXs in edible tissue of mollusks, echinoderms, tunicates, and marine gastropods shall be 1 mg YTX equivalent/kg.

3. The mouse and rat bioassay are the preferred methods of analysis for the toxin. In the event of discrepancies between these two methods, the mouse bioassay must prevail.

Ireland - In 1984, the Biotoxin Monitoring Programme was initiated. This was based on screening samples for the presence of DSP toxins. Recently, they included other toxins for monitoring like Azaspiracids and the monitoring is done now on the weekly basis. A website information system is being developed to have easy access to this information.

Turkey - Regulation and detection is based on mouse assay.

North America

Canada - In 1995, Hallegreaff et. al reported that in Canada, monitoring for Dinophysis and Prorocentrum spp. is done and when the DSP toxin exceeds the tolerable level, closure of harvesting areas is implemented (5).

United States - There is no monitoring of DSP in the United States simply because there is no confirmed DSP. The FDA (Food and Drug Administration) is the primary agency on seafood safety and marine biotoxin. The National Marine Fishery Service of the National Oceanic and Atmospheric Administration has several programs geared on fishing and wildlife.

Internationally, it is the FDA that sets up the memoranda of understanding with other countries to regulate imported seafood products.

Central and South America

Argentina - Argentina has a national monitoring program for mussel toxicity and its fixed station is in Mar del Plata.

Brazil - Brazil had one pilot monitoring initiative for one year but does not have a national monitoring program.

Chile - The Chile government has two monitoring agencies. The National Health Service is responsible for monitoring toxicity using bioassay at 40 stations on a monthly basis. and the Fisheries Research Institute which is university based. The Chile Minister of Health through its regional Health Services is responsible for assessment and the closure of harvesting areas. The National Fish Service (NFS) is the agency for seafood export. It has a memorandum of agreement with the US and Europe to permit shellfish export. Before 2001, PSP and DSP severely affected the public health and economy of Chile.

Uruguay - Uruguay has national monitoring programme for mussels and toxic phytoplankton. Their analysis is based on mouse assay.

Venezuela - Regulation is based on the mouse assay.

Asia

China - China has no regulatory program for DSP toxin and no regulatory program for algae biotoxins. One major program on red tide was funded that includes regular monitoring in two areas. This monitoring is biweekly for shellfish and plankton.

Japan - Japan has the Prefectural Fisheries Experimental Station in major shellfish areas that periodically collect plankton samples and conducts cell count of Dinophysis species. Shellfish are also collected and assayed. Maximum level of DSP toxin for human consumption is set at 5 MU/100 gram of whole meat detected by mouse assay. Japan has a well-defined network for easy dissemination of information connecting government agencies, fisheries cooperatives, fishers, mass medias and the public.

The Republic of Korea- The Republic of Korea has the National Fisheries Research and Development Institute (NFRDI) that assesses plankton samples in key areas biweekly. It has over two hundred stations. They assess these samples for PSP, DSP, and ASP. They set up a tolerance limit for DSP at 5 MU / 100 grams.

Thailand - Regulation is based on mouse assay.

Oceania:

Australia - Their recommended limit for DSP is 16-20 µg OA (Okadaic acid) eq/100 grams of shellfish meat.

New Zealand - The New Zealand government has the Biotoxin Monitoring Programme that assesses phytoplankton and performs regular shellfish testing.

NEUROLOGIC SHELLFISH POISONING (NSP)

Neurotoxin shellfish poisoning are caused by ingesting shellfish contaminated with *Brevetoxin* which are produced by unarmored dinoflagellates. Cases of NSP are usually associated with shellfish harvested during red tide blooms. This type of poisoning may not be as serious as the PSP or DSP, but some cases required hospitalization. There are no reported fatalities so far for NSP. Patients suffering from NSP develop gastrointestinal, respiratory, dermatological, and neurological symptoms.

This poisoning is caused by Polyether Brevetoxin produced by unarmored dinoflagellates. This is tasteless, odorless, heat and acid stabile, lipid-soluble, cyclic polyether neurotoxin. This poison is toxic to humans, birds, marine mammals, and fish, but not to shellfish (oysters, clams, mussels).

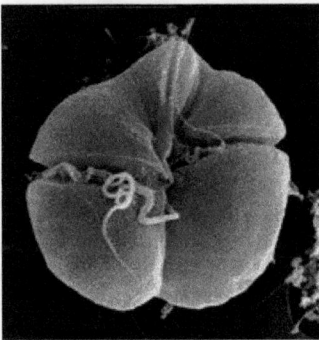

Fig. 12-27 *Karenia brevis (Formerly called Gymnodinium brevis)*[38]

CASES and OUTBREAKS

Figure 12-28. Occurrence of NSP toxins in coastal waters of North American ICES countries from 1991 to 200 (represented by blue dots)

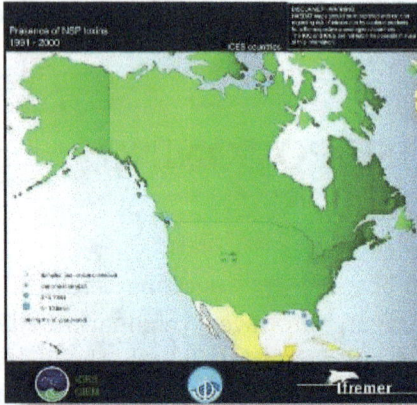

Source: Food and Agriculture Organization of the United Nation, 2004, Neurologic Shellfish Poisoning http://www.fao.org/3/y5486e/y5486e0o.htm#bm24 Reproduced with permission

Europe

France - In October 1991, *F. japanica* was reported in the Channels coast of Normandy. [39]

Germany - *H. Akishino* was detected in the German Wadden Sea beginning in the summer of July1995. F. japonica was observed near Sylt in summer of 1997. In March to May of 1998 2000, and 2001, an extensive algae bloom of *Chattonella* occurred killing fish.

Netherland - *F. japonica, Chattonella antiqua* and *Chattonella marina* were reported in 1991 to 1993 in the Wadden Sea, North Sea and/or Delta area south of Rhine. In summer of 1997, *F. Japonica* was found along the Dutch coast from Noordwijk to Borkum.

The Russian Federation - There were reports of fish deaths in the Amurskii Bay caused by *Chattonella*.

The United Kingdom of Great Britain and Northern Ireland - There were reports of red tide of *H. akashimo* from England and Bermuda that caused fish deaths.

South Africa - In the coastal resort of Hermanus in Walker Bay, episodic respiratory problems were seen on residents in the summer of 1995 and 1996. This was due to the bloom of the toxic dinoflagellate species, *Gymnodinium.*

Canada - Report of red tide due to *H. akashiwo* has caused mortality of cultured fish in Canada (Van Apeldoorn, et al., 2001).

United States of America:

East Coast - From 1987 to 1988, there were reports of bottlenose dolphin deaths (*T. truncatus)* attributed to Brevetoxin along the mid-Atlantic coast. In November 1987 till early 1988, there was an algae bloom due to *G. breve* in North Carolina coast resulting in forty-eight cases of NSP in humans. Red tide due to *H. akashimo* have caused mortality on cultured fish.

Florida and Gulf of Mexico -

As early as 1946, there were already reports of bottlenose dolphin mortalities in Southwest Florida due to an etiologic agent which later was identified as Brevetoxin. In 1987 and 1988, there were reports of bottlenose dolphin deaths associated with Brevetoxin.

On June 16, 1996, three patients developed symptoms of NSP according to the Sarasota County Health Department. They had eaten clams (*Chione cancellate*) and whelks (unknown species) harvested from areas closed for harvesting clams. The clams were cooked until they opened. From March to April 1996, there was a significant red tide dinoflagellate bloom in the South coast of Florida resulting in the deaths of at least 149 manatees.

West coast - In California, brevotoxin was identified as the cause of summer mortality in common murres. There were reports of mortality of cultured fish by *H. akashimo* (https://www.fao.org/3/y5486e/y5486e0o.htm)

Mexico - In the Gulf of Mexico, along the coasts of Veracruz and Tamaulipas, *G. breve* developed huge blooms every year in the autumn killing fish. Beginning in 1994, this algae bloom increased in permanence, resulting to in fish kills. Residents were affected by exposure to sea sprays or immersion in seawater.

Hongkong - In March 1987, Hong Kong Special Administrative Region reported the first harmful bloom of raphidophytes in Yim Tin Tsai. In 1991, a bloom of *Chattonella marina* occurred, killing fish, and posing a serious threat to finfish cultures.

Japan - In 1972, a red tide of *F. japonica* was reported from coastal areas of Ehime Prefecture causing heavy mortalities of caged young yellowtail. This red tide was later reported in Atsumi Bay, the Seto Inland Sea and Harima Nada.

Republic of Korea - Red tide due to *H.akashimo* was reported in embayments, killing cultured fish.

Australia - In Boston Bay, Southern Australia, an algae bloom of *Chattonella marina* occurred, affecting the tuna industry. Elevated levels of a breve-like toxin were found the liver of these farmed bluefin tuna. Another finding is the epithelial swelling of the tuna gills with copious mucus. In January 1994, mussels in the Gippsland coast of Victoria were reported to have prominent level of NSP toxin.

New Zealand - There were 186 cases of NSP recorded (Van Apeldoorn et al, 2001). From September 1994 to July 1996, about 0.2% of weekly samples of shellfish along the coastline of New Zealand showed NSP toxin beyond regulator limits.

From mid-February to April 1998, there was a severe toxic outbreak in Wellington Harbor that decimated all marine life (including seaweeds). This outbreak killed eels, flounders, pelagic fish, and other marine invertebrates. Eighty-seven people developed a respiratory illness ranging from dry cough, severe sore throat, runny nose, and skin irritation. The people affected were beach goers, swimmers, and windsurfers. The algae bloom was dominated by *Gymnodinium sp.*

CHEMICAL STRUCTURE OF BREVETOXIN:

Brevetoxin is a heat and acid stable, lipid soluble substance produced by marine dinoflagellates. There are four analogues of Brevetoxin isolated from contaminated shellfish: BTX- B1 analyzed from cockles and the BTX- B2, BTX-B3, BTX - B4 analyzed from mussels. (https://www.fao.org/3/y5486e/y5486e0o.htm) -

(Moroshi,A et al.1995. Brevetoxin B3, a new brevetoxin analog isolated from the greenshell mussel *Perna canaliculus* involved in neurotoxic shellfish poisoning in New Zealand. Tetrahedron Letters. Volume 36. Issue 49. 4 Dec.1995. 8995-8998.)

Fig 12-29

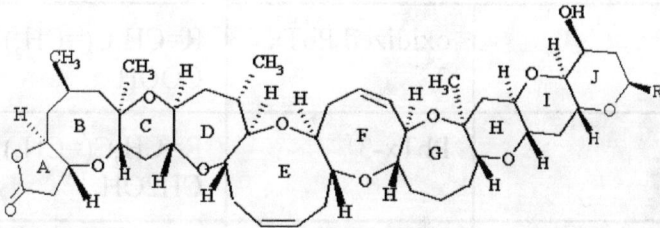

TABLE 12-2-R GROUPS IN BREVETOXINS

Type 1 (A) Brevetoxins:	PbTx-1,	R = CH$_2$C(=CH$_2$)CHO
	PbTx-7,	R=CH$_2$C(=CH$_2$)CH$_2$OH
	PbTx-10,	R=CH$_2$CH(CH$_3$)CH$_2$OH

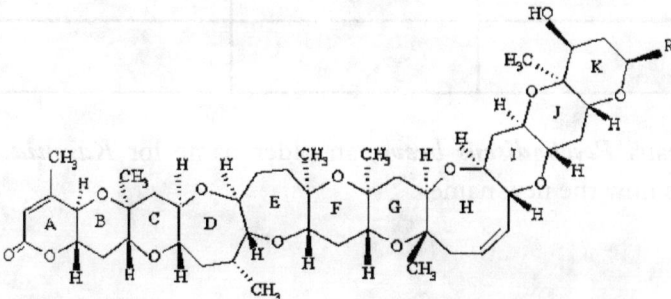

TABLE 12- 3

Chemical structures of type A and B Brevetoxins

(Hua,Y, et al 1996.On line liquid chromatography-electrospray ionization mass spectrometry for determination of the brevetoxin of the brevetoxin profile in natural "red tide" algal blooms. J of Chromatography Volume 750, Issues 1-2 25 Oct 1996 pp.115-125)

Type 2 (B) Brevetoxins:	PbTx-2	$R = CH_2C(=CH_2)$ CHO
	oxidized PbTx-2	$R=CH_2C(=CH_2)$ COOH
	PbTx-3	$R=CH_2C(=CH_2)$ CH2OH
	PbTx-8	$R=CH_2COCH_2Cl$
	PbTx-9	$R=CH_2CH(CH_3)$ CH2OH
	PbTx-5	the K-ring acetate of PbTx-2
	PbTx-6	the H-ring epoxide of PbTx-2

Pb -Means *Ptychodiscus breve*, an older name for *Karenia brevis* which is now the new name.

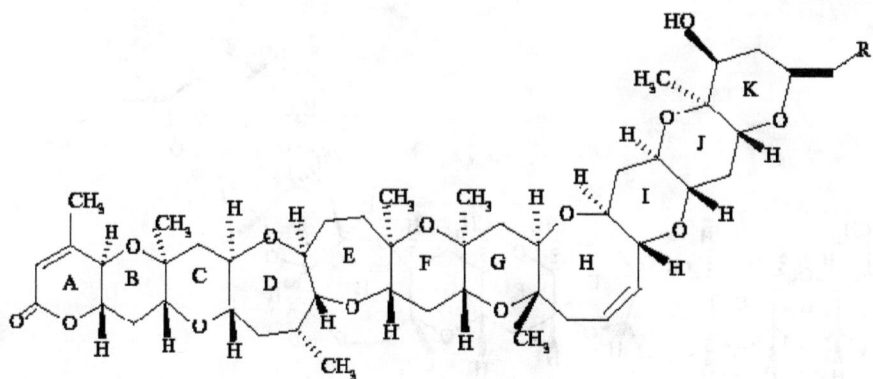

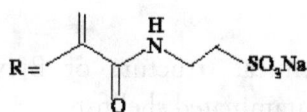

BTX-B1   R =

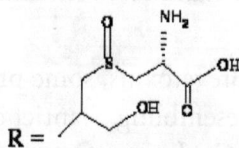

BTX-B2   R =

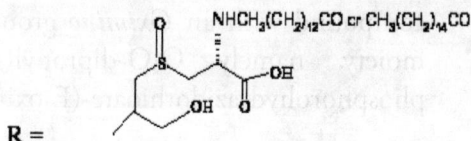

BTX-B4   R =

Source: (https://www.fao.org/3/y5486e/y5486e0o.htm)

Figure Fig 12-30: Chemical structures of Brevetoxin analogues BTX-B1, -B2 and -B4 isolated from contaminated shellfish

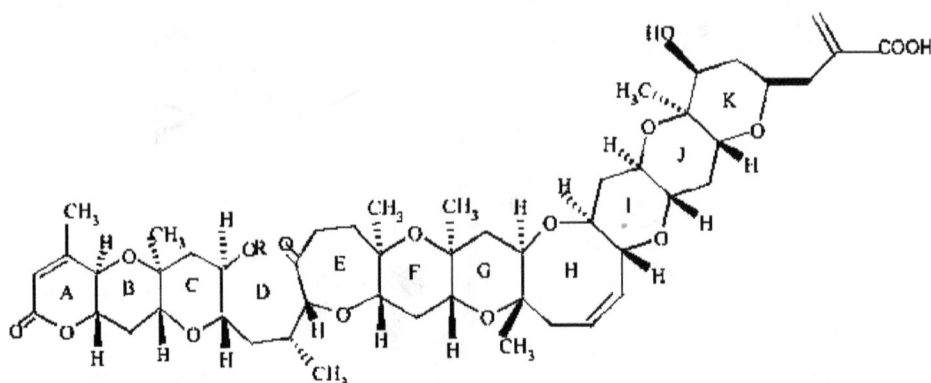

Fig.12-31`: Chemical structure of Brevetoxin analogue BTX-B3 isolated from contaminated shellfish

In addition to Brevetoxins, some phosphorus containing ichthyotoxic compounds resembling anticholinesterases, have also been isolated from **G. breve**. One example is an acyclic phosphorus compound with an *Oximino* group in addition to a thiophosphate moiety, namely O,O-dipropyl(E)-2-(1-methyl-2-oxopropylidene) phosphorohydrazidothioate-(E)oxime (Van Apeldoorn et al., 2001).

Fig. 12-32: Phosphorus containing ichthyotoxic toxin isolated from G. breve.

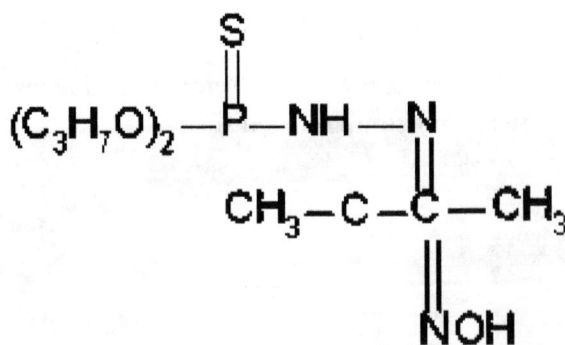

Phosphorus containing ichthyotoxic toxin isolated from G. breve

Source: Food and Agriculture Organization of the United Nation, 2004, Neurologic Shellfish Poisoning, http://www.fao.org/3/y5486e/y5486e0o.htm#bm24

SOURCE ORGANISMS:

Dinoflagellates (**Gymnodinium breve** aka syn. **Ptychodiscus breve**)

Algae species belonging to class Raphidophyceae like *Chattonella antiqua*

Algae species **Fibrocapsa japonica**

Algae species **Heterosigma akashiwo**

Shellfish Containing NSP Toxins

Most common shellfish that contain Brevetoxin are the oysters, clams, mussels, and welks. These shellfish are not susceptible to the toxin, but fish, birds and other mammals are.

Some copepods (**Temora turbinata, Labidocera aestiva, Acartia tonsa**) have been traced to have Brevetoxin; however, in experiments when these copepods were combined with juvenile fish, the fish were not killed.

Brevetoxin were also found in some tuna in Australia, in menhaden and mullet from the coast of Florida, and Muir birds from the coast of California.

Predisposing Condition for Growth

G. breve blooms in the west coast of Florida from summer to winter, more frequently in autumn.

They initially appear in summer when the winds are weakest and continue into the autumn when the winds are strong. With these intense winds, these algae are transported to the shore. The growth of these algae are further enhanced from additional human made sources.

In the great flood of Mississippi in 1993, massive amounts of agricultural nutrients from Midwest farms were poured into the Gulf of Mexico, creating a so called "dead zone" of water with low oxygen leading to the growth of this Dinoflagellates. It was documented in 1987 and 1988 that wind coming from Gulf of Mexico carried **G. breve** to the east coast of Florida and north to North Carolina.

C. marina belonging to raphidophytes produces Brevetoxin in coastal areas rich in organic materials.

H. akashiwo blooms, which are found in the Pacific and Atlantic coasts, requires metals such as iron and manganese, and nitrogen, phosphorous and vitamin B12 for growth. These are supplied by water with low oxygen content plus the wind induced turbulence of water sediments.

The growth of *Chattonella antiqua* are supported by nitrate, ammonium and, to a lesser extent, urea. Iron and vitamin B12 promote growth. Maximum/terminal growth is 25 degree centigrade, at salinities between 25 and 41%, and under light intensities above 0.04 ly/ min.

Pathogenesis

Brevetoxin opens voltage gated sodium ion channels in the cell membrane. This will enhance the influx of sodium ions altering the cell membrane potential, making it less negative. This will result in slowing of flow of impulse or action potential. Moreover, because these sodium channels are opened, there are persistent activations and repetitive firings.

Neuro-excitation from continuous nerve membrane depolarization leads to spontaneous firing. In most cases, aside from depolarizing the nerves, the muscles are also depolarized. (9) Since the toxin is aerosolized, once inhaled it can cause bronchospasms leading to different respiratory symptoms. Not only that, but Brevetoxin can cause the release of the neurotransmitter ACH from the autonomic nerve endings. This will cause contractions of smooth muscles in organs, specifically the lungs, stomach, and the uterus. In the lungs in particular, contractions of the smooth muscles of the trachea and bronchi cause further bronchial constriction and bronchial spasms, resulting in breathing difficulty.

Also, there is an increased histamine release which causes the capillaries to dilate and increases permeability. This will allow fluid in the plasma and cells to squeeze between the endothelium of the capillary walls to migrate to interstitial space. This leads to the symptoms of an allergic reaction that an exposed person experiences, such as conjunctival irritation, rhinorrhea, and a non-productive cough.

Since histamine causes vascular dilation and an increase in permeability, it will decrease the intravascular volume, leading to lower fluid volume and low blood pressure. Histamine is one of the substances that can stimulate hydrochloric acid secretions by the gastric mucosa.

Clinical Symptoms/Treatment

Oral exposure: Eating raw or cooked contaminated shellfish may cause a toxic syndrome like PSP and ciguatera poisoning, but to a lesser degree. The latency period from the intake of shellfish to the appearance of symptoms is usually 30 minutes to three hours and these symptoms last for three to four days The gastrointestinal symptoms are diarrhea, abdominal pain, nausea, and vomiting. In severe cases, the patient may develop hypotension due to increase vascular permeability, arrhythmias, sweating, chills, numbness, and paresthesia of the lips, face, and extremities. Respiratory symptoms such as breathing difficulty follows. If no immediate treatment is initiated, this may lead to paralysis, seizures, or coma. So far, no mortality has been reported from NSP .

Treatment of NSP is more symptomatic and supportive as the patients recover spontaneously. An antispasmodic is given for abdominal pain. If vomiting is severe, an injectable anti-emetic drug can be administered.

Particular attention must be given to diarrhea. If the patient shows signs and symptoms of dehydration, fluid administration is necessary and electrolyte imbalance, if any, must be corrected. The neurologic manifestations, such as *Paresthesias*, numbness, or weakness are usually self-limiting. Respiratory symptoms from oral exposure are milder compared to those from aerosol exposure.

Aerosol exposure - Inhalation of aerosolized surf or its red tide causes respiratory distress and eye and nasal allergy symptoms.

Respiratory symptoms include rhinorrhea, non-productive cough, and difficulty breathing due to *Bronchoconstriction*. Allergic reactions including conjunctival irritation, copious catarrhal exudates, mucus secretions, and skin rashes are present. Asthmatics and the elderly with chronic pulmonary disease whose respiratory status are already compromised, are particularly susceptible. Brevetoxin - 3 has been implicated as the primary toxin responsible for respiratory symptoms in humans.

Treatments for respiratory symptoms are usually supportive. In normal individuals, respiratory symptoms are usually temporary. The individual should leave the beach and enter an air- conditioned room, and the symptoms will subside.

Dermal exposure - Since the **G. breve** organism has no outer shell covering like other dinoflagellates, it is easily broken open in the surf releasing the toxin. Direct contact with swimmers causes eye irritation, mucous membrane irritations, and skin irritations. These are usually temporary and will subside after leaving the beach area and entering an air-conditioned room.

Regulations and Monitoring in Some Countries

Europe

Denmark - Monitoring program for selected species exists. For **Gymnodinium** species anything above 5×10^5 cells per liter causes the fish harvesting areas to close.

Italy - Harvesting areas for fishery products are closed once there is both the presence of algae in water and toxin in mussels.

United States of America - A level of 80 mg of PbT-2 in 100 grams of shellfish tissue analyzed by mouse bioassay in shellfish would necessitate regulatory action by FDA.

Florida and the Gulf of Mexico - Since mid-1970, the Florida Department of Natural Resources has enacted a control program. In 1984, the *G. breve* concentration exceeded 5,000 cells/liter leading to the closure of shellfish beds. The closure would last from few weeks to six months.

This measure above prevents NSP from eating shellfish, but not respiratory irritation associated with exposure to aerosolized red tide toxins. Unlike Texas, Florida does not close beaches for recreational and occupational activities, even during an active near shore algae bloom.

Central and South America

Argentina - This country has a national monitoring program for mussel toxicity and a fixed monitoring station is in Mar de Plata.

Brazil – Has a pilot monitoring initiative but does not have a national monitoring program.

Uruguay - Has national monitoring program on mussel toxicity and toxic phytoplankton.

Oceana

New Zealand - With the advent of the NSP in 1993, New Zealand created a management strategy of weekly sampling of commercial and non-commercial shellfish harvesting areas, which monitors for phytoplankton.

AZASPIRACID SHELLFISH POISONING (AZA)

Azaspiracid was identified in 1995, when eight people in the Netherlands became sick after consuming mussels (*Mytilus edulis*) that originated from Killary, Ireland. The symptoms were like diarrheic shellfish poisoning but investigators found that the outbreak was not caused by DSP toxins (dinophysistoxin and okadaic acid). Furthermore, samples collected were different from the known samples that cause DSP. The causative toxin was later identified as Azaspiracid. Since then, it has been isolated in shellfish from countries all around the world.

Cases/Outbreaks

Europe

Ireland - The first recorded case of Azaspiracid Shellfish Poisoning (AZP) that led to the identification of Azaspiracid as the causative toxin occurred in Ireland in 1995. Eight people became ill after consumption of mussels (Mytilus edulis) from Killary Harbor, Ireland. They presented with gastrointestinal symptoms similar to DSP, but it was noted that their okadaic acid and dinophysistoxin levels were low enough to not cause symptoms. They later found out that the toxin was Azaspiracid. In the following years, several ASP incidents were reported in the Arranmore Island region of Donegal, Northwest Ireland, among other countries. They also noted that it was not only mussels that carries the toxin but other bivalves like oysters.

In 1999, they assessed 1,800 samples and 5% assessed positive for Azaspiracid.

Norway - Azaspiracid has been identified in mussels.

Portugal - There were some reports of toxicity like AZP from cockles (*Cerastoderma edule*).

United Kingdom - Azaspiracid has been found in mussels.

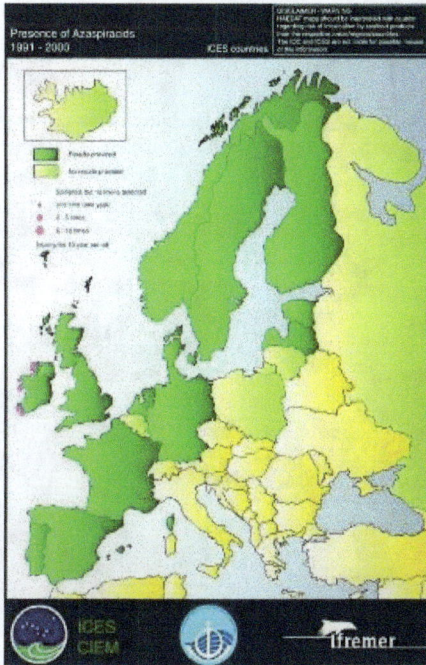

Fig. 12-33: Occurrence of AZP toxins in coastal waters of European ICES countries from 1991 to 2000 (represented by purple dots)

Source: Food and Agriculture Organization of the United Nation, 2004, Marine Biotoxin, Azaspiracid Paralytic Shellfish Poisoning, Reproduced with permission

Causative Organism

Azaspiracids are produced by the small dinoflagellates *Azadinium spinosum* and *Protoperidinium crassipes*. These dinoflagellates are eaten by bivalves such as oysters, mussels, and scallops, also by marine crabs.

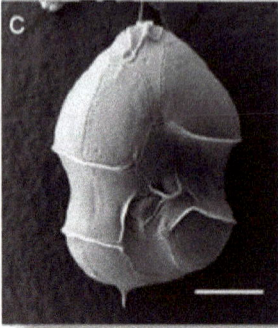

Fig 12-34- Azadinium (SEM) [41]

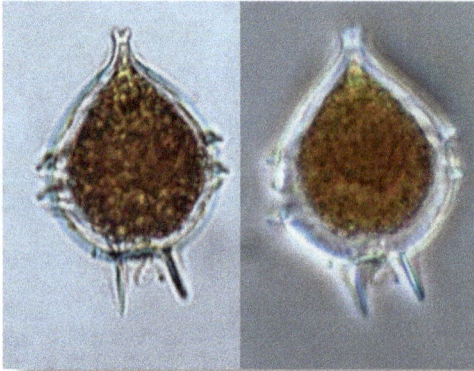

Fig 12-35 Protoperidinium stein. Photo by Rajashree Gouda[42]

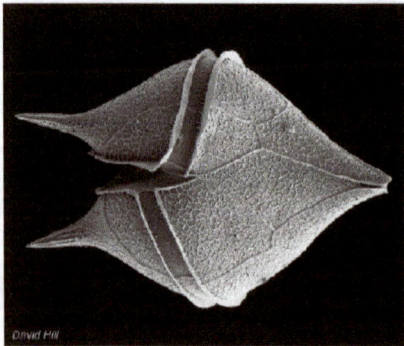

Fig 12-36- Protoperidinium (SEM) Photo by David Hill[43]

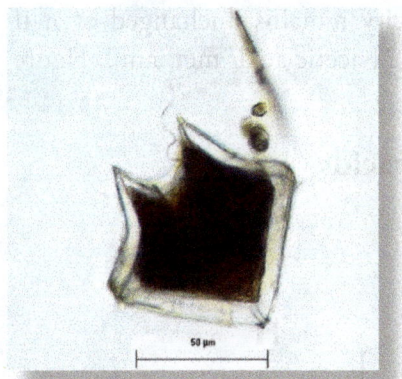

Fig 12-37- Protoperidinium [44](*LM light micrograph*)

Mechanism of Action: There is no data for mechanism of action for Azaspiracid; however, research has determined that the lethal dose of AZA is 150 ug/kg. Injection of this lethal dose in mice causes the swelling of the liver due to fatty accumulation and vacuole formation in the hepatocytes. The stomach becomes swollen leading to erosion and bleeding. It causes pyknosis in the pancreas leading to fragmentation of the nucleus and cell apoptosis. Dead lymphocyte debris were present in the thymus and spleen. No histological changes were reported in the heart, lung, and kidney. In another study, they injected acetone extract from contaminated mussels to the mice intraperitoneally causing "neurotoxin-like" symptoms such as respiratory difficulties, sluggishness, spasm, progressive paralysis, and death within 20-90 days. These pathologic changes are considered different from those changes induced by toxins of other shellfish poisoning.

Chemical Structure

Azaspiracid is a colorless, amorphous substance with no UV absorption maxima above 210 nm. Azaspiracid (AZA) has four analogues, AZA 2 to AZA 5. AZA 4 and AZA 5 which are oxidized metabolites of AZA 3. So, AZA, AZA 2, and AZA are the genuine substances that cause Azaspiracid Shellfish Poisoning. This compound is considered stable because its toxicity remains unchanged even if heated at 5°C for 150 minutes in 1.0 N acetic acid/ methanol. Figure 2 is the chemical structure of AZA.

Chemical structures of Azaspiracids

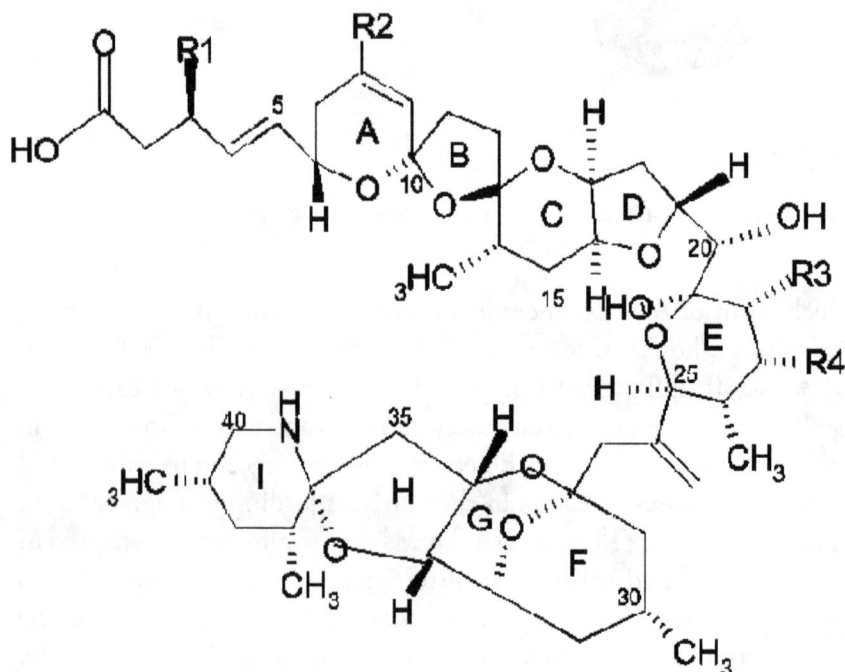

Fig 12-38

Source: Food and Agriculture Organization of the United Nation, 2004, Marine Biotoxin, Azaspiracid Paralytic Shellfish Poisoning, **http://www.fao.org/docrep/007/y5486e/y5486e0p.htm#bm25**

Reproduced with permission

Clinical Manifestation

Symptoms of AZP are like those from diarrheic shellfish poisoning such as nausea, vomiting, stomach cramps, and diarrhea. Investigators initially thought that these were DSP cases, but the toxins responsible for DSP were exceptionally low, so it was thought that these cases were of a different entity. These symptoms usually appear within hours of ingestion and persisted for two to three days, after which there was a full recovery.

As to the long-term effects, studies showed AZA can cause damage to epithelial cells of the intestinal tract explaining the diarrhetic symptoms in humans. Because of these findings, it is possible that AZA may lead to chronic gastrointestinal conditions such as Crohn's disease, ulcerative colitis, or even cancer of the gastrointestinal tract, but it remains to be established.

For AZP management, for persons suffering from generalized manifestations, the treatment is symptomatic such as antipyretic for fever and antispasmodic for abdominal pain. Since patients may develop diarrhea and vomiting, make sure that their electrolytes are balanced by giving intravenous fluid if necessary. No mortality has been reported to date.

Regulation and Prevention

In 2001, the EU Commission for Consumer Health and Protection sponsored a workshop and established a regulatory limit of 160 ug AZA/ kg whole shellfish flesh. Since this regulatory limit was established, there were only five reported AZP events between 1995 and 2000, a period of over seven years. This proved that a regulatory mechanism is especially important in prevention of diseases.[45]

Depuration has been suggested as another way to prevent AZP. This is a procedure by which marine or freshwater animals are placed into a clear water environment for a specified period to purge them of biological contaminants and impurities. This preventive method was mentioned as a way of freeing shellfish from the toxin AZA. Some researchers opined that this might be a slow process considering that the toxin is in the digestive tract and other tissues of mussel.

PALYTOXIN POISONING

Palytoxin is a very potent and dangerous marine toxin that can be acquired by inhalation. ingestion, and skin exposure. This toxin has been sourced from some marine corals in the aquarium at home. This toxin can be found also in certain marine bacteria and some fish and seafoods in tropical countries. Here are some reported palytoxin poisoning.

A case of a man together with his wife and daughter went to the hospital due to shortness of breath, and a worsening nonproductive cough. These were followed by fever, chills, and myalgia. These symptoms started after the man cleaned exotic corals in the aquarium. His wife was in the basement while his daughter was upstairs. All three persons in the family had the same symptoms with degrees of severity commensurate to the distance from the aquarium. The cause of their symptoms were later sourced from soft marine corals belonging to the Zoantharian (zoanthids) genera known as *Palythoa*.[46]

In May 2017 in Adelaide, Australia, a family member cleaned the saltwater aquarium inside the house by removing and scrubbing unwanted growth in the rocks and corals. Unknowingly, the aerosolized palytoxin was released. All the family members in the house developed severe shortness of breath ending up in the hospital.[47] Between 2000 and 2004, The US National Poison Data System reported 171 + phone calls from people who have inhaled palytoxin.

Eighty percent of these cases were all from the residences leading them to surmise that the cause was from the home aquarium. In August 2019, a family of five from Shropshire, England was poisoned and hospitalized after cleaning a tropical aquarium. In 2018 and 2019, two poisonings involving the handling of corals were reported in Quebec. In 2017, seven family members and their dog were treated for breathing difficulties after their saltwater aquarium was cleaned.

Reports of the marine bacteria, fish, and seafood as the source of the poisoning were reported. One was in 2005, when 200 beach going tourists developed respiratory symptoms ending up in the hospital instead of enjoying themselves on the Italian beach. Ten Percent of the patients required intensive care.[48] In the Philippines, one person died after eating a crab species *Demania reynaudii* containing palytoxin.[49]

Causative Organisms

Palytoxin is produced by species of **Palythoa** and **Zoanthus** soft coral. These are the corals used in the house aquarium. People underestimate the health risk of these corals because there are only few reported cases.

Palytoxin can be found also in species of the phytoplankton genus *Ostreopsis,* a dinoflagellate and other bacteria. Palytoxin can be found in sponges, mussels, starfish, cnidaria and fish and crabs by the process called *Biomagnification.* The thing is that this toxin does not kill the fish, crustaceans, worms, and other marine inhabitants. Humans acquired the poisoning by ingesting these marine inhabitants. It is noteworthy to know that palytoxin destroy the stony hard corals but not the soft corals so it is not advised to put them together in the aquarium.

Toxicology

Palytoxin is a huge molecule weighing around 2,680 Daltons and has a backbone of 115 Carbon Atoms long. A 1989 C&EN story considered palytoxin the Mount Everest of chemistry for the monumental effort required to understand and synthesize the complex molecule. Palytoxin is not heat labile so boiling water cannot remove the toxin.

Here is the Chemical Structure of Palytoxin.[50]

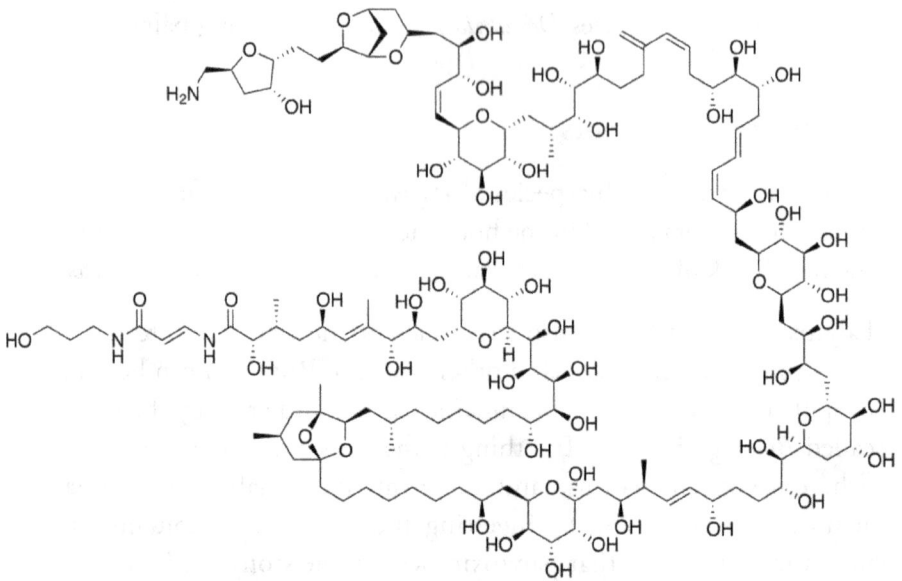

Fig 12-39

Mechanism of Action

Palytoxin binds with the Na+/K+-ATPase, a trans membrane protein enzyme found in cell membranes of every vertebral cell. This enzyme is necessary to start Na+/K+ pump. This pump is particularly important in maintaining homeostasis of the cells. It pumps Na+ out of the cell and pumps K+ ion into the cell to maintain the concentration gradient between the intracellular fluid and extracellular fluid. This is necessary for viability of all cells. Normally, there are more Na+ outside the cell and there are more K+ inside the cell. When a stimulation happens, these Na+ channels open and there is sudden rush of Na+ into the cells. To counteract this event, the K+ rush out of the cell to go to the extracellular fluid. Eventually this Na+ inside the cell (as a result of rushing in) must be brought back outside the cell and the K+ outside the cell (because of rushing out) must go back into the cell. This is accomplished by Na+/K+ pumps. Disruption of the pump would cause free diffusion of Na+ and K+ in and out of the cell eventually leading to loss of the normal electrolyte gradients between the inside and outside of the cell. In RBCs, this will result in hemolysis of blood cells while in the heart and muscle cells, it causes continuous contraction of the cells.[51]

Palytoxin can also affect the eyes. It causes corneal edema, corneal ulcer leading to blurring of vision. The posterior surface of the cornea of the eye is lined by simple squamous epithelium with Na-K Pump keeping sodium and water out of the cornea to make the cornea clear. Disruption of this cellular pump causes more water to remain in the cornea because it cannot be pump out. This hydrated cornea becomes edematous and later develop bulla and ulcers leading to blurring of vision

Clinical Manifestations:

Palytoxin is acquired by ingestion, inhalation, skin exposure (skin wounds) and ocular exposure. Among these routes, the skin and respiratory routes pose grave problems for human health. The thing is that the palytoxin does not kill these marine inhabitants but is harmful to humans. Upon ingestion, the toxin accumulates in the bloodstream and spreads to the different organs of the body causing the systemic symptoms. These symptoms appear suddenly within minutes to hours after exposure.

The respiratory symptoms are cough, sore throat, runny nose shortness of breath with fever and in severe cases, accumulation of fluid in the lungs (hydrothorax) that may result to respiratory failure. The two hundred patient-tourists (see introduction above) in Italy presented with fever, cough, sore throat, and dyspnea. Ten percent of these patients were brought to intensive care units.

In another reported case in Virginia, a patient developed respiratory symptoms after cleaning a zoanthid coral with boiling water. Physical examination revealed a febrile, tachypneic and tachycardic patient.

Those patients who acquired the palytoxin from ocular exposure, develop conjunctivitis, photophobia, blurring of vision, and worse corneal ulceration. Those who acquired the toxin from skin exposure develop rashes, itchiness, numbness, and Dermatitis.

Those patients who acquired palytoxin by ingestion, developed abdominal cramps. nausea, vomiting, diarrhea. There might be some oral symptoms such as distortion of taste called dysgeusia, or paresthesia of the perioral area. Once the toxin enters the bloodstream, patients may develop systemic symptoms affecting the nervous system such as dizziness, speech disturbance, tremors, and numbness of the extremities. The toxin can cause damage to the cardiac muscles manifested as irregular, slow or fast heart rate, low or high blood pressure.

Severe cases can develop multi organ failure such as renal failure, heart failure and respiratory failure.

Diagnosis

Diagnosing Palytoxin Poisoning is based on clinical presentations and history of exposure. There is no confirmatory test nor a laboratory marker for this type of poisoning. Other types of poisoning may present similar manifestations making it hard to diagnose this condition. Presented with this typical symptoms of poisoning, the Palytoxin poisoning is considered more of a differential diagnosis than a primary diagnosis. The only item that will pinpoint to Palytoxin poisoning is the exposure to aquariums. Public and even medical professionals have little awareness even though it is a serious condition.

A medical professional can request a chest x-ray to check for lung conditions, especially Pulmonary edema and Congestion, and an EKG to check the status of the heart. Laboratory examinations like CBC is important because cases of Palytoxin poisoning usually cause *Leukocytosis* (increased numbers of WBC). Arterial blood gas determination (ABG) can be requested to check for hypoxemia. For severe cases, the major organs of the body must be checked like requesting liver function tests to check the liver status, kidney function and tests to check for the renal status.

Treatment

No definite treatment nor antidote for Palytoxin poisoning is available. The treatment is supportive and symptomatic. Giving antipyretics for fever, anti-tussive medications for cough, bronchodilators for dyspnea, antidiarrheal and anti-emetic for gastrointestinal symptoms and anti-arrhythmia for heart involvement. Oxygen inhalation for hypoxemia and IVF administration for hypotension and electrolyte imbalance.

Serious cases however must be hospitalized. There was one case in Danville, PA who was hospitalized due to fever and dyspnea. His condition deteriorated and he was transferred to the ICU. In this case, the blood oxygen level spiraled down and he was connected to mechanical ventilators. In serious gastrointestinal symptoms with diarrhea and vomiting, electrolyte imbalances are the usual complications. The electrolyte level in the blood must be monitored and corrected right away for any imbalances. If the palytoxin was taken within three to four hours, gastric lavage is done to bring out the toxin from the stomach. Those patients with cardiac involvement develop arrhythmias. Such patients must be sent to the ICU and hooked up to a cardiac monitor. Treating the arrhythmias is necessary to prevent heart failure.

Those patients with ocular symptoms are usually given artificial tears and corticosteroids. In *Dermal Exposure*, an antihistamine is given. Topical corticosteroids can be given, too. There is no antidote for Palytoxin poisoning currently. Palytoxin however, being a very potent vasoconstrictor can be treated with vasodilators as an antidote. Examples of these *Vasodilators* are *Papaverine* and *Isosorbide dinitrate*. These drugs must be injected immediately upon exposure.

Preventions

There are no official guidelines from the CDC for Palytoxin Poisoning as quantitative data are still inadequate and further research is still to be done. It is worthwhile though, to give some advice to people overseeing or maintaining aquariums. These corals indeed add color and beauty to your aquarium, but they also bring naturally occurring toxins.

When cleaning an aquarium, one must wear long gloves, eye goggles, and face mask to prevent exposure to aerosolized palytoxin. Better soak the rocks and corals with ten percent bleach solution for thirty minutes before cleaning. This will neutralize the remaining palytoxin. If possible clean the rocks and corals outdoors to disperse the aerosolized palytoxin. If this is not possible, you can clean them indoors with open windows or run an exhaust fan. Do not use pressurized stream water in cleaning because this will aerosolize the palytoxin from the rocks and corals.

While rare, eating contaminated fish and crustaceans can be fatal. This is most likely the reason why there are no official CDC guidelines. There are no regulations and toxicity determination on Palytoxin in shellfish. There is no way for us consumers to know if the fish or crabs we buy from the market have Palytoxin. The fact is that the toxin is thermostable and cannot be removed by boiling and they remain stable in aqueous solution for a longer period. The good thing is that its toxicity is lost in acidic or alkaline solution. Having said that, it will be better to mix the fish, crabs, and shellfish in an acidic solution like vinegar before cooking.

MICROCYSTIN POISONING (MCYST)

Introduction

When water bodies are stagnant, warm, rich in phosphorus and nitrogen from sources such as agricultural land and runoff from sewage and septic tanks, the surface water appears in different colors raging from red to brown. They are also called P*ond Scum*. These colors are due to the rapid multiplication of algae called *Algae Blooms*. These algae may appear in surface waters or at the bottom of the water. They emit a foul odor described as rotten plant odor. A particular alga that commonly causes this bloom is the Blue Green Algae called *Microcystis*, a *Cyanobacterium*. They are called as such because they share common characteristics with bacteria. This algae bloom can occur anytime of the year but more commonly during late summer or early fall. These blue green algae produce several toxins called cyanotoxins and the most common among them is *Microcystin*, which will be discussed in this section.

If you happen to wade on this stagnant pond with blue green algae bloom, you may develop skin rashes or respiratory symptoms. Worse if you happened to drink water contaminated with microcystin, you may develop renal or liver damage.

Microcystin producing algae blooms are a worldwide problem especially in Australia, Brazil, China, South Africa, United States and European countries. The Hartbeespoort Dam in South Africa is believed to be the most contaminated site in the world with cyanobacteria[52].

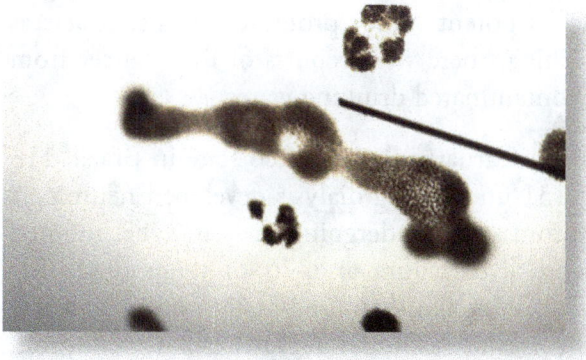

Fig 12-40 *Microcystis aeruginosa* Photo by David Arieti

Toxicology

Microcystin is a toxin produced by the freshwater cyanobacteria of Microcystis species but is also produced by other species which include . **Anabaena, Hapalosiphon, Nostoc, Planktothrix, and Phormidium.** This toxin is lethal and is one of most studied toxins by scientists, biologists, and ecologists.

Microcystin is stable and resists *Hydrolysis, oxidation*, and other common chemical reactions. Ordinary cooking will not eliminate the toxin. It breaks down at extreme pH. There are some bacteria that produce proteases (enzymes that breakdown proteins) and these enzymes break down microcystin. These types of bacteria however are seldom found in the water bodies.

Microcystin LR has several non-proteinogenic amino acids which are covalently bonded and inhibit protein phosphatase PP1 and PP2A. We know that these phosphatase enzymes are necessary for cellular phosphorylation. This biological process is an important part in mitosis, signal transduction, cell growth, protein synthesis, activation, or deactivation of some cellular enzymes. Failure of this biological process can lead to abnormal proliferation, differentiation of cells, cancer, or apoptosis (cell death). Although microcystin causes gastrointestinal symptoms, neurologic symptoms, the main effect of the toxin is in liver cells. The liver cells enlarge, bleeds and become congested, eventually leading to liver cell necrosis. It is interesting to note that microcystin is a potent tumor promoter in rats. It acts as a tumor initiator. In China, there were reports of liver cancer from ingesting microcystin contaminated drinking water.

In February 1996, in Caruaru, Pernambuco state in Brazil, 116 (89%) patients out of 131 undergoing dialysis developed neurologic and hap ototoxic symptoms after undergoing dialysis. One hundred patients developed acute liver failure and 76 of them died. The microcystin originated from a lake they used as the source of water due to severe drought. This lake was found out later to have blue green algae bloom.

Chemical structure of Microcystin LR[54]

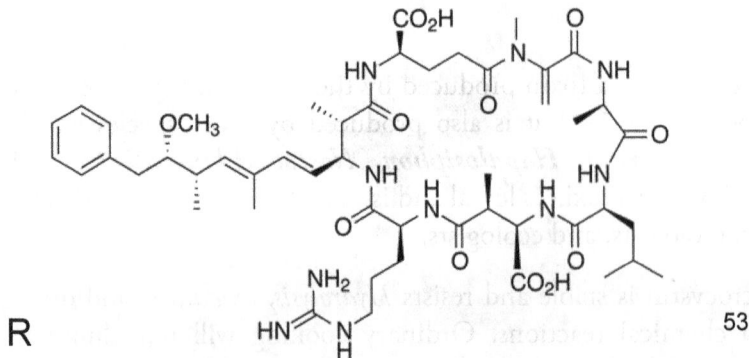

R

53

Fig 12-41

Symptoms:

The microcystin can be acquired through direct skin contact as you wade intentionally or accidentally in a pond with blue-green algae. It can be acquired by inhalation of airborne droplets containing the toxin in boating or waterskiing and acquired accidentally by swallowing water from this pond. Animals that are fond of wading or swimming in the pond are vulnerable because they usually lick their body to dry off after leaving the water.[55] Symptoms of Microcystin poisoning may appear several hours to days after exposure but the usual incubation period is within the first week after exposure.

Those individuals who had skin contact with the toxin, present erythema on the skin, with blisters, hives, rashes, irritations and other allergic-like manifestations. They are more frequently located on the lips and under the swimsuits.

Those individuals who acquired the toxin by inhalation would present respiratory symptoms, runny nose and eye, dyspnea, sore throat, cough, other asthma-like symptoms, and allergic-like symptoms.

Those individuals who accidentally drink water with the toxin would develop nausea, vomiting, diarrhea, fever, headache. Ingestion of greater amounts of toxin can cause liver damage called acute hepatic necrosis which will eventually lead to hepatic failure which is usually manifested as pain in the right upper quadrant, jaundice, elevated liver enzymes levels and neurologic symptoms such as disorientation.

Wildlife, livestock and even pets are more vulnerable than humans as they swim. wade and drink water in the ponds and lakes. Symptoms of microcystin poisoning in the animals include vomiting, anorexia, diarrhea, excessive salivation, and difficulty in breathing. In severe cases, they may develop seizures or death.

Diagnosis

Presented with sudden onset of gastrointestinal symptoms such as nausea, vomiting and diarrhea with neurologic symptoms of weakness, headache, and disorientation followed by symptoms of acute liver failure, one must consider Microcystin poisoning. This is further supported if the patient has a history of exposure to stagnant water bodies or has accidentally drunk water from those water bodies.

Water samples can be assessed by microscopy for the presence of Blue Green Algae. It can be analyzed for the presence of Microcystin by the ELIZA test or Liquid Chromatography- Triple Mass Spectrometry. These examinations are offered in California Animal Health and Food Safety Laboratories in Davis, California.

(http://cahfs.ucdavis.edu) and Auburn University CyanoPros, (http://www.cyanopros.com). [56]

If there are elevations of these liver enzymes (alkaline phosphatase, ALT, AST) in blood then it will further boost your diagnosis of microcystin poisoning. Other tests to determine Microcystin are Kidney Function Tests (Blood Urea Nitrogen Level, Creatinine Level), Serum Electrolytes and Chest X-rays, if Respiratory Symptoms are more pronounced.

Treatment

There is not an antidote for Microcystin poisoning. The treatment is symptomatic and supportive. For Gastrointestinal symptoms, the patient must be put on NPO (Nothing Per Orem –nothing by mouth) and if necessary, give the patient fluid and electrolytes intravenously. If the intake is within three to four hours, gastric lavage is of utmost importance to remove the toxin from the stomach.

Those patients with more severe poisoning develop liver damage manifested as jaundice, fever, confusion, or disorientation. They are monitored in the hospital. Treatment is geared to prevent hepatic encephalopathy, which is hepatic failure. The damaged liver cannot breakdown the toxin in the body which eventually accumulates to affect the central nervous system causing confusion and disorientation. In worse case scenarios there may be a coma.

The goal of treatment is to eliminate the toxins from the body. They are first given lactulose. It function as a laxative to draw the toxin to the large intestine and defecated out. They also soften the stools. Antibiotics such as neomycin is given to eliminate bacteria that create the toxin from the digested food thereby decreasing the toxin in the body. Low protein diet is recommended to decrease the production of ammonia.

For those patients with a history of inhaled aerosolized toxin, movement of the patient to non- contaminated areas is recommended. Bronchodilators may be given for patients with difficulty breathing.

If you encounter water contaminated with microcystin, remove your clothing, jewelry and other accessories and wash with fresh water and soap for 10-15 minutes as soon as possible. Antihistamines or steroid cream may be applied for skin rashes, irritation, and itchiness.

For eye exposure, wash your eyes with normal saline for 15 minutes. Remove contact lenses, if symptoms persists after washing, consult an ophthalmologist.

Prevention

For the public especially children and pregnant women, avoid wading, playing, or swimming in the lake, pond and other water bodies that appear with distinct colors because they are likely to have algae blooms. Refrain from drinking or swallowing any recreational water especially those coming from lakes or streams.

For livestock and other animals, they should be prevented from staying close to water bodies with algae blooms. When they get out of water after wading or swimming, they lick their body to dry off, thereby acquiring the toxin. If there is no other water area for drinking, these animals should drink on the shore of the lake only. Placing barriers such as logs, floating plastic to keep the algae bloom from reaching the shore were proven to be ineffective.

From the government side, several water treatment techniques have been used. Among them, are chlorination and ozonation. Chlorine and ozone are oxidants that kill cyanobacterium cells. Some environmentalists are skeptical of these techniques because as the cells are killed by the chemicals, they release the mycotoxin in the water. Unless another process is added to remove these toxins, these techniques are not recommended.

Another chemical compound they used is copper sulfate ($CuSO_4$) or any copper-based compound. It prevents the formation of cyanobacteria algae blooms. The formula of the mixture of copper and water is based on gallons of water or acres of water bodies. This technique is different from ozonation or chlorination because copper sulfate prevents the formation of cyanobacteria.

Of particular attention is addition of chemicals that bind and remove Phosphorous from the Eutrophic Water. This type of water is rich in chemicals which serve as nutrients for the algae. These waters are usually run off from chemical factories, agricultural land using pesticides and fertilizers and from commercial establishments. Phosphorous is one of the nutrients for algae formation. Among the chemicals that bind with phosphorous are aluminum sulfate, ferric chloride, and some particles in clay. Another chemical that binds and removes phosphorous from water bodies is Lanthanum. This is a naturally occurring earth element that is commercially used to increase the water quality in ponds, lakes ,and reservoirs. This compound has been patented in USA since 2010.[57]

PALM ISLAND MYSTERY DISEASE (CYN)

In November 1979, an outbreak of a "hepatitis-like illness" (associated with dehydration and bloody diarrhea) was reported in Palm Island, northern Queensland, Australia. These involved 148 people (10 adults and 138 children) of Aboriginal and Torres Strait Islander descent. Investigators found that all the patients drank water from the same source, the Solomon Dam. Residents who had other sources of water were not affected. A few days before the incident, the

Solomon Dam was treated with copper sulfate to control the algae bloom. Copper sulfate at 1ppm (1 ug/ ml) breaks the cyanobacteria and releases its toxic components into the water.

Investigators received samples of water from the dam, cultured the organism, and the results were administered to mice. The mice later developed tissue injury on their gastrointestinal organs, kidneys, and liver. It was determined that the causative algae was *Cylindrospermopsis*. They later identified the cyanobacteria as *Cylindrospermopsis raciborskii*. Eventually, the toxin produced by this cyanobacterium was found out to be Cylindrospermopsin (CYN).

Fig 12-42 Palm Island from Wallaby Point, Queensland , Australia[58]

The affected people manifested anorexia, malaise, headache, and vomiting with initial constipation followed by bloody diarrhea at varying degrees of dehydration. Laboratory results showed elevated liver enzymes indicating liver damage. Most of these patients needed intravenous fluid while those with a severe condition developed hypovolemic/acidotic shock. The stools and food were examined, eliminating common infectious organisms and toxins as probable causes of the outbreak. All patients eventually recovered after treatment. There was no mortality.

Causative organism: *Cylindrospermopsin raciborskii* which produces a toxin called *Cylindrospermopsin*.

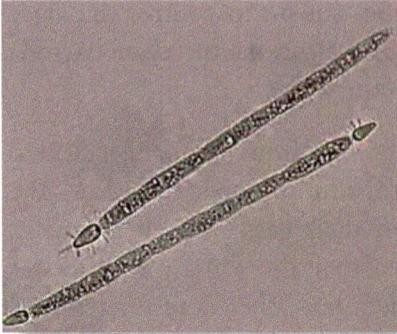

Fig 12- 43 Cylindrospermopsis sp.[59]

Cylindrospermopsis raciborskii can form toxic cyanobacterial blooms

Figure 12:44 Here is the chemical structure of Cylindrospermopsin (CYN)[60]

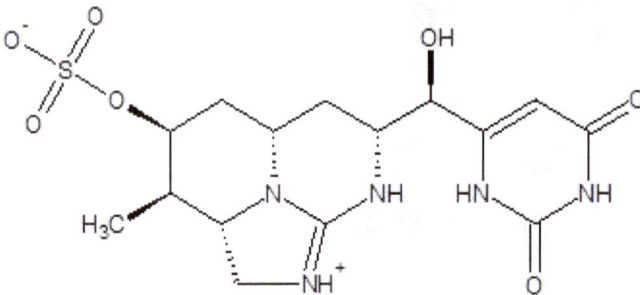

Source: https://natoxaq.ku.dk/toxin-of-the-week/
cylindrospermopsins/

Cylindrospermopsin is a potent inhibitor of protein synthesis produced by cyanobacteria, most commonly by *C. rackborskii*. This toxin can cause cell death with the liver as the main target organ. It has an uracil moiety attached to guanidino moiety suggesting that it may have a carcinogenic effect. CYN is not easily degraded in water because of its high-water solubility and stability to a wide range of heat, light, and pH.

Clinical Manifestations

The mode of transmission for most of the documented cases of Palm Island Mystery Disease is drinking water. The main manifestations of the patient involved is in the gastrointestinal tract. These symptoms are abdominal pain, nausea and vomiting, bloody diarrhea, and headache. Another mode of transmission is by direct contact. The patient may develop skin or eye irritation. If inhaled, it may cause a sore throat, dry cough or worse, atypical pneumonia.

Treatment

As used in any other type of poisoning, if the case is of mild to moderate severity, the management is symptomatic or supportive. For gastrointestinal symptoms, fluids and electrolytes balance is very particularly important. If needed, the patient should have intravenous therapy.

For severe cases, aside from the supportive management, the patient must be monitored for the possibility of liver toxicity since the main target of the toxin is the liver. Have a liver enzyme determination especially for a patient with persistent symptoms. Some patients may develop atypical pneumonia and/or other respiratory symptoms. The respiratory functions must be well monitored in these cases.

PFIESTERIOSIS

Pfiesteriosis is the general name for any or all symptoms a human may experience after exposure to *Pfiesteria piscicida* (Pp). *Pfiesteria* poisoning is a general name for any or all manifestations of fish infected with *Pfiesteria piscicida*.

In 1988, researchers at the College of Veterinary Medicine at North Carolina State University were startled to find a considerable number of dead fish in their aquaria. It was later discovered that the appearance of dead fish occurred simultaneously with an increased number of a specific microalga. The microalga was later identified as a freshwater dinoflagellate named *Pfiesteria piscicida* in honor of the late Lois Pfiester, who had done extensive research on this organism.

This organism is found in the Albermarle - Pamlico Estuarine System which is the second largest estuary in the United States. Sometime in summer of 1997, this toxic dinoflagellate that causes illness in fish was the hottest topic in local media in the area. They coined this microalga as *Cell from Hell* or *Fish Killer*. At present the common habitat of the *Pp* are the areas from the Gulf of Mexico to the Atlantic estuarine waters, including Florida, North Carolina, Maryland, and Delaware. There were some reports of the presence of *Pp* in the Mediterranean Sea.[61]

Fig 12-45 https://commons.wikimedia.org/wiki/File:PamlicoSound-EO. JPG viewed 31 Aug 2023

The location of the Albemarle-Pamlico Sound Estuary System relative to the State of North Carolina and the southeastern United States

Causative Organism.

The causative organisms are dinoflagellate species of genus *Pfiesteria (Pp)*. Although many dinoflagellates act like Pfiesteria, there are only two species that have toxin producing capabilities. These are *Pfiesteria piscicida* and *Pfiesteria shumwayae*. The latter was named after renowned scientist Sandra Shumway, who also did extensive research on these toxic dinoflagellates.

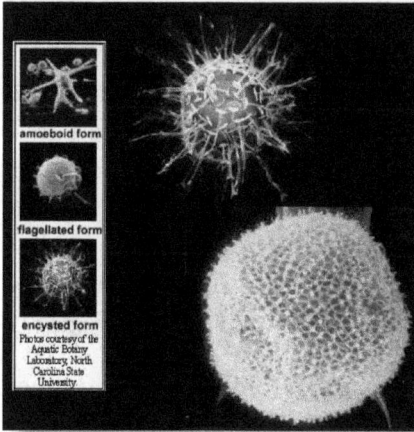

Figure 12-46. Scanning electron micrograph showing a toxic zoospore.

Pfiesteria [62]

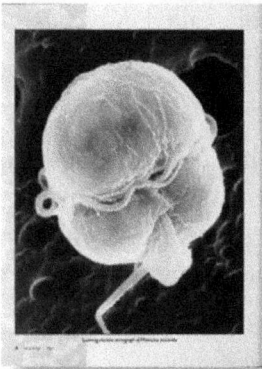

Fig 12-47 Pfiesteria piscicida. Photo by US Sea Grant College program[63]

Pfiesteria piscicida has been found to be active at a temperature range of 9-33°C although most of the growth is at 18°C. It is also active in brackish water of estuaries with a salinity of 2-20, parts per thousand (ppt) (note: fresh water is less than one and marine water is 35ppt on the salinity scale).

Pfiesteria piscicida has a remarkable life cycle. It was the first toxic dinoflagellate to attack fish prey. Unlike other toxic dinoflagellates that are plant-like organisms with pigments that may color water, the *Pp* are translucent unless they consumed pigmented fish prey. It is difficult to monitor its presence as it has more than twenty forms or stages in the water column or benthic sediments. This is also the first dinoflagellate to have dormant stages in a form of a cyst that can last for years. When the environment is conducive to their growth, they transform to a motile form that swarm the water and produce toxins.[64]

Pp has versatile nutrition. Its food ranges from bacteria and other algae to mammalian tissue, live or dead. They consume dissolved organic substances found in poorly treated sewage and animal waste. At the initial stage, **Pfiesteria piscicida** may be harmless to the fish but further contact will cause these dinoflagellates to multiply and become lethal. This will support the assumption that the growth of Pp is stimulated by the presence of the fish tissue itself, its secretions, and excretions. Furthermore, they can survive even without organic substances because they have a large food vacuole called *Kleptoplast* that stores the chloroplasts they acquired from other algae. These serve as their nutritional supplement.

Several **Pfiesteria piscicida** outbreaks in North Carolina happened in estuaries with a high concentration of anthropogenic origin such as animal waste, poorly treated human sewage, cropland, and lawn fertilizer runoff with a large phosphorous content.

Toxin of *Pfiesteria piscicida*

The lethal toxin produced by *Pp* has not been identified yet. There were several toxicity studies done and all of them were not universally conclusive. As of now, what is definite is that the compound responsible for epidermal damage to the fish is a fat-soluble compound. Another compound obtained from a dinoflagellate was a water-soluble compound with neurotoxin-like properties. Collaborative efforts between three (National, Oceanic and Atmospheric Administration) scientists and researchers in North Carolina State University and University of Miami are trying to characterize and identify the toxin.

Mechanism of Action

Although the mechanism of action has not been established yet because the toxin has not been identified, several studies were done on the effect Pp toxin on fish and humans. Data shows that the Pp toxin lowers the fish white blood cells to 40-60% of the normal, compromising its immunity and lowering their resistance making them susceptible to other opportunistic pathogens. The toxin also destroys the osmoregulatory system located in the epidermis of the fish making them vulnerable to the high salinity of the estuary water. The toxin of *Pp* was observed to impair the reproductive system of some fish. Commercial fish such as striped bass (*Morone saxatilis*) and killifish (*Fundulus heteroclitus*) do not hatch after exposure to toxin of Pp.

Upon contact with toxic *Pp* or even indirectly with only the toxin, the fish became inactive, moving wayward and sluggishly. In acute exposure, the fish develop focal or diffused skin lesions ranging from ulcerations and hemorrhages leading mostly to death. These skin lesions appear within two to 12 hours upon exposure.

There is no significant research as to the effects of Pp toxin in humans, but studies done in rats suggest that the toxin targets the N-methyl-p-aspartate (NMDA) receptors. As we know, the NMDA receptor is particularly important in the body. It controls the synaptic activity and memory functions. The toxin inhibits the NMDA receptors in the brain leading to cognitive and memory impairment. The intensity and duration of exposure to produce the symptoms is still unknown, but there is a dose-response relationship as more neurocognitive impairments were seen in people who have had longer exposures to water with fish killers.

Clinical Manifestations

It is surprising to know that the potential effect of Pp toxin was first noticed in a laboratory setting when marine scientists developed symptoms while studying the effects of the toxin in fish health. One of them was hospitalized due to severe symptoms. Similar sickness was reported among fishers in Maryland. They observed that those fishers who had repeated contact with estuary water with fish kill developed the illness. [65]

The usual modes of acquiring the disease are through prolonged skin water contact or inhaled aerosols where fish were diseased or dying and where actively toxic *Pfiesteria* populations were present. Unlike poisoning associated with other marine biotoxins. Pp associated syndrome is not acquired by oral routes.

The medical complaints of those exposed to the Pp toxin are respiratory irritation, eye irritation, gastrointestinal symptoms such as stomach cramps, nausea and vomiting, fatigue, and headache. These symptoms usually last for two weeks depending upon the time span of contact. Other individuals manifest neurologic symptoms such as altered mental status, confusion, disorientation, mood and personality changes, dyskinesia, and ataxia. Some individuals' manifest amnesia with impairment of antegrade memory, less concentration, and impaired dexterity. In some individuals, lumbar tap was performed to assess the infection in the brain. They found elevated protein and immunoglobulins in the cerebrospinal fluid. These symptoms disappear two to three weeks after exposure.

Skin irritation is manifested in the form of a burning sensation, itchiness, red bumps, or sores erythematous papules usually in the extremities and trunk. If symptoms persist, individuals may experience numbness and a tingling sensation of extremities. These lesions are suggestive of allergic reaction, inflammatory, or toxic process.

How about oral route of Pp toxin? At this time, Pp does not adversely affect people who ate seafood originating from areas with known toxic Pp. Was this because of the high acidity of our gastric environment that will neutralize the toxicity? Studies show that it is not because of the inherent human protector in the stomach but with the characteristics of the toxin. The toxin has been observed to be labile and unstable in water. Studies were done when the water taken from estuaries with dead fish were filtered to remove the Pp strain, the remaining toxin in the water will kill fish only within 8 hours. Beyond this time the fish survive. Some consider the water safe for human use after 24 hours. As a precautionary measure in Maryland, their policy was to reopen the areas after three days without fish deaths.

More research needs to be done on the impact of Pp toxins to human health. Potential effects on humans during Pp outbreaks cannot be ruled out until more studies are done on the toxin in the fish tissue. The greatest obstacle standing in the way of more research on the *Pp* toxin to humans is the lack of an assay for toxin identification.

Diagnosis

Having the symptoms as mentioned above are not enough to have a diagnosis of Pfiesteria associated syndrome. These symptoms are not reliable to consider this syndrome as these lesions can be caused by fungal infections. A more reliable basis is the history of prolonged exposure to water known to have the Pf toxin. Although neurocognitive tests are nonspecific, positive results on these tests will support the diagnosis of Pfiesteria associated syndrome. At present, neuroscientists are working in collaboration with the CDC to determine basic testing batteries for persons inclined to develop the *Pp* associated syndrome.

Regulation and Monitoring

Since Pp proliferate in an environment with high concentrations of anthropogenic origin such as animal waste, poorly treated human sewage, cropland, and lawn fertilizers, preventative guidelines are geared towards minimizing these factors. [66]

1. Local, state, and federal governments should enact regulations to lower and hopefully end water pollution, especially in estuarine systems.

2. Hog farms should be placed under strict regulations.

3. Companies whose chemical waste is leaking should be forced to meet exacting standards of containment for the waste.

4. Conduct further research. At this point, many details of *Pfiesteria piscicida*'s way of life, origins, and effects remain unknown. Federal agencies have allocated $750,000 for **Pfiesteria piscicida** research at North Carolina State University. This is a substantial step in the right direction, but until **Pfiesteria piscicida** is fully understood, research efforts should continue.

5. Educate the public. Spread the word about *Pfiesteria piscicida*.

6. Inform farmers and industrial companies about their effects and impacts on the watershed. Teaching people about the potentially hazardous consequences involved with their actions may help to terminate unsafe behavior.

7. Alert the public regularly and frequently about water conditions and warn them about the causes and symptoms of Pfiesteriosis (**Pfiesteria**).

8. Protect wetlands and marshes which function as buffers from pollutants to rivers and streams. Create and enforce regulations to preserve these fragile and valuable ecosystems.

Regarding the prevention of Pp associated syndrome in the laboratory setting, there are no federal guidelines or regulations at this time. Federal agencies have recommended standard laboratory practices when managing toxic algae, which includes the use of laboratory coats, gloves, and other personal protective equipment. Universities engaged in Pp research have constructed a laboratory with BSL 3 (Biological Safety Level 3) protocols, described below.

BSL-3

A BSL-3 Laboratory typically includes work on microbes that are either indigenous or exotic (Non-Native) that can cause serious or potentially lethal disease through inhalation. Examples of microbes worked with in a BSL-3 includes *Yellow fever*, *West Nile virus*, and the bacteria that causes *Tuberculosis*.

The *Microbes* are so serious that the work is often strictly controlled and registered with the appropriate government agencies. Laboratory personnel are also under medical surveillance and may receive immunizations for the microbes that they work with.

For a better understanding of Biosafety levels 1-4 , log on the website:

https://www.cdc.gov/orr/infographics/00_docs/biosafety.pdf

Common requirements in a BSL-3 Laboratory

- Standard personal protective equipment must be worn, and respirators might be required.

- Solid-front wraparound gowns, scrub suits or coveralls are often required.

- All work with microbes must be performed within an appropriate BLS-3.

- Accessible hands-free sink and eyewash stations are available near the exit.

- Sustained directional airflow to draw air into the laboratory from clean areas towards potentially contaminated areas (Exhaust air cannot be re-circulated).

- A self-closing set of locking doors with access away from general building corridors.

- Access to a BSL-3 laboratory is restricted and always controlled.

CIGUATERA FOOD POISONING (CFP)

Ciguatera food poisoning is the most reported poisoning from seafoods. This is an illness caused by eating reef fish that contain the toxin *(ciguatoxin)* produced by marine microalgae called "Gambierdiscus toxicus". Symptoms include nausea, vomiting, diarrhea, and neurologic symptoms such as tingling of the fingers and toes, dizziness, and weakness. There are more than four hundred species of fish, including edible fish such as sea brass, snapper, perch that are contaminated with ciguatoxin. These fish are typically inhabitants of coral reef that consumed the Marine Microalgae.

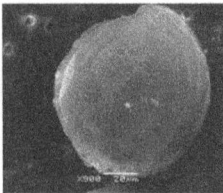

Fig 12-48 *Gembierdiscus toxicus*[67.]

Incidence

Ciguatera poisoning has been reported way back in 1511 by Peter Martyr de Anghera in West Indies. It was also reported in 1601 by Harmansen in the islands of the Indian Ocean. By 1606, there were reports of Ciguatera Poisoning in several archipelagos of the Pacific Ocean. The word Cigua was referring to a certain univalve mollusk *(Turbo pica)* in Cuba which was believed to contain the toxin. In 1787, Antonio Parra in Cuba transferred the word to describe the intoxication caused by eating the mollusk.[68]

In the old days, the incidence of Ciguatera Food Poisoning was limited to coastal areas, islands communities of Aboriginal people but now because of increased universal food consumption, commercialism and increased travel, this poisoning has been present worldwide.

Ciguatera food poisoning affects all age groups with males and females in equal numbers. It is interesting to note that ciguatoxin was detected in semen that can be transferred to females after intercourse. It was also detected in breast milk, so it is possible to transfer the toxin from the nursing mother to the babies. [69]

Toxicology/Pathogenesis

Ciguatera toxins are odorless, tasteless compounds undetected by simple means. It is heat stable and lipid soluble. It is composed of 13-14 rings fused by ether linkages into a ladder structure. It cannot be destroyed by cooking and exposure to mild acidic and alkaline conditions.

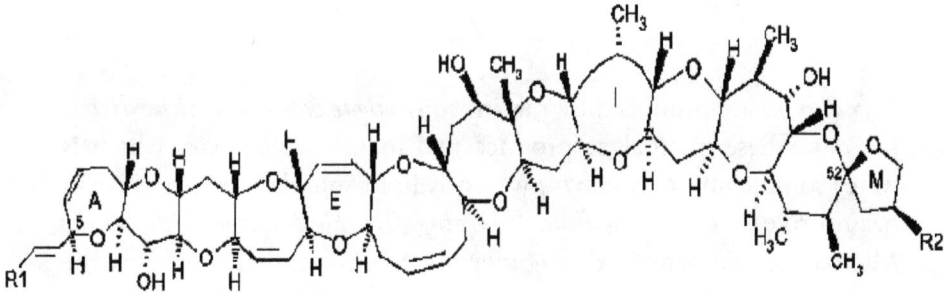

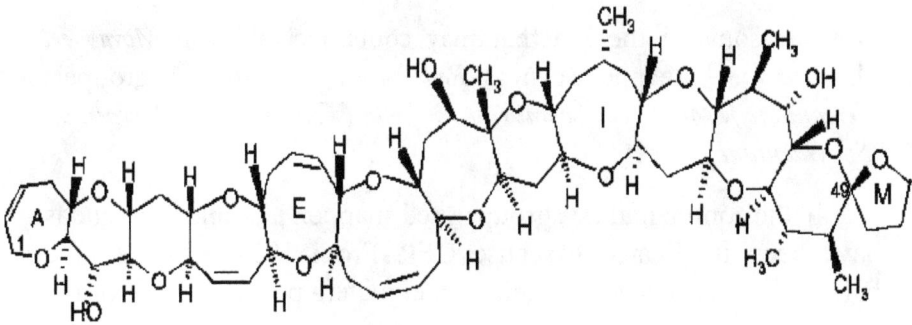

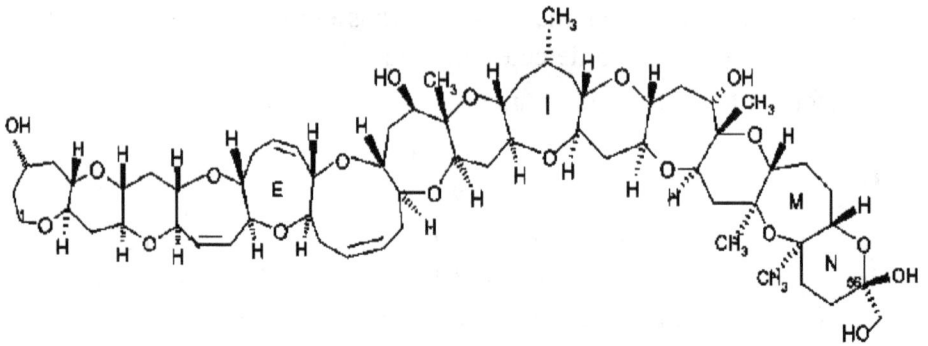

Fig 12.49

Structure of Pacific (P) and Caribbean (C) ciguatoxins (CTXs) [70]

Ciguatoxins are produced by marin*e microalgae called* **Gambierdiscus toxicus.** These microalgae produce two toxins: *Maitotoxin,* which is water soluble and *Ciguatoxin,* which is lipid soluble. The Maitotoxin has no role in *Ciguatera Food Poisoning. The Herbivorous Fish eat the* Microalgae that contain the *ciguatoxin and are* eventually eaten by the *Carnivorous Fish.* The ciguatoxin has been found in the liver, muscle, skin, and bones of the large carnivorous fish which are eventually eaten by humans.

Here are some of the fish that may contain ciguatoxin: *Moray eel (Muraenidae),* snapper such as *Red Bass (Lutjanidae),* groupers *(Serranidae), Mackerel (Scombridae), Jacks (Carangidae), Barracudas (Sphyraenidae).*

In the continental US, grouper, red snapper, jack and barracudas have been implicated to cause CFP. From 1954 to 1992 the barracudas has been associated with ciguatera poisoning in Florida.

The ciguatoxin affects the excitability of the cell membranes of the nerve and muscle. It specifically binds to voltage gated sodium channels in the cell membrane causing influx of Na even at resting membrane potentials. This causes imbalances among the electrolytes inside and outside the cell. This will result in increased Calcium entry into the cell that will cause increased muscle cell contraction. This will result in significant slowing of nerve conduction and prolongation of absolute refractory period. In the heart it causes increased cardiac muscle contraction (increase inotropic effects). In the intestinal mucosa, the increase of Calcium into the smooth muscle causes more fluid secretions from the digestive glands leading to diarrhea.

Symptoms

The symptoms of *Ciguatera* poisoning start as early as 30 minutes after consumptions of the contaminated fish for severe symptoms. For mild form, the symptoms may appear 24-48 hours after consumption. The initial symptoms are numbness of the lips and tongue followed by tingling sensation of the hand and feet. Gastrointestinal symptoms follows which are nausea , vomiting, diarrhea, and abdominal cramps. Neurologic symptoms may ensue such as generalized weakness, restlessness, dizziness, blurring of vision and even coma. Cardiac symptoms may appear in the form of *Hypotension* and *Bradycardia*. Gastrointestinal symptoms last for days but neurologic symptoms lasts for several days. Severe cases can develop hypotension, bradycardia and respiratory difficulties but death is uncommon. The reason death is uncommon is that the toxin in the fish rarely reach a level lethal to humans. The fish are vulnerable to elevated levels of toxin. Researchers found out that ciguatoxin can be stored in the adipose tissue. Any condition that causes increased lipid metabolism such as in times of stress, exercise, weight loss may result in the toxin re-entering the blood stream and subsequently appearance of the symptoms and signs. This is the reason the symptoms may last for weeks and months in some patients.

Treatment:

At present, there is no antidote for Ciguatoxin. Fortunately, most of the cases are mild and managed symptomatically such as hydration for diarrhea and vomiting, *Paracetamol* (acetaminophen) or *Nifedipine* for headache, *Gabapentin* for neuropathic pain. For severe poisoning, a Gastric Lavage can be administered if the contaminated fish was eaten within three to four hours earlier. This means that food eaten is still in the stomach that can be brought out by lavage. Food eaten usually stays in the stomach for three to four hours. Severe gastrointestinal symptoms can be managed by continuous intravenous fluid infusion with monitoring of the *Electrolytes* and kidney functions. Patients rarely develop respiratory failure and coma that require intubation and assisted ventilation. For those patients with more pronounced neurologic symptoms such as coma, *Mannitol* is given as infusion. Mannitol being an osmotic substance will attract fluid as it pass through the brain decreasing the cerebral and neuronal edema that cause the neurologic symptoms. Aside from that, Mannitol attracts free radicals brought about by the ciguatoxin eventually decreasing the action of the toxin to the voltage gated sodium channels. Note that Mannitol serves also as an osmotic diuretic that may decrease the total fluid volume so it is recommended that patients must be hydrated first before administering Mannitol.

In some areas in New Caledonia, Western Pacific, inhabitants use traditional herbal medicine to treat ciguatera poisoning. These are extracts from *Argusia argentea leaves* or *Davallia* , but there is no scientific evidence of their efficacy.[71]

Prevention

It is hard to prevent *Ciguatera* poisoning because of the physical and chemical characteristics of the toxin. It is odorless and tasteless so it is hard to detect contaminated fish. It is heat stable so it cannot be eliminated by boiling, frying, salting, marinating and even freezing. It is interesting to note that most of the fish causing the Ciguatera Poisoning are those fish caught by sport fishing, not from commercial fishing. It is advised not to consume bigger fish because the bigger the fish is, the greater is the toxin content. This toxin can remain in these fish throughout their life span. It is also advised not to eat the visceral parts of the fish such as the liver and head because these parts contain more toxin than in the muscles of the fish.

Prevention is more on community outreach and education of residents especially in endemic areas. They should be made aware of the kinds of fish that harbor the toxin. Here are some examples of those fish:[72]

Moray eel

Barracuda

Grouper

Kingfish

Amberjacks

Snapper

Sturgeon fish

Parrot fish

Hogfish

Narrow barred Spanish mackerel

Coral trout

Flowery cod

Red emperor

Any case of CFP, whether it is confirmed or suspected must be reported to authorities. In Florida, there is a law that requires medical practitioners such as physicians, chiropractors, naturopathy practitioners and veterinarians to report any incidence of CFP. Confirmed cases are recorded while suspected cases are reviewed to gather more data as well as obtaining fish samples for analysis by the Food and Drug Administration (FDA). When a report is done, immediate warnings are issued to the residents as to what fish to avoid and areas not to fish.

Government agencies maintain a website for providing information to residents about the CFP. In Florida for example, there is such thing as Waterborne Disease Surveillance Program that serve as coordinating network among the medical practitioners, Florida Poison Information Center, and other government agencies to review data and give recommendations to the affected residents.

The Florida Poison Information Center - Miami has a 24-hour hotline

(1-888-232-8635) where anyone can report CFP.

BMAA (Beta-Methylamino-L-alanine)

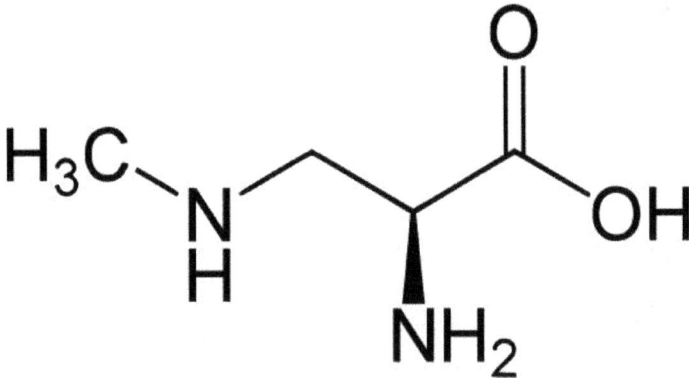

Beta Methyl-Amino-L-Alanine

There are many chemicals that can cause neurodevelopmental and neurodegenerative disorders. We are aware of POPs (persistent organic pollutants), bisphenol and phthalates from plastic, low molecular weight hydrocarbons and aromatic hydrocarbons. A chemical of particular interest recently is BMAA (β-methylamino-L-alanine). This chemical has been associated with degenerative disorders of the brain like Amyotrophic Lateral Sclerosis (Lou Gehrig's disease), Alzheimer's disease, Parkinson's disease and other neurodegenerative disorders.

β-methylamino-L-alanine (BMAA) is a naturally occurring neurotoxin. It is a non-essential amino acid that can be produced by the harmful algae like cyanobacteria, diatoms and dinoflagellates. It biomagnifies in the food chain, and can be found in common human commodities such as seafoods and shellfish. Ingestion of these foodstuffs can lead to the neurological consequences such as impairment of long term learning, memory deficits, eventually leading to neurodegenerative disorders such as Alzheimer's disease (AD), Parkinson's disease (PD), and Amyotrophic Lateral Sclerosis (ALS).

Fig 12-50 Guam-Photo by NASA.[73]

Fig 12-51 Chamorro people in Guam[74]

Discovery of BMAA

After the Japanese war in 1944, a US Navy neurologist who happened to visit the Chamorro people in Guam, noticed a peculiar disorder of the natives characterized by dementia, paralysis and tremors. This was locally called "lytico- bodig" . Other neurologists were fascinated by the news and subsequently flocked to the area to investigate this condition which was dubbed later as ALS-PDC (**Amyotrophic Lateral** Sclerosis- Parkinsonism Dementia Complex) . Recently, it has been said that the incidence of ALS in the Chamorro people is about 100 times higher than elsewhere in the word. [75]

In 1967, BMAA was first isolated in Guam by Marjorie Whiting , a nutritional anthropologist and Arthur Bell, a plant biochemist and director of the Kew Botanic Garden. They were researching on the cause of lathyrism, a progressive, irreversible spastic paralysis of the legs, previously prevailing in Europe, North Africa, Middle east but at that time restricted in China, India, Bangladesh... This condition has been linked to excessive consumption of a certain seed called " cycad seed " believed to contain BOAA. In their experiments however, they were not able to isolate **Beta-oxalylaminoalanine (BOAA)** instead they isolated BMAA. With these findings, the researchers now diverted their attention to BMAA as the cause of lathyrism....

Another researcher by the name of "Cox" went to Guam in the late 1990s and discovered that the Chamorros made tortillas from the

Cycad seeds. They were aware of the possible toxicity of these seeds. As a precaution, they washed the seeds thoroughly and gave the wash water to the chicken for drinking. If the chicken did not die, the cycad seeds were believed to be safe for making tortillas. Likewise they ate bats named "Marianas flying fox" that feed on cycad seed. In 2002, Cox with another researcher, Oliver Sack, a neurologist from Columbian University Medical Center hypothesized that long term consumption of the cycad seed created a reservoir of BMAA in the brain of the Chamorros, causing the " lytico-bodig disease. Their research was focused on the Marianas flying fox bats as well as the cycad seeds.

It was however Sandra Banack, a biology professor at California State University Fullerton, a colleague of Cox, who found out that the BMAA were produced also by an algae called cyanobacteria which thrive symbiotically with the roots of the cycad seeds. Cyanobacteria thrive alone in water or have symbiotic relationships with other organisms. Given favorable conditions for their growth (fertilizer run off, sewage runoff and warming climate change), they rapidly multiply to form toxic blooms. BMAA biomagnifies (moves up the food chain) in fish and shellfish which are eventually consumed by humans.

In 2005. Cox and other colleagues tested 30 laboratory strains of Cyanobacteria and found out that 95% of these strains produced BMAA. With this discovery, the researchers were then focusing on the environmental toxicant rather than genetics as the cause of ALS. In the course of time, Walter Bradley , an ALS expert and former chairman of Neurology at the University of Miami, Miller School of Medicine. mentioned that " Only 5-10% of ALS, AD and Parkinson's cases are due to inherited genetic mutations." They concentrated more on the environmental toxicants.

Pathophysiology

BMAA enters the body through the oral route. It may also enter through inhalation. The gastrointestinal tract is the interface between the dietary metabolites from the food and drinks and the ENS (enteric nervous system), the nervous system in the gastrointestinal tract. The ENS are nerve plexuses in the muscular wall (myenteric nerve plexuses and submucosal walls) of the gastrointestinal tract (GIT). This ENS is connected to the Central nervous system by the vagus nerve. The GIT is also colonized by microbes that are beneficial to the body. These microbes regulate gut motility, maintain intestinal mucosa for absorption of nutrients, metabolism of drugs and prevent absorption of toxic substances like BMAA. Inequality of these normal microbes due to infection, stress ,antibiotic and even inappropriate diet affect the homeostasis allowing toxic substances like BMAA to be absorbed along the intestinal mucosa. This substance ended on the dendrites of the sensory neurons of the ENS. They affect the neuronal excitability and membrane action potential of the neuron, and are transported to the central nervous system by retrograde direction (Retrograde transport)

In the central nervous system, BMAA passes through the blood brain barrier (a structure specific only to the capillary of the brain that selectively permeates substances and chemicals to enter the brain tissue). In the neurons this chemicals affect the Translation step in the Protein synthesis (there are two steps in protein synthesis of the cells: (1) Transcription that involved the formation of mRNA in the nucleus and (2)Translation that involves the formation of amino acid chains to form the protein. This second step happens in the ribosomes of the cytoplasm. BMAA which is a non-proteinogenic amino acid is said to be mis-incorporated in place of the amino acid serine in the process of protein synthesis, forming an abnormal protein that tends to misfold and not function correctly. Further formations of these abnormal proteins lead to accumulation of these abnormal proteins that causes neuron cell death called apoptosis. The accumulation of these protein aggregates is the classic hallmark of ALS, PD/AD. The problem of the neuron is that it does not multiply. It does not undergo mitosis. The amount of neurons at birth are approximately the same amount in old age, barring neurons lost in injuries. There is no way for these accumulated abnormal proteins to be diluted as what happened in one cell before it divides to two cells. The increase in the amount of cytoplasm as a cell prepares for cell division will in some way dilute the abnormal proteins, thereby decreasing their toxicity. Since the neuron does not multiply, these abnormal proteins accumulate to toxic levels that will result in death of the neuron. Such an event will lead to paralysis when the motor neurons are affected. This has been postulated as the pathophysiology of some of the neurodegenerative disorders like Amyotrophic Lateral Sclerosis and Parkinson's Disease.

Amyotrophic Lateral Sclerosis (ALS)

Amyotrophic Lateral Sclerosis is also known as Lou Gehrig disease or Charcot disease. This disease was identified by Jean-Martin Charcot in 1869. It was recognized internationally in the early 1940s when it led up to the retirement of a well decorated baseball player named Lou Gehrig of the New York Yankees. He was considered the first person diagnosed with ALS that is why it was named after him. That was on June 10 1939 when he visited the Mayo Clinic in Rochester Minnesota after suffering from weakness and loss of coordination while playing baseball.[76]

In medical terminology, AMYOTROPHIC means " no muscle nourishment. ("A" means without, MYO means muscle, TROPHIC means nourishment). When the body cell is deprived of nourishment, they eventually die and their death causes sclerosis to the area. The word "Lateral" refers to the lateral side of the spinal cord. This is the area where these motor neurons pass through. This disease usually affects the motor neuron of the Central Nervous System. These are the neurons that stimulate and control the movement of the muscles. Affection of these neurons lead to weakness , loss of coordination of the muscle movements and worse paralysis of the muscles. These motor neurons can be divided into two types; (1) Upper Motor Neurons that transmit impulse from the brain to the spinal cord, (2) Lower Motor neurons that transmit impulse from the spinal cord to the muscles. If it affects your skeletal muscles, you develop weakness and loss of coordination. If it affects the muscle of speech, you develop aphasia. ALS is incurable. There are some medications that you can avail of but these drugs don't treat ALS but delay its progression, thereby prolonging the patient's life. In advanced cases, it may affect the neurons supplying the respiratory muscles or breathing muscles. Paralysis of such muscles results in death.

Causes of ALS

The real cause of ALS is still unknown but there are several hypothesis:

(1) **Environmental Toxic Exposure**. Of particular focus here is the BMAA that is produced by the Cyanobacteria. This is well explained in the above section of this article. This is the explanation why the Chamorro people in Guam have about 100% more incidence of ALS compared to the world. This was also one of the explanations why some of the soldiers from the Gulf war suffered this ALS/ PD syndrome. They might be exposed to the cyanobacteria and other microorganisms in the desert sand while patrolling in the area. They also hypothesize that BMAA may be acquired by longer exercise and trauma This is probably the possible cause why there are some incidences of ALS in athletes, just like what happened to Lou Gehrig.

Fig 12-52 *Nostoc pruniforme* forming amorphous semi-filaments.[77]

Photo by Mathew Larson

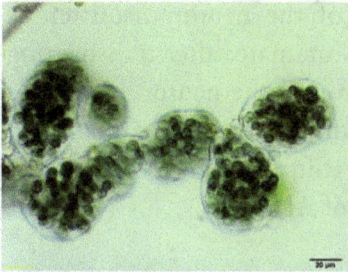

(2) **Genetic factor** . 10-15% of cases are genetically linked. A person may not have the disease but in his lifetime acquired the disease by genetic mutation, he would surely transfer this disease to his children.

(3) **Oxidative Stress**. In ALS, there is increased production of a molecule that contains oxygen. These are toxic byproducts of cellular metabolism. They could lead to protein modification, precipitation and eventually lead to neuron apoptosis. This will cause the muscles supplied by these neurons to shrink , atrophy and lose its functions. A drug, edavarone, targets this oxidative stress. It alleviates the effects of these oxidants and preserve the function of the neuron, thereby prolonging the life of the muscles.

(4) Mitochondrial Dysfunction. One of the organelles in the cytoplasm is the Mitochondria. This is the powerhouse of the cell because it produces ATP which is the energy unit of the cell. Mitochondria has its own DNA and ribosomes. It has been observed that patients with ALS have shown aberration in the neuronal mitochondria.

(5) Abnormalities of the Immune System. There are two types of cells in the nervous tissue. The neurons that transmit impulses and the neuroglia that protect, nourish , and support the neurons. One of the neuralgias is the Microglia which has the immune function. They are there in the nervous tissue to protect the neurons from toxic substances and invading microorganisms. It has been observed that Microglia may be beneficial to the neuron against ALS initially but harmful at a certain stage of ALS.

(6) Glutamate toxicity. In the nervous tissue, there is such a thing as Neurotransmitters. These are chemicals that transmit impulses from one neuron to the other neuron. One of the neurotransmitters is Glutamate. It has been found out that Glutamate, after transmitting signals to the neuron, would build up around the neuron instead of dispersing away from the neuron. Accumulation of Glutamate will cause death of the neuron. A drug called Riluzole reduces the level of Glutamate that can slow the progression of ALS.[78]

Parkinson's disease (PD) (Paralysis agitans)

Another neurodegenerative disorder linked to BMAA is Parkinson's Disease. This is a neurological disorder characterized by uncontrollable movements as tremors, rigidity , stiffness, slowing of movements, with loss of coordination, abnormality of gait and postural balance leading to difficulty in walking and slurred speech. The face may show no or little expression, the upper extremities may not move in coordination with walking. Other signs and symptoms of Parkinson's disease are depression, derangement of sleep, fatigue, memory difficulties and bladder problems . These signs and symptoms deteriorate over time.

This condition usually starts at the fifth and sixth decades of life with men more affected than women. Presently it affects approximately two percent of Americans in their lifetime[79]

These signs and symptoms of Parkinson's Disease are due to inadequate chemicals in the brain called "Dopamine". A certain area in the midbrain called Substancia Negra (a basal ganglia) produces this chemical. The neurons in this basal ganglion deteriorate progressively, thereby failing to produce adequate dopamine. This chemical communicates with other basal ganglia (caudate nucleus, putamen, globus pallidus and nucleus accumbens) in the corpus straitum to coordinate and refine the movement of the human beings. Deficiency of this Dopamine, leads to uncoordinated movements.[80]

What causes the failure of the neurons of the subtancia nigra to produce dopamine:

(1) Familial Cause – There are at least seven different genes associated with Parkinson disease. Examples are **LRRK2, PARK7, PINK1, PRKN, or SNCA, SLC7A11** genes . Being familial, it can be transferred from the parents to the children. This however comprises only 10% of cases. We will not discuss this familial cause in the book as we are focused on the BMAA.

(2) Toxic substances and poisons. It has been said that some chemicals like BMAA can cause the formation of Lewy Bodies, an aggregate form of protein Alpha -Synuclein , which is the hallmark of neuro degenerative disrorder. Just like in ALS, they have the same mode of transmission by oral route to the ENS of the intestine , eventually ending up in the CNS through retrograde transmission via Vagus nerve. These chemicals cause degeneration of the cells of the Substantia negra, resulting in deficiency of Dopamine.

Another hypothesis is that when BMAA reaches the Subtancia nigra, it targets a certain gene SLCA11 which is necessary in the synthesis of an antioxidant Glutathione. BMAA will slows down the production of Glutathione resulting in the build up of free radicals which are toxic to the neurons that produce dopamine in the substancia negra .

Other possible causes of PD are:

(3) Brain damage from injuries. There is such thing as post traumatic PD which are seen in athletes in contact sports such as boxing and football who have repeated head trauma as exemplified by Muhammad Ali. There was a study done among veterans for a period of 12 years. Their subjects were 1462 veterans diagnosed with PD. 949 of these veterans had traumatic brain injury. They studied their medical records and developed a criterion for mild and moderate to severe traumatic brain injury. Mild was defined as loss of consciousness less than 30 minutes and memory loss less than 20 hours while moderate and severe was defined as memory loss more than 30 minutes and loss of consciousness for more than 24 hours. They found out that there was a 56% risk of developing PD in mild cases of traumatic brain injury while there is a risk of 83 % for moderate and severe cases.[81]

(4) Infection. Inflammation of the brain **(Encephalitis)** can cause parkinsonism

(5) Medications. There are medications that can cause Parkinson-like effects. If continuously taken can result in permanent Parkinson symptoms.

Alzheimer's Disease (AD)

Another neurodegenerative disease that BMAA is implicated in is Alzheimer's disease. This is a neurodegenerative disorder characterized by impairment of cognitive functions resulting in dementia. The main symptoms of AD are memory loss, cognitive dysfunction, abnormal personality changes, and impaired judgment. If you ask an Alzheimer's patient what to do if there is fire outside, she would say I better watch TV. One of the prominent personalities who had Alzheimer's was President Reagan. In his eulogy, his son said that every time he visited his father, he was recognized not as his son but the person who visited before. According to the Alzheimer's Association, there are about 50 million AD patients worldwide and its incidence doubles every five years after the age of 65 years . AD accounts for 60.80% of dementia cases.[82]

There have been many hypotheses as the cause of AD but the most accepted among these hypotheses are the beta -amyloid protein and tau protein hypotheses. They postulated that there is an increased amount of abnormal tau protein in the part of the brain involved in memory and there are beta amyloid plaques in between neurons. It has been said that BMAA can bind with the Amyloid beta protein or the BMAA replacement of amino acid serine can result in accumulation of the amyloid beta protein.

Recently however, there is research that contradicts this hypothesis. They found out that there was no binding between BMAA and amyloid beta protein. They concluded that incorporating BMAA in place of serine amino acid in the protein synthesis, would not alter the conformational dynamics of the beta Amyloid peptide.[83]

In some analytic research, they did not find BMAA in the brain of former ALS and AD patients and in some patients where they found BMAA, the quantity of BMAA was not significant enough to induce the disease . They cannot conclude though that BMAA cannot cause AD. All they can say is that the BMAA hypothesis as the etiology of these neurodegenerative diseases has still to be substantiated .[84]

Having said that, it is not in the realm of our book to discuss AD as we are only dealing with the conditions that implicate BMAA as the cause.

I PUT CHARTS OF TOXINS CAUSED BY THE CYANOBACTERIA BEGINNING ON PAGE 626 BECAUSE THEY ARE A BIG PROBLEM DUE TO THEIR TOXINS. THEY POSE CHALLENGES DUE TO HARMFUL BLOOMS IN THE OCEANS AND IN FRESHWATERS.

TYPES AND ACTIVITIES OF TOXINS ASSOCIATED WITH CYANOBACTERIA

HEPATOTOXINS [85] (Toxic to the Liver)

TABLE 12-4 MICROCYSTINS

ACTIVITY	GENERA THAT PRODUCE THESE TOXINS
Hepatotoxic	*Microcystis*
Protein phosphatase inhibition	*Anabaena*
Membrane integrity	*Nostoc*
Conductance disruption	*Planktothrix*
Tumor promotors (Cancer)	*Anabaenopsis*
	Hapalosiphon

TABLE 12-5 NODULARIN

ACTIVITY	GENERA THAT PRODUCE THIS TOXIN
Same as with Microcystins	Nodularia spumigena Nodularia sphaerocarpa

TABLE 12-6 CYLINDROSPERMOPSIN

ACTIVITY	GENERA THAT PRODUCE THE TOXIN
Necrotic injury to liver	Cylindrospermopsis
Spleen	Aphanizomenon
Kidneys	Anabaena
Lungs	Raphidiopsis
Intestines	Umezakia
Protein synthesis inhibition	
Genotoxic	

NEUROTOXINS

TABLE 12-7 ANATOXIN-A

ACTIVITY	GENERA THAT PRODUCE THESE TOXINS
Postsynaptic,depolarizing neuromuscular blockers	*Apanizomenon*
	Anabaena
	Raphidiopsis
	Oscillatoria
	Planktothrix
	Cylindrospermum

TABLE 12-8 ANATOXIN –a(S)

ACTIVITY	GENERA THAT PRODUCE THESE TOXINS
Acetylcholinesterase inhibitor	*Anabaena*

TABLE 12-9   SAXITOXINS

ACTIVITY	GENERA THAT PRODUCE THESE TOXINS
Sodium channel blockers	Aphanizomenon
	Anabaena
	Planktothrix
	Cylindrospermopsis
	Lyngbya

DERMATOTOXINS

TABLE 12-10 LYGBYATOXIN-A AND APLYSIATOXINS

ACTIVITY	GENERA THAT PRODUCE THE TOXIN
Inflammatory agents	Lyngbya
Protein kinase C activators	Schizothrix
Oscillatoria	

ENDOTOXINS

TABLE 12-11 LIPOPOLYSACCHARIDES

ACTIVITY	GENERA THAT PRODUCE THE TOXIN
Inflammatory agents	Many cyanobacteria
Gastrointestinal irritants	

TABLE 12-12 SYMBOLS ASSOCIATED WITH TOXIC ALGAE

Symbol	Disease	Symbol	Disease
ASP	Amnesiac shellfish poisoning	NSP	Neurotoxic shellfish poisoning
DA	Domoic Acid	YTX	Yessotoxins
OA	Okadaic Acid	AZA	Azaspiracids
DSP	Diarrhetic shellfish poisoning	CYN	Cylindrospermopsin
MCYST	Microcystin	BMAA	Beta-methyl-amino-L-Alanine
BTX	Brevetoxin	PSP	Paralytic shellfish poisoning
CFP	Ciguatera food poisoning	STX	Saxitoxin

1   Jonsson,P.R. et al 2009. Formation of harmful algal blooms cannot be explained by allelopathic interactions.

Proc Natl Acad Sci U S A. 2009 Jul 7; 106(27): 11177–11182. doi: 10.1073/pnas.0900964106

https://www.ncbi.nlm.nih.gov/pmc/articles/PMC2708709/

2   Kinkaid, C 2014. Toxic Algae: How to Treat and prevent Harmful Algal Blooms in Ponds, Lakes Rivers, and Reservoirs.P14. Solardyne.com

3   https://commons.wikimedia.org/wiki/File:Mixed_zooplankton_sample.jpg

4   https://commons.wikimedia.org/wiki/File:Red_tide.jpg Viewed 18 May 2021.

5   https://hab.whoi.edu-Viewed 20 April 2022.

6   https://hab.whoi.edu/impacts/impacts-ecosystems/ viewed 20 April 2022.

7   Kinkaid, C.2014.Toxic Algae: How to treat and prevent harmful algal blooms in ponds, lakes, rivers, and reservoirs. P4 Solardyne ,LLC.

8   https://en.wikipedia.org/wiki/Cyanobacteria#/media/File:Morphological_variation_within_cyanobacterial_genera.jpg Viewed 21 April 2022.

9   Sato, N.2021Are Cyanobacteria an Ancestor of Chloroplasts or Just One of the Gene Donors for Plants and Algae?

Genes 2021 Jun, 12(6):p823 doi: 10.3390/genes12060823

10   https://hab.whoi.edu/species/species-life-cycle/cyanobacteria/ Viewed 22 April 2022.

11   https://commons.wikimedia.org/wiki/File:Diatoms.png Viewed 234 April 2022.

12   https://hab.whoi.edu/species/species-life-cycle/diatom/ Viewed 22 April 2022.

13   Bahls, L. (2016). Cymbella hantzschiana. In Diatoms of North America. Retrieved May 18, 2021, from https://Diatoms.org/species/cymbella_hantzschiana With permission

14   https://commons.wikimedia.org/wiki/File:Toxic_Algae_Bloom_in_Lake_Erie.jpg

Viewed 2 Dec 2022.

15   Marine Biotoxin, Amnesic Shellfish Poisoning, Food and Agricuture Organization of the United Nations, Rome, 2004 http://www.fao.org/docrep/007/y5486e/y5486e0n.htm#bm23

16   http://cfb.unh.edu/phycokey/Choices/Bacillariophyceae/Pennate/biraphes/biraphe_colony/PSEUDONITZSCHIA/Pseudonitzschia_Image_page.html#pic01   Viewed 31 May 2021.

Phycokey - Alexandrium images (unh.edu) Viewed 2 Dec 2022.

17   http://cfb.unh.edu/phycokey/Choices/Rhodophyceae/Macroreds/CHONDRIA/Chondria_image_page.htm#pic01

Viewed 13 Jan 2022.

18   Amnesic Shellfish Poisoning, California Poison Control System.   Lasoff D, Ly B. https://calpoison.org/news/amnesic-shellfish-poisoning

19   https://www.marinemammalcenter.org/science.org/science/top-research-projects/domoic-acid-toxicity.html

20   *http://www.fao.org/docrep/007/y5486e/y5486e0n.htm#bm23 Viewed 26 April 2022.*

21   Saxitoxin (STX) has also been found in Bullfrog (Rana catesbeiana)plasma

22   Harmful Algae, Amnesic Shellfish Poisoning, Fleming L, https://www.whoi.edu/redtide/page.do?pid=15159&tid=523&cid=27686

23   Saxitoxin is one of the most common toxin that causes PSP

24   Dinoflagellates and cyanobacteria, diatoms and dinoflagellates. Many algal species produce similar toxins.

25   There are many species of Alexandrium which also produce STX. They are A. tamaransis, A. fundyense, A.catenella. Species of Gymnodinium include G. catenatum.

26   Mesquita, M.C.B.2019. Combined Effect of Light and Temperature on the Production of Saxitoxins in Cylindrospermopsis raciborskii Strains. Toxins 2019 11(1): 38. https://doi.org/10.3390/toxins11010038

27   http://cfb.unh.edu/phycokey/Choices/Dinophyceae/PS_dinos/ALEXANDRIUM/Alexandrium_Image_page.html#pic02 Viewed 15 Feb 2022.

28   Marine Biotoxin,Azaspiracid Paralytic Shellfish Poisoning, Food And Agriculture Organization of the United Nation, Rome, 2004,

29   http://www.fao.org/3/y5486e/y5486e0c.htm#TopOfPage

30   Epidemiologic Notes and Reports , Paralytic Shellfish Poisoning, Massachussetts   and Alaska, 1990, CDC, MMWR March 15, 1991 / 40 (10) ; 157-161,

https://www.cdc.gov/mmwr/preview/mmwrhtml/00001927.htm

31   PSP in the Philippines: three decades of monitoring a disaster (2006), Farida F, Juan B,   Relox Jr, Yasumo F, **http://agris.fao.org/agris- search/search.do?recordID=AV20120103502**

32   2 dead,40 hospitalized over red tide poisoning, Desacada M, (The Philippine Star) - December 27, 2016, https://www.philstar.com/nation/2016/12/27/1656993/2-dead-40-hospitalized-over-red-tide-poisoning.

33   Marine Biotoxin, Diarrhoeic Shellfish Poisoning, Food and Agriculture Organization of the United Nations, Rome,   2004   http://www.fao.org/docrep/007/y5486e/y5486e0f.htm#TopOfPage

34   https://commons.wikimedia.org/wiki/File:Dinophysis_acuminata.jpg Viewed 26 April 2022

35   http://cfb.unh.edu/phycokey/Choices/Dinophyceae/PS_dinos/PROROCENTRUM/Prorocentrum_Image_page.html#pic03 Viewed 26 April 2022

36   http://www.fao.org/3/y5486e/y5486e0e.htm#bm14, Viewed 26 April 2022

37   Diarrhetic Shellfish Poisoning, Washington, USA, 2011, Lloy J, Duchin J, Borchert J, Quitana, H.Robertson, A. https://www.ncbi.nlm.nih.gov/pmc/articles/PMC3739508//

38   https://commons.wikimedia.org/wiki/File:Karenia_brevis.jpg-Viewed 31 Dec 2022

39   Marine Biotoxin, Neurologic Shellfish Poisoning, Food and Agriculture Organization

 of the United Nation, Rome, 2004,,http://www.fao.org/docrep/007/y5486e/y5486e0o.htm

40   Marine Biotoxin, Azaspiracid Paralytic Shellfish Poisoning, Food and Agricuture Organization of the United Nations, Rome, 2004, http://www.fao.org/docrep/007/y5486e/y5486e0p.htm#bm25

41   https://upload.wikimedia.org/wikipedia/commons/3/36/Microalgal_species_from_the_Gulf_of_Naples.jpg

Viewed 1 June 2021.

42   http://cfb.unh.edu/phycokey/Choices/Dinophyceae/PS_dinos/PROTOPERIDINIUM/Protoperidinium_Image_page.html#pic05 Viewed 31 July 2021.

43   http://cfb.unh.edu/phycokey/Choices/Dinophyceae/PS_dinos/PROTOPERIDINIUM/Protoperidinium_Image_page.html#pic05 Viewed 1 June 2021.

44   http://cfb.unh.edu/phycokey/Choices/Dinophyceae/PS_dinos/PROTOPERIDINIUM/Protoperidinium_Image_page.html#pic05 Viewed 1 July 2021.

45   Twiner M, Rehmann N, Hess P, Doucette G, Azaspiracid Shellfish Poisoning: A Review on the Chemistry, Ecology, and Toxicology with an Emphasis on Human Health Impacts.

https://www.ncbi.nlm.nih.gov/pmc/articles/PMC2525481/

46   https://ncceh.ca/content/blog/palytoxin-potent-poorly-understood-marine-toxin-found-aquarium-coral

47   https://cen.acs.org/articles/96/i2/Palytoxin-danger-hidden-tropical-aquariums.html chemical structure of palytoxin

48   https://cen.acs.org/articles/96/i2/Palytoxin-danger-hidden-tropical-aquariums.html chemical structure of palytoxin

49   https://www.fragglereef.co.uk/palytoxin-symptoms

50   https://en.wikipedia.org/wiki/Palytoxin#:~:text=Palytoxin%20is%20a%20polyhydroxylated%20and,over%201021%20alternative%20stereoisomers.

51   https://www.ncbi.nlm.nih.gov/pmc/articles/PMC5099280/

52   https://en.wikipedia.org/wiki/Microcystin

53   https://www.sciencedirect.com/topics/pharmacology-toxicology-and-pharmaceutical

54   https://commons.wikimedia.org/wiki/File:Microcystin-LR.svg Viewed 31 Dec 2021.

55   https://www.ncbi.nlm.nih.gov/pmc/articles/PMC91088/

56   https://www.sciencedirect.com/topics/pharmacology-toxicology-and-pharmaceutical-science/microcystis-aeruginosa

57   https://www.merckvetmanual.com/toxicology/algal-poisoning/overview-of-algal-poisoning

58   https://commons.wikimedia.org/wiki/File:View_of_Palm_Island_from_wallaby_point.JPG

Viewed 2 June 2021.

59   http://cfb.unh.edu/phycokey/Choices/Cyanobacteria/cyano_filaments/cyano_unbranched_fil/tapered_filaments/CYLINDROSPERMOPSIS/Cylindrospermopsis_Image_page.htm#pix01

Viewed 2 June 2021.

60   https://natoxaq.ku.dk/toxin-of-the-week/cylindrospermopsins/

Viewed 27 April 2022.

61   State of the Art Clinical Article, 1999, Glenn, M. , Pfiesteria, " The cell from hell " and Other toxic Algal Nightmare , Morris Glenn Jr. Department of Medicine, Division of Hospital Epidemiology, University of Maryland School of Medicine and VA Medical Center, Baltimore, Maryland . https://watermark.silverchair.com/28-6-1191.pdf?token=AQECAHi208BE49Ooan9kkhW_Ercy7Dm3ZL_9Cf3qfKAc485ysgAAAlMwggJPBgkqhkiG9w0BBwagggJAMIIC

62   https://commons.wikimedia.org/wiki/File:Pfiesteria_large.jpg Viewed 11 July 2021

63   https://commons.wikimedia.org/wiki/File:Coast_watch_(1979)_(20471959890).jpg

Viewed 14 Sept. 2021.

64   Bioscience, October 2001, Burkholder J.M. et al, History of Toxic *Pfiesteria* in North Carolina Estuaries from 1991 to the Present: Many toxic *Pfiesteria* outbreaks have plagued the Albemarle-Pamlico Estuarine System, including events both before and after the 1997 outbreaks in Chesapeake Bay https://doi.org/10.1641/0006-3568(2001)051[0827:HOTPIN]2.0.CO;2

65   Bioscience, October 2001, Grattan, L. M. et al Human Health Risks of Exposure to *Pfiesteria piscicida*: Environmental exposure to toxic *Pfiesteria piscicida* produces a distinct clinical syndrome in some persons; efforts to identify the toxin(s) responsible for the syndrome and to increase understanding of how it disrupts the central nervous system have important implications for public health https://doi.org/10.1641/0006-3568(2001)051[0853:HHROET]2.0.CO;2

https://academic.oup.com/bioscience/article/51/10/853/245245

66   Pfiesteria piscicida: A Toxic Dinoflagellate Plaguing North Carolina, Solution to stop Pfiesteria pisccida.

67   https://www.mtholyoke.edu/~akpeters/solutions.html

68   https://en.wikipedia.org/wiki/Ciguatera_fish_poisoning#:~:text=The%20current%20name%2C%20introduced%20in,the%20univalve%20mollusc%20Cittarium%20pica. Viewed 2 March 2023

69   https://rarediseases.org/rare-diseases/ciguatera-fish-poisoning/

70   http://www.fao.org/3/y5486e/y5486e0q.htm#bm26

71   https://www.ncbi.nlm.nih.gov/pmc/articles/PMC2579736/53

72   IAMAT (International Association for Medical Assistance to Travelers) Ciquatera Poisoning. https://www.iamat.org/risks/ciguatera-fish-poisoning#:~:text=Any%20reef%20fish%20can%20cause,intestines%2C%20heads%2C%20and%20roe. https://www.iamat.org/risks/ciguatera-fish-poisoning#:~:text=Any%20reef%20fish%20can%20cause,intestines%2C%20heads%2C%20and%20roe.iewed 1 Jan 2023.

73   https://commons.wikimedia.org/wiki/File:Guam_ali_2011364_lrg.jpg Viewed 21 Nov 2022.

74   https://commons.wikimedia.org/wiki/File:Chamorro_people_in_1915.jpg Viewed 21 Nov 2022.

75   https://alsnewstoday.com/news/study-suggests-mechanism-als-caused-by-toxic-bmaa-compound-algae/

76   https://www.als.org/understanding-als/lou-gehrig

77   http://cfb.unh.edu/phycokey/Choices/Cyanobacteria/cyano_filaments/cyano_unbranched_fil/untapered_filaments/heterocysts/vis_sheath/NOSTOC/Nostoc_Image_page.html#pic03 Viewed 4 Dec 2022.

78   https://www.nia.nih.gov/health/parkinsons-disease#:~:text=Parkinson's%20disease%20is%20a%20brain,have%20difficulty%20walking%20and%20talking.

79   https://www.northshore.org/neurological-institute/centers-and-programs/parkinsons-disease-movement-disorders/?gclid=Cj0KCQjw4omaBhDqARIsADXULuVl70TXUrlSdam_-olcRrhzwYYLWtzALc6EQhWzl-7XZ5pe8sjbxCkaAvt9EALw_wcB&g

80   https://www.ncbi.nlm.nih.gov/books/NBK6271/

81   https://my.clevelandclinic.org/health/diseases/8525-parkinsons-disease-an-overview

82   https://www.frontiersin.org/articles/10.3389/fneur.2019.01312/full

83   https://pubmed.ncbi.nlm.nih.gov/29649877/#:~:text=Abstract-,%CE%B2%2D%20N%2DMethylamino%2Dl%2Dalanine%20(BMAA),monomeric%20and%20protein%2Dbound%20form.

84   https://cfpub.epa.gov/si/si_public_record_report.cfm?Lab=NHEERL&dirEntryId=345

85   Blaha,L et al.2009. Toxins produced in cyanobacterial water blooms – toxicity and risks.Interdisc Toxicol. Vol 2 (2):36-41

Section 3

Chapter 13

Solutions That Must Be
Implemented to
Prevent the End of
Humanity?

Epilogue

Glossary

Index of genera and species
discussed in the book

General Index

Chapter 13

SOLUTIONS THAT MUST BE IMPLEMENTED TO PREVENT THE END OF HUMANITY

Assuming that you the reader have read everything in this book, you would have learned the following about Biology: the six kingdoms of organisms which include the Archaebacteria, Eubacteria, Animal, Fungus, Plant, Protist as well as genetics, DNA and of course the basics of Environmental Science. In addition, you have learned how humans make the conditions suitable for new organisms to run wild and cause diseases. Furthermore, you become enlightened about the ways that these organisms mutate. In fact, you have read throughout this book about a long list of diseases, many of them quite horrifying, which you are undoubtedly hoping to avoid.

So, here is the question we must ask: Can the human race do anything to prevent these diseases or stop new dangers that are not on the radar yet? The answer is an absolute, definite MAYBE. How do we go about doing this? Pay attention to this chapter.

Let me suggest ways in which we may help prevent these diseases, and simultaneously maintain a livable environment. There are measures we can take to reduce, and hopefully eliminate to address the activities and attitudes that are harming our planet. Additionally, there are other positive changes we can make to protect and improve life on Earth. This will be immensely beneficial to all of humanity.

Solutions that Must Be Implemented To Prevent Disease and Other Catastrophes

- **PRESIDENT JOE BIDEN**
- **PRESIDENT BARACK OBAMA**
- **IS THE INCREASE IN QUARANTINE NECESSARY?**
- **END WARFARE SO WE CAN CONCENTRATE ON MORE IMPORTANT ISSUES**
- **MAKE POOR COUNTRIES LIVABLE**
- **HUMAN POPULATION CONTROL**
- **GETTING RID OF ALL FOSSIL FUELS FOR ENERGY PRODUCTION, INCLUDING THE ECONOMIC INCENTIVE**
- **SOLARIZE THE PLANET**
- **EAT LOWER ON THE FOOD CHAIN, NO MEAT**
- **STOP USING ANTIBIOTICS IN ANIMAL FEED**
- **GETTING RID OF MONEY AS WE KNOW IT**
- **HEALTH CARE FOR EVERYONE**
- **CHANGE ATTITUDES TOWARDS LIFE AND ALL ORGANISMS ON THE PLANET**
- **ENVIRONMENTAL EDUCATION FOR EVERYONE**
- **LOWER THE ENVIRONMENTAL FOOTPRINT OF PEOPLE WITH TOO MUCH STUFF (BECAUSE THEY HAVE LOTS OF MONEY)**
- **IMPLEMENTATION OF THE EARTH CHARTER**
- **CHANGE THE WAYS IN WHICH HUMANS THINK**

- **HAVE BILLIONAIRES USE SOME OF THEIR MONEY TO SAVE ENTIRE COUNTRIES**
- **GIANT CARBON SUCKING MACHINES**

Unfortunately, Earth and all of its inhabitants are presently facing serious issues. As mentioned earlier, Climate Change threatens our very survival. There are too many problems to count. Also, most are caused by people and their destructive interactions with the planet. For example, fuel emissions from automobiles are wreaking havoc with the atmosphere. Moreover, the difficulties on our radar all interrelated. It is my intention here to list some of the reasonable solutions to the environmental mess that we are now facing. Perhaps some readers may not agree with me, but I am confident that my ideas are based on sound rationales.

PRESIDENT JOE BIDEN

Now as I write these words, we have a new administration in the United States. This comes as a relief from the horrid past four years under the misguided leadership of an incompetent, privileged, narcissistic man, Donald Trump. Now that Joe Biden has assumed the presidency, the planet can breathe a collective sigh of relief. Allow me to clarify that this will not be a political rant. Rather, it is an opportunity to discuss the many positive environmental changes that Joe Biden made immediately after becoming president:[1]

- Bringing the United States back into The Paris Climate Accord
- Cancelling the Keystone XL Pipeline
- Placing leaders who respect climate:
  - Energy Secretary Jennifer Granholm
  - Brian Deese as National Economic Council Director
  - Deb Haaland as Secretary of the Interior.

Secretary Haaland revoked a series of Trump Era orders that allowed drilling on federal lands and she established a Climate Task Force that incorporates the social cost of carbon.

- Biden also has set serious goals for carbon reductions

As we approach 2022, we expect more environmental changes for the benefit of the planet's ecosystems. This is just of the advantages of having more environmentally aware people in the Biden Administration.

PRESIDENT BARACK OBAMA

First, we have taken a positive, major step toward solving some of our environmental problems with the election of Barack Obama as the 44th President of the United States for two terms. The fact that we do not have an oil company supporter, as we did with former Republican President George W. Bush, is in our favor. Previously, climate change and humanity worsened when Republican Administrations were in charge. Specific examples include the Iraq War with a total disregard for all types of life, and feigning ignorance with a total disregard of scientific reports about Earth getting warmer, along with the deaths of people, animals, and plants. Additionally, bacteria was spreading and that put lives at risk. Currently, we are now headed in the right direction environmentally. We hope it will last.

Our dependence on fossil fuels does not need any additional reinforcement from highly elected officials whose personal, family, and professional wealth is dependent on oil. This was the case for President Bush and Vice President Dick Cheney. When President Obama was elected, my blood pressure went down because he understands and accepts science. Furthermore, Mr. Obama and Democratic Congressional Members helped pass the Affordable Health care ACT (ACA).

As a result, millions of Americans could now afford health care.[2] If Mr. Obama encourages other sources of energy which do not involve burning something, then the rest of the world may follow suit.

Also, let's face it; from the second that Mr. Obama took office, our worldly image of the United States quickly rose into positive territory. Additionally, the stock market finally improved last year. Other countries regard our new President with much more respect and optimism than they viewed some of his predecessors.

What more proof do we have that he alone may set the impetus for change? He won the Nobel Peace Prize for two major reasons. One is that Mr. Obama won the award just for being there, as in the movie (*Being There*, 1979, starring Peter Sellers, directed by Hal Ashby), and the second is that many of his ideas were recognized by the Nobel Peace Prize committee as promoting world peace and other improvements.

As I write these words it appears that Mr. Obama is in a fight with the Republicans over health care. I believe that if we can solve the crisis here in The United States, then we may be able to prevent catastrophe. Please see the section on health care for America below. Mr. Obama is also partly the savior of our planet's environment.

The first edition of this book was written in 2011. Prior to this, we saw emerging diseases such as AIDS, which quickly spread from chimpanzees to humans; Ebola, believed to have originated in bats; and Lassa fever from rodents. Other diseases such as Lyme disease spread by ticks and of course diseases like SARS (Sudden Acute Respiratory Syndrome), MERS (Middle East Respiratory Syndrome). Since then, many unpleasant things have occurred such as fires in the Amazon, world droughts, warfare in various countries and of course COVID19. As indicated in this chapter in the first edition I suggested preventing infectious diseases by quarantining Africa.

However, as of now, I do not believe this is the only solution. It may mean that we may have to quarantine every country in order to prevent infectious diseases. Obviously, this cannot be done. So, what can we do?

IS QUARANTINE NECESSARY?

Most infectious diseases may emerge from Tropical regions of the planet. Is it fair to single out individual countries or continents to be quarantined such as Africa as is mentioned in the first edition of this book, now that we have COVID 19? In the previous edition of this book, I singled out Africa to quarantine because of the possibility of having disease organisms spreading such as the Ebola virus, but since then it appears that Asia or other areas including the tropics might be harboring other disease organisms.

SHOULD AFRICA BE QUARANTINED?

If there is one continent that illustrates the potential for disastrous spread of disease, it is Africa. For the past 25 years I have been extremely interested in Africa and the poverty, internal strife and disease located there. A good description of these problems is explained in a book written by Richard Preston.[3,4] It is titled, *Panic in Level 4*. In it the author describes the situation in Kikwit in the Congo, formally under the weak leadership of President Mobutu Sese Seko.

According to Preston, Kikwit has no sewage system, no running water, no telephone network, no newspaper, no radio station; the city's main hospital has one X-ray machine, and there are very few paved roads.[5] It is incredible to see that African countries lack the basic necessities that we in the West always take for granted.

Perhaps Ross Donaldson, in his latest book, *The Lassa Ward*, sums up their situation most profoundly: *"Lassa affected the most disposed people in one of the poorest countries in the world--it was an affliction of the unfortunate amidst the deprived."* [6] This statement perhaps summarizes Africa as a whole.

Map 5-1 (in Chapter 5) shows deaths from vector-borne diseases around 2002. Notice that of all the continents, Africa takes the lead with approximately 500-1900 deaths per million people while the United States and Europe have between one and 20 deaths per one million people.

The Indian sub-continent has between 50 and 200 deaths per one million.

Now what is the point that I am trying to make? Simple. We see that Africa has many vector- borne diseases. People travel to and from Africa every day. Viruses and other organisms such as bacteria and protists may travel with them and cause an outbreak in the United States, Europe, or some other location where the traveler goes. Since Africa has thousands of pathogens. It seems obvious that the possibility exists that an infectious agent can make its way elsewhere.

If we look at many of these African countries, we realize that many of the diseases are taking a large toll. In Sierra Leone, for instance, it is assumed that AIDS has infected almost half of all soldiers.[7] Perhaps this fact alone will prevent war. If soldiers are infected maybe they will be too weak to fight. Wishful thinking? Perhaps.

It appears to me that there are three possible solutions to prevent disease that involve Africa.

1. Do we leave things as they are?

2. Should Africa be Quarantined?

3. Do we invest a trillion dollars to help build infrastructure in the countries there?

1. Do We Leave Things as They Are in Africa? (All this was mentioned in the first edition)

Many Americans and Europeans, with monetary means, probably want to leave things the way they are. Why? Because they are living well with all their financial, medical and housing needs met, and they do not care about others. Now let us discuss the state of health care in many African villages. Sub-Saharan Africa has millions of people suffering from preventable and treatable diseases. The difference between the situation in Africa now and that of many Americans is the fact that the United States has the equipment and doctors. However, what this nation lacks is access for everyone who needs it. Africa has neither the equipment nor proper access. I mention this comparison because as I write this, the United States is debating health care because insurance companies have lobbied the government to ensure that their astronomical profits remain intact to the detriment of patients. See health care later in this chapter for more on this topic.

Africa has one quarter of the disease burden of the world but has 3% of all health care workers.[8]

It is estimated that 820,000 doctors and nurses are needed to provide adequate health care. Six hundred medical and nursing schools would be required to meet these needs.

I will never forget one picture that I saw in a book. The photo, taken in Kikwit, Zaire showed ten beds in one room and a dead Ebola virus patient lying on the floor while another patient was surrounded by four people who were all wearing masks. Would a situation like this ever occur in the United States or in a European country? Of course not. Therefore, rich countries cannot just sit by and watch this happen.[9] Some may be preoccupied with the profits of Wall Street.

However, wealthier countries need to instead shift our focus to the main streets or primary dirt pathways of the world.

Of course, the old expression "Out of Sight, out of Mind" applies here. If most people do not see or know that this is happening, how can they be aware of it or do something about it?

However, if the status quo remains, then we are perhaps looking at a human-induced disaster the likes of which the world has never seen. And what if something happens in Africa that leads to a new disease outbreak?

2. Quarantine Africa?

What exactly does this mean? Simple. No one either will be allowed in or out of Africa. Why? Because that continent must contain the disease or it could easily spread to other countries or entire continents. In fact, the disease may not have been identified yet. This can conceivably happen as a possible consequence of evolution. (See

Chapter 2). With airplane travel, everyone on the plane, or even at the airport, can be infected. Today anyone can come on a plane with a disease during its latency or non-infective period. Then the passengers and crew who pick up the infectious agent can help it spread just by doing their jobs.

So, theoretically, one solution to the problem would be to simply prevent Africans from leaving their continent and forbidding residents of other continents to travel there. Obviously, this would have many practical disadvantages, not the least of which would be the short-sightedness of assuming that diseases are *their* problem, not *ours*.

3. Do We Invest a Trillion Dollars to Help Build Infrastructure in Africa?

In order to prevent a catastrophe, those of us in the rich part of the world, must make every effort to make poor countries livable for their own people. We cannot let these countries disintegrate into total chaos. Countries like Liberia and Sierra Leone which are on the verge of total collapse because of internal warfare may be helping to spread disease. With poor sanitation and virtually no infrastructure, such as roads, running water, or hospitals, it is very possible that many people will not go for medical care, and diseases like Lassa Fever and other viral diseases may spread. It is imperative that we in the rich countries spend billions of dollars on helping these African countries get their infrastructure rebuilt.

It is important to note that there are approximately fifty-four countries in Africa, many of which are in a similar state as Sierra Leone and Liberia. If we continue this process of ignoring these countries and allowing their populations to struggle in extreme poverty, then we are looking at total elimination of the human race. How? If a person carrying a disease flies across the Atlantic and winds up in New York, he may help spread a disease. Then that disease will spread among the populations of New York State and other states. The destructive possibilities are limitless. This may sound like a science fiction movie. But it is not.

In fact, it is unconscionable that numerous affluent CEOs and others ignore Africa just so they can have whatever they want to enrich their bank accounts. Maybe if the media, whether it be the internet, newspapers, TV, radio or whatever, would report Africa's plight, then perhaps we can raise the money needed to help their infrastructure. Africa is ripe with uncountable diseases. Just imagine if we here in the United States would begin to come down with many of the highly contagious illnesses mentioned in Section 2. What would life be like? Now try to envision what the local African citizens go through every day.

END WARFARE SO THAT WE CAN CONCENTRATE ON MORE PRESSING ISSUES

It may not be so bad if disease were the only problem. Unfortunately, Africa not only contends with plagues, but with rampant, bloody warfare. During the past 50 years Africa had wars, civil or otherwise, as well as real plagues caused by disease organisms. Some of the countries involved with these vicious battles are Angola, Liberia, Sierra Leone, the Democratic Republic of the Congo, and others. See Table 13-1. These countries are rich in diamonds and other minerals. So, these wars are fought over control of these valuable, natural resources.

Additionally, wars exist here because of the lack of value for human lives. Finally, the environment here is not a priority to many, with little concern for the trees, animals, bacteria, or fungi which are innocent victims of these brutal wars.

TABLE 13-1

AFRICAN COUNTRIES PRESENTLY AT WAR OR HAD RECENT WARS

Angola	Mozambique	Ethiopia	Eritrea	Sudan

Liberia	Sierra Leone	Democratic Republic of the Congo	Chad	Algeria
Rwanda	Tanzania	Uganda	Senegal	Guinea
Ivory coast	South Africa-apartheid	South Sudan		

REASONS FOR WARS:

There are three basic reasons for wars in Africa.

1. Struggle to gain independence from colonial rule.
2. Religious, ethnic, or racial identity.
3. Perceived wealth, such as diamonds and other minerals.

Anyone who has read about history or current events knows that some variation on these themes has led to wars not only in Africa, but throughout the world, century after century. Political and territorial disputes, conflicting beliefs and identities, desires for more resources, profits, or control are all reasons for battles. Now, does it really matter what the causes of war are? No, it does not, As a friend of mine once said, "*The only sure thing about war is that somebody is going to get killed.*" As you can see, he was 100% right.

WHY THE CONCERN?

Those who fight these wars generally forget one thing: "When you a make war on humans you make war on nature." This is especially true in Africa where there is a tremendous richness in biodiversity, which maintains the planet's life support systems.

Without biodiversity we would not eat food or breathe oxygen. When we make war, we are destroying the infrastructure of these countries. This includes water treatment facilities, water wells, hospitals, schools, and roads. Without infrastructure people are more stressed; ultimately, this will result in a diminished quality of life, as well.

As of May 19, 2021 the amount of money spent on wars since 2001 was $5,372,599,342,146 at 22;40:11 (10:40PM). At 22:47 May :19 (10:47) PM 5,372,602,014,381 the amount of money increased by $2,672,235 in roughly seven minutes. This means, assuming that the web site is correct, the amount spent during seven minutes is over two million dollars.[10] It is hard to fathom. But it is the truth.

Think of what can be done with this money in poor countries. Wouldn't it be better to have invested this money in helping many of Africa's countries such as the Congo, Rwanda, Uganda, and many other areas such as the Gaza Strip in Southern Israel? In my opinion this money should have been used to build infrastructure such as schools, water distribution systems, sewage systems, and local clinics loaded with state-of-the-art medical equipment. The latter would include X-ray machines and heart monitoring equipment, on-site labs to check blood for diseases, and all other medical amenities that wealthier nations take for granted.

MAKE POOR COUNTRIES LIVABLE FOR THEIR OWN PEOPLE

For many people who live in the West (another name for developed countries) the plight of the average African, Asian, and South American (Latin American) is never on their minds. Although many in the West are relatively wealthy compared to the less fortunate in other countries, they do not seem to care about the plight of the people in the developing nations. Furthermore, why does the United States sit idly watching Mexico, as it implodes? In fact, the United States government was about to place a fence between Mexico and The USA, when Trump was president to keep illegal Mexican immigrants out of the United States. Many poor men, women, and children risk their lives to come to the United States? Why would they do this?

It is because they want the better opportunities that America offers. At least they believe that a finer life is possible here.

I grew up never having to experience what it is to be hungry, cold, or homeless. This helped me to be grateful and to think of other people. Now after seven decades plus, I continue to genuinely care about the plight of poverty-stricken nations. In fact, I get a little angry when I see endless waste, especially when restaurants throughout the U.S. throw out perfectly good food every day. They do this because they are afraid of being sued if they feed hungry men and women and someone gets sick from it. As a result, there are thousands of people who regularly eat out of garbage dumpsters in the wealthiest country on the planet.

During the 1950's when I was growing up, I remember watching TV. They showed people in poor countries trying to make a living by doing whatever they could. The situation may be even more grave now because of constant wars in Africa and greedy dictators in Indonesia. This leads to some of the poorest men and women stealing.

For example, in Africa, there are some who will poach animals, plants and whatever else they can get their hands on. We see this all the time when poachers are causing the extinction of various wild animals in order to sell their body parts. Specifically, Ivory Tusks are often poached. These used to sell for $125/pound.[11] However, poachers would not receive much money for their efforts because they sell to intermediaries who make the most money.

One can see that if the planet's biodiversity is poached to extinction, life as we know it may also become extinct. Suppose we have a big erector set. We built a tower with all the metal parts. Once it is completed, we take away one screw at a time. Notice that whenever we do this, the structure is weakened. Also, if we take out too many screws, the tower collapses. We can say the same thing with the ecosystems of the Earth. Every one of the millions of organisms has a function. We may not know what every function is, but we will always be on point if we say that they have an ecological function.

This is precisely why we must prevent people from poaching? Speaking of this, we are all currently witnessing a legal form of poaching right now and that is the overfishing of the oceans. This is driven by greed. Fishers want as much money as they can get.

This is why it is imperative that we help stop this reckless activity. What if we could ensure that everyone has the basic necessities of life. How could we achieve this? Have rich people donate a major part of their fortunes to help poor countries rebuild and sustain their communities, infrastructure, jobs, food, clean water, and other necessities that the average citizen needs. At the very least, pass legislation requiring the wealthiest people to pay their fare share of income taxes. Multi-millionaires and billionaires can afford it. After they pay their taxes, they will remain affluent. In 2021 the number of billionaires has increased to 2,755 with a net value of US$13.1 trillion.[12]

Funding from the fair share of the wealthiest one percent's taxes and wealthy governments will ensure safe water distribution systems, life-saving hospitals, good schools, farms, roads, and energy production. Instead of the wealthiest members of our society buying luxury hotels, huge cruise ships and golf courses, they should invest in humanity.

Speaking of golf, as of this writing (Dec. 15, 2009) Tiger Woods has been outed for extramarital affairs. It seems that every woman he has cheated on his wife with is speaking up. Throughout the past eight years, Woods was the highest paid athlete. earning approximately $110 million last year.[13] That means he probably made over one billion dollars. It is a mind-blowing amount. Now there are over seven hundred million people living in poverty worldwide. What if Tiger Woods became a philanthropist, donating $25 million to help build infrastructure in Haiti, South American countries, and some in Africa? Many people would forgive and forget about his infidelity and respect him for helping humanity, we hope.

In addition to Tiger Woods, there are many other billionaires that could help impoverished nations. According to *Forbes*, the top twelve all have $22.5 billion or more. Perhaps the quintessential example of overpaid people is David Tepper, a hedge fund manager. He made a record amount of $4 billion.[14] How incredible it would be if he would donate a large portion of his earnings to countries Haiti and Kenya to develop water supply systems? As of July 2022, we can add the following super rich billionaires to the list. The first five are Elon Musk (worth $237.9 Billion); Bernard Arnault and Family (worth only $149.3 B); Jeff Bezos (worth $141.2B); Bill Gates (worth $124.6B) and Gautam Adani and Family (worth$105.0B).[15]

We should ask the ultra-rich to provide hundred of millions of dollars for developing a healthier Earth. The money taken would be used to supply jobs locally so that local populations do not have to poach their natural resources (animals, plants) in order to survive.

Much of the underdeveloped countries also are involved with wars mainly due to economic reasons. If these countries had their basic needs met, then they would not have to cut, chop, dig and fight simply to live.

If you remember the situation in Somalia during the Reagan administration, the United States had to send in troops with food. They were met with some resistance. Perhaps if Somalia became self-sufficient, they would not have to keep killing and viciously maiming each other in frequent battles. Of course, the root of the problem here was fighting the local war lords.

In another part of the world extreme wealth is prevalent. It is a city called Dubai, in the United Arab Emirates (UAE). Dubai is loaded with excesses. One example is a large indoor ski resort. Yes, an *indoor* ski resort that rivals Aspen, Colorado for snow. However, in Dubai, snow is produced artificially. They also have the only seven-star hotel in the world, the Burj Al Arab.

Also, Dubai has the tallest building in the world at 2640 feet or one-half mile. It has 160 stories and it is named The Burj Dubai.[16]

In fact, developers in Dubai are planning to refrigerate the sand in front of the Hotel Palazzo Versace so that tourists will not burn their feet.[17] In addition, Dubai has created man made islands. Needless to say, living there will be exorbitant. When will this stupidity end?

In the United States, there should be a twelfth cabinet position, which will be responsible for helping developing countries with the necessities of life. This will benefit all of their populations, rather than only the wealthiest.

HUMAN POPULATION CONTROL

In 2010 the world's human population was around 6.8 billion with a net gain of three humans every second to add approximately seventy million new people every year. This comes out to roughly the size of three- and one-half Mexican cities every year. Now in 2021, the world's population is around 7.8 billion.[18]

Overpopulation contributes to increased erosion of the Earth's environment, further depletion of precious natural resources and increased poverty. Al Gore's book *The Earth in the Balance* states that population control is the major solution to the global environmental crisis.[19] This is only common sense, because the majority of environmental problems are caused by people. A historical milestone occurred when the sixth billionth person was born on October 12, 1999.[20] This number is symbolic because it brings further emphasis on the strain on Earth's carrying capacity. Carrying capacity is defined as ***the maximum population an area can sustain without bringing in outside resources.***

Solving this problem requires a more responsible action guided by common sense, something which is lacking in government circles. Countries like the United States can make a major contribution to alleviate overpopulation in the developing world. Yet, instead of common-sense dictating policies toward these areas, such decisions are often made by ideology based on certain religious beliefs that will ensure reelection of the politicians involved in decision making. Now in July 2022 SCOTUS (Supreme Court of the United States) just got rid of Roe v Wade which means that they made abortion illegal. This means more overpopulation. The reader must realize that most of the men on the Supreme Court are right wing idealogues with little knowledge about the environment (authors opinion).

Developed countries like the United States should strive for a far more humane approach that would encourage birth control, contraception devices such as condoms, (now both male and female condoms) family planning, sex education, employment, and access to medical care in many parts of the world.

Perhaps one of the best ways to stop population growth is to only have one child as China has tried to implement. In his wonderful book, *Maybe One*, author Bill McKibben[21] suggests that maybe one child is enough. He maintains that children without siblings are in fact psychologically stable and not necessarily as spoiled as previously thought. Having one child makes sense from an economic standpoint because of less stress on the parents and on the environment. This is essentially a numbers game which benefits the Earth by less consumption, less pollution, and more chance of a sustainable ecosystem as well as a sustainable future.

Pope Francis succeeded Pope Benedict XVI in March of 2013. He has taken a stance against many sensible solutions to alleviating overpopulation such as denouncing birth control.

According to an article in the British paper, The Guardian, *"Indeed, Francis in the encyclical explicitly rejects the idea of population growth as a strain on global resources.*

"Demographic growth is fully compatible with an integral and shared development," [22]. It is obvious that he never took a class in environmental science with me. It appears that the Pope is out of touch with many of his members who suffer from his teachings regarding reproductive rights. This is primarily because adequate birth control is not easily available in developing countries. Instead of preventing such measures by abstinence, the Pope should encourage birth control as a humane solution which alleviates health problems, early death, and poverty, especially of women.

It is for these very reasons that birth control is a necessity for the developing countries. Unfortunately, it is not only limited access that prevents birth control from becoming a reality in underdeveloped regions, but also the lack of health and environmental education. Moreover, there is the widespread problem of illiteracy and adherence to religious doctrine.

Now what does human population growth have to do with disease and Climate Change? Most people like to eat meat, which comes from cows, pigs, and goats. Animals raised for meat put out lots of methane gas (CH_4) from their digestive tracts. In addition, in order to raise meat-laden animals, farmers must raise lots of feed such as corn, hay and other vegetable animal foods. In addition to raising crops, trees are being cut down in the rainforests of the world for human use. These include land for farming, even though rainforest soil is not fertile. They use the land for tree plantations and for cattle raising. When trees are chopped down, termites gobble up the wood residues and they too expel methane gas from their digestive tracts which is produced by protists in their digestive tracts. It is estimated that termites produce up to 20,000,000 tons of methane per year.[23] Do not forget that methane accounts for roughly 18% of greenhouse gas emissions.

These crops consume a lot of water, and that makes food more expensive for those people living on the margins of survival. It would be better to feed poor people the food that we raise for animals, such as corn. If we had fewer people to feed, we would not have many of these environmental problems. Furthermore, utilizing water for crops is not the only problem. In order to increase yields, farmers add fertilizers and pesticides which wind up as food residues. As a result, this causes unwanted growth of algae in lakes and rivers which leads to a process called ***Eutrophication.*** Also, this makes humans sick with pesticide poisoning. When poisons or other xenobiotics, non-natural substances, accumulate in bodies of organisms, including people, it is called *bioaccumulation.* [24]

Fig 13-1 shows the dramatic correlation between population growth and oil production. Because of the oil spill in the Gulf of Mexico in April 2010 many organisms such as crabs and oysters bioaccumulated toxic products from the oil.

If you look at the chart below closely, you will notice that oil production continues rising with population growth. More people need more energy. More oil-based energy means more carbon dioxide. More carbon dioxide means more climate change. More oil produced means more oil spills.

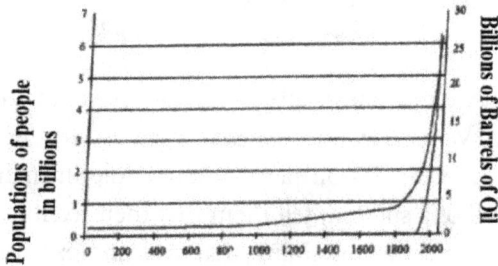

Fig 13-1
Population and Oil Production

Figure 13-1

(Illustration by Kevin Bley)

X axis represents Barrels of Oil
Y axis represents Population in Billions

GET RID OF ALL FOSSIL FUELS FOR ENERGEY PRODUCTION, INCLUDING THE ECONOMIC INCENTIVES

How do we get rid of fossil fuels? We have to convince the whole world that they are a major cause of Global Warming and Climate Change. In April 28, 2010, there was a terrible oil spill going on in the Gulf of Mexico thanks to British Petroleum or BP. This caused immeasurable damage in the waters off of Louisiana. In fact, it was estimated that there were thousands of barrels of oil leaking into the world's waters every day. One barrel of oil equals forty-two gallons. BP is not doing much to prevent these dangerous oil spills that kill so many fish and mammals. Could it be that BP is more interested in saving the Oil pipelines and increasing its profits rather than protecting the planet's ecosystems?

A previous oil spill in the Gulf of Mexico occurred on June 3, 1979 when an oil well named ITOX 1 blew. This occurred in Campeche Bay (Bahia de Campeche). It was owned by PEMEX (Petroleos Mexicans). A total of 140,000,000 gallons (140 million) of oil was spilled before it was brought under control in March 1980.

If we do not change to solar or some other non-carbon dioxide-producing energy source, the world will be imperiled and perhaps made

657 | David Arieti

uninhabitable for everyone due to incredibly high temperatures. This includes rich oil barons and their minions. *Specifically, All Offshore Drilling must be outlawed worldwide permanently. Is anyone listening?*

So, how can we prevent instability in economies that depend on oil like Saudi Arabia, Venezuela, and Nigeria? The answer is not easy, but we must try. In many countries oil plays a major role in maintaining their economy. If we, all of a sudden, tell them that their fossil fuels are useless, then one can imagine the instability that will ensue: bankruptcy, riots, warfare, and total misery for everyone involved.

Since most Middle Eastern countries have lots of sand (silicon dioxide, SiO_2) they can now become rich without using their carbon-based fossil fuels. They literally have a gold mine within the sand because it can be used to manufacture photovoltaic cells. Yes, the very thing that converts solar energy (light, photons) into energy. With electricity, there is power to run almost everything from lights to cars and yes, possibly in the future, spaceships. The latter could beam down electricity from space from large cells absorbing photons.

Campaign financing and oil spills

As we have seen since April 20th, 2010, oil has been spewing out of a pipe due to deep well drilling in the Gulf of Mexico. One reason for this is that the government was hesitant in outlawing *all* offshore drilling because of the economics involved. Oil companies contributed over six billion dollars over the past 20 years to federal campaigns of both major political parties. [25] Specifically, members of Congress were paid off to ensure vast profits in the billions of dollars for the oil industry. As long as the United States allows any form of campaign financing, disasters like oil spills will occur. Anyone in their right mind would not drill in a body of water due to potential accidents like oil drill blowouts. The oceans contain an incredible wealth of life which are always destroyed in spills. Accidents have occurred in Australia and other places ever since deep well drilling has occurred. By now we should know better.

In addition, chemical and pharmaceutical manufacturers have been known to contribute large sums of money in order to be guaranteed large profits at the expense of the environment as well as the average citizen. It is for this reason that all private campaign financing should be abolished rather than upheld by the Supreme Court of the United States. The consequences of their decisions are too dire to contemplate.

Some congressional members, as well as citizens who never took a class in environmental science, are still in favor of drilling because they cannot wait to get to their money-maker. Politicians try to spin it, saying that oil drilling will employ lots of people. They even oppose the sensible moratorium on offshore drilling for six months which President Obama proposed.

However, they are forgetting one essential thing: if the environment goes, then we all go. Why do they not view this from a healthier perspective? Funding solar means creating millions of environmental or green jobs. Don't people realize that every accident involving fossil fuels, including coal mine disasters, are all related to not going green? Is it really necessary to poison the air, land, and water of the planet for profit?

On one of the cable news there is a mother of an oil rig worker killed in an on-site accident. In one breath she says that she hopes that a mishap like this never happens again. However, in her next breath, she states that she still favors more drilling. Didn't anybody learn a lesson from this disaster? Evidently not.

SOLARIZING THE PLANET

What does *solarizing* the planet mean? It involves utilizing solar energy as the main source of energy on the planet. It is ironic that most organisms, especially animals and humans, are a result of solar energy. How could this be? Everyone alive eats food, whether it be from animals or plants. Plants are grown in sunlight. The animals that we eat are mainly vegetarians such as cows. When we eat cows (beef), for example, we eat plants that have been converted to meat. So, we are all already a part of solar energy. Why can't we take the next step and use solar as an energy source to run our machines?

Below are examples of how solar power can be generated. At the present time there are many homes and other facilities worldwide that use solar power, especially photovoltaic cells. Below are examples of how solar voltaic cells work.

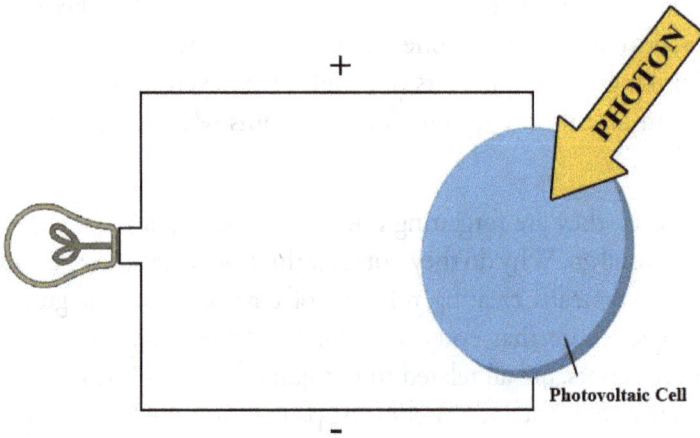

Fig. 13-2. Electricity generated by photovoltaic cells (Illustration by Kevin Bley)

Simply stated, photons (particles of light) hit a silicon wafer made with other elements, and then electrons are emitted, thus generating electricity. Many countries use solar power for many purposes, the most important of which is for the generation of electricity.

In Kenya, 100,000 homes are solarized, enough to produce sufficient power for one television and four light sources.[26] Why can't the rest of the world solarize all of its homes? Maybe it is because the fossil fuel-based industries, specifically oil and coal companies, are hindering development of solar energy in order to hold on to their billions.

EAT LOWER ON THE FOOD CHAIN

Essentially, eating lower on the food chain means becoming a vegetarian. Animals that consume plant material contribute to the depletion of land by adding manure and methane gas to both soil and air. In addition, cattle grazing decimates large areas of rainforest which is essential to the world's climate and oxygen content. Additionally, it poisons waterways with added nutrients, which causes the growth of toxic aquatic organisms such as ***Pfiesteria piscicida*** which was made famous after it's overgrowth in North Carolinas Neuse River. Other than carbon dioxide (CO_2), methane (CH_4) is the second most common gas involved in the greenhouse effect; it is emitted from intestinal tracts and mouths as burps from many animals. See Chapter 4 (page 154-ANIMAL'S CONTRIBUTION TO GREENHOUSE GASES)

Obviously, food is essential to human survival. However, consider the consequences of consuming meat products, which have been a staple of wealthy nations for many years. In order to gain one pound, a person would have to eat ten pounds of food. If you eat this food in the form of vegetable material such as grains, then all you need is ten pounds. However, if you eat ten pounds of cow, you are eating the equivalent of one hundred pounds of grain. Therefore, if we would eat plants (vegetables) directly, we can save 90% of the grain or plant material for human consumption. As a result, we could possibly alleviate starvation throughout the world.

Food chains consist of basically four levels:

1. Primary producers--these are basically plants

2. Primary consumers –these are animals which eat plants (including humans)

3. Secondary consumer--an animal that eats primary consumers

4. Tertiary consumer--an animal that eats secondary consumers--humans, bears and sharks are examples.

By eating lower on the food chain, we save fuel, electricity, seeds for crops, labor, valuable land, farm equipment and other important limited resources. See Figure 1-4, a typical Aquatic Food Chain. Eating less meat is not only logical for the environment, but it is also beneficial for people. In fact, it is especially good for those who live in western or wealthier countries. High cholesterol and heart attacks are common in meat-eating countries. I believe that eating less meat means fewer health problems, especially heart problems.[27] One must remember that the heart is the hardest- working organ in humans. Someone once asked the following question: *If meat is no good, why did God make it so tasty?*

The following animals are raised in factory farm situations: cows, chickens, fish, turkeys, ducks, and geese. There may be others as well. Do not forget: When these animals are raised in terribly horrible conditions they suffer. Many never see daylight. Many breathe contaminated air due to the quantity of flatulence. Some never go outside. It is vile. These animals are born just to die weeks or months later so that they can become a meal for people. For those of you interested in animal factories, which is what they are, go to the People for the Ethical Treatment of Animals (PETA) Website.

STOP USING ANTIBIOTICS IN ANIMAL FEED AND FOR HUMANS

The use of antibiotics in ever increasing amounts has led to the resurgence of many disease organisms immune to old and newer antibiotics. See Chapter 2, go to evolution. Antibiotic sales reached a total of $26 billion in 2002.[28] Because of plasmids, microorganisms are gaining resistance to many antibiotics. (See page 51). This is because of genetic ecology. Microbes exchange plasmids, many containing genes for resistance. One of the main causes of antibiotic resistance is due to animals in feeding lots, better known as Concentrated Animal Feeding Operations (CAFOs).

A good example of the possible spread of an antibiotic-resistant organism is *Campylobacter*.[29] These species are part of the normal flora in many avian species. *Campylobacter* are recognized as being the main causes of bacterial foodborne diseases in many developed countries. Could it be that a strain of *Campylobacter* can mutate to be more virulent than the ones that exist now?

Other examples of organisms that can be spread by feedlots are *Salmonella* species. The Scottish *Salmonella* Reference Laboratory had records that showed that 137 different kinds of **Salmonella** caused disease in humans and cattle between 1974-1976. It was pointed out that only four kinds of **Salmonella** were responsible for 52 percent of the epidemics.[30]

In another example it was shown that by using human antimicrobials such as *lincomycin and bacitracin* in chicken feed, the colonization of Salmonella in the chickens increased. One can just imagine what could happen if these organisms mutate more due to some unforeseen environmental change.

GETTING RID OF MONEY BY GOING TO ANOTHER SYSTEM

As I mentioned in my first book, one of the solutions that I believe should be implemented was that of devaluing money. I did not know how to do it and I still do not, but as long as the world cares more about the stock markets instead of its life support systems, we are all doomed. With so many poor people and the rich who only care about keeping their wealth, is it any wonder that the elite allow so many inequities? These inequities include immense poverty and the old adage:*out of sight, out of mind.*

As I write these words Haiti just had a 7.0 earthquake. In 2004 Hurricane Jeanne passed north of the country, killing three thousand people. In 2008 four storms: Fay, Gustav, Hanna, and Ike, dumped heavy rains on Haiti. The Haitians' plight was made much worse than it should have been because 98% of their rugged hillsides were stripped of their trees. Floods killed hundreds of people due to soil erosion, and there was no natural barrier to stop the water.[31]

How come the wealthy countries of the world did not do much to alleviate their problem when the other hurricanes hit? All they had to do was increase building standards to make them earthquake resistant and reinforce present buildings. If the United States and other wealthy nations had chipped in after the other storms then the damage would have been much less, including the death toll.

Apparently, the United States and other rich countries did not see fit to help Haiti before the earthquake because there simply was no money in it; or put another way, it was not in our interest. I know that this is a cynical attitude, but I believe this to be true.

The United States spent over two billion dollars for the 2008 election, which included money both for the presidential and congressional campaigns. Now here is the point. It would make more sense to have used that money to help poor countries. But what if the world economies did not depend on money? Wouldn't that be great? I ran across an article about this very issue.[32]

It discusses the difference between a doctor and a house cleaner. Should the children of each of those professions have an equal chance in life? I say yes. So how do we do it? I really do not know, but I am sure that we can.

In our world, most people from underdeveloped countries are born into poverty and they will die in it, unless they are extremely lucky. How can everyone get their basic needs met without using money? This is a good and rather perplexing question. We will always have selfish, powerful, obscenely wealthy corporations and people. This means that the poor will always be at the bottom and the rich will always be on top.

The poor will destroy their own nest or home, just to eke out an existence. The previous examples of poachers and people who mistakenly try to clear and farm the rainforest illustrate this. People in Africa and other impoverished areas will kill animals for money. Or they will cut down old and irreplaceable trees for fuel. They do these things in an effort to survive. Some misguided people may think that they are improving their way of life; but they are actually doing the opposite. Without trees, the Earth will lose its lungs.

HEALTH CARE FOR EVERYONE

Why should we be concerned with health care for everyone? As I write this, the United States Congress, without the help of a single Republican, passed a much-needed health care bill. Proper medical care is a basic human right. Health care should begin at birth for all citizens in the county. However, it does not. At least not for millions of people. Perhaps this is why many are so worried. For the uninsured, one major injury or illness that involves a hospital stay and/or surgery can lead to bankruptcy.

If a person remains healthy or practices preventive medicine, then he/she may not get a disease that infects many other people. Being a teacher, I have learned that students or virtually anyone can spread diseases inadvertently. Now here is the problem. What if the hypothetical disease could spread and devastate a population? What if the disease was prevented in the first place by having the hypothetical carrier have an annual checkup at virtually no cost? A disease may be prevented. The situation is so bad here in the United States that sick people who do not have insurance must have fund raisers to treat expensive diseases or face bankruptcy. I myself witnessed one when the husband of a secretary who collaborated with me got electrocuted and had to lose one arm and leg. He needed expensive treatment and they held a fundraiser for him. This is what happens when you have a ridiculous health care system that does not cover many Americans. This includes many adjunct faculty (part time) at colleges in the United States, as well.

Due to an inadequate medical system, some Americans go to other countries to have an operation (medical tourism) or some other procedure performed. This is due to the fact that the cost of everything including the flight, hotel and food would be less than just the procedure alone in the United States. Let us say that an American flies to India for an operation that he cannot afford in the United States because his insurance company will not pay for it. An Indian doctor will prefer to take on an American medical case because even though the American is paying "only" a few thousand dollars, it would be a lot of money for the doctor compared to what he would get from treating the average citizens in India.

This is immensely unfair to the people that were born and raised in poverty ravaged India. The doctor should be practicing his or her trade on the local Indigenous population and not on some wealthy American. Just think what an American would feel like if he lived in

Chicago and needed a highly specialized operation from a surgeon. Let us say that that doctor is busy treating a person who came to the United States just for that particular operation, thereby denying the American citizen his operation. That neglected patient will not be too happy. I believe an Indian citizen would feel the same way if a doctor treated a foreign American instead of him or her.

India has over 1.1 billion people and will soon exceed the population of China. India will need every doctor they have for the local population. Now extrapolate this to any other country and you will see what I mean. This by the way will *not* increase our prestige in any way in other countries. One other possibility may happen. What if the American who just had an operation gets bitten by some mosquito, tick or some other disease carrying vector? The American arrives at Kennedy airport. When he leaves the plane, another insect bites him and picks up the disease. The insect that just bit him will bite someone else and spread the disease. If it is one of the diseases in Section 2, that's unbelievably bad news.

CHANGE ATTITUDES ABOUT LIFE AND ALL ORGANISMS ON THE PLANET

One of the traits of Americans who are reality challenged is the fact that they have strange attitudes towards animals, and the things that they hear about them. In many instances, they have never even seen these creatures. Let me give you an example. When Sarah Palin (former governor of Alaska) was running for vice president of the United States with John McCain, there was an ad put out that could be seen on the internet. It mentioned that Palin advocated killing wolves in Alaska from airplanes because the wolves would kill some of the livestock that belonged to farmers. The ad got quite graphic when it showed a wolf being hit with a bullet in his back. The wolf was in severe pain and screaming. I showed this ad to my environmental science class. I mentioned that it is horrible to kill wolves in such a manner. One female student replied: "*Oh, they are just wolves.*" So, I asked the student, "How would you like to be shot in the back from an airplane?" No reply.

Notice the attitude of the student who said that "they are just wolves." Why should we care? Now she probably never saw an Alaskan wolf in her life. The point here is that Americans and other citizens of wealthy countries have no idea about other animals, people, cultures, feelings, or even other ecosystems. Some people are only concerned with themselves and are trying to get through the day with as much pleasure as possible.

Another Story to Think About

On January 15, 2009, a US Airways passenger airliner landed in the Hudson River because birds got sucked into the engines. Fortunately, everyone survived. The pilot, C.B. Sullenberger, was the hero of the moment. I admit he is pretty impressive. But now comes the interesting part. Remember the news reported that birds were sucked in the engines. Who weeps for the birds?

Are we humans so immersed in ourselves that we forgot that birds-- yes, members of the phylum Chordata like everyone who reads these pages-- are organisms with feelings? The answer is unequivocally *yes*. Maybe someone should have had a memorial for our feathered friends just to make a point. I mentioned this story to my class in another school and another student said: *"They are just birds."*

Now ask yourselves: What functions do these birds have? Well, imagine that the birds who were sucked into the engines of the US Airways flight that landed in the Hudson eat virus- infected insects. Let us say that one of the sucked-in birds was ready to eat an infected mosquito. What if that the mosquito was infected with a virus and instead of being eaten by a bird, bit a person and spread the infection. It sounds simple, doesn't it? Now you know the importance of the birds.

An exceptionally good example of these situations is mentioned in the book *On A Beam of Light* written by Gene Brewer. It is a sequel to his book *K-Pax* which was made into a movie in 2001 starring Kevin Spacey and Jeff Bridges. In a section of the book entitled "The wisdom (or craziness) of Prot," the protagonist is the hero of the book who has made brilliant statements while making an appearance on TV. Three of his many quotes are as follows: [33]

1. *Do not blame the politicians for your problems. They are merely a reflection of yourselves.*

2. *Many humans feel sorry for the dolphins who are trapped in tuna nets. Who weeps for the tuna?*

3. *Hunting is not a sport; it is cold-blooded murder. If you can outwrestle a bear or chase down a rabbit, then you can call yourself a true athlete.*

Needless to say, I was astounded when I read these words. They pretty well sum up attitudes of some people. I bet if the people with these attitudes had to walk two or three miles a day just to get water for their needs, as people in many parts of the world do, including Africa, they would have a vastly different outlook.

ENVIRONMENTAL EDUCATION FOR EVERYONE

Perhaps the best way to help end the environmental crisis with its associated problems is environmental education. Children *and* adults *must* be educated on this topic in order to assure the future of humanity. This education must be a priority all over the world. If people understand that there is more to life than merely pursuing wealth, eating, sleeping, being entertained, and going to the store to buy things, then they will have a better environmental consciousness. Most of the packaging that comes with their purchases can easily be recycled. First, however, they must realize that everything they buy is manufactured from raw material that comes from nature. In fact, this includes some products engineered by human beings, such as plastics and synthetics. As consumers who use these items in massive numbers, this illustrates one more way that we are dependent on petroleum. Therefore, it is imperative for people to learn how everything in nature is not only connected, but fragile and finite. Courses in Ecology and

Biology will lead to a greater understanding of the role of humanity in nature and the importance of preserving our planet, rather than pillaging from it.

Even now, without the benefit of widespread environmental education, many children are aware of the importance of recycling. Additionally, they know about wasteful consumerism (culturally influenced), and the depletion of natural resources. Occasionally I encounter this awareness with young adults while teaching my classes. Children have been known to educate their parents on environmental matters. It is this attitude that must be further encouraged and incorporated into our education system, even as early as kindergarten or grade school.

LOWER THE ENVIRONMENTAL FOOTPRINT OF PEOPLE WITH TOO MUCH STUFF (BECAUSE THEY HAVE LOTS OF MONEY)

Being an American I have grown up in both the city and suburbs. In the suburbs one sees many car garages. Most of these garages are filled up with lots of stuff such as tires, drawers, lawnmowers, old newspapers, books, bookcases, chemicals such as paint and paint thinner, oil for car engines and basically a whole mess of things. Americans are prone to hoard lots of stuff that becomes unused except for being stored in closets or garages. Needless to say these items are never used again. (See Fig 13-3)

Fig. 13-3 Compulsive hoarding in an apartment[34]

How do you lower the footprint of rich people all over the world? Tough question. Maybe ask them why they need so much stuff. From Figures 13-4a and 13-4b one can see the discrepancies between the rich and poor in Los Angeles, California in the United States. Just look at the photos and figure out who owns more stuff which overwhelms much of the planets land masses by cutting, chopping and digging to get raw materials for building large homes without giving it a thought about how oversized houses weaken the Earth's environment. I believe the two photos below says it all.

Fig.13-4a Tents in Los Angeles Skid row. Photo by Russ Allison Loar[35]

Fig 13-4b An affluent house in Los Angeles.[36] Photo by Atwater Village Newbie (Eric Beteille)

IMPLEMENTATION OF THE EARTH CHARTER

What exactly is the Earth Charter? "The Earth Charter is a declaration of fundamental principles for building a just, sustainable, and peaceful global society in the 21st Century."[37] The Earth Charter was written by an international committee and approved at a UNESCO meeting in Paris in March of 2000. The Earth Charter was influenced by contemporary science, international law, Indigenous cultures, and various religious and philosophical traditions.

According to the brochure, the importance of the Earth Charter is that it: "…challenges us to examine our values and to choose a better way. It calls on us to search for common ground in the midst of our diversity and to embrace a new ethical vision that is shared by growing numbers of people in many nations and cultures throughout the world.[38]"

Finally, it is reassuring that there are many others who want to save the environment. The fact that there are large UN-based organizations on our side should give us encouragement and the realization that there is hope.

The following is a list of the sixteen principles of the Earth Charter:

1. Respect Earth and life in all its diversity.
2. Care for the community of life with understanding, compassion, and love.
3. Build democratic societies that are just, participatory, sustainable, and peaceful.
4. Secure earth's bounty and beauty for present and future generations.
5. Protect and restore the integrity of Earth's ecological systems, with special concern for biological diversity and natural processes that sustain life.
6. Prevent harm as the best method of environmental protection and when knowledge is limited, apply a precautionary approach.

7. Adopt patterns of production, consumption, and reproduction that safeguard Earth's regenerative capacities, human rights, and community well-being.

8. Advance the study of ecological sustainability and promote the open exchange and wide application of the knowledge acquired.

9. Eradicate poverty as an ethical, social, and environmental imperative.

10. Ensure that economic activities and institutions at all levels promote human development in an equitable and sustainable manner.

11. Affirm gender equality and equity as prerequisites to sustainable development and ensure universal access to education, health care, and economic opportunity.

12. Uphold the right of all, without discrimination, to a natural and social environment supportive of human dignity, bodily health, and spiritual well-being, with special attention to the rights of Indigenous peoples and minorities.

13. Strengthen democratic institutions at all levels, and provide transparency and accountability and governance, inclusive participation in decision making, and access to justice.

14. Integrate into formal education and life-long learning the knowledge, values, and skills needed for a sustainable way of life.

15. Treat all living beings with respect and consideration.

16. Promote a culture of tolerance, non-violence, and peace.[39]

CHANGE THE WAYS HUMANS THINK

Since the publication of the first edition, I was becoming more upset and disenfranchised with the human race. Therefore, I want to focus on four issues that merit our prompt attention. They are the following:

- Warfare
- Deforestation
- Dying coral reefs
- Destroying wildlife for profit

Warfare

We all know the degree of damage to the planet after our major world wars. We saw mass destruction of virtually everything in some countries with nothing, except rubble. You would think that as members of the human race, we have seen enough of this carnage. However, as I write this (June 21, 2021), havoc continues. So, what places on Earth have seen constant bombardment for the past 75 years? Israel and The Gaza Strip. In fact, Israel was established by the United Nations in 1948; but the Arab countries refused to recognize Israel as an Independent state. So, even after three major wars, and a war of attrition, the fight goes on. Figures 13-5 and 13-6 shows what Gaza can look like without constant tit- for tat by both the Israelis and the Palestinians.

President Putin of Russia invaded The Ukraine in February of 2022 with the intention of destroying it and claiming that it shouldn't exist. Today (16 July 2022) the news mentioned that an apartment house in Vinnytsia, a city in west-central Ukraine was hit by a missile. The bombing killed many people including Liza, a girl with Down's syndrome on her way to a speech therapist with her mother. This is just one of horrible, sad stories occurring on our planet since humans were created on the Earth. The fact that we still have this insane attitude about warfare does not bode well for the future. Because of the war Ukraine, which is one of the world's top agricultural producers and exporters, is unable to export its food products to countries located in Africa and Western and Central Asia. Fig 13-7 shows a bombed building thanks to Russian airstrikes. (Positively shameful)

Fig 13-5 Gaza before Israeli Bombardment.[40]

Fig 13-6 Parts of Gaza after Israeli bombardment.[41]

Fig. 13-7 A building destroyed by Russian missiles in Mariupol, Ukraine Viewed 9 Jan 2024.[42]

Deforestation

During the past 50 years or so deforestation has occurred worldwide, especially in tropical rain forests because of money. One of the primary reasons for deforestation is for production of palm oil and for raising cows. With this destruction comes the loss of oxygen producing life support systems (trees) and making thousands of animals homeless, especially orangutans in Borneo and Sumatra. (There are now around 107,700 orangutans in Borneo and 7,500 orangutans in Sumatra)[43]

Palm oil is found in the following food and non-food products: bread, soap, shampoo, chocolate ice cream, makeup, cooking oil and numerous other products. However, most humans would rather have oxygen than palm oil.

Fig 13-8 Deforestation[44]

Fig 13-9 –Intact Rainforest[45]

Dying Coral Reefs

Look at these two photos of Coral Reefs. Now imagine what it will look like if the deterioration and neglect of our planet continues.

Fig 13-10 Coral Reef in good shape[46]

Fig 13-11 Bleached coral reef[47]

Destroying Wildlife for Profit

Perhaps the most egregious example of killing animals for profit is the desire by some countries for Ivory. As of 2015 the five major countries interested in Ivory are the following: China, Philippines, Thailand, United States, and Vietnam.[48]

Fig 13-12 Elephants in good shape. Photos by Y .S. Krishnappa, T. Breuer, and M. M. Karim[49]

Rather than show pictures of dead elephants I will show you artwork made with Ivory. Please consider that in order to make this art, it resulted in the death of an elephant.

Fig 13-13. Ivory Plaque with Enthroned Mother of God (Stroganoff Ivory), 950-1025 AD, Byzantine, Constantinople, ivory - Cleveland Museum of Art - DSC08411.JPG[50]

Fig 13-14 Ivory amulet from Alaska, Tlingit, 1820-50, Metropolitan Museum of Art, 1979.206.518.JPG[51]

The importance of these principles cannot be ignored because they are the concern of many environmentally conscious people like myself. Sharing these principles provides a foundation for solving the world's environmental problems in a practical manner. These principles are sound and can be practiced by everyone, from corporate executives and politicians to average people throughout the world, in order to save the Earth.

My list of solutions, many of which concur with the principles of the Earth Charter, do require massive changes to our infrastructure, social behavior, and consumption patterns. By educating the world's public to be more environmentally conscious, these solutions can be gradually implemented before catastrophic events which are now occurring become even more severe.

HAVE BILLIONAIRES USE SOME OF THEIR WEALTH TO SAVE ENTIRE COUNTRIES

According to the 35th Annual Forbes List, the world has 2,755 billionaires.[52] During recent years, I have emailed Jeff Bezos and Elon Musk about donating at least $1 billion to save the Amazon Jungle. Specifically, I suggested that they give the funds directly to individuals, rather than government. Similar to the Buy Back Weapons Programs in the United States, I specified that these billionaires should go to citizens directly, buying back every chain saw for $1000 and then destroy them. This will have the stipulation that they never cut down another tree. Otherwise, there could be the punishment of life imprisonment.

It is incumbent on billionaires to help save the planet by utilizing their money in a more productive way. After all, their wealth is greater than the populations of entire countries. In addition, those who have multiple billions should help save the planet's ecosystem. It is a shame to continue with excess while the planet burns. I am sure that there are many billionaires who do give. However, in order to make a true difference, more needs to be done when saving this planet and all who inhabit it.

Many readers agree with me.

INVENTING CARBON DIOXIDE SUCKING MACHINES[53]

In June of 2017 the world's first commercial plant for capturing carbon dioxide from the air opened in Switzerland. It's called the Climeworks AG facility near Zurich. It can capture 900 tons of CO_2 Annually. The plant is powered by waste heat from a recovery facility. Fans push air through the filter that collects CO_2. When the filter is saturated CO_2 is separated at temperatures at 100°C.

I believe that we can all assume that there will be bigger and better CO_2 sucking machines soon. No picture that we can use was available. Hang in there and hope.

[1] What Biden's Done for the Climate in His First 100 Days https://www.outsideonline.com/2424442/haaland-suspends-anwr-oil-leases

[3] Richard Preston is perhaps one of the best writers at the end of the year 2000. I consider him the John Steinbeck of the new century.

[2] He may also be good for the doctors who work on Republican patients because when Mr. Obama won the presidency their blood pressure went up, and they had to see doctors about that problem.

[4] Author of *The Hot Zone*. It is about the arrival of the Ebola virus in The United States.

[5] Preston, R. *Panic in Level 4*. Random House, 2008, p.72.

[6] Donaldson, Ross, *The Lassa Ward*, 2009, St. Martin's Press, p.68.

[7] Op. cit. Ross, p.101.

[8] http://blogs.driversofchange.com/globalvillage/2007/11/health care in africa.html Viewed October 4, 2009.

[9] Brookesmith, Peter. Biohazard: *The Hot Zone and Beyond: Mankind's battle against deadly diseases*. Barnes and Noble, 1997, p.94.

[10] https://www.nationalpriorities.org/cost-of/?redirect=cow

[11] http://www1.american.edu/ted/elephant.htm viewed 18 Nov. 2009.

[12] https://en.wikipedia.org/wiki/The World%27s Billionaires Viewed 19 May 20, 2021

[13] http://www.forbes.com/2009/06/17/top-earning-athletes-business-sports-top-earning-athletes.html. Viewed 14 December 2009.

[14] http://business.timesonline.co.uk/tol/business/industry sectors/banking and finance/article7085137.ece viewed May 27, 2010

[15] Forbes, 2022. The World's Real-Time Billionaires. https://www.forbes.com/real-time-billionaires/#64efedc53d78 Viewed 9 July 2022.

[16] Boyle, C. Burj "Dubai, now the world's tallest building at over 2,600 feet, to open in United ArabEmirates"
http://www.nydailynews.com/news/world/2010/01/02/2010-01-_02_burj_dubai_now_the_worlds_tallest_building_at_around_.html

[17] Kantar,James. "GlobalTourism and a chilled beach."
http://greeninc.blogs.nytimes.com/2009/01/02/global-tourism-and-a-chilled-beach-in-dubai/ viewedJanuary 2, 2010.

[18] https://www.worldometers.info/world-population/ viewed 19 May 2021

[19] Al Gore, *Earth in the Balance*. New York: Houghton/Mifflin, 1992, p.307.

[20] Bradley Keoun, "Of Earth's 6 Billion.0.11% call this Home," *Chicago Tribune,* October 12,1999: 1,12.

[21] McKibben, Bill. *Maybe One.* Simon and Schuster, 1998.

[22] https://www.theguardian.com/world/2015/sep/24/popes-climate-stance-is-nonsense-rejects-population-control-_says-top-us-_scientist#:~:text=Indeed%2C%20Francis%20in%20the%20encyclical,development%2C%E2%80%9D%20the%20pope%20wrote 2015. The Guardian. Viewed 9 July 2022.

[23] Holland,D. **HOW A TERMITE'S MOUND FILTERS METHANE (AND WHAT IT MEANS FOR GREENHOUSE GASES). University of Melbourne. Pursuit**
https://pursuit.unimelb.edu.au/articles/how-a-termite-s-mound-filters-methane-and-what-it-means-_for-greenhouse-gases#:~:text=Globally%2C%20it%20is%20estimated%20that,ends%20of%20these%20humble%20insects. Viewed 14 July 2022.

[24] Bioaccumulation is the process whereby toxins accumulate in the fatty tissues of organisms including humans. They may lead to illnesses such as cancers and mimic the effects of hormones.

[25] http://www.grist.org/article/2010-05-05-bps-donations-to-congress-more-worrying-than-donations-to-obama/ viewed June 2,2010.

[26] Kinyanjui, Kui. Kenya: Chinese Firm to Invest Sh9 Billion in Solar Power Plant. In *Kenya London News.* http://kenyalondonnews.co.uk/index.php?option=com_content&task=view&id=3878&Itemid=44viewed 9 November 2009.

27 http://www.goveg.com/heartdisease.asp viewed March 31, 2010.

28 http://www.asianresearch.org/articles/2717.html Viewed March 31, 2010.

29 Iovine, N. and Blaser, M.L., Antibiotics in Animal Feed and Spread of Resistant *Campylobacter* from Poultry to Humans. CDC. http://www.cdc.gov/ncidod/EiD/vol10no6/04-0403.htm. Viewed April 6, 2010.

30 Lappe, Marc. *Germs that will not die.* Anchor Press/Doubleday, 1982, pp.133-134.

31 Masters, Jeffry. "Hurricanes and Haiti: A Tragic History." http://www.wunderground.com/education/haiti.asp viewed Jan 24, 2010.

32 http://anthologyoi.com/blogish/beyond-the-socialist-dream-a-money-less-society-part-i.html. Viewed March 29, 2010.

33 Brewer, Gene, 2001. *On a Beam of Light,* St Martin's Press, p.two hundred.

34 Compulsive Hoarding. Wikipedia. https://commons.wikimedia.org/wiki/File:Compulsive_hoarding_Apartment.jpg Photo by Grap Viewed 17 July2022.

35 https://commons.wikimedia.org/wiki/File:Tenting_in_Los_Angeles_Skid_Row.jpg Viewed 17 July 17 2022.

36 https://en.wikipedia.org/wiki/Economic_inequality#/media/File:The_Manor,_Holmby_Hills,_Los_Angeles,_in_200_8.jpg Viewede 17 July 2022.

37 The Earth Charter brochure, from the Earth Charter International Secretariat. n.d. San Jose, Costa Rica.

38 The Earth Charter brochure, from The Earth Charter International Secretariat. nd. San Jose, Costa Rica.

39 Ibid. For more details about these environmental principles see www.earthcharter.org or write to themat The Earth Charter Fund, P.O.B 648, Middlebury, VT 05753.

40 https://commons.wikimedia.org/wiki/File:Gaza_City.JPG Viewed 6/11/21
41 https://commons.wikimedia.org/wiki/File:20140805_beit_hanun7.jpg Viewed 6/11/21

[42] https://en.m.wikipedia.org/wiki/File:Mariupol_destruction.jpg

[43] WWF (World Wildlife Fund) https://www.worldwildlife.org/species/orangutan#:~:text=A%20century%20ago%20there%20were,was%20announced%20in%20November%2C%202017.
Viewed 16 July 2022.

[44] https://upload.wikimedia.org/wikipedia/commons/a/a9/Illegal_Deforestation_Plantation.jpeg Viewed 13 June 2021
https://commons.wikimedia.org/wiki/File:Borneo_rainforest.jpg Viewed 13 June 2021

[45] https://www.youtube.com/watch?v=0fts6x_EE_E

[46] https://commons.wikimedia.org/wiki/File:Coral_reef_98.jpg Viewed 13 June 2021.

[47] https://commons.wikimedia.org/wiki/File:Bleached_coral,_Acoropora_sp.jpg Viewed 13 June 2021

[48] https://www.nationalgeographic.com/pages/article/150812-elephant-ivory-demand-wildlife-trafficking-china-world Viewed 13 June 2021.

[49] https://commons.wikimedia.org/wiki/File:Elephant_Diversity.jpg

[50] https://commons.wikimedia.org/wiki/File:Ivory_Plaque_with_Enthroned_Mother_of_God_(Stroganoff_Ivory),_950 -1025_AD,_Byzantine,_Constantinople,_ivory_-_Cleveland_Museum_of_Art_-_DSC08411.JPG Viewed 13 June 2021.

[51] Ivory amulet from Alaska, Tlingit, 1820-50, Metropolitan Museum of Art, 1979.206.518.JPG Viewed 13 June2021

[52] Dolan,K. 2021. Forbes World Billionaires List-The Richest in 2021.

[53] https://www.science.org/content/article/switzerland-giant-new-machine-sucking-carbon-directly-air Viewed 19 July 2022.

Epilogue

To begin with the epilogue, it is incumbent on me to show some of the recent headlines that typify the times that we live in:

Most of the US should get ready for a warm winter.[1]

The Climate Crisis is Here Now, Experts Warn, as Death Tolls From Summer Disasters Mount[2]

Terrifying climate change map pinpoints 3 MILLION homes to be flooded by 2050[3]

Climate crisis worst-case scenario laid bare - Expert outlines impacts hitting globe now[4]

End of the world: David Attenborough sends 'catastrophic' warning: 'Not false alarm!'[5]

End of the world as 'triple planetary crisis' threatens humanity: 'Quick action needed'[6]

Vapor Storms are Threatening People and Property[7]

One of the most frightening headlines was released just as I was sending this chapter to the publisher.

Congo to Auction Land to Oil Companies: 'Our Priority Is Not to Save the Planet'[8]

(Peatlands and rainforests in the Congo Basin protect the planet by storing carbon. Now, in a giant leap backward for the climate, they're being auctioned off for drilling.)

I want to point out that this is just a small percentage of the number of articles written about climate change. But this should alert the reader that this is just the tip of the iceberg. That means that if we don't do something soon, like tomorrow, the planet's life including humans are doomed. A very pleasant and optimistic thought, isn't it?

As I write these sentences the COP26 just wrapped up in Glasgow, Scotland.

This was a conference which included 197 states and was attended by 20,000-30,000 delegates, NGO (Non-governmental organizations), clean energy business leaders, climate activists, 120 world leaders including our own President Biden and finally **503 fossil fuel lobbyists**. The oil and gas companies sent out around 24 people more than the largest delegation had.[9]

The UK wanted those countries attending to be more ambitious than COP 21 held in Paris in 2015.

The goals of the Glasgow Conference (COP26) were as follows:[10]

1. Reach net-zero carbon emissions[11] by the year 2050 and keep global temperatures below 1.5° C increase

2. Protect Ecosystems and habitats

3. Have developed nations mobilize $100 billion for poor nations

4. Collaborate to finalize the Paris Rulebook which sets out rules for the Paris Agreement (2015).

To reach the temperature, increase of no more than 1.5°C. the following actions had to be taken:

- Phase out coal (I say eliminate it totally)

- Prevent deforestation

- Use Electric vehicles

- Renewables such as wind and solar

There were two major agreements which to me are the most important for the future of most life on the Planet Earth:

End Deforestation and methane (CH_4) emissions.[12]

The percentage of global emissions by each country is presented in TABLE E-1 below. The four main emitters of CO_2 are China, The United States, India and Russia. China and Russian leaders were absent from COP26. (Both nations taken together are responsible for 33% of CO_2 emissions.)

Below is a Table E-1 showing the amount of CO_2 in were emitted to the atmosphere by country.

Table E-1 Tonnage and percentage of CO_2 emission by country[13] (2020)

COUNTRY	TONS OF CO_2	PERCENTAGE
China	10.4 billion	29.18 %
United States	5 billion	14%
India	2.5 billion	7%
Russia	1.7 billion	5%
Japan	1.2 billion	3%
Germany	.775 billion	2%
Canada	.675 billion	2%

South Korea	611 million	2%
Saudi Arabia	582 million	2%
Canada	576 million	2%
South Africa	478 million	1%
Brazil	466 million	1%
Mexico	438 million	1%
Australia	411 million	1%
Turkey	405 million	1%
UK	370 million	1%
Italy	337 million	1%
France	323 million	1%

I am sorry to say that the outcome of the meeting was not optimistic and the leader of the meeting, Alok Sharma (an Indian born British minister who was secretary of State for Business in the UK), was visibly upset and apologized after last-minute changes were made to the wording on coal [14](Yes that black carbon containing rock that is the worst form of energy used on the planet) at the end of the conference because he knew that the results were not good.

India wanted the words "Phase Down" instead of Phase Out" when it referred to coal despite the fact that India's capital, New Dehli is one of the most polluted cities in the world, if not the most polluted and yet the Indian government is very adamant about doing something to mitigate air pollution.

NOW FOR THE IRONY OF THE COP26 MEETING (Maybe Hypocrisy), 400 PRIVATE JETS

According to news reports there were 400 private jets containing leaders, presidents, celebrities and others who arrived in Glasgow to attend COP26.

Based on reports, when taken together, these jets spewed out 13,000 tons of CO_2.[15]

Private jets emit 9.57 kg CO_2 per gallon (3.87 L) burnt[16] (2.47kg of CO2 for each liter of fuel consumed.) Irony or hypocrisy. Let's do a little calculating here. According to the article just mentioned, a three hour flight on a Boeing 737 commercial jet will burn 2250 gallons of fuel per hour while producing 21,533 kg (47,472 LBS) of CO_2.

Now let's look at private jet, the Cessna Citation, a very popular jet. It burns 210 gallons of fuel an hour but produces 6,030 kg of CO_2 in a three-hour flight.

Fig E-1 Cessna Jet. Photo by Markus Eigenheer [17]

Israel has invented an airplane that is fully electric. Maybe soon all planes will be electric. It's called the Eviation Alice.[18] (yes, Eviation is spelled correctly)

Fig E-2 The Israeli, Eviation Alice electric passenger airplane.[19] Photo by Matti Blume

So here is the irony. The main purpose of the meeting is to have the 200 countries who participated limit CO_2 emissions. Just going there in jets created more CO_2 which is what the COP26 was trying to prevent.

GRETA THUNBERG

Fig E-3 Greta Thunberg[20](Photo by [Cholérine Derilinoseau](...))

The world's most famous teenager, Greta Thunberg, the Swedish environmental activist, said that COP26 was a failure and turned into a PR event. She also used her favorite group of words, Blah, Blah Blah to describe the facts that delegates were just talking without making real efforts to solve the Climate Change problem. And I'm afraid that she was correct.

According to Greta,

"The COP has turned into a PR event, where leaders are giving beautiful speeches and announcing fancy commitments and targets, while behind the curtains governments of the Global North countries are still refusing to take any drastic climate action."

She also said the following: *"the most exclusionary COP ever,"* *saying those at the sharp end of the climate crisis remain unheard. She added that the event could be considered a "two-week-long* celebration of business as usual and blah, blah, blah."[21]

I believe that the conference should have formulated plans to eliminate 100% the use of fossil- based fuels and encourage the use of solar rather than wind or some other renewable resources which do not do physical or environmental damage to the planet. If the use of solar power dominates as an energy source, it is almost 100% certain that temperature rise of the planet will be kept to a minimum. It is my opinion that they didn't specify solar energy by itself because they were under economic pressures from the oil industry. This is to be expected because of their in-place infrastructure. The April 20th, 2010, oil spill in the Gulf of Mexico just proves my point. See Chapter three of this book where I mention OIL CEOs and certain present- day politicians which are very dangerous because of their lack of environmental education and their loyalties to their political parties.

Currently, we have fantastic scientific achievements such as the internet and sending satellites to photograph other planets such as the New Horizons interplanetary probe which took nine years to get to Pluto in 2015.

See Figure E-4.

Fig. E-4 Photo of Pluto by Dr. Alex Parker [22]

Headline from BBC News:

COP26: Climate deal sounds the death knell for coal power-PM [23].

The headline sounds good, but it is not ideal. We are still headed for doom because the countries only agreed to "Phase down coal" instead of 'phase out coal". Here is what was supposed to happen. Pledges of Greenhouse Gas emissions before COP26 were 52.4 Gt (Gigatons of CO_2); Pledges at COP26 get to this 41.9GT; according to what is needed we are far off. By 2030 we should be at 26.6 GT of CO_2.[24]

It appears that although we are trying very hard with all these COP meetings nothing is really getting done. We try and try and try. If we keep bickering because of our obsession with money and doing things the old way nothing will get done.

India was one of the countries to "Phase Down Coal" instead of Phasing out Coal". The Air Quality Index (AQI) for New Delhi for PM2.5 was **551 (hazardous range)** on Dec 20, 2021.

Considering that clean air contains PM2.5 less than 50 the AQI is way off ; 11 times the AQI for clean air to be exact[25].

In fact, the idea that we still have dirty air due to fossil fuels and we have the ability to get to clean renewable energy sources and don't emphasize them is pure lunacy. It is lunacy taken to the exponent of infinity (L^∞).[26]

Why am I bringing this up? Simple. Because to use the old phrase: Nero fiddles while Rome burns; In this case, world leaders are fiddling around while the Earth (Rome in the phrase) is literally burning up (Global Warming). Hence, the part of the first edition's subtitle…The possible end of humanity? *The possible end of humanity?*

The world is loaded with idiots, especially with political idiots. They don't understand that every yoctosecond (a septillionth of a second 1×10^{-24})[27] we argue about what is going on with our climate, due mainly to fossil fuel companies and their apathetic CEOs, the environment deteriorates. If we don't do anything we move closer to death by some dreaded disease mentioned in Section 2 or some other as yet-to-be-discovered terrible disease-causing organism. (This part was written 12 years ago in 2009 and now we see that a disease is plaguing the world, Covid19). COVID 19 is bad but perhaps not as bad as a future horrible viral or bacterial disease.

In an updated report from The United Nations Climate press release from 25 Oct 2021 there is a dire warning:

"At the same time, the message from this update is loud and clear: Parties must urgently redouble their climate efforts if they are to prevent global temperature increases beyond the Paris Agreement's goal of well below 2C – ideally 1.5C – by the end of the century.

Overshooting the temperature goals will lead to a destabilized world and endless suffering, especially among those who have contributed the least to the GHG emissions in the atmosphere. This updated report unfortunately confirms the trend already indicated in the full Synthesis Report, which is that we are nowhere near where science says we should be," [28]

In another article written by C.J. Polychroniou, Robert Pollin and Noam Chomsky share their insights into the Glascow COP26. In the article there is another warning:

The urgency for action at COP26 cannot be overstated. We are running out of chances to save the planet from a climate catastrophe. But in order for the stated goals of COP26 to be attained, it is imperative that narrow views of national interest be put aside, and great powers steer clear of geopolitical confrontations. Indeed, without international cooperation, the continued use of fossil fuels is set to drive societies across the globe into climate chaos and collapse. [29]

Both Chomsky and Pollin know what must be done on a global basis to save humanity and all organisms on the planet from a climate catastrophe.

During the week that I am writing this paragraph there was an article in "Express" a webpage with news. The article has this shocking headline:

"Horror maps pinpoint EXACTLY where on earth people will die first from climate change". [30]

In this article, the British MET Office (United Kingdom's National Weather Service) published research that showed that one billion people will be affected by extreme heat stress if the rise in global temperatures reaches 2°C. It showed that in almost all continents the impacts of climate change will be expressed as heat stress that will affect all those people.

Most scientists agree that Global Warming is occurring as we speak. Yet there are some naysayers who believe that the mention of Global Warming is a hoax. One person comes to mind. His name is James Inhofe. The senator from Oklahoma. He is trying to convince the American public that Global Warming is one of the greatest hoaxes perpetrated on humanity.[31] He was chairman of the Senate Committee on the Environment and Public Works during the Bush administration. (Inhofe passed away 3 Jan 2023)

Let's assume that Global Warning is not a human made event. It still would be wise to treat it as such and limit our carbon output. The adage, *"It is better to err on the side of caution rather than do nothing."* Perhaps it is wishful thinking that the people in power would do something. I would assume that there are many James Inhofe's in other countries as well. They don't want to recognize the fact of global warming because they fear that if they do something, the economy will suffer. Here is the ridiculous part. Most people don't realize that if you don't have a decent pollution free environment there won't be any need for an economy because everyone will be dead. Remember, the premise of this book is death by disease caused by climate change, and other events caused indirectly by human involvement.

As mentioned earlier, there exists an incredible amount of misfortune and plain bad luck. These misfortunes include the following: disease, environmental refugees, crime, starvation, warfare, homelessness, ridiculous costs of medicine, and perhaps the biggest misfortune of all: STUPIDITY. Remember the statement by author Harlan Ellison,

"The two most common elements in the universe are hydrogen and stupidity."[32]

But perhaps a better statement attributed to Einstein," *The Difference between Genius and Stupidity is that genius has it's limits"*

It seems that the present powers and political leaders (perhaps they should be called *misleaders*) are afraid to take climate change seriously. Perhaps they think that death is not so serious.

Lastly but perhaps the most important concept and idea that can save the planet it is this.

According to Mike Malloy a brilliant commentator, Fascism, war, global warming, hurricanes, death, famine, destruction are all connected. We as members of the human race should realize this and end the destroyers of life on this planet such as nuclear weapons and the continued use of fossil fuels as if they are harmless.

They should be outlawed and abandoned permanently. Will this happen? Probably not because we humans are trapped with our outdated ideas such as the misinterpretation of the second amendment in the United States which states that *"the right to bear arms shall not be infringed"* or that we must use fossil fuels even though we know that they are destroying the planet because of the mega fossil fuel companies. Let's hope that we humans (specifically the world's politicians) wake up and stop this nonsense before we all get destroyed.

One aspect of famine is being played out now as Vladimir Putin is attacking the Ukraine. He doesn't realize that he is jeopardizing the health and nutrition of much of Africa and other countries because they depend on the Ukraine for wheat, barley, and corn.

The area of Russia is 6.602 million square miles (17,099,101.5 square kilometers) where the area of the Ukraine is 233,031 square miles (603,547.5 square kilometers). Ukraine is slightly smaller than the combined areas of France and Germany with a combined surface area of 348.081 thousand square miles or 901,530 square kilometers.

Why does Russia want so much land? The entire planet has 57,308,738 square miles of land. Russia already has around 11% of the earth's land mass. The continental United States makes up 6.6% of the Earths land mass.

Russia is only 28.33 times bigger than Ukraine. Why does Russia want so much land? I know. Because of Putin's ego.

...UNLESS!!

What's an unless? This is the question that a boy asks the Once-ler (who represents the greedy CEOs of polluting corporations) in the 25 minute movie, "The Lorax" as he passes a ring of bricks with a sign, **UNLESS**. "The Lorax," an animated thriller (by Dr Seuss) is about the destruction of all three major areas of the environment: the land, the water, and the air.

It is a brilliant document about the destruction of the environment and its consequences on the biota of the planet. The basis of the cartoon is as follows: A strange creature (whose name is Onceler, pronounced "wonce-ler") and who represents the greedy CEOs and Wall Streeters, whose head we never see, comes to town, and finds a certain tree called a Truffula tree. He realizes that he can make something called a THNEED. (Notice the word *THNEED* where the word _need_ appears). A thneed is a shirt, a sock, a hat, or essentially any item that everyone needs. A monkey-like creature called a *Lorax*-- who represents someone like me, the Worldwatch Institute and all the environmental organizations in the world-- pops out of a tree stump and tries to convince the Onceler to stop cutting down the Truffula trees. As the reader can guess, the end of the movie/story shows a smog-filled planet where the Lorax has moved elsewhere. The only hope left remains in the hands of a boy who comes along. The Onceler gives the boy the last seed in the world of a Truffula tree which is symbolic of all the life support systems of the planet and encourages him to plant it. Hence the following: [33]

"The boy: What's an unless?

The Onceler: Just a faraway word; Just a faraway thought;

Unless someone like you cares a whole lot, nothing is going to get better--it's not." The following is from me, David Arieti:

If world leaders don't care about the sick, poor of the world, the homeless, people who own nothing, or all the biota of the world to include all the bacteria, fungi, plants, animals, protists and viruses (some are probably good), the situation will continue to decline... Unless we take action NOW!!!!

Just as this book was being readied for publication a storm named Daniel hit a Libyan town named DERNA and flooding killed over 20,000 people.

Politicians considered it a natural disaster, but experts claimed the extensive damage and deaths were due to years of corruption, poor maintenance of public structures and years of political infighting.[34]

But wait. Now comes the quintessential reason why the human race needs readjustments and why it should win the DARWIN AWARD*. Not only did the residents of Derna try to escape the floods from broken dams but **THEY HAD TO AVOID DISPLACED LANDMINES LEFTOVER FROM YEARS OF CIVIL CONFLICT BETWEEN 2011-2021.**[35]

Maybe this should be a wakeup call for the world to stop these idiotic conflicts such as wars and concentrate on protecting the planet with more environmentally friendly activities geared towards saving the planet.

*(The Darwin Award is given to people who contribute the most to humanity by removing their genes from the gene pool).[36]

We better not wait for a third edition of this book.......

---

[1] https://www.audacy.com/wccoradio/news/national/most-of-the-us-should-get-ready-for-a-warm-winter Viewed 22 Aug 2022

[2] Tigue, Kristoffer. 2023. The Climate Crisis Is Here Now, Experts Warn, as Death Tolls From Summer Disasters Mount. Inside Climate news. 18 Aug 2023. https://insideclimatenews.org/news/18082023/climate-crisis-here-now-experts-warn-as-death-tolls-from-summer-disasters-mount/

[3] Murrison, P 2021 Terrifying climate change map pinpoints 3 MILLION homes to be flooded by 2050 Express https://www.express.co.uk/news/uk/1510402/climate-change-map-flood-risk-locations-evg Viewed 24 Dec 2021.

[4] Langton,K. 2021 Climate crisis worst-case scenario laid bare - Expert outlines impacts hitting globe now https://www.express.co.uk/news/world/1509532/climate-crisis-global-warming-temperatures-impact-EVG Viewed 24 Dec 2021

[5] Hoare,C 2021 End of the world: David Attenborough sends 'catastrophic' warning: 'Not false alarm!' https://www.express.co.uk/news/science/1511588/end-of-the-world-david-attenborough-catastrophic-warning- climate-change-cop26 Viewed 24 Dec 2021.

[6] Paul, J. 2021 End of the world as 'triple planetary crisis' threatens humanity: 'Quick action needed'. https://www.express.co.uk/news/science/1509212/climate-change-uk-boris-johnson-biodiversity-treeconomy-triple- planetary-crisis-cop26

Viewed 24 Dec 2021. https://www.scientificamerican.com/article/vapor-storms-are-threatening-people-and- property/ Viewed 24 Dec 2021.

[7] Francis, J 2021. Vapor Storms Are Threatening People and Property.

[8] https://www.nytimes.com/2022/07/24/world/africa/congo-oil-gas-auction.html **Viewed 28 July** 2022.

[9] Mellor, S. 2021. **COP26 is 'crawling with fossil fuel lobbyists,' and they're watering down negotiations, climate hawks warn. Fortune from 12 Nov 2021.** https://fortune.com/2021/11/12/cop26-fossil-fuel-lobbyists- climate-un-shell-exxonmobil-bp/. **Viewed 12 Nov 2021.**

[10] Aljazeera. 2021. **Infographic: COP26 goals explained in maps and charts** Infographic: COP26 goals explained in maps and charts | Infographic News | Al Jazeera

[11] Net Zero refers to balance between greenhouse gases released and those gases removed from the atmosphere.

[12] Baker, S.2021. **2 agreements from the COP26 conference pledge major cuts to deforestation and methaneemissions, as leaders scramble to address the climate crisis. Business Insider, India** 2 agreements from theCOP26 conference pledge major cuts to deforestation and methane emissions, as leaders scramble to address the climate crisis | Business Insider India

[13] https://www.worldometers.info/co2-emissions/co2-emissions-by-country/

[14] 'I am deeply sorry': Alok Sharma fights back tears as watered-down Cop26 deal agreed – video |Environment | The Guardian Viewed **24 Dec 2021.**

[15] Fergueson, J.2021. Private Jets Flying To COP26 Will Spew More CO2 Than 1,600 Scots In A Year. Private Jets Flying To COP26 Will Spew More CO2 Than 1,600 Scots In A Year - Climate Change Dispatch

[16] Coffey,H. 2021. How bad are private jets for the environment**?Independent. (1 Nov 2021.)** How bad are private jets for the environment? | The Independent | The Independent Viewed 7 Nov 2021.

[17] https://commons.wikimedia.org/wiki/File:N682A_Cessna_680A_C680_(18231022444)_(cropped).jpg Viewed 10 Nov 2021.

[18] https://en.wikipedia.org/wiki/Eviation_Alice Viewed 13 Nov 2021.

[19] https://commons.wikimedia.org/wiki/File:Eviation_Alice,_Paris_Air_Show_2019,_Le_Bourget_(SIAE8856).jpg https://commons.wikimedia.org/wiki/File:Eviation_Alice,_Paris_Air_Show_2019,_Le_Bourget_(SIAE8856).jpgiew ed 13 Nov 2021.

[20] https://commons.wikimedia.org/wiki/File:Greta_Thunberg,_March_2020_(cropped).jpg Viewed 12 Nov 2021.

[21] Meredith, S. 2021. **'COP26 is a failure': Greta Thunberg says climate summit has turned into a PR event**https://www.cnbc.com/2021/11/05/greta-thunberg-says-cop26-climate-summit-is-a-failure-and-a-pr-event.html Viewed 7 Nov 2021.

[22] https://en.wikipedia.org/wiki/File:Pluto-01_Stern_03_Pluto_Color_TXT.jpg Viewed 9 Nov 2021.

[23] Faulkner,D. 2021. COP26: Climate deal sounds the death knell for coal power-PM. https://www.bbc.com/news/uk-59284505

[24]Rincon,P . COP26: New global climate deal struck in Glasgow. 2021.https://www.bbc.com/news/world-59277788. Viewed 15 Nov 2021.

[25] https://www.iqair.com/us/india Viewed 20 Dec 2021.

[26] L∞ means Lunacy taken to power of infinity

[27] Another way to look at yoctoseconds is showing the reader this number. One yoctosecond (YS) is equal to aseptillionth of a second or 10^{-24}s

[28] United Nations.2021. Updated NDC Synthesis Report: Worrying Trends Confirmed Updated NDC Synthesis Report: Worrying Trends Confirmed | UNFCCC

[29] Polychroniou, C.J. 2021. Chomsky and Pollin: COP26 Pledges Will Fail Unless Pushed by Mass Organizing. Truthout. 28 oct 2021. https://truthout.org/articles/chomsky-and-pollin-cop26-pledges-**will-fail-unless-pushed-by-mass-organizing/** **Viewed 7 Nov 2021.**

[30] Whitfield, K. 2021. Horror maps pinpoint EXACTLY where on earth people will die first from climate change.Horror maps pinpoint EXACTLY where on earth people will die first from climate change | Nature | News | Express.co.uk. Viewed 9 NOV 2021.

[31] http://en.wikipedia.org/wiki/Jim_Inhofe

[32] https://quotefancy.com/quote/1057200/Harlan-Ellison-The-two-most-common-elements-in-the-universe-are-hydrogen-and-stupidity Viewed 2 Feb 2022

[33] Geisel, T. and A. Geisel. Dr. Seuss. *The Lorax*. 1971, Random House, last page.

[34] https://www.aljazeera.com/news/2023/9/13/libyans-search-for-families-after-catastrophic-flood-kill-thousands?traffic_source=KeepReading Viewed 17 Sept 2023

[35] Libya flooding: Landmines pose new threat, death toll rises. https://www.dw.com/en/libya-flooding-landmines-pose-new-threat-death-toll-rises/a-66837101

[36] https://www.urbandictionary.com/define.php?term=Darwin%20Awards

GLOSSARY

ABIOTIC FACTOR- A nonliving factor in an ecosystem such as air, rocks, gravity, atmospheric pressure, sunlight,etc.

ACID RAIN- Rain with a pH below 5.6.

ADULT (Insects)- The stage in insects when they stop molting.

ALBEDO-Fraction of light that gets reflected from the Earth's surface.

ALGAE- Mainly single cell eukaryotic organisms that lack stems, leaves and roots. Most contain chlorophyll although there are many species that don't photosynthesize.

ALGAEMIA-Is a rare disease in which green algae enter the blood stream. It is observed in cattle, dogs, cats and humans. It is also called Protothecosis.

ALGAL BLOOM- Excessive growth of algae where there can be millions of cells per liter of water.

ANTHROPOCENTRISM-The idea that humans are at the center of the universe and they are in control of everything. Also that everything on the planet is for humans.

ANTHROPOGENIC-This is a term used in environmental science which means caused by humans.

ANTHROPONOSES- Diseases spread from humans to humans.

ANTIBODY- A glycoprotein produced in response to an antigen.

ANTIGEN-A foreign substance that when introduced into a body is recognized by host's immune system. White blood cells may produce antibodies against antigens.

ARBOVIRUS-Viruses spread by Arthropods that cause disease.

ARTHROPODS-These are organisms that have three basic characteristics. They are the following: Jointed legs, a chitinous exoskeleton, and highly developed sense organs. Examples are ticks, mosquitoes, crabs and all insects.

ATMOSPHERE- The technical name for the part of the biosphere containing air. The major components of an unpolluted atmosphere are nitrogen gas (N2) accounting for 78.09%, and Oxygen (O2) accounting for 20.94% of the atmosphere, and Carbon dioxide accounting for roughly .04%. The atmosphere is composed of five main parts differentiated by altitude. They are in order starting from sea level: Troposphere, stratosphere, mesosphere, thermosphere and exosphere. (See Table 1-2)

ATMOSPHERIC RIVER- Regions of the atmosphere that carry water vapor. They are long and flowing.

AUTOTROPHIC- This is a type of nutrition in which organisms photosynthesize. They make their own food from inorganic sources such as carbon dioxide (CO2) and water, nitrates and phosphates. These include plants and algae.

AZASPIRACIDS-These are a group of algal toxins which causes gastrointestinal problems after eating mussels (a type of shellfish). It is not fatal.

BARREL OF OIL (bbl) = 42 gallons (US).

1 BECQUEREL - 1 radioactive decay per second, 1 gram of radium is equivalent to 0.037 TBq (terabecquerel).

BIOACCUMULATION-This means the buildup of chemicals in bodies of organisms. Example: DDT accumulates in fatty tissues of humans. See Biomagnification.

BIODIVERSITY- This term includes the wide variety of genes, species and ecosystems. As of 2001 there were approximately 1.8 million known species of organisms which include bacteria, single celled organisms, plants, animals and fungi.

BIOINVASIONS- A tremendous increase in certain species of plants, animals and other organisms which have a negative effect on ecosystems.

BIOMAGNIFICATION-This is like bioaccumulation except that chemicals tend to move up the food chain. An example is the fact that a chemical may be found in a small algae cell. The algae cell is then eaten by a zooplankter, a little animal, then a fish will eat the zooplankter, then a bigger fish will eat the other fish etc. thus increasing the concentration of the chemical in organisms higher up in the food chain.

BIOME- A large geographical area that has a similar climate, soil, vegetation and animal life. Examples are the tropical rain forest, deserts and grasslands.

BIOSAFETY LEVEL 4- This is a facility in a laboratory where a set of biocontainment precautions are required. These facilities are used to study highly deadly viruses such as Ebola. Biosafety levels range from level -1 (BSL-1), the lowest level like those in a college bio lab to those called Biosafety level 4 like those in institutions like the CDC in Atlanta Georgia and United States Army Medical Research Institute of Infectious Diseases (USAMARIID) in Fort Ditrick, Maryland..

BIOSPHERE- The area of the planet which includes the land, water, and air where most life survives. Technical names are the lithosphere, hydrosphere, and atmosphere.

BMAA-Beta-Methylamino-L-Alanine is a non-proteinogenic amino acid produced by Nostoc, a cyanobacterium. It has been implicated in neurodegenerative diseases such as ALS and Parkinson's disease and Dementia. It was noticed in people on

Guam from seeds of cycad trees.

BOD- The term means Biochemical Oxygen Demand. It is used to describe the cleanliness of waste water before and after sewage treatment. It is expressed as milligrams of oxygen per liter of water. The higher the BOD, the worse the water quality.

BIOTIC FACTOR- Living things in an ecosystem. It includes plants, animals, fungi, bacteria and protists. See Abiotic factor.

BYCATCH- Aquatic animals that are caught by mistake in nets and are usually thrown away and left to die. These include birds, sharks, starfish, and a host of other organisms. A form of collateral damage.

CARBON DIOXIDE- The main gas involved in climate change (CO_2).

CARBON FIXATION- Conversion of carbon into organic compounds by photosynthesis.

CARRYING CAPACITY- The maximum population that an area can sustain without importing outside resources.

CDC- Centers for Disease Control and Prevention, located in Atlanta, Georgia.

CERCLA- Comprehensive Environmental Response Compensation and Liability Act also known as Superfund.

CFCs-Class of compounds containing Carbon (C)Hydrogen (H),Chlorine (Cl) and Fluorine(F).Used in refrigerants.

CHEMOSYNTHETIC- Organisms that get their energy from the oxidation of Hydrogen sulfide gas (H_2S).

CHITIN- A nitrogen containing substance found in fungus, arthropods, and protists. It is the chemical that exoskeletons of Arthropods are made of.

CHLOROPHYLL-Chlorophyll is the main pigment that plants use during photosynthesis to make food and give off oxygen.

CHRYSALIS- A pupal stage in butterflies.

CIGUATERA FISH POISONING-A foodborne illness by eating reef fish contaminated with certain toxins.

CLEAN AIR ACT- The first attempt to reduce air pollution. It was first established in 1963 but was amended in 1970 to reduce emissions from cars and industries.

CLEAN WATER ACT- The Clean Water Act (CWA) is the primary federal law in the United States governing water pollution.

CLIMATE CHANGE- A global change in regional climate patterns mainly noticed after the 1950's caused by increasing CO_2 concentrations.

COPEPOD- A member of the zooplankton and are probably the most abundant animal on the planet. There are around 11,500 known species. They live in salt and freshwater.

CURIE- (Ci) 37 billion disintegrations per second. A unit of radiation exposure.

CYANOBACTERIA- A form of prokaryotic algae which are related to bacteria. They are believed to be the ancestor of the chloroplast. These bloom sometimes and produce toxins that kill humans and wildlife.

DALTON- A unit of mass equal to 1/12 the mass of a carbon-12 atom which weighs $1.660\ 538\ 86\ X10^{-27}$ Kilograms. It's used in Biochemistry.

DEMOGRAPHIC FATIGUE- The growing inability of developing countries with burgeoning populations to cope with new threats to society...such as AIDS, malaria, tuberculosis, land degradation, warfare and other miseries.

DIATOMS- A type of algae that have two valves made of silicon. They are very ornamental. They come in many shapes.

DINOFLAGELLATES-A type of algae that has two flagella and blooms into huge numbers. Many are responsible for potent algal toxins. These are the main culprits responsible for red tides.

DOMOIC ACID- Causes life threatening illness manifested by neurological and gastrointestinal symptoms. It's produced by the diatom *Pseudo-nitzchia.*

EBOLA- An RNA virus. Named after the Ebola River in the Democratic Republic of the Congo.

ECOLOGY-The study of organisms and their physical environment.

ECOSPHERE- Same as biosphere.

ECOSYSTEM- The area where there is an interaction between the living and nonliving.

Living meaning organisms such as bacteria, animals, plants, fungi and bacteria and non-living meaning air, water, atmospheric pressure etc.

ECOTONE- A sharp boundary between two different types of communities. An example is the sharp difference between where a forest ends and a grassland begins.

EFFLUENT- Water leaving a sewage treatment plant. It includes treated and untreated sewage.

ENDEMIC- Found only in one location on the planet. This refers to plants, animals and other organisms which are native to specific parts of the planet. An example is the large tortoises found on the Galapagos Islands and giant redwood trees found only in California.

ENVIRONMENTAL POLLUTION-1) The contamination of Mother Nature by Human Nature 2) The effluence of affluence 3) The contamination of an ecosystem by the addition or removal of something.

ENVIRONMENTAL SCIENCE- The study of the environment

to include all aspects of the environment and their interactions.

ENVIRONMENTAL SUSTAINABILITY- Maintaining ecosystems for future generations.

ERMS -(European Register of Marine Species). An authoritative list of species occurring in the marine environment of Europe. (See WoRMS in the glossary)

ETIOLOGY-The cause of the disease.

EUKARYOTIC ORANISM-Includes all organisms with a membrane-bound nucleus.

EUTROPHICATION- The term used when a body of water, especially a lake, has too many nutrients causing growth of aquatic plants and algae which eventually die and decompose. The body of water will eventually turn into dry land.

GLACIER-a large body of ice that moves slowly down a slope or valley.

GLOBAL WARMING POTENTIAL (GWP) - A measure of energy needed to heat up the environment where Carbon dioxide has a GWP of one and other gases such as methane have a GWP of around 86 which means that methane absorbs heat 86 times that of Carbon Dioxide (CO_2).

GONOCYTES are long-lived precursor germ cells responsible for the production of spermatogonial stem cells (SSCs). Gonocytes relate to both fetal and neonatal germ cells from the point at which they enter the testis primordial until they reach the base membrane at the seminiferous cords and differentiate.

HABs- Hazardous Algal Blooms.

HAEMOPHAGOUS- Feeding on vertebrate blood such as mosquitoes.

HETEROTROPHIC- A type of nutrition where organisms eat preformed organic material such as proteins, fats and carbohydrates.

HFRS- Hemorrhagic fever with renal syndrome. This is associated with the Puumala virus.

HPS-Hantavirus pulmonary syndrome.

HYDROCARBONS- This is a class of chemicals like components of petroleum made up of only hydrogen and carbon.

HYDROSPHERE- The area of the planet that contains water.

IMAGO- The last stage of development of an insect.

INFLUENT- Water entering a sewage treatment plant.

INORGANIC CHEMICALS- These are chemicals that are composed of elements other than carbon. Examples include sodium chloride (NaCl) table salt, H_2SO_4, sulfuric acid, and H_2O (water).

INSTAR- A stage of arthropods, especially in insects between molts.

INVERTEBRATE- An animal without a backbone. (See vertebrates).

IPCC- Intergovernmental Panel on Climate Change. This is an intergovernmental body that evaluates the risk of global warming and climate change.

LARVA- A juvenile form of an organism before undergoing metamorphosis.

LEACHATE- The liquid containing dissolved chemicals. Leachate is formed when water dissolves chemicals, which are usually from landfills or from rain in contact with dissolvable substances.

LITHOSPHERE- The term used to describe the land mass of the planet.

MALACOLOGY- The study of mollusks such as clams and oysters.

MICROCYSTIN-a toxin produced by the cyanobacterium *Microcystis*. It causes digestive problems, liver problems and neurologic symptoms.

MOUSE UNIT - A mouse unit (MU) is the amount of toxin required to kill a 20g mouse in 15 minutes via Intraperitoneal injection. It is used to determine toxicity of algae.

NATC- North Atlantic Thermohaline Circulation / gulf stream.

NEMATODA- These are unsegmented roundworms with about 30,000 species described but it is estimated that there may be thousands more undiscovered. This is a (phylum) group of worms which range in size from less than a mm to more than a meter. They are very important in maintaining the proper functioning of ecosystems. Some are parasitic on plants, animals, and humans but the majority are beneficial to the planet.

NICHE- The role that an organism plays in its environment.

NITROGEN FIXATION-Where nitrogen is converted into ammonia.

NOAA-National Oceanic and Atmospheric Administration.

NON-POINT SOURCE POLLUTION- Refers to pollutants where the source is not identified.

NOXs- This is the term used to identify oxides of nitrogen.

There are three oxides of nitrogen: N_2O, (nitrous oxide); NO, (nitric oxide); and NO_2, (nitrogen dioxide).

NYMPH- Juvenile insects that resemble adults without wings.

OIL-See petroleum.

ORGANIC CHEMICALS- Those chemicals consisting of carbon and hydrogen in covalent linkage. Examples are sucrose (table sugar), ethyl alcohol and starch. There are millions of organic chemicals.

OSHA- Occupational safety and health Administration.

OZONE- A natural gas which filters out Ultra-violet light. It has the formula O_3.

PAN-Peroxyacetyl nitrate-A form of photochemical smog.

PARALYTIC SHELLFISH POISONING- A food borne disease acquired by eating contaminated shellfish.

PARTICULATE MATTER- A particle generally smaller than 100 micrometers in the air.

PETROLEUM-A fossil fuel like coal and natural gas made when algae, zooplankton, animals and plants are covered with sedimentary rock for prolonged periods of time under prolonged heat and pressure. It is mainly composed of hydrocarbons which contain hydrogen, carbon, smaller amounts of Nitrogen, Oxygen, Sulfur and metals.

PFIESTERIA- A toxic dinoflagellate (Alga) which has been associated with fish kills from Delaware to North Carolina (Neuse River).

pH-Measure of the acidity of a substance. pH ranges between 0 to 14. pH below 7 is acidic; pH above 7 is alkaline or basic.

PHOTOCHEMICAL SMOG- Pollution in the atmosphere where many chemicals such as gases and hydrocarbons react together, using the sun as an energy source.

PHYTOPLANKTON- These are photosynthetic single celled algae that make up the plankton community. (See plankton).

PLANKTON- These are generally marine drifters which include both small algae and very small animals such as copepods.

PLASTIC-These are products made of polymers of various elements composed of carbon, hydrogen, oxygen, nitrogen, sulfur and chlorine.

PLUME- A concentrated mass of pollution such as from a

smokestack or a sewage outfall pipe.

POINT-SOURCE POLLUTION- This is a term mainly used with water pollution and air pollution. It refers to an identifiable source of pollution such as a sewage outfall pipe or a smokestack.

PRION-Proteinaceous Infectious particle-responsible for Mad Cow Disease and Creutzfeldt-Jakob disease.

PROCHLOROCOCCUS- A newly discovered photosynthetic cyanobacterium considered to be the most abundant photosynthetic organism on earth. They are found by the trillions in the marine environment.

PRODUCTIVITY-Rate at which organic matter is produced.

PROKARYOTIC ORGANISM- An organism that does not have a membrane bound nucleus. It includes only bacteria.

PROTISTA-Single celled Eukaryotic organisms which are found in both terrestrial and aquatic biomes. Algae and Paramecium are examples.

PUPA- A stage found in insects that undergoes complete metamorphosis.

RED TIDE- This is an algal bloom generally caused by algae known as dinoflagellates.

RICKETTSIA-These are nonmotile , gram-negative, non-spore-forming bacteria that cause many diseases such as Rocky mountain spotted fever, Typhus, Ehrichiosis and other diseases. These are obligate parasites.

RIPARIAN- Refers to areas adjacent to rivers and streams.

SALINIZATION- The buildup of salt in soil due to too much irrigation of fields.

SAXITOXIN- An algal toxin which causes paralytic shellfish poisoning after eating clams and other mollusks.

SEAWEED- This is a term which includes thousands of species of large multicellular algae which are mainly marine. Large seaweed such as kelp provide nursey habitats for fish and other aquatic organisms. The major groups of seaweeds are the red, (Rhodophyta), Brown (Phaeophyta) and the green algae (Chlorophyta).

SEROTYPE- A group of organisms or viruses based on their surface antigens.

SHELLFISH-This is a general fisheries term which includes crustaceans such as lobsters, crabs, molluscs such as clams and oysters and many other species without backbones.

SUBSIDENCE - Downward sinking of land masses.

SUPERFUND - Also known as CERCLA.

Superfund was set up to investigate and clean up hazardous waste sites.

SYNERGISM - The multiplicative effects of two substances taken together rather than each taken separately. Put in another way, one could make the statement 1+1=5. An example is the health effects of smoking and working with asbestos. If one smokes, he may have a 15% chance of developing lung cancer. If a person works with asbestos and doesn't smoke, he may have 12% chance of getting lung cancer, but if the person smokes and works with asbestos his chances of getting cancer can shoot up 92 times.

TROPHIC-DYNAMIC RELATIONSHIPS - Transference of energy by food chains.

TROPHIC LEVEL - Place in the food chain.

USAMARIID - United States Army Medical Research Institute of Infectious Diseases in Fort Ditrick, Maryland..

VECTOR - Something that spreads a disease. A good example are mosquitoes which can spread Malaria.

VERTEBRATE - An animal with a backbone. All vertebrates belong to the phylum Chordata.

VIRUS - A structure that has a core of either RNA or DNA and a protein coat. Many are disease-causing agents.

WoRMS-(World Register of Marine Species)-The aim of WoRMS is to provide an authoritative list of marine organisms. (See ERMS in the Glossary)

ZOONOSIS - Diseases that can be transmitted from animals to humans.

ZOOPLANKTON - Heterotrophic (Those that eat) animals that drift in both freshwater and saltwater.

ZOOXANTHELLAE- A yellowish-brown symbiotic dinoflagellate found in corals and in other invertebrates.

ORGANISM INDEX

A

B

Dracunculus medinensis 206, 207, 402, 403, 404, 413

Dreissena polymorpha 118

D. rotunda 560 (D stands for Dinophysis)

D. tripos 560 (D stands for Dinophysis)

D. variabilis 266 (D stands for Dermacentor)

E

Ebola 41, 53, 54, 55, 137, 138, 140, 142, 150, 168, 205, 216, 319, 320, 321, 322, 323, 324, 325, 327, 328, 399, 643, 644, 646, 680, 701, 705

Ebola virus 41, 53, 54, 138, 140, 319, 320, 321, 322, 323, 324, 325, 644, 646, 680

E. coli 222, 223, 224

Ehrlichia chaffeensis 204, 311

Ehrlichia ewingii 204, 311

Elephantiasis 38, 207, 409, 410, 412, 413, 414, 415, 416, 417, 430

Encephalitis virus 392

Endocrinopathies 495

Equus sp 201

Escherichia coli 37, 222, 223, 225, 226, 248

F

Fibrocapsa japonica 575

Filovirus - Ebola virus (Ebola Hemorrhagic Fever) 319

F. japonica 568, 570

Flavivirus 319, 335, 341, 362, 393

Flexal virus 319

Francisella holarctica 266

Francisella tularensis 265, 266, 268

Fundulus heteroclitus 605

G

Ostreopsis 587

P
(The P stands for Pseudo-nitzchia for the following)

P. australis 529, 530, 531

P. fraudulenta 531

P. multiseriata 531

P. pungens 531

P. seriata 531

P. turgidula 531

Palythoa 586, 587

Panstrongylus geniculatus 198

Papilloma virus 217, 374, 375, 376, 379, 400

Paramyxoviridae 206, 384, 385

Paresthesia 262, 367, 538, 577, 578, 590

Pasteurella multocida 382

Pasteurella tularensis 265

P. concavum 560 (P stands for prorocentrum)

Pediastrum simplex 40

Pediculosis humanus 259

Pediculus capitis 207, 301

Pediculus humanus corporis 301

Peliosis hepaticus 258

Penicillium notatum 39

Peridinium gatunense 525

Perna canaliculus 571

Perna viridis 559

P. falciparum 141, 439, 440, 441, 442 (P stands for Plasmodium)

Pfiesteria piscicida 153, 601, 602, 603, 604, 608, 634, 635, 661

Pfiesteria shumwayae 602

R

Rhipicephalus sanguineus 207, 208, 304, 308

Rhodnius prolixus 449

Rickettsia africae 207, 291

Rickettsia conorii 207, 208, 291, 308

Rickettsia felis 207, 297

Rickettsia prowazekii 207, 300, 301

Rickettsia rickettsii 207, 290, 303, 305

Rickettsia typhi 297, 298

Rifampicin 244, 264, 287, 296, 314

Rift Valley Fever 319, 388, 389, 390

Rochalimae hensalae 257

Ross River Virus 206, 217, 366, 368

Rousettus aegyptiacus 325

S

Sabia virus 319

Sabethes 363, 395

Saccharomyces cerevisae 39

Salmonella 662, 663

Salmonella parathypii 227

Salmonella typhi 227

Sarcopsylla 474

Sarcopsylla penetran 474

Saxidomus gigantea 551

Schistosoma haematobium 419, 424

Schistosoma japonicum 419, 424

Schistosoma mekongi 419, 424

Schizothrix 626

Scombridae 612

Serranidae 612

Shigella boydii 247, 422

Trypanosoma brucei rhodesiense 208, 445, 447, 448, 449

T. truncatus 569 (T stands for Tursiops)

Tularensis holarctica 268

Tunga penetrans 474, 475

Turbo pica 610

U

Ulva lactuca 40

Umezakia 209, 626

V

Vibrio cholera 102, 281,

Vibrio parahaemolyticus 102

W

West Nile Virus 121, 135, 205, 216, 341, 342, 343, 344, 345,

Wuchereria bancrofti 207, 412, 413, 415

X

Xenopsylla cheopis 200, 210

Y

Yellow fever virus 362,

Yersinia pestis 270, 271, 289

Z

Zaire 53, 137, 140, 319, 320, 646

Zalophus californianus 529

Zaire ebolavirus 319

Zooxanthellae 117

Zosimus aeneus 545

INDEX

A

Agar gel immunodiffusion test 371

Ague 436

AIDS 41, 242, 246, 249, 282, 294, 346, 423, 424, 427, 432, 433, 461, 464, 510, 643, 645, 704

Albedo 106, 110

Albemarle-Pamlico Estuarine System 634

Albendazole 417, 471

Albuterol 486

Algaemia 700

Alok Sharma 688

Alphonse Laveran 436

ALS 209, 617, 618, 619, 620, 621, 622, 625, 702

Alzheimer's disease 535, 616, 624

Amantadine 372

Amastigote 450, 458

Amberjacks 615

American Plague. 361

Aminosidine 461

Amla (amalaka) 357

Amnesiac shellfish poisoning 130, 208, 219, 518, 525, 628

Amocarzine 411

Amoxicillin 237, 278

Amphibian 119, 451, 516

Amphotericin B: 435

Amu Darya 81, 109, 179, 189

Amyotrophic lateral sclerosis 616, 617, 620

Anaplasma bacteria 311

Anaplasmosis 216, 311, 312, 313, 314

Andrew Jackson 280

Atovaquone and Proguanil HCl (Malarone) 443

Atoxyl 444

ATP 154, 622

Australia 7, 83, 108, 112, 293, 349, 354, 366, 384, 392, 393, 478, 490, 505, 530, 537, 539, 546, 553, 559, 567, 570, 575, 586, 593, 598, 599, 658, 688

Australian X disease 392

Autotrophic 10

Avian Influenza 217, 368, 370, 371

Ayurvedic 357

Azaspiracid 208, 219, 518, 536, 565, 580, 581, 583, 584, 631, 632, 633, 701

Azaspiracid shellfish poisoning 208, 219, 518, 580, 583, 633

Azithromycin 465, 296

B

Babesiosis 461, 462, 466, 467

Bacillary Angiomatosis 256, 258

Backpack Diarrhea 467

Bacteria 31, 32, 33, 37, 38, 207, 258, 261, 289, 310

Bacteriophage 42

Baheda 357

Baleri 218

Bangkok 5

Bangladesh 104, 105, 132, 140, 156, 161, 186, 280, 385, 386, 457, 458, 617

Barbeiro (The Barber in Brazil) 449

Barbituates 535

Barking pig syndrome 386

Barracuda 612, 615

Basal cell carcinoma 509, 510, 511

Bayticol 293

Beaver Fever 467

Beclomethasone. 486

Bedbugs 47

Bees 21, 46

Beetle 22, 28

Benzodiazepines 535

Bermuda triangle 107, 183

Bernard Arnault 653

Beta-methylamino-L-alanine 219, 519, 616

Beta-oxalylaminoalanine (BOAA) 617

BHP Billiton 78

Biltricide 422

Bill Gates 653

Bioinvasions xxi, 118, 180

Bioinvasive species 118, 165

Biomagnification 587, 702

Biome 6, 7, 8, 702

Biosphere 2

Biota 9, 58

Biotic factor 703

Biting midges 47, 383

Black death 269

Black fever 457

Blackflies 46, 408

Black Sea 118

Black typhus 331

Black Vomit, 361

blood diathesis 319, 320

Blooms xx, xxi, xxiv, 14, 26, 130, 213, 516, 517, 518, 519, 522, 528, 551, 558, 567, 569, 572, 575, 576, 592, 593, 597, 598, 600, 618, 629, 636, 705, 706

Blue Ear Pig Disease 379

Blue Tongue disease 217

BMAA 209, 219, 519, 618, 628, 702

Bolivian Hemorrhagic fever (See also Machupo) 146, 205, 216, 319, 330, 331, 332, 333

Bolsonaro, Jair 65, 69

Borneo 545, 675, 683

Borrelial Lymphocytoma. 235

Boutonneuse Fever 208, 216, 307

BP 78, 337, 657, 681, 693, 698

Brazilian Hemorrhagic fever 319

Breakbone fever 336

Brevetoxin 567, 569, 571, 572, 573, 574, 575, 576, 577, 578, 628

Brill-Zinsser disease 301

Broecker, Wallace 59, 75, 84

bronchopneumonia 332

Brown dog tick 208, 304, 308

Brunzinski Sign 286

BSL-3 Laboratory 609

BT 150, 151

Bubonic plague 193, 270, 271, 272, 273

Buchanan, Pat 92

Budeso Bunyavirus –Hemorrhagic Fever with Renal Syndrome 319, 707

Burj Al Arab 653

Burj Dubai 653

D

376, 378, 488, 622, 639, 712

Dobson units 173

Dodo bird 168

Domiciliary type 450

Domoic acid 208, 525, 526, 531, 532, 533, 534, 535, 536, 628, 705

Dopamine 623, 624

DOTS 244

Dourine 218, 456

Doxycycline 237, 254, 278, 293, 296, 298, 310, 314, 443

Dracunculias (Dracunulosis) 217, 402, 405

Drought 111, 112, 183, 184

Dubai 202, 653, 681, 716

Dumdum fever 457

Dyskinesia 606

Dyspnea 241, 371, 422, 464, 484, 589, 591, 595

E

Earth Charter 671, 678, 682

Ebola Bundibugyo (Uganda) 137

Ebola Cote d'Ivoire 137

Ebola 41, 53, 54, 55, 137, 138, 140, 142, 150, 168, 205, 216, 319, 320, 321, 322, 323, 324, 325, 327, 328, 399, 643, 644, 646, 680, 702, 705

Ebola fever 216

Ebola Hemorrhagic fever 319, 323, 327, 328

Ebola Sudan 53, 137

Ebola Zaire 53, 137

E. coli 222, 223, 224

Ecosystem 27, 65, 181, 212, 518

Ecthyma migrans 311

hepatic necrosis 332, 595

Hepatitis 205, 216, 244, 345, 346, 347, 348, 349, 399

Hepatitis A 216, 345, 346, 347, 348, 349, 339, 564

Hepatorenal Syndrome 364

Hepatosplenomegaly 229, 295

Hepatotoxins 518, 522, 626

Herbicides 178

Heterotrophic 10, 524, 712

Hippocampus 531, 532, 534, 535

Histoplasmosis 433

His-Werner Disease 258

Hogfish 615

Hoof and Mouth Disease 205, 216, 349, 351

Horse 384, 386, 387

Hotel Palazzo Versace 653

Human babesiosis 218, 461, 466

Human granulocytic anaplasmosis (HGA) 311

Human monocytic ehrlichiosis (HME) 311

Human Papilloma virus infection 217, 374, 375, 376, 379

Hunger 182

Hydrosphere 2, 8, 70

Hydrothermal vents 12

Hydroxychloroquine 237, 513

hyponatremia 305, 314, 492

hypovolemia 282, 301, 323, 327, 334, 336, 492

Hypovolemic shock 282, 322, 326, 329

I

Icteric syndrome 277

Icthyotoxic 574,575

Immunochromatographic test 410, 441

Immunochromatographic test (ICT) 410

Immunodeficiency Syndrome 282

Indonesia 140, 156, 162, 280, 369, 418, 651

Influenza 322, 336, 338, 364, 368, 369, 370, 371, 372, 373, 374, 400

Inhofe, James xxiv, 26, 694

Insects xxiv, 45, 431, 700

Intermediate yellow fever 362

International Criminal Court in the Hague (ICC) 70

Intervet Fort Dodge Animal Health 398

IPCC iii, 60, 61, 84, 126, 707

ipratropium 486, 487

Iraq National Oil Company 78

Iraq war 642

IRS 443

Isoniazid 244

Isoproterenol 487

Isosorbide dinitrate 591

Itraconazole (Sporonox) 435

Ivermectin Stromectol 410, 477

Ivory Coast 53, 140, 320, 649

Ixtoc I 79, 87

J

Jail fever 300

James Polk 280

Jeff Bezos 653, 678

Jennifer Granholm 641

Japanese encephalitis 139

Jellyfish 14, 27, 45, 117, 118

Jenga 13

Joe Biden 640, 641

John McCain 666

Joseph Bancroft 413

Junin virus 319, 333, 719

Juvenile Onset Recurrent Respiratory Papillomatosis 377

K

Kala-azar, or Dumdum fever. 457

Kaodzera 457

Kaposi sarcoma 509

Katayam disease 420

Katrina 129, 228

Kazakhstan 81, 88, 109, 179

Ketoconazole 461

Keystone XL Pipeline 641

Kikwit 644, 646

Kingfish 615

Kiyoshi Shiga 444

Kleptoplast 604

Korea 293, 553, 566, 570, 688

Korsakoff syndrome 535

Kudzo 118

Kunjin virus 392, 393, 394

Kupffer cells 347

Kuru 43, 151

Maculopapular Rashes 292, 359

Mad cow disease 43, 151

Maitotoxin 612

Malaria xiv, 40, 90, 116, 121, 139, 142, 144, 146, 147, 166, 184, 188, 192, 208, 214, 217, 320, 322, 323, 326, 330, 332, 338, 360, 364, 436, 437, 439, 441, 443, 465, 467, 712

Malaysia 140, 152, 293, 354, 385, 386, 539, 545, 553

Malignant melanoma 505

Maltese Cross 465

Mannitol 501, 614

Marayo fever 217, 395

Marburg fever 216

Marburg hemorrhagic fever 205, 319, 324, 327

March fever 436

Mariupol 674

Mary Robinson xx

May Abigail Fillmore 280

McLaughlin, John 92

MDCs 156

Mechanical Respirator 487

Meclizan 410

Mediterranean spotted fever 208, 216, 290, 291, 307, 308

Mefloquine 443

Mefloquine (Lariam) 443

Megaloblastic anemia 503

Meglumine antimoniate 461

Melarsoprol IV 448

Meningitis belt 284

Meningococcal Meningitis 216, 284, 285, 286

Merkel cell carcinoma 509

MERS 643

Murine typhus 297, 298

Murray valley encephalitis 392, 393, 394

Mutation 49, 506

Mycophenolic acid 340

Myocarditis 272, 295

N

Nagana 208, 218, 455

Naproxyn 357

Narrow barred Spanish mackerel 615

National 26, 56, 59, 69, 77, 78, 181, 232, 256, 358, 375, 485, 536, 537, 532, 564, 565, 566, 579, 586, 604, 641, 693

National Fisheries Research and Development Institute (NFRDI) 566

Natural selection 49

Natures services 18

necrotizing enteritis 332

Nedocromil 486

Nematode 45, 206, 207, 401

nervousness 295

Netherland Meningitis Cohort Study 286

Neurological shellfish poisoning 219, 519

Neurotoxic shellfish poisoning 130, 571, 628

Neurotoxins 518, 522, 627

Nifedipine 613

Nigeria 140, 156, 328, 474, 658

Nipah virus 152, 384, 385, 386

Niridazole 477

NMDA N-methyl-p-aspartate (NMDA) 605

N-methyl-p-aspartate 605

NOAA 27, 85, 181, 604

Noam Chomsky 693

Nobel Peace Prize 92, 643

Norway 92, 269, 396, 463, 523, 540, 556, 580

NPO 596

NSAIDS 237, 368, 441, 489

nuchal rigidity 295, 313

Nyama 389

O

Obama, Barak 92, 640, 642

obtundation 295

Obtunded mental status 498

OEPA 411

Ohara fever 265

Oil xxi, 64, 76, 77, 78, 79, 95, 180, 190, 477, 658, 685

Oil spills xxi

Okadaic acid 209, 560, 562, 567, 580

Oligotrophic 169

Onchocerciasis 144, 207, 217, 406, 407, 409, 410, 411, 412

onchocercomas 408

O'nyong-nyong virus infection 358, 359, 360

ookinete 438

Orange Roughy 166, 167

Orangutan 675, 683

Oropouche fever 383, 384

Oroya fever 260,

Oseltamivir 372

OSHA (Occupational Safety and Health Administration) 245, 247

Osteoarticular sporotrichosis 432, 433, 434

Overhunting 168

Overpopulation 654

Oxygen 3, 170, 182, 252, 340, 487, 500, 591, 701, 703, 709

O'nyong-nyong fever 216, 356, 358, 359, 360, 361

Oysters 102

Ozone xxi, 170, 173, 174, 175, 176, 190

P

Palm Island Mystery Disease xxi, 209, 219, 519, 598, 601

Palytoxin 219, 518, 586, 587, 588, 589, 590, 591, 592, 633

Palin 666

Papaverine 591

Paracetamol 357, 613

Paranasal Sinuses. 432

Parinaud oculoglandular syndrome 261

Paralytic shellfish poisoning xxi, 130, 209, 219, 518, 537, 538, 539, 541, 547, 553, 562, 581, 584, 626, 631, 632, 709, 711

Parasitemia 438, 440, 446, 451, 455, 465, 466

Parasites 457

Parasympathetic Nerve Endings 484

Parkinson's disease 616, 620, 622, 623, 702

Parrot fish 615

Patrick Mason 413

Paul Ehrlich 160, 444

Peabody Energy 77

Pectenotoxin 560, 561

Peliosis hepatis 256

Pemex 77, 78, 79, 657

Penicillin 278

Pentamidine IV 448, 461

pentavalent antimonials 461

Perioral edema 538

Perivascular lymphohistiocytic 312

Permafrost 64, 107

Q

Ross River virus infection 366, 367

Rowland., F. Sherwood 174, 175

Royal Dutch Shell 77

Russia 109, 156, 265, 267, 280, 293, 456, 673, 687, 695

Rwanda 121, 159, 161, 176, 649, 650

S

Salmeterol 486

Sandflies 195, 196, 458

Sao Paulo fever 303

Saprotrophic 11

SARS 399, 643

Saturated potassium iodide solution treatment 435

Saudi Aramco 77, 78

Saxitoxin 209, 537, 538, 546, 547, 548, 549, 627, 630, 631, 711

Schistosomiasis 146, 148, 188, 214, 217, 418, 420, 421

scrub typhus 294, 295

Scrub typhus 293, 294

Scrub Typhus 207, 216, 295, 296, 297

Seagrass 19, 28

Sea Lions 529, 532

Septic Shock 252

Septicemic Plague 271

Serous Filariasis 217

severe pain 248, 272, 405, 492, 666

Sharp fever 300

Shigellosis 215, 224, 246, 247, 248, 249

Shinbone Fever 258

Ship fever 300

Sir Ronald Ross 436

Shinbone Fever 258

Syphilis 38, 236, 434

Syr Darya 89, 109, 179, 189

Systemic lupus Erythematosus 236, 513

T

Tahaga (camel disease in Algeria) 457

Taliagos 218

Tamiflu 372

tenesmus 248

Termites 96, 142

Tetracycline 254, 264, 273, 283, 293, 296, 297, 298, 299, 302, 303, 306, 310, 311, 465, 504

Tetrad of Merozoites 465

Thailand 5, 140, 148, 293, 369, 539, 546, 566, 676

The Hot Zone 53, 57, 137, 185, 187, 320, 680

THNEED 696

Thermosphere 5, 6

Thermohaline circulation 127, 128, 185, 708

Thunberg, Greta 690, 698, 699

Tibialgia Fever 258

Tick paralysis 218, 478, 479

Ticks 192, 207, 291, 307, 462, 463, 466

Tick Typhus 303, 308

Tiger Woods xxii, 652

Tillerson, Rex 65, 73, 75, 78, 87

Tinidazole 471

Timothy Lewis 413

Titanic 11

Tobia fever 303

TOMS satellite 173

Total SA 78

Uveitis 276, 277, 326

UV light 49, 171, 173, 175

V

Vaqta 349

Vectors xxiii, xxv, 7, 9, 134, 141, 191, 233, 259, 449, 450

venereal warts 377

Venezuelan Hemorrhagic Fever 319

Verruga peruana 261, 262

Vinchuca (Argentina, Bolivia, Paraguay) 449

Viral Hemorrhagic fever 216, 319, 328

Viroids 43, 44, 56

Viruses xxii, 41, 317, 363, 375, 645, 700

Vomiting 498

W

Warfare xxi, 13, 673

War Fever Camp fever 300

washerwoman's hand 282

Waste water 21

Waterhouse-Friderichsen Syndrome 286

Weil-Felix Agglutination Reaction 298

Weil-Felix test 263, 302

Weil's disease 274

Western Equine Encephalitis 139

West Nile virus infection 216, 341

Wildfires 74, 95, 108, 490

Wolhynia fever 258

Woods, Darren 76

Woods, Tiger xxii, 652

Wright-Geimsa stain 314

Y

Yellow fever 144, 361, 362, 364, 365, 436, 609

Yellow Jack, 361

Yessotoxins 560, 628

Z

Zafirlukast 486,

Zaire 53, 137, 140, 319, 320, 646,

Zanamivir 372

Zanzarin 477

Zileuton 486

Zoanthus 587

Zooxanthellae 117

Zooplankton 102, 118, 281, 516, 517, 629, 704, 709, 712

CONTACTING DAVID ARIETI

darieti@comcast.net

David's Web Page

Davids-earth.org

David's Youtube Channel

https://www.youtube.com/@DavidsEarth/videos

Contacting Jacob Nieva

jnieva@oakton.edu

www.ingramcontent.com/pod-product-compliance
Lightning Source LLC
Chambersburg PA
CBHW070711190326
41458CB00004B/941